T0419839

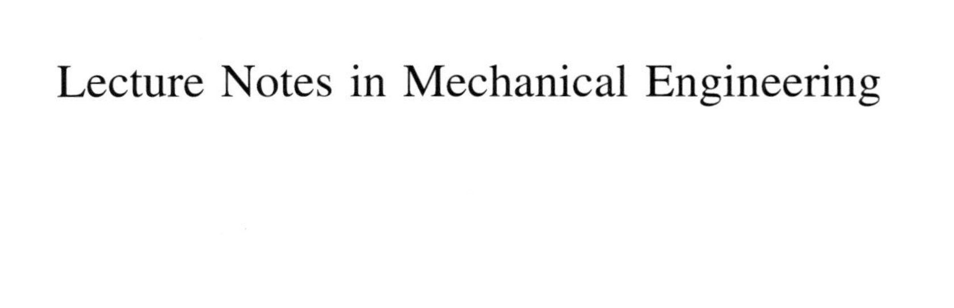

Lecture Notes in Mechanical Engineering

For further volumes:
http://www.springer.com/series/11236

Amaresh Chakrabarti · Raghu V. Prakash
Editors

ICoRD'13

Global Product Development

Part I

Editors
Amaresh Chakrabarti
Centre for Product Design
and Manufacturing
Indian Institute of Science
Bangalore, Karnataka
India

Raghu V. Prakash
Mechanical Engineering
Indian Institute of Technology Madras
Chennai, Tamil Nadu
India

ISSN 2195-4356 ISSN 2195-4364 (electronic)
ISBN 978-81-322-1049-8 ISBN 978-81-322-1050-4 (eBook)
DOI 10.1007/978-81-322-1050-4
Springer New Delhi Heidelberg New York Dordrecht London

Library of Congress Control Number: 2012954861

Printed on acid-free paper

Springer is part of Springer Science+Business Media (www.springer.com)

Preface

Design is ubiquitous, yet universal; it pervades all spheres of life, and has been around ever since life has been engaged in, purposefully changing the world around it. While some designs that matured many centuries ago still remain in vogue, there are areas in which new designs are being evolved almost every day, if not, every hour, globally. Research into design and the emergence of a research community in this area has been relatively new, its development influenced by the multiple facets of design (human, artefact, process, organisation, and the micro- and macro-economy by which design is shaped) and the associated diversification of the community into those focusing on various aspects of these individual facets, in various applications. Design is complex, balancing the needs of multiple stakeholders, and requiring a multitude of areas of knowledge to be utilised, with resources spread across space and time.

The collection of papers in this book constitutes the Proceedings of the 4th International Conference on Research into Design (ICoRD'13) held at the Indian Institute of Technology Madras in the city of Chennai, India during 7–9 January 2013. ICoRD'13 is the fourth in a series of biennial conferences held in India to bring together the international community from diverse areas of design practice, teaching and research. The goal is to share cutting edge research about design among its stakeholders; aid the ongoing process of developing a collective vision through emerging research challenges and questions; and provide a platform for interaction, collaboration and development of the community in order for it to address the challenges and realise the collective vision. The conference is intended for all stakeholders of design, and in particular for its practitioners, researchers, teachers and students.

Out of the 201 abstracts submitted to ICoRD'13, 175 were selected for full paper submission. One hundred and thirty-two full papers were submitted, which were reviewed by experts from the ICoRD'13 International Programme Committee comprising 163 members from 127 institutions or organisations from 32 countries spanning five continents. Finally, 114 full papers, authored by over 200 researchers from 91 institutions and organisations from 23 countries spanning five continents, have been selected for presentation at the conference and for publication as chapters in this book. ICoRD has steadily grown over the last three editions, from a humble beginning in 2006 with 30 papers and 60 participants, through 75 papers and 100 participants in ICoRD'09, and 100 papers and 150 participants in ICoRD'11. This is also the first time that ICoRD has taken place outside Bangalore, with the Indian Institute of Technology Madras and Indian Institute of Science Bangalore jointly sharing the responsibility for its organisation.

The chapters in this book together cover all three major areas of products and processes: functionality, form and human factors. The spectrum of topics range from those focusing on early stages such as creativity and synthesis, through those that are primarily considered in specific stages of the product life cycle, such as safety, reliability or manufacturability, to those that are relevant across the whole product life cycle, such as collaboration, communication, design management, knowledge management, cost, environment and product life cycle management. Issues of delivery of research into design, in terms of its two major arms: design education and practice, are both highlighted in the chapters in this book. Foundational topics such as the nature of design theory and research methodology are also major areas of focus. It is particularly encouraging to see in the chapters the variety of areas of application of research into design—aerospace, healthcare, automotive and white goods sectors are but a few of those explored. The theme of this year's conference is Global Product Development. The large number of chapters that impinge on this theme reflects the importance of this theme within design research.

On behalf of the Patron, Steering Committee, Advisory Committee, Local Organising Committee and Co-Chairs, we thank all the authors, reviewers, institutions and organisations that participated in the conference and the Conference Programme Committee for their support in organising ICoRD'13 and putting this book together. We are thankful to the Design Society and Design Research Society for their kind endorsement of ICoRD'13. We thank the Indian Institute of Technology Madras and the Indian Institute of Science, Bangalore for their support of this event. We also wish to place on record and acknowledge the enormous support provided by Mr. Ranjan B. S. C., Ms. Kumari M. C. and Ms. Chaitra of IISc in managing the review process, and in preparation of the conference programme and this book, and the group of student-volunteers of Indian Institute of Technology Madras led by Swostik, Suraj and Sahaj in the organisation and running of the conference. Finally, we thank Springer India for its support in the publication of this book.

Amaresh Chakrabarti
Raghu V. Prakash

Contents

Part II Design Creativity, Synthesis, Evaluation and Optimization

Part III Design Aesthetics, Semiotics, Semantics

Part IV Human Factors in Design

Part V Eco-Design, Sustainable Manufacturing, Design for Sustainability

Part VI Design Collaboration and Communication

Part VII Design Management, Knowledge Management and Product Life Cycle Management

Organizing Committee

Patron

Ramamurthi, Bhaskar, Director, Indian Institute of Technology Madras, Chennai, India

Chairs

Prakash, Raghu, Indian Institute of Technology Madras, Chennai, India
Chakrabarti, Amaresh, Indian Institute of Science, India

Co-Chairs

Blessing, Lucienne, University of Luxembourg, Luxembourg
Cugini, Umberto, Politecnico di Milano, Italy
Culley, Steve, University of Bath, UK
McAloone, Tim, Technical University of Denmark, Denmark
Ray, Gaur H., Indian Institute of Technology Bombay, India
Taura, Toshiharu, Kobe University, Japan

Steering Committee

Athavankar, Uday, Indian Institute of Technology Bombay, India
Gero, John, George Mason University, USA
Gurumoorthy, B., Indian Institute of Science, India
Harinarayana, Kota, National Aeronautical Laboratories, India (Chair)
Lindemann, Udo, Technical University of Munich, Germany

Advisory Committee

Arunachalam, V. S., Center for Study of Science, Technology and Policy, India
Das, Anjan, Confederation of Indian Industry, India
Dhande, Sanjay G., Indian Institute of Technology Kanpur, India
Forster, Richard, Airbus Industrie, Toulouse, France
Gnanamoorthy, R., IIIT D&M, Kancheepuram, India
Horvath, Imre, Delft University of Technology, The Netherlands
Jaura, Arun, Eaton India Engineering Center, Pune, India
Jhunjhunwala, Ashok, Indian Institute of Technology Madras, Chennai, India
Leifer, Larry, Stanford University, USA
Pathy, Jayshree, CII National Committee on Design, New Delhi, India
Mitra, Arabinda, International Division, Department of Science and Technology, India
Mohanram, P. J., Indian Machine Tool Manufacturers Association, India
Mruthyunjaya, T. S., Chairman Emeritus, CPDM, Indian Institute of Science, India
Nair, P. S., ISRO Satellite Centre, Indian Space Research Organisation, India
Pitroda, Sam, National Innovation Council, India
Saxena, Raman, USID Foundation, India
Subrahmanyam, P. S., Aeronautical Development Agency, India
Sumantran, V., Hinduja Automotive, UK
Lt. Gen. Sundaram, V. J., National Design and Research Forum, India

International Programme Committee

Albers, Albert, Karlsruhe Institute of Technology, Germany
Allen, Janet K., University of Oklahoma, USA
Anderl, Reiner, Technical University of Darmstadt, Germany
Arai, Eiji, Osaka University, Japan
Aurisicchio, Marco, Imperial College London, UK
Badke-Schaub, Petra, Delft University of Technology, The Netherlands
Bernard, Alain, Ecole Centrale, Nantes, France
Bhattacharya, Bishakh, Indian Institute of Technology Kanpur, India
Blanco, Eric, Institut Polytechnique de Grenoble, France
Bohemia, Erik, Northumbria University, UK
Boks, Casper, Norwegian University of Science and Technology, Norway
Bolton, Simon, Cranfield University, UK
Bonnardel, Nathalie, University of Provence, France
Bordegoni, Monica, Politecnico di Milano, Italy
Borg, Jonathan C., University of Malta, Malta
Braha, Dan, University of Massachusetts, USA
Brazier, Frances, Delft University of Technology, The Netherlands
Brissaud, Daniel, Institut Polytechnique de Grenoble, France
Bruder, Ralph, TU Darmstadt, Germany
Burvill, Colin, University of Melbourne, Australia

C. Amarnath, Indian Institute of Technology Bombay, India
Caillaud, Emmanuel, Universite de Strasbourg, France
Cascini, Gaetano, Politecnico di Milano, Italy
Cavallucci, Denis, INSA Strasbourg, France
Chakrabarti, Debkumar, Indian Institute of Technology Guwahati, India
Childs, Peter, Imperial College London, UK
Clarkson, John P., University of Cambridge, UK
Das, Amarendra K., Indian Institute of Technology Guwahati, India
Deb, Anindya, Indian Institute of Science, India
Dong, Andy, University of Sydney, Australia
Dorst, Kees, University of Technology Sydney, Australia
Duffy, Alex, University of Strathclyde, UK
Duflou, Joost, Katholieke Universiteit Leuven, Belgium
Duhovnik, Jožef, University of Ljubljana, Slovenia
Echempati, Raghu, Kettering University, USA
Eckart, Claudia, Open University, UK
Eckhardt, Claus-Christian, Lund University, Sweden
Eynard, Benoit, Universite de Technologie de Compiegne, France
Fadel, Georges, Clemson University, USA
Fargnoli, Mario, University of Rome "La Sapienza", Italy
Friedman, Ken, Swinburne University of Technology, Australia
G. K. Ananthasuresh, Indian Institute of Science, India
Ghosal, Ashitava, Indian Institute of Science, India
Girard, Phillippe, University of Bordeaux, France
Goel, Ashok, Georgia Institute of Technology, USA
Goldschmidt, Gabriela, Technion, Israel
Gooch, Shayne, University of Canterbury, New Zealand
Grimhelden, Martin, Royal University of Technology, Sweden
Gu, Peihua, University of Calgary, Canada
Hanna, Sean, University College London, UK
Hekkert, Paul, Delft University of Technology, The Netherlands
Helander, Martin, Koln International School of Design, Germany
Heskett, John, Hong Kong Polytechnic University, China
Hicks, Ben, University of Bath, UK
Horne, Ralph, RMIT University, Australia
Hosnedl, Stanislav, University of West Bohemia, Czech Republic
Howard, Thomas, Technical University of Denmark, Denmark
Ijomah, Winifred, University of Strathclyde, UK
Ion, William, Strathclyde University, UK
Iyer, Ashok Ganapathy, Manipal University—Dubai Campus, UAE
Jagtap, Santosh, Lund University, Sweden
Johnson, Aylmer, University of Cambridge, UK
Kailas, SatishVasu, Indian Institute of Science, India
Karlsson, Lennart, Lulea University of Technology, Sweden
Kasturirangan, Rajesh, National Institute of Advanced Studies, India

Rohmer, Serge, Universite de Technologiede Troyes, France
Roozenburg, Norbert, Delft University of Technology, The Netherlands
Rosen, David W., Georgia Institute of Technology, USA
Roy, Rajkumar, Cranfield University, UK
Roy, Satyaki, Indian Institute of Technology Kanpur, India
Saha, Subir Kumar, Indian Institute of Technology Delhi, India
Salustri, Filippo A., Ryerson University Canada
Sarkar, Prabir, Indian Institute of Technology Ropar, India
Sato, Keichi, Illinois Institute of Technology, USA
Seliger, Guenther, Technical University of Berlin, Germany
Sen, Dibakar, Indian Institute of Science, India
Shah, Jami, Arizona State University, USA
Shimomura, Yoshiki, Tokyo Metropolitan University, Japan
Shu, Li, University of Toronto, Canada
Sikdar, Subhas, Environmental Protection Agency, USA
Singh, Mandeep, School of Planning and Architecture New Delhi, India
Singh, Vishal, Aalto University, Finland
Sotamaa, Yrjo, University of Art and Design Helsinki, Finland
Srinivasan, V., Technical University of Munich, Germany
Srinivasan, Vijay, National Institute of Standards and Technology, USA
Sriram, Ram, National Institute of Standards and Technology, USA
Storga, Mario, University of Zagreb, Croatia
Subrahmanian, Eswaran, Carnegie Mellon University, USA
Sudarsan, Rachuri, National Institute of Standards and Technology, USA
Sudhakar, K., Indian Institute of Technology Bombay, India
Thallemer, Axel, Universitaetfürindustrielle und kuenstlerische Gestaltung, Austria
Thompson, Mary Kathryn, Technical University of Denmark, Denmark
Tiwari, Ashutosh, Cranfield University, UK
Tomiyama, Tetsuo, Delft University of Technology, The Netherlands
Torlind, Peter, Lulea University of Technology, Sweden
Tripathy, Anshuman, Indian Institute of Management Bangalore, India
Tseng, Mitchell, The Hong Kong University of Science & Technology, China
Umeda, Yasushi, Osaka University, Japan
Vancza, Jozsef, MTA SZTAKI, Hungary
Vaneker, Tom, Univeristy of Twente, The Netherlands
Vasa, Nilesh, Indian Institute of Technology Madras, India
Verma, Alok K., Old Dominion University, USA
Vermass, Pieter, Delft University of Technology, The Netherlands
Vidal, Rosario, UniversitatJaume I, Spain
Vidwans, Vinod, FLAME School of Communication, India
Vijaykumar, A. V. Gokula, University of Strathclyde, UK
Voort, M. C. van der, University of Twente, The Netherlands
Wiegers, Tjamme, Delft University of Technology, The Netherlands
Yammiyavar, Pradeep, Indian Institute of Technology Guwahati, India
Yannou, Bernard, Ecole Centrale Paris, France

Zavbi, Roman, University of Ljubljana, Slovenia
Zeiler, Wim, Technical University Eindhoven, The Netherlands
Zwolinski, Peggy, Institut Polytechnique de Grenoble, France

Local Organising Committee

Swaminathan, Narasimhan, Indian Institute of Technology Madras, India
Siva Prasad, N., Indian Institute of Technology Madras, India
Krishnapillai, Shankar, Indian Institute of Technology Madras, India
Seshadri Sekhar, A., Indian Institute of Technology Madras, India
Balasubramaniam, Krishnan, Indian Institute of Technology Madras, India
Varghese, Susy, Indian Institute of Technology Madras, India
Ganesh, L. S., Indian Institute of Technology Madras, India
Ramkumar, Indian Institute of Technology Madras, India
Srinivasan, Sujatha, Indian Institute of Technology Madras, India
Sivaprakasam, Mohan Sankar, Indian Institute of Technology Madras, India

Student Organising Committee

Das, Swostik, Indian Institute of Technology Madras, India
Gullapalli, Suraj, Indian Institute of Technology Madras, India
Anand, Sahaj Parikh, Indian Institute of Technology Madras, India
Dhinakaran, S., Indian Institute of Technology Madras, India
Arun Kumar, S., Indian Institute of Technology Madras, India
Sarkar, Biplab, Indian Institute of Science, India
Devadula, Suman, Indian Institute of Science, India
Kumari, M. C., Indian Institute of Science, India
Madhusudanan, N., Indian Institute of Science, India
Ranjan, B. S. C., Indian Institute of Science, India
Uchil, Praveen T., Indian Institute of Science, India

Organised By

Indian Institute of Technology Madras, Chennai, and Indian Institute of Science, Bangalore, India

Endorsed By

The Design Society and The Design Research Society

Part I
Design Theory and Research Methodology

How I Became a Design Researcher

Gabriela Goldschmidt

Abstract No one trajectory leads to becoming a design researcher. This paper is an interim overview of one case of a designer who became a design researcher. Through concerns derived from experience as a designer and a design teacher, and learning from observations of children designing, some fundamental questions were formulated under the common theme of design cognition. Two major lines of research were undertaken over the years: first, modeling and measuring design reasoning. This is done with linkography, which is a system for the notation and analysis of design activities based on protocols. With linkography light can be shed on the structure of the design process and its quality, especially its creativity. Second, visual thinking in design and primarily the generation of sketches and the use of visual stimuli are investigated. Experimental work confirms the importance of visual thinking in designing.

Keywords Cognition · Creativity · Linkography · Sketching · Visual thinking

1 Introduction

When I was an architecture student, in three different schools, I was absolutely certain that my future is in architectural design. I saw practice as the only career I would ever want to follow. Once out of school I first worked in established firms, later in my own small independent practice. Little by little I realized something was missing. Practice was exciting but also very tiring and frustrating, and work

G. Goldschmidt (✉)
Technion-Israel Institute of Technology, Haifa 32000, Israel
e-mail: gabig@technion.ac.il

A. Chakrabarti and R. V. Prakash (eds.), *ICoRD'13*, Lecture Notes in Mechanical Engineering, DOI: 10.1007/978-81-322-1050-4_1, © Springer India 2013

posed too many practical difficulties and too few intellectual challenges. I started teaching architectural design part time and found it rewarding, in that students dealt with hard core issues that forced me to struggle with those same dilemmas myself. My frustrations as a designer and observations of students' progress and difficulties started raising fundamental questions: how do we actually think when we design? How do we learn to design? What does it mean to be a 'good' designer and why isn't every designer's performance excellent? How and what does experience contribute? What is visual about our thinking in design, and why is that important, if it is? Questions of this nature occupied my mind but I was not sure how to go about answering them until I decided one day, quite intuitively, to look at children designing. Much to my surprise I discovered that children could design, and some of what they do does not seem to differ greatly from what my students or I do when we design. This was when I started talking to psychologists and soon I made two discoveries: First, I learned that cognitive psychology deals with some of the questions I was interested in, though it does not focus on designing. Second, I found there were a handful of others in the design world who were interested in the same questions and who had already started researching them. It was then that I turned into a design researcher; first in parallel to my design activities and later in their stead, although I continued to teach design for a long time thereafter. Certain insights regarding design thinking and its research have matured over the years and this paper presents them and traces the trajectory that led to them—a personal journey toward design research, which has become more than a career: it is a love and a passion.

2 Experiences as a Designer

Ever since I was a student design was not something that came easily to me. Every new project, at school and later in practice, required a fairly long and sometimes tortuous search until a good enough idea crystalized. Even though the final result was usually satisfactory, I envied those of my peers who were visibly more relaxed and easy-going about developing their design ideas. I wanted to be like them but did not know what would make that possible. Things got better with time and I concluded that having accumulated experience helped. But I still did not know what about that experience was responsible for a more direct path to a good solution. Have I become a better thinker? How so? Or was it just a matter of self-confidence? And then there were design reviews. How was it that for the most part reviewers agreed on what constitutes a good project and what was a less successful project, even when reasons were not elaborated?

When I started working with my own clients I had to convey design ideas to them. Things that now seemed obvious to me turned out to be far from obvious to them and sometimes I failed to gain their understanding of why something was important. Why couldn't I reach out to them? How could one ensure seeing design issues eye to eye with someone else, whether a professional or a layperson?

Once again this boiled down to design thinking, to what we know, to the nature of expertise and the contribution of experience to it. It also posed questions about creativity, communication, and what I was much later able to label as the nature of one's *design space*, and shared *mental models*. But these terms—and their significance—were not part of my terminology and my agenda until many years later. In the meantime I just continued to struggle and tried my best to connect with clients; some would call it: educating the client.

3 Experiences as a Design Teacher

When I started teaching architectural design studios I was a practitioner with a few years of experience in reputable international architectural firms. I was enthusiastic and wanted all of my students to produce wonderful work. As it were, some did but others did not do as well. I would spend hours on end with weaker students with the hope of assisting them; I tried all strategies, such as citing examples and precedents, modeling and discussing alternatives with them, eliciting thumb rules and conventions, and more. Sometimes it helped, sometimes it did not. I invented exercises that were geared at seeing the task from a different perspective: starting the design from sections as opposed to plans, starting from a whole to be divided into components, as opposed to starting with components to be assembled into a whole (or vice versa); continuing a project started by somebody else; developing a story that the design would illustrate; and more. Again, this was helpful, but not always and not as much as I had hoped it would be.

I wanted to know why some students 'get it' and others do not, although their starting point was apparently the same. Why some students needed only a subtle hint and could then leap forward on their own, whereas others were unable to develop ideas even when they were shown explicit solutions. Why some students were able to use feedback to make incremental progress from week to week and others were unable to incorporate feedback and started anew after every feedback session ('crit'). These were burning questions and they arose with students of different standing, from complete novices, i.e. first year students, to nearly professionals, that is fifth year students. I started realizing that what had weighed on me as a practitioner and what bothered me as a teacher were questions of the same kind, but I still did not know how to begin to explore them.

4 Learning from Children's Designs

One day I played with a young child who had spent the morning in a swimming pool, or rather a toddler's pool, as she did not yet know how to swim. This was an extraordinary experience which she enjoyed a lot, as she came from a small new settlement were there was no swimming pool. That afternoon she played with building blocks and a few other items and announced she was going to build a 'model' of her settlement. The settlement had a ring-road around which there were two rows of houses; the back row was served by pedestrian access paths, leading from the road to each second-row house. These paths were favorite play areas for young children. Other than single-family houses and a small number of community facilities, the settlement had semi-underground bomb-shelters scattered among the houses; their sloped roofs touched the ground and children used them as slides.

The child's model consisted of houses and a row of wooden blocks—"the path", leading to a circle of wooden blocks, "a swimming pool". Nearby there were a few bigger wooden blocks designated as "shelters". A few small plastic human figurines were scattered about, "children". I realized that the child was using her knowledge of the familiar home environment—houses, paths, shelters, to which she added a desired feature—a swimming pool, a newly acquired concept that she considered appropriate and desirable as an addition to the existing environment. It dawned on me that what this child was doing was not essentially different from what we designers do. Don't we introduce new features to old and existing ones to produce new syntheses?

With this insight in mind I conducted a workshop with older children (aged 9–10), who were given a big crate full of materials and were reminded of a children's story they all knew and loved, Pippi Longstocking. In the story Pippi and two friends run away from home, with a horse and a monkey, and have various adventures as runaways. Since they have no home now, the children's task in the workshop was to design a house for them, which they did by building models from the materials in our magic crate (cardboard sheets and rolls, wooden blocks and round pieces, wire, paper, and so on, plus glue, tape, scissors etc.). The workshop took place at school and the children were so enthusiastic that the principal decided to cancel the remaining classes for the day to allow them to complete their projects. They worked alone or in groups of two or three, and got adult help only with technicalities like cutting a piece to their specifications with a sharp cutting knife. At the end of the day they were interviewed about their projects wherein they were able to explain their acts and their choices [1]. Amazingly enough all children completed the task. It was again apparent that they used knowledge about the concept of 'house', which they were familiar with of course, and synthesised it with knowledge about Pippi's personality that was available to them because they knew the story. They were able to translate personality traits into design features (for example: Pippi is very strong and playful so she can enter the house by jumping from a seesaw and grabbing a rope hanging from the window). The main lesson from the workshop was that different children construed the task differently; the main focus was either on 'house' or on 'Pippi', leading to different design priorities.

The same workshop was repeated with first and fourth year architecture students, with far less interesting results. The first year students were held up by the fact that they were not given a 'program' and were not told at which scale they should work. Some fourth year students thought this was not a serious enough exercise. In addition to confirming the ability to use knowledge that is not design-specific, we learned about design education and pedagogy, on which we shall not dwell here.

5 Questions that Beg Answers

Having had the experiences described above in both the professional arena and the educational studio, it became clear that there are some persistent mega-questions regarding the process of designing that were wide open and invited exploration. Some of them were outlined above, and in addition the contribution of visual thinking to design, at least in some fields (e.g., architecture, industrial design, graphic design, mechanical engineering), became intriguing. Rudolf Arnheim's seminal *Visual thinking* [2] was a major influence that in addition to shedding light on visual thinking also served as an introduction to Gestalt psychology. With hindsight, the main questions I was able to formulate ran approximately as follows:

- How are different knowledge items synthesised into a design solution?
- How do we reason on the fly about issues, knowledge and acts related to a design entity that is in the process of being generated?
- How do we construe ill-defined design problems, and turn them into solvable problems? Why are there individual differences in this process?
- How are design skills acquired? What is teachable and learnable about such skills?
- How can we characterize creative designing? Can we measure creativity?
- What is the role of visuals in designing? This question may be divided into two:
- On the one hand, there is the 'consumption' of external visuals which act as inspirational agents. What mechanisms are involved in being 'inspired'?
- On the other hand there are self-generated sketches. Why do designers sketch?

When the accumulated questions gained enough coherence it became obvious that they are all of a kind; in fact they can be grouped under the heading of design cognition. This was a new concept that required some clarification.

6 Design Cognition

I had already written a few papers on the process of designing [3–5] drawing primarily on my own experiences when I felt ready to become a 'real' researcher. I realized that to succeed, I needed to belong to a research community, but which community? I was not yet aware of other designers who were interested in questions similar to the ones that intrigued me, but I found that psychologists were in fact quite interested in discussing the questions I asked. First, through the children's design study, these were developmental psychologists; through them I met other psychologists who were more interested in certain cognitive issues, in particular creativity. I started to educate myself in the field of cognitive psychology and in parallel I came across writings by designers that echoed my interests; I was delighted to find out that indeed a community of design thinking researchers was coming into being and within this community there were some prominent design thinking students like Nigel Cross, Bryan Lawson, Omer Akin, John Gero and others, whose pioneering work was of great value e.g., [6–9]. Schön [10], in essence not really a design researcher, was of great influence in that community. I eagerly joined those forerunners and before soon many more young researchers joined in: a community of design thinking researchers, interested mostly in design cognition, became an established fact.

I put together a course for graduate students called Cognitive Aspects of Architectural Design, which was taught for many years during which it continually changed and grew. This led to students knocking on the door, wanting to work on theses and dissertations in the area of design thinking. It was time to learn how to conduct empirical research and exposure to cognitive psychology paved the way to experimental work. Protocol analysis became the major methodology for our work, and it continues to serve us well to this very day. Two main and complimentary research directions emerged: modeling and measuring design reasoning, and visual thinking in design.

7 Modeling and Measuring Design Reasoning

When I first discovered protocol analysis I tried to apply it as prescribed [11, 12] to protocols we have generated. I failed to find coding schemes of categories that could reveal anything of value or interest and thought that there must be another way to use protocols of on-line design session recordings to derive information that would lead to new insights regarding the design process. My interest in cognition led to the wish to find out not only what designers were thinking about, which coding protocols aptly reveals, but also how they think, that is, how do they form ideas and concepts. This prompted the notion that what we should be looking at is links among units of verbalization, into which protocols are parsed. These units, also called segments, can be anything from a few words to long stretches of verbalization. If the thinking scale one is interested in is cognitive, the units should perforce be rather short, in the order of approximately one sentence. I decided to borrow the term *move* from chess, and here it means a step, an act, an operation, that transforms the design situation somewhat relative to the state in which it was prior to that move. The protocol is now transformed to a numbered, sequential list of design moves. These moves represent the universe in which the designer acts during the vignette that is captured in the protocol.

Design theories tell us that the crucial outcome of the preliminary design process is synthesis. A synthesis is manifest in a partial or complete design proposal that reflects the best possible balance among considered options, given requirements and goals, analysis of the givens and the situation (constraints, opportunities), the designers' knowledge and experience, and their values and personal propensities. A synthesis cannot be arrived at 'in toto' in one go: a synthesis, that is, a candidate solution, is constructed step by step in a search process that is typical of ill-structured/ill-defined problem solving, of which design is a paramount example. At the cognitive scale we equate the search steps with design moves. To arrive at a synthesis design moves must be integrated in a network such that the ensuing solution would be sure to take into account all relevant aspects of the situation and the problem. My suggestion was that this process is captured in a network of links among design moves, based on contents and arbitrated by common sense, using disciplinary knowledge. Once this understanding was reached, the task became clear: we need a system of notation and analysis of links among design moves.

Such a system was developed in the framework of *linkography*, which consists of the theory behind the relevance of links among design moves and the technical method of notating and analyzing such links e.g., [13, 14]. The linkograph is where a link is notated, if one exists, between every pair of moves in the sequence, if one exists. For n moves, the number of possible links is $n*(n-1)/2$. For each pair of moves, consecutive or far apart, we ask: is there a link between these two moves? The answer is yes, or no; if positive—a link is notated in the linkograph. There are two types of links: backlinks and forelinks.

Backlinks: Starting with move 2, we ask about a possible link with each of the preceding moves; therefore such links are designated as backlinks.

Forelinks: If move 2 links back to move 1, then we may say, with hindsight, that move 1 links forward to move 2. This is a virtual link of course as at the time move 1 is generated we cannot know whether or not it will link to move 2 (and subsequent moves). However, we are interested in forelinks because they are indicative of steps that future design activities refer back to.

Critical Moves: Not all design moves are of the same magnitude; some are more significant than others in terms of developing the final outcome. We call the most important moves critical moves (CM) and distinguish between CMs due to a large number of backlinks (<CM) and CMs due to a large number of forelinks (CM>). For each investigation we determine a threshold that indicates the minimum number of links necessary to qualify a move as critical. If, for example, the number is 5 links, the threshold of 5 is indicated (CM^5). Figure 1 depicts a linkograph of a short design vignette (20 moves).

We carried out a large number of linkographic studies and other researchers joined us in using linkography, primarily in order to investigate the structure of design reasoning e.g., [15–18]. The linkograph shows clearly where moves are concentrated and where they are sparse and far between. In as far as a linkograph displays fairly autonomous chunks of interlinked moves, with few links to neighboring chunks, we may refer to such chunks as micro-phases and the initial and last moves of a chunk are of particular interest, especially if they are critical moves (as they often are). In a way, looking at critical moves—their positioning in the sequence of moves and their contents—gives us a very good overview of the

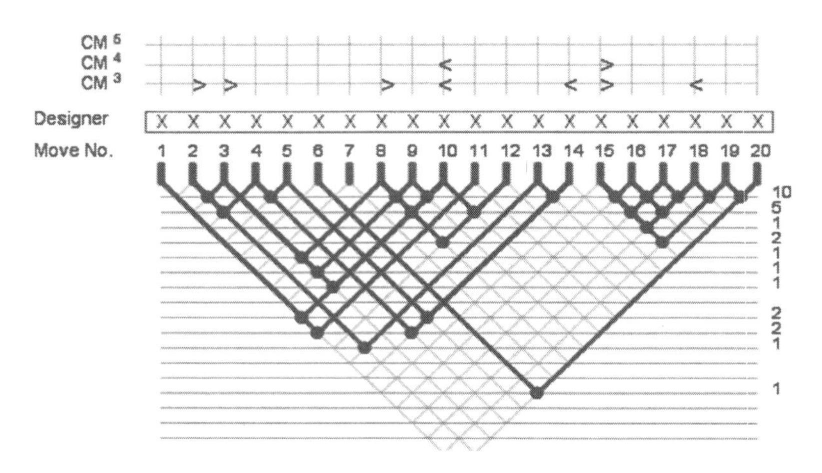

Fig. 1 Linkograph. <CMs and CMs> are indicated at thresholds of 3 and 4 links

design process is question. I propose that moves with more forelinks are roughly reflective of divergent thinking: new ideas come up in them, which further moves refer back to. Similarly, moves with primarily backlinks represent convergent thinking: they test, evaluate, confirm or question preceding moves in which a new proposal had been put forth. If we accept this notion, then the proportion and interplay among critical moves tell us something about the creativity of the process under scrutiny. Needless to say, we are not talking here about creativity of the kind that changes the world; instead, we detect acts of small-scale, quotidian creativity (that Boden [19] calls P-Creativity), which are typically marked by shifts between divergent and convergent thinking e.g., [20–22].

Linkographic studies have been able to establish a correlation between independent creativity assessments and the proportion of critical moves, especially CMs>, in a variety of settings. They also taught us a lot about problem framing and problem solving reasoning which is noticeable in the structure of the linkograph. This served as a basis for comparisons between novices and advanced students, which taught us something about learning and the role of experience. Other linkographic studies looked at communication patterns in the design studio, topic shifts in design, the dynamics of team work, and much more. Linkography was also used to look at phenomena at non-cognitive scales such as prolonged design periods and other problem-solving processes.

8 Visual Thinking

Architects and other designers are in the habit of making rough, free-hand sketches. This habit continues since industrially produced paper became available in Europe in the last quarter of the 15th century, and it persists today despite the digital revolution. The sketches we are talking about are not after-the-fact renderings but quick freehand drawings made on the fly during the search in the preliminary conceptual phase of the design process. Sketches serve a number of purposes [23]: prescription, communication, and thinking. In the framework of design cognition it is the latter we are interested in. In the hands of fluent sketchers, sketching has many cognitive advantages: it off-loads memory; is rapid and therefore requires minimal cognitive resources, stop rules are flexible, it is only minimally rule-bound, it is reversible and transformable at any stage, it is tolerant of incompletion, inaccuracy and lack of scale, it provides stimuli and cues, and finally, it supports feedback loops between external and internal representation (visual imagery).

Feedback loops are of particular interest, as they establish the 'cooperative' dynamics between imagery which we know is a valuable cognitive capacity in creative work, and what is being laid down on paper. Thus sketching does not merely record mental activity, it is part and parcel of it [24]. In fact designers discover things in their own sketches [15], sometimes much more so than anticipated. Since the early 1990s a host of studies have shown that sketching is a high impact design strategy.

Our sketching studies led to a wider interest in the mode in which designers exploit visual stimuli while developing design solutions. One of the conclusions was that images in stimuli may serve as bases for analogy, unconsciously or consciously, especially when one is explicitly instructed to use analogical thinking [25]. Like visual imagery, analogy is a helpful cognitive strategy in creative endeavors, and in both cases sketching aids the relevant cognitive operations. Of course, stimuli may also cause designers to fixate on them, thus barring them from developing original solutions. This is more likely to happen when the stimulus is an actual example of a solution to the problem in hand. Generally we were able to show that visual stimuli, with or without suggestions of analogies, are beneficial in the design process.

I propose that visual thinking is so important in designing because when the final outcome is a physical entity that needs to be considered in terms of form and function, a good synthesis can be achieved only if in the process of reasoning about candidate partial and whole solutions the designer shifts back and forth between embodiment and rationale, such that all design aspects are covered in a final well-integrated design outcome [26]. Visual thinking helps represent and assess embodiments, and therefore it is invaluable, whether sketches are hand-drawn or simulations are created digitally.

9 What Next?

Two important hallmarks of design cognition have been established in our studies: the power of interlinking moves, and the need to shift between embodiment and rationale. These traits of design cognition may not be unique to designing, but they are particularly important in this activity that deals with form and function, and that must bring about a well synthesised artifact. Important as these findings are, they are not enough; there is more to design thinking, and we must discover additional fundamental characteristics of such thinking if we are to build tools to effectively impact them. The design research community is invited to partake in this valuable, exciting and rewarding project.

References

1. Goldschmidt G (1994) Development in architectural designing. In: Franklin MB, Kaplan B (eds) Development and the arts: critical aspects. NJ Lawrence Erlbaum Associates, Hillsdale, pp 79–112
2. Arnheim R (1983) Design thinking. University of California Press, Berkeley
3. Goldschmidt G (1983) Doing design, making architecture. J Architect Educ 37(1):8–13
4. Goldschmidt G (1988) Interpretation: its role in architectural designing. Des Stud 9(4):235–245
5. Goldschmidt G (1989) Problem representation versus domain of solution in architectural design teaching. J Architect Plann Res 6(3):204–215
6. Cross N (1984) Developments in design methodology. Wiley, Chichester
7. Lawson B (1980/2005) How designers think: the design process demystified. Butterworth/ Architectural Press, Oxford
8. Akin Ö (1978) How do architects design. In: Latombe JC (ed) Artificial intelligence and pattern recognition in computer aided design. IFIP. North Holland Publishing, Amsterdam, pp 65–104
9. Gero JS, Maher ML (1990) Theoretical requirements for creative design by analogy. In: Fitzhorn P (ed) Proceedings of the 1st international workshop on formal methods in engineering design, manufacturing and assembly, Colorado State University, Fort Collins, pp 19–27
10. Schön DA (1983) The reflective practitioner. Basic Books, New York
11. Ericsson KA, Simon HA (1984/1993) Protocol analysis: verbal reports as data. MIT Press, Cambridge, MA
12. van Someren MW, Barnard YF, Sandberg JA (1994) The think aloud method: a practical guide to modeling cognitive processes. Academic, London
13. Goldschmidt G (1990) Linkography: assessing design productivity. In: Trappl R (ed) Proceedings of the 10th European meeting on cybernetics and systems research. World Scientific, Singapore, pp 291–298
14. Goldschmidt G (Forthcoming) Linkography: unfolding the design process. MIT Press, Cambridge, MA
15. Suwa M, Tversky B (1997) What do architects and students perceive in their design sketches? A protocol analysis. Des Stud 18(4):385–403
16. Kan JWT, Gero JS (2008) Acquiring information from linkography in protocol studies of designing. Des Stud 29(4):315–337
17. Kvan T, Gao S (2006) A comparative study of problem framing in multiple settings. In: Gero JS (ed) Design computing and cognition'06. Springer, Dordrecht, pp 245–263
18. van der Lugt R (2003) Relating the quality of the idea generation process to the quality of the resulting design ideas. In: Folkeson A, Gralén K, Norell N, Sellgren U (eds) Proceedings of 14th international conference on engineering design, Design Society, Stockholm (CD: no page numbers)
19. Boden MA (1994) What is creativity? In: Boden MA (ed) Dimensions of creativity. MIT Press, Cambridge, pp 75–117

20. Plucker JA, Renzulli JS (1999) Psychometric approaches to the study of human creativity. In: Sternberg R (ed) Handbook of creativity. Cambridge University Press, Cambridge, pp 35–61
21. Gabora L (2010) Revenge of the 'neurds': characterizing creative thought in terms of the structure and dynamics of memory. Creativity Res J 22(1):1–13
22. Dong A (2007) The enactment f design through language. Des Stud 28(1):5–21
23. Ferguson ES (1992) Engineering and the mind's eye. MIT Press, Cambridge
24. Goldschmidt G (1991) The dialectics of sketching. Creativity Res J 4(2):123–143
25. Casakin H, Goldschmidt G (1999) Expertise and the use of analogy and visual displays: implications for design education. Des Stud 20(2):153–175
26. Goldschmidt G (2012) A micro view of design reasoning: two-way shifts between embodiment and rationale. In: Carroll JM (ed) Creativity and rationale in design: enhancing human experience by design. Springer, London, pp 41–55

Why do Motifs Occur in Engineering Systems?

A. S. Shaja and K. Sudhakar

Abstract Recent years have witnessed new research interest in the study of complex systems architectures, in domains like biological systems, social networks etc. These developments have opened up possibility of investigating architectures of complex engineering systems on similar lines. Architecture of a system can be abstracted as a graph, wherein the nodes/vertices correspond to components and edges correspond to interconnections between them. Graphs representing system architecture have revealed motifs or patterns. Motifs are recurring patterns of 3-noded (or 4, 5 etc.) sub-graphs of the graph. Complex biological and social networks have shown the presence of some triad motifs far in excess (or short) of their expected values in random networks. Some of these over(under) represented motifs have explained the basic functionality of systems, e.g. in sensory transcription networks of biology overrepresented motifs are shown to perform signal processing tasks. This suggests purposeful, selective retention of these motifs in the studied biological systems. Engineering systems also display over(under) represented motifs. Unlike biological and social networks, engineering systems are designed by humans and offer opportunity for investigation based on known design rules. We show that over(under) represented motifs in engineering systems are not purposefully retained/avoided to perform functions but are a natural consequence of design by decomposition. We also show that biological and social networks also display signs of synthesis by decomposition. This opens up interesting opportunity to investigate these systems through their observed decomposition.

A. S. Shaja (✉) · K. Sudhakar
Department of Aerospace Engineering IIT Bombay, Powai, Mumbai 400076, India
e-mail: aerocomputer@gmail.com

K. Sudhakar
e-mail: sudhakar@aero.iitb.ac.in

A. Chakrabarti and R. V. Prakash (eds.), *ICoRD'13*, Lecture Notes in Mechanical Engineering, DOI: 10.1007/978-81-322-1050-4_2,

Keywords Motifs · Engineering systems · Synthesis by decomposition

1 Introduction

This section gives an introduction to the field of our research and an overview of insights proposed in this paper.

1.1 Complex Systems Architecture

Recent years have witnessed a growing interest in the study of complex systems architectures, in domains like biological systems and social networks [1]. Unifying principles have emerged [2]. Literature has commented on the hesitation of researchers in complex engineering systems, to look at their problems, in the light of emerging ideas in complex systems in general. "Engineering should be at the centre of these developments, and contribute to the development of new theory and tools" [3]; "Engineers seem a little bit indifferent as if engineering is at the edge of the science of complexity" [4].

Architecture is the fundamental structure of components of a system—the roles they play, and how they are related to each other and to their environment [5]. A layman definition of complexity refers to interconnected/interwoven components. Complexity of a system scales with the number of components, number of interactions, complexities of the components and complexities of interactions [6]. Complex engineering systems are synthesized from a large number of components coupled to each other, giving them a physical architecture. Architecture of a system (from any domain—say engineering, biology, sociology) can be abstracted as a network/graph, where the nodes/vertices correspond to components in the system and edges correspond to interconnection between them.

1.2 Literature Survey

In biology, over-represented motifs have led to interesting insights in the areas of protein–protein interaction prediction [7, 8]. For instance, in sensory transcription (protein–protein interaction) networks of biology the over-represented motif has been theoretically and experimentally shown to perform signal-processing tasks. This has led to the belief that over-represented motifs are simple building blocks of complex networks and can help understand the basic functionality of a system [7]. Importance of ideas related to motifs has recently become research interest in other domains.

Incremental ideas related to motifs have also been proposed in recent literature. Paulino et al. [9] proposed a different type of motif named 'chain of motifs' (that is, sequence of connected nodes with degree 2). They divided chains into subdivisions named cords, rings etc. depending on the type of their extremities (e.g. open or connected). The main difference between these chain motifs and the motifs by Milo et al. [7] is that the former may involve a large number of vertices and edges. They calculated the statistics of chain of motifs for few biological networks and reported the appearance of chain motifs in these networks [9].

Milo et al. [10] proposed an approach to study similarity in the structure of networks based on the Motif Significance Profile (MSP) of their graphs. These profiles are seen to be highly correlated across systems of the same family (i.e., MSPs for all systems of same type are highly correlated, e.g. Sensory transcription network of *E. coli* and Yeast of Biology family are highly correlated). Due to the distinct motif signature indicated by systems, motif significance profile signatures have also been proposed as a classifier for systems [11]. In this paper, we proceed to investigate motifs and possible reasons for its occurrence in engineering systems [12].

1.3 New Insights

Unlike biological and social networks, engineering systems are designed by humans and offer opportunity for investigation based on known design rules. We show that over(under) represented motifs in engineering systems are not purposefully retained/avoided to perform functions but are a natural consequence of design by decomposition. We also show that biological and social networks also display signs of synthesis by decomposition.

2 Theoretical Background About Motifs

"Motifs are recurring sub-graphs of interactions from which the networks are built" [7]. If a graph/network representing a system has N nodes there are NC3 3-node 'triads' in it. Some of these triads need not be connected and the rest that are connected are sub-graphs of the graph. Each 3-node sub-graph will correspond to one of 13 possible motifs (Fig. 1).

Each of the ${}^{N}C_3$ triplets, if a sub-graph, will assume the pattern of one of the 13 motifs and one can count the occurrence of each motif in a graph and define a vector, of size 13 $n = \{n_i, i = 1 \text{ to } 13\}$. In a network, the count for a particular motif may be high, which by itself is not considered important. It is possible that such high count for that motif is unavoidable for a network synthesized using the N nodes that preserve the degree distribution of the real network. To investigate this, randomized networks are created [7] using same N nodes, i.e., the number of nodes and their degree distribution is preserved. A large number of randomized networks (i = 1 to m) will define a vector of means, $\mu = \{\mu_{\mathrm{i}}, \mathrm{i} = 1 \text{ to } 13\}$ and a vector of standard deviations of motif counts, $\sigma = \{\sigma_{\mathrm{i}}, \mathrm{i} = 1 \text{ to } 13\}$. For the real network one can check the motif significance profile (MSP) of all the 13 motifs by a vector $\mathrm{Z} = \{\mathrm{Z}_{\mathrm{i},} \mathrm{i} = 1 \text{ to } 13\}$

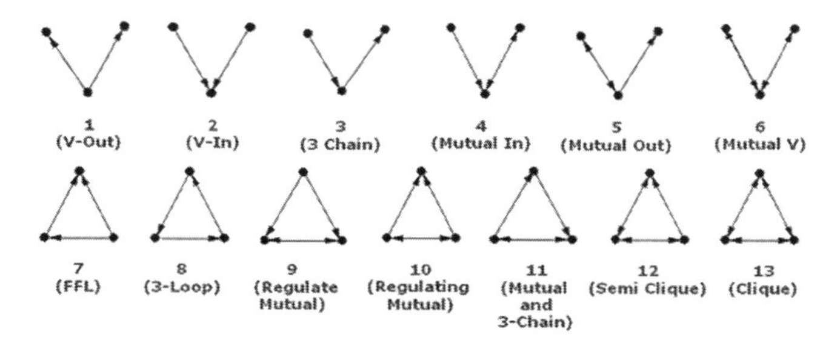

Fig. 1 All 13 patterns (motifs) for 3 node sub-graphs. The numbers are motif-ids and within brackets are nomenclature

$$[Z = (n_i - \mu_i)/\sigma_i \quad \text{for } i = 1 \text{ to } 13] \tag{1}$$

In simple words, Motif Significance Profile (MSP) is the vector (of size 13 for 3-node motifs) of the extent of over(under)-representation of all 13 motifs in a system. For a normally distributed random number, $-3 \leq Z_i \leq 3$ implies a rare occurrence (beyond $\pm 3\sigma$ limit). Any motif with its $Z_i > 2$ is considered an over-represented motif and any motif with its $Z_i < -2$ is an under-represented motif [7].

Milo et al. [10] argue that Z is influenced by the size of the network and propose normalisation of Z to make it largely independent of network size. Thus, the normalised significance profile vector, W is defined as $W = \{W_i,\ i = 1 \text{ to } 13\}$ where $W_i = Z_i/|Z|$.

3 Motifs in Engineering Systems

This section gives an introduction to the field of our research and an overview of insights proposed in this paper.

3.1 Details of Systems

We have gathered architecture data for more than 100 diverse engineering systems ranging from mechanical, software and electronic circuits. In this paper we consider 38 arbitrarily chosen systems from literature and study their architectures. Systems considered range from aircraft engine [13], softwares [14], electronic circuits [15, 16], robot [17], refrigerator [18], bacteria *E. coli* [19], yeast *S. cerevisiae* [19], language networks [19]. These 38 systems are of vastly different sizes (ranging from minimum 16 components to maximum 23,843 components). We extracted architecture data from these datasets by developing some tools for parsing/filtering from the raw data. Table 1 briefly identifies each of the 38 systems for the data.

Table 1 38 systems considered for study

S.no	System name	Nodes	S.no	System name	Nodes
1	Digital fractional multiplier (s208)	122	20	Traffic control system (s400)	186
2	Digital fractional multiplier (s420)	252	21	PLD (s820)	312
3	Digital fractional multiplier (s838)	512	22	Traffic control system (s382)	182
4	*E.coli*	423	23	ALU (74181)	87
5	Yeast	688	24	PLD (s832)	310
6	Apword	1,096	25	ECAT (c499)	243
7	Linux	5,420	26	ALU (c880)	443
8	Mysql	1501	27	ALU (c7552)	3,718
9	Vtk	778	28	PLD (s641)	433
10	Xmms	1,097	29	ECAT (c1908)	913
11	Traffic control system (s444)	205	30	ALU (c3540)	1,719
12	PLD (s713)	447	31	Traffic control system (s562)	217
13	ALU (c2670)	1,350	32	Aircraft Engine	54
14	ECAT (c1355)	1,355	33	Refrigerator	16
15	Forward logic chips (s9234)	5,844	34	Robot	28
16	Forward logic chips (s13207)	8,651	35	English	7,724
17	Forward logic chips (s15850)	10,383	36	French	9,424
18	Forward logic chips (s38417)	23,843	37	Japanese	3,177
19	Forward logic chips (s38584)	20,717	38	Spanish	12,642

In electronic circuits, nodes represent component gates and edges represent the flow of digital signals between gates. In case of software systems, nodes represent classes and edges represent directed collaboration relationships between classes. In mechanical systems, nodes represent physical components and edges represent exchange of energy, material or signal between components. In case of biological systems, nodes represent genes and edges represent direct transcription interactions. In case of languages, each word in a passage is a node and each edge represents word adjacency in the passage.

3.2 Motif Experiment and Results

We create 1,000 random networks for each considered system using same N nodes, i.e. number of nodes and their degree distribution is preserved (we have proposed a method named 'switching method' for doing this. The details of this method along with its comparison with existing method from literature are archived in our website [20]). We estimate the μ and σ of motifs of these random networks based on 1,000 random networks. For 10 arbitrarily chosen systems we create 10,000 and then 1,00,000 random networks to confirm that μ and σ of motifs counts based on 1,000 random cases are converged values. The further observations and analysis made in this paper is based on 1,000 random networks for each system.

For each real network we compute the significance of each of the 13 motifs of 3-noded sub-graphs, $Zi = (ni-\mu i)/\sigma i$ for i = 1 to 13. For example, the digital fractional multiplier s838 has n = [860, 1100, 0, 401, 0, 0, 0, 0, 40, 0, 0, 0, 0], μ = [856.9, 1213.7, 0, 397.9, 3.1, 0, 0, 0, 1.1, 0, 0, 0, 0], σ = [1.8, 3.6, 0, 1.8, 1.8, 0, 0, 0, 1, 0, 0, 0, 0] and therefore Z = [1.72, −31.89, 0, 1.72, −1.72, 0, 0, 0, 37.1, 0, 0, 0, 0]. One shall note the under-representation of motif id 2 and the over-representation of motif id 9 in the above example. We found all the systems that we studied had some or the other motif over-represented or under-represented. Z vectors are in fact computed for 3-noded, 4-noded and 5-noded sub-graphs. The results of 3-noded are available at [21] and the results of 4-noded, 5-noded are available at our website [20]. (It may be noted that the size of Z vector for 4-noded is 199 and for 5-noded is 9,364). Further study in this paper is restricted to 3-noded sub-graphs only.

3.3 What Causes Over(Under)-Represented Motifs?

All systems studied, including engineering systems, display over(under)-represented motifs; i.e., counts of some motifs in the real system are far in excess or short of their expected counts (beyond $+3\sigma$) in random graphs created using the same nodes. Such motif counts represent highly improbable events. In naturally evolving biological or social systems such motif presence can be attributed to deliberate retention to create useful functionality. But engineering systems are designed and the design process does not address functionality through motifs. So, why do over(under)-represented motifs appear in engineering systems?

4 What Causes Over-Represented Motifs

Engineering systems are designed by humans and offer opportunity for investigation based on known design rules. All engineering systems display over (under) represented motifs and they are rare events as per accepted interpretations. If designers of engineering systems explicitly retain/avoid motifs for the purpose of meeting system design requirements or system design objective, the rarity would have got explained. But we know these motifs are not retained/avoided by designers for any specific purpose. This prompts us to look for an interpretation that renders the motifs counts in a system as probable events. Thus we look for design rules that are responsible for the motifs counts in engineering systems. The motif counts when viewed without regard to those design rules will appear as rare events, but when viewed with regard to those design rules will appear as probable events. Artzy-Randrup et al. [20, 23] have argued that motifs can arise by various mechanisms other than evolutionary selection for function and highlighted for the first time that a rule in synthesis can influence motif counts in a system. They showed that a rule like "the probability of preferential connection to other nodes falling off with the physical distance between nodes" can explain the over-represented motif in neural-connectivity map of a nematode *Caenorhabditis elegans*. But that design rule was unable to reproduce the full motif significance profiles [7].

One major design rule in complex engineering systems is 'design by decomposition' that is invoked to conquer complexity. System is decomposed into sub-systems (and recursively so for very complex systems) such that nodes within each sub-system are densely inter-connected and nodes from across sub-systems are sparsely inter-connected. We investigate impact of design by decomposition on motif counts in engineering systems. Consider an arbitrarily chosen engineering system—digital fractional multiplier s832 [16]. It has $N = 512$ nodes with each node having specific in-degree and out-degree and has a motif count vector of n, i.e. $n = n_i$, $i = 1$ to 13 is the count of 13 motifs in s832. We first study expected motif counts, of random graphs synthesized monolithically, i.e. without decomposition, from these 512 nodes. This is referred to as single cluster configuration and designated by $c = 1$. Large number of such randomized graphs are created by inter-connecting all node pairs such that the degree distribution of nodes and the count of 2 node sub-graphs as in the real network are retained in the random graphs. A vector of means of motif counts, $\mu_1 = \mu_{1,i}$ where $i = 1$ to 13 and a vector of standard deviations, $\sigma_1 = \sigma_{1,i}$ where $i = 1$ to 13 are defined. Here the subscript 1 of μ and σ refers to $c = 1$. The motif significance profile (MSP) vector [10] which we have defined in Sect. 2 as, $Z_1 = (n-\mu_1)/\sigma_1$ is computed. Some elements of Z_1 have values outside of ± 3 (From the picture (1) of Fig. 2 it can be seen that $Z_{1,2} < -3$ is under-represented and $Z_{1,9} > +3$ is over-represented). With regard to these over(under) represented motifs we can take a stand that a rare event is being witnessed. But such a stand becomes not justifiable when similar rare events are witnessed for all systems. So we take the alternate stand, that the event witnessed does not belong to configuration $c = 1$ and proceed to investigate other configurations.

We then create two cluster configurations out of same 512 nodes to represent two sub-systems. Each cluster has roughly $N/2 = 256$ nodes. We create large number of random graphs by inter-connecting edges of node pairs within a cluster with higher probability ($p = 0.9$) than node pairs across clusters ($p = 0.1$) along with preserving degree distribution of nodes and the count of 2 node sub-graphs as in the real network. Vector of means of motif count, μ_2 and vector of standard deviations, σ_2 are estimated. We now define MSP as $Z_2 = (n-\mu_2)/\sigma_2$ for this

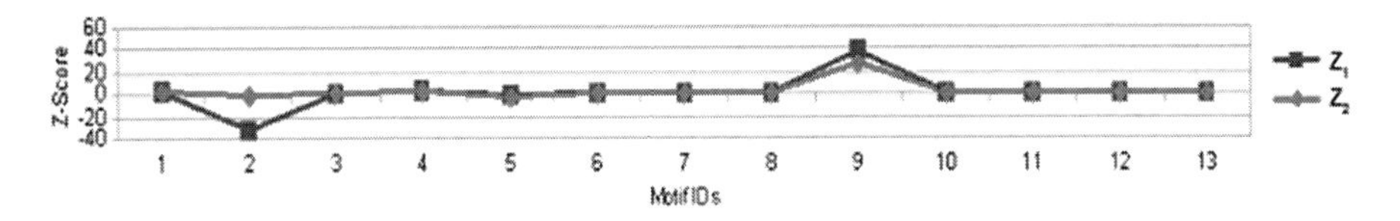

Fig. 2 MSP for Z_1 and Z_2 of digital fractional multiplier s832

c = 2 configuration. Motif significance profile vector Z_2 for clustered case (configuration c = 2) is significantly different from Z_1 (configuration c = 1). Motif id 2 ceases to be under-represented, while motif id 9 is continues to be over-represented whilst all other motif ids continue to be near zero. We similarly study cluster numbers c = 3, 4, 5, etc. and observe a clear dependence of motif significance profile vector, Z_c to clustering.

Let us assume that the real system is synthesized by the designer with k subsystems. Since k for the s832 system is not known we use the following approach: We first use Walktrap Community Detection algorithm by Pons and Latapy [23] to find the best possible sub-systems grouping for a given k, from k = 1 to k = N. In order to choose the best k out of this, we use the system modularity index proposed by Newman and Girvan [24]. The modularity index calculates how modular is a given division of a graph into subgraphs. The system modularity index for clusters k = 1 to k = N is computed and shown in Fig. 3.

When k = 1 all nodes are in one subsystem and have same probability to be connected to each other. When k = N each node is a separate cluster and has same probability to get connected to each other node. The similarity of modularity index for k = 1 and k = N is explained. Modularity index is highest for k = 38 suggesting that s832 is designed with k = 38 sub-systems. We show MSP for k = 38 as $Z_{38} = (n-\mu_{38})/\sigma_{38}$, in comparison with Z_1 in the Fig. 4. Z_{38} has no over(under)-represented motifs and hence no rare events.

We now repeat the process for aircraft engine [25] for which N = 54. The number of clusters present is discovered as k = 5 (Fig. 5).

Sosa et al. [25] have reported the number of modular sub-systems in aircraft engines as 6, which is close to what we discover here. Z_1 and Z_5 are computed for aircraft engine and compared in the Fig. 6. It can be seen that extent of over(-under)-represented motifs in Z_5 as reduced significantly compared to Z_1. We have repeated this exercise for other engineering systems to confirm the above

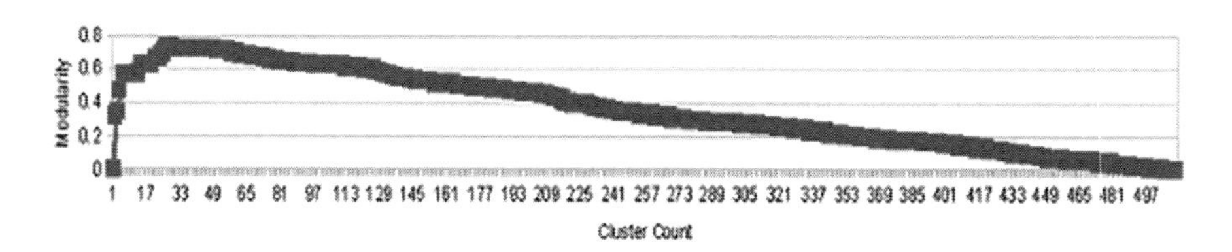

Fig. 3 System modularity index for various clusters sizes of s832

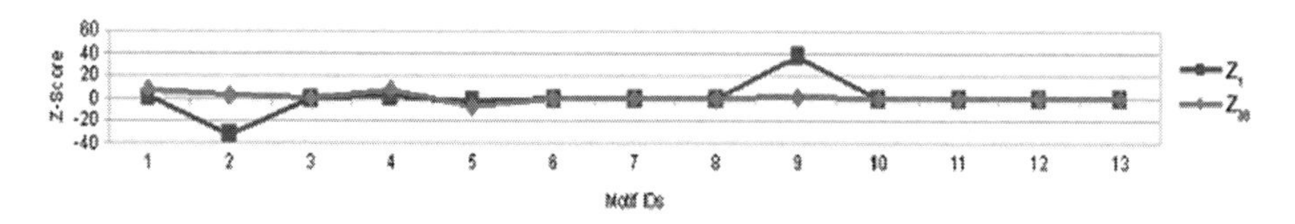

Fig. 4 MSP for Z_1 and Z_k (here k = 38) of digital fractional multiplier s832

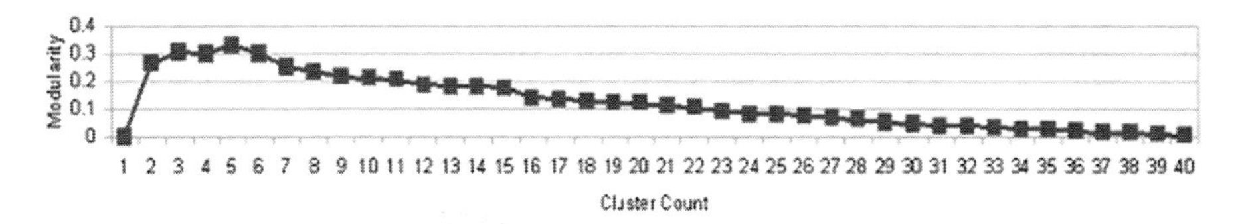

Fig. 5 System modularity index for aircraft engine [12] peaks at k = 5

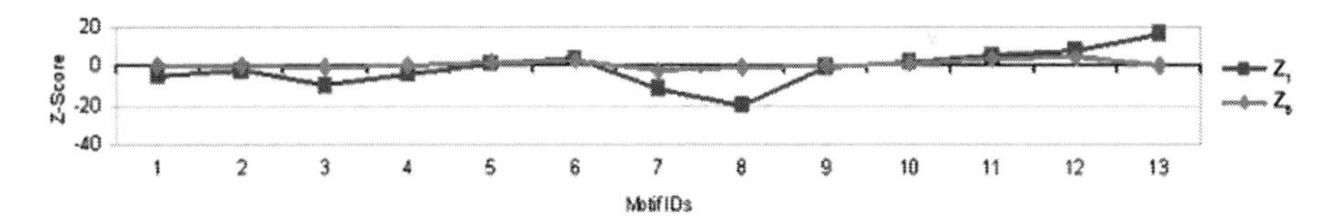

Fig. 6 MSP for Z_1 and Z_k (here k = 5) of aircraft engine

observation (the results are archived in our website [20]). We conclude that over(under) represented motifs observed are merely an outcome of comparing motif counts in a real system synthesized by decomposition to mean motif counts of random networks synthesized monolithically. Over(under) represented motifs do not show up if motif counts in the real system are compared to mean motif counts of random networks synthesized by decomposition. Randomization does not try to mimic exact nodes that go into each cluster or even exact number of nodes in each cluster, but has roughly equal number of nodes randomly picked in each cluster. But such randomization still shows remarkable likeness in motif count to real system.

5 Impact of Our Observations on Biological and Social Networks

Engineering systems are invariably designed through decompositions and it is evident that observed motif counts are a natural consequence of design by decomposition. With this backdrop of understanding for engineering system we now investigate biological systems and social networks.

We first investigate *E. coli* [19] for clustering and discover that it is not a connected graph and actually a collection of 28 sub-graphs not connected to each other. We investigate this collection of 28 sub-graphs to discover 49 subgraphs[1] (Fig. 7). We estimate Z_k for k = 28 and 49 and compare it with Z_1 (Fig. 8) and find a reduction in the extent of over(under) representation of the significant

[1] Out of the 28 sub-graphs, the big enough ones are decomposed further to discover 49 sub-graphs.

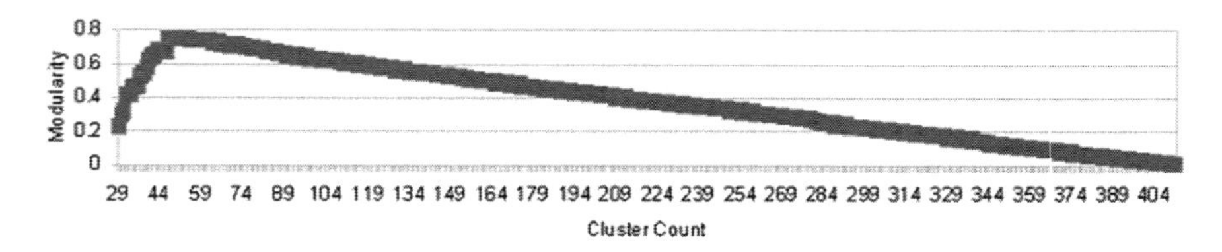

Fig. 7 System modularity index for *E. coli* [19] peaks at k = 49

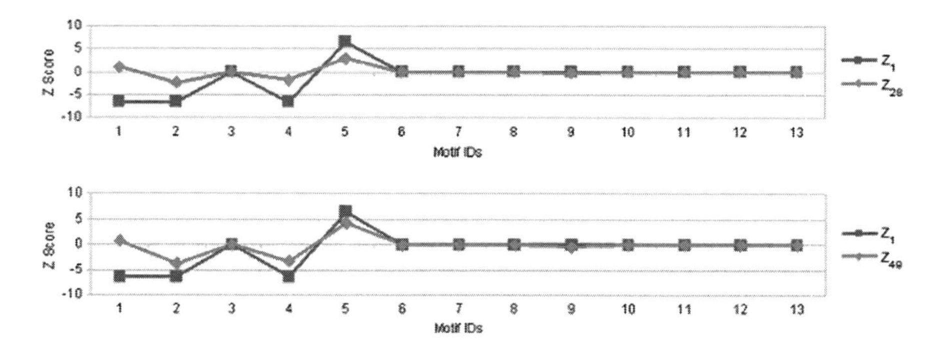

Fig. 8 MSP for Z_1 and Z_k (here k = 28, k = 49) of *E. coli*

motifs, though the reduction is not as dramatic as in engineering systems. There could be other rules apart from clustering that are present in these systems that may further reduce the extent of over(under) representation.

It is not clear why a bio-logical system must have sub-systems (clusters). Previous researchers have studied the role of over-represented motifs in a biological system. We feel it could be more revealing to investigate role of clustering. What function do clusters of specific nodes with dense interconnections perform in biological system may lead to interesting and useful findings.

We finally investigate a social network, representing games played between American (NCAA) college football teams during the year 2000. Radicchi et al. [26] have reported the number of modular teams in football system under study as 9, which is same as what we discover here k = 9 (Fig. 9). We estimated Z_9 and compared it with Z_1 (Fig. 10). It can be seen that extent of over(under)-represented motifs in Z_9 has reduced significantly (almost close to zero indicating no rare

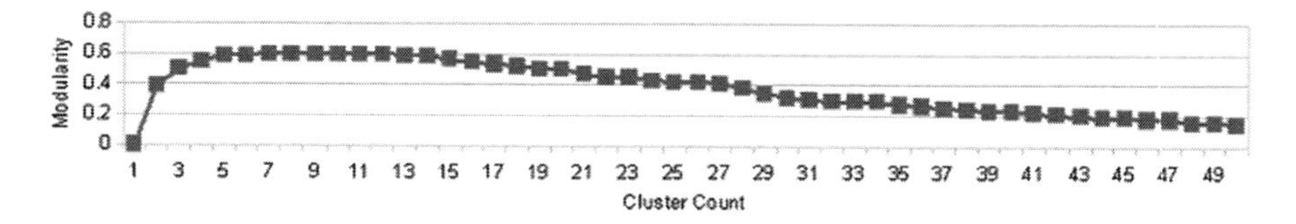

Fig. 9 System modularity index for football [20] peaks at k = 9

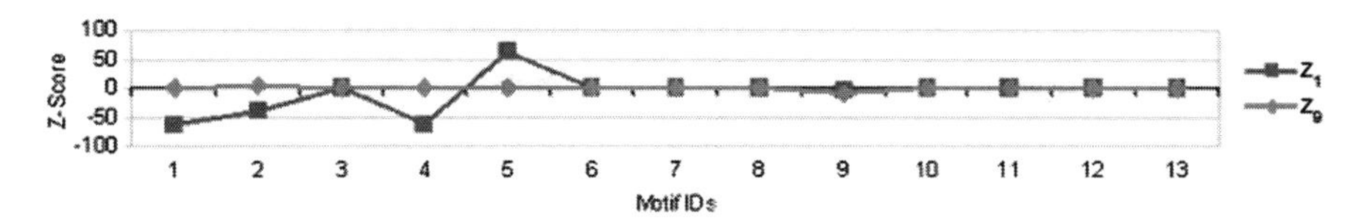

Fig. 10 MSP for Z_1 and Z_2(here k = 9) of football

events) compared to Z_1. This again implies that over(under) represented motifs observed are merely an outcome of comparing motif counts in a real system synthesized by decomposition to mean motif counts of random networks synthesized monolithically.

6 Computational Aspects

All the code that we have developed as part of this research as a software framework named CASMot. Some computational aspects about this framework are mentioned in Table 2.

Table 2 Computational aspects related to CASMot

Features	
Functionalities supported by CASMot framework	Automated scripts to convert raw domain specific data to network, discover over(under)-represented motifs, create MSP, Create CCP, perform decomposition Analysis.
Software	
Operating system	Debian Linux with kernel version 2.6
Programming language	Statistical R, Erlang, Bash shell scripting
Lines of code	48539
Main software paradigm	Functional programming using map-reduce architecture
Hardware	
CPU1	Eight core CPU 2 nos with a processing speed of 1.5 GHz
CPU2	Dual core CPU 5 nos with a processing speed of 1.5 GHz
RAM1	2 GB in the dual core machines
RAM2	16 GB in the eight core machines
Hard disc	80 GB in the dual core machines 500 GB in the eight core Machines
Computational effort	
After harnessing the computing capacity of both the hardware computational effort to run motif experiments	Computations required to generate MSPs of the 38 systems took roughly 850 h (approximately 35 days)

The reader is requested and encouraged to refer to our webpage [20] for the algorithms used for producing clustered random graphs, how to use our distributed software framework named CASMot for doing motif experiments etc.

7 Conclusion

Ideas related to complex system architectures may give insight into previously complex and poorly understood phenomena in engineering domains. Barabasi [27] argues that, "The science of networks is experiencing a boom. But despite the necessary multi-disciplinary approach to tackle the theory of complexity, scientists remain largely compartmentalised in their separate disciplines". The application of this complex system architectures theory is still in infancy and has very recently entered into study of engineering systems or their design. We have shown that over(under) represented motifs in engineering systems are not purposefully retained/avoided to perform functions but are a natural consequence of design by decomposition. We also have shown that biological and social networks also display signs of synthesis by decomposition. This is shown by considering 38 arbitrarily chosen systems ranging from—biology systems, languages, electronic circuits, software systems and mechanical engineering systems. This study has thrown some new insights about Classification of Systems from Component Characteristics.

Acknowledgments We thank Centre for Aerospace Systems Design & Engineering, IIT Bombay, India for the excellent research environment and Aeronautics Research and Development Board (ARDB), India for this project grant. We are also thankful to Mr. Mahesh for his support in plotting few graphs.

References

1. Watts DJ (2004) Six degrees: the science of a connected age. Norton and Company, New York
2. Boccaletti S, Latora V, Moreno Y, Chavez M, Hwang D (2006) Complex networks: structure and dynamics. Phys Rep 424(4):175–308. doi:10.1016/j.physrep.2005.10.009
3. Ottino JM (2004) Engineering complex systems. Nature 427:399. doi:10.1038/427399a
4. Jiang Z-Q, Zhou W-X, Bing X, Yuan W-K (2007) Process flow diagram of an ammonia plant as a complex network. AIChE J 53(2):423–428. doi:10.1002/aic.11071
5. ANSI IEEE Standard 1471. http://www.iso-architecture.org/ieee-1471/
6. Crawley E, de Weck O, Eppinger S, Magee C, Moses J, Seering W, Schindall J, Wallace D, Whitney D (2004) The influence of architecture in engineering systems. MIT Eng Syst Div (Monograph)
7. Milo R, Shen-Orr S, Itzkovitz S, Kashtan N, Chklovskii D, Alon U (2002) Network motifs: simple building blocks of complex networks. Science 298:824–827

8. Ronen M, Rosenberg R, Shraiman BI, Alon U (2002) Assigning numbers to the arrow: parameterizing a gene regulation network by using accurate expression kinetics. In: Proceedings of the national academy of sciences vol 99, no 16, pp 10555–10560
9. Villas Boas PR, Rodrigues FA, Travieso G, Da Fontoura Costa L (2008) Chain motifs: the tails and handles of complex networks. Phys Rev E—Stat, Nonlinear Soft Matter Phys 77(2):026106
10. Milo R, Itzkovitz S, Kashtan N, Levitt R, Shen-Orr S, Ayzenshtat I, Sheffer M, Alon U (2004) Superfamilies of evolved and designed networks. Science 303(5663):1538–1542
11. Ahnert SE, Fink TMA (2008) Clustering signatures classify directed networks. Phys Rev E—Stat, Nonlinear Soft Matter Phys 78(3):036112
12. Blanchard BS, Fabrycky WJ (1998) Systems engineering and analysis, 3 edn. Prentice Hall, Upper Saddle River
13. Sosa ME, Eppinger SD, Rowles CM (2003) Identifying modular and integrative systems and their impact on design team interactions. J Mech Des 125(2):240–252. doi:10.1115/1.1564074
14. Software graph data for specified software systems. http://www.tc.cornell.edu/~myers/Data/SoftwareGraphs/index.htm
15. ISCAS High level models. http://www.eecs.umich.edu/~jhayes/iscas.restore/
16. ISCAS'89 benchmark data. http://www.pld.ttu.ee/~maksim/benchmarks/iscas89/bench/
17. Farid AM, McFarlane DC (2006) An approach to the application of the design structure matrix for assessing, reconfigurability of distributed manufacturing systems. In: Proceedings of the IEEE workshop on distributed intelligent systems: collective intelligence and its applications, art no. 1633429, pp. 121–126. doi: 10.1109/DIS.2006.10
18. Pimmler TU, Eppinger SD (Sep 1994) Integration analysis of product decompositions. ASME design theory and methodology conference, Minneapolis
19. Datasets for Bacteria *E-coil*, yeast *S. cerevisiae* and language networks http://www.weizmann.ac.il/mcb/UriAlon/groupNetworksData.html
20. Artzy-Randrup Y et al Casmot: motif related tool made at centre for aerospace systems design and engineering. Indian Institute of Technology. https://www.casde.iitb.ac.in/complexsystems/motif/results
21. Shaja AS, Sudhakar K (2009) Overrepresented and underrepresented patterns in system architectures across diverse engineering systems, in 19th annual INCOSE international symposium, (Singapore)
22. Artzy Randrup Y, Fleishman SJ, Ben Tal N, Stone L (2004) Comment on network motifs: simple building blocks of complex networks and superfamilies of evolved and designed networks. Science 305(5687):1107c
23. Pons P, Latapy M (2006) Computing communities in large networks using random walks. J. Graph Algorithms Appl 10(2):191–218
24. Newman M, Girvan M (2004) Finding and evaluating community structure in networks. Phys Rev E 69:026113
25. Sosa ME, Eppinger SD, Rowles CM (2003) Identifying modular and integrative systems and their impact on design team interactions. J Mech Des 125(2):240–252
26. Radicchi F, Castellano C, Cecconi F, Loreto V, Parisi D (2004) Defining and identifying communities in networks. In: Proceedings of the national academy of sciences of the United States of America, vol 101(9), pp 2658–2663
27. Barabási A-L, Complexity Taming (2005) Taming complexity. Nat Phys 1:68–70. doi:10.1038/nphys162

Thinking About Design Thinking: A Comparative Study of Design and Business Texts

Marnina Herrmann and Gabriela Goldschmidt

Abstract In the past 10 years design thinking has become a popular buzzword in the design and business communities alike. While much has been written on the subject in both academic and popular literature no consensus has been reached as to its actual definition and nature. The following study employs a semiotic analysis in order to identify lexical patterns that can provide us with insight into many underlying principles of design thinking.

Keywords Design thinking · Language · Semiotics · Business

1 Introduction

It was in the 1960s that the stage was first set for the design thinking revolution. Individuals such as Simon [1], McKim [2], Rittel and Webber [3] first introduced the world to concepts such as design as a way of thinking and planning, systems thinking, and the solving of wicked problems. As these ideas developed throughout the second half of the last century and helped establish a practice of human-centered design, they gained popularity and began to receive the attention of other disciplines.

Today a quick Amazon search of the term reveals its sudden burst of popularity in the past few years. Unfortunately, reading the literature only reveals just how ambiguous the term actually is. There are those who define design thinking in terms

M. Herrmann (✉) · G. Goldschmidt
Israel Institute of Technology, Haifa 32000, Israel
e-mail: marnina@technion.ac.il

G. Goldschmidt
e-mail: gabig@technion.ac.il

A. Chakrabarti and R. V. Prakash (eds.), *ICoRD'13*, Lecture Notes in Mechanical Engineering, DOI: 10.1007/978-81-322-1050-4_3, © Springer India 2013

of its goals, others who define it in terms of process and methodology and others still who define it in terms of ideology. While there does not appear to be a consensus as to its meaning explicitly stated in any of the current literature, when one digs a little deeper into the language design thinkers use to discuss their process, certain patterns rise to the surface. A deeper understanding of these patterns is crucial in order for the design thinking discourse to move forward in a meaningful way.

Design thinking has not only gained popularity in the design world, but in the business world as well. Business leaders did not adopt a design thinking ideology or methodology overnight. Rather, it was the result of a long process that began with the acknowledgment that design had a positive impact on the bottom line, as well as the more recent shift from designer as solitary practitioner to an integral part of a collaborative business team [4]. This shift came about as a result of business leaders' realization that to stay competitive in the modern world, efficiency was no longer enough. In today's innovation-based economy consumers want more varied options and richer experiences. Design thinking assists businesses in delivering these options and experiences [5].

With businesses turning to designers with new challenges and responsibilities, designers have been given their long awaited permission to expand the reach of their own practice. To effectively do this they used design thinking as a way to expand their definition of design. In his book "Design Thinking" Lockwood, president of the Design Management Institute describes design thinking as:

> ...a human-centered innovation process that emphasizes observation, collaboration, fast learning, visualization of ideas, rapid concept prototyping, and concurrent business analysis, which ultimately influences innovation and business strategy. [6, p. xi]

While none of these are new ideas, by combining these concepts and verbalizing them designers are able to more effectively communicate and exchange ideas with collaborators from the business world. It is this exchange of ideas that will help to bring design to the front and center of the business world and further propel businesses into the 21st century.

While the term design thinking had no one single definition or description of its nature it is clear from the literature that there are specific commonalities and differences in how the two communities described it. This study brings those commonalities and differences to light by examining the choice of language used by authors of design thinking texts.

2 Methodology

The study employed an in-depth literature survey, semi-structured interviews with design thinkers from various different fields, a modified semiotic analysis of design thinking texts and a comparison of the results garnered from the analysis using a two factor design—factorial analysis. The examination of both formal and informal sources of information will allow for a broader understanding of the evolution of the term design thinking. Because the study is primarily qualitative it can take on a more flexible nature. Rather than develop a hypothesis and try to prove it, the conclusions presented arose inductively through the classification and examination of the gathered data in a manner similar to grounded theory.

2.1 *Semiotic Analysis*

The bulk of the study consists of a modified semiotic analysis. Semiotics is a branch of linguistics that seeks to understand how signs convey meaning in context [7]. Semiotics can be applied to anything which signifies something [8] be that a word, picture, gesture, etc. Semiotic analysis uses various analytical tools to determine the meaning of signs in different contexts. Semiotic analysis enables texts of all types of media to be analyzed in the same way. For this study the primary tool used was syntagmatic analysis, which refers to the contextual relationship between two or more linguistic units.

An assortment of both business and design texts from a variety of media were collected. The inclusion criteria were that the texts were written no earlier than the year 2000 (as it was only after 2000 that the term design thinking gained its popularity in the business world), that their primary purpose was to discuss design thinking as a whole and that the term design thinking was actually used. Texts were categorized by community (business or design) and medium. The community was determined primarily according to the background of the author, but content was taken into account as well. Exact definitions for the term design thinking (when explicitly stated) were extracted from the texts. Terms that were syntagmatically related to the term design thinking were then extracted from the text and categorized according to their use as describing: goals and outcomes, process, ideology, comparable terms, qualities, and context. Not all texts contained terms that fit into all six categories. As well, some texts had words that fit into multiple categories. Each list of terms was then alphabetized and repetition of terms within the individual lists was removed. Then, all terms were combined into two lists: design and business, and the terms in each of these lists were counted and organized according to frequency of appearance.

2.2 *Factorial Analysis*

Following the semiotic analysis a two factor design—factorial analysis was applied. 50 terms were used that were most prominent in the semiotic analysis. Texts were then reviewed for the appearance of these 50 terms. Each term was then given a value representing the proportion X/N where X is the number of times a term appears and N is the number of texts in each category (in our case 9). This information was then graphed.

3 Results

Results were culled from an analysis of 18 texts,[1] nine design and nine business. Within each category there was two journal articles, three magazine or newspaper articles, one book, one essay and two blog posts. The number of extracted terms was not limited to a specific number and the final number of terms was somewhere in the hundreds.

3.1 Semiotic Analysis: Commonalities

Over all the design thinking process looks fairly similar within both communities. Both communities see design thinking as a repetitive (Fig. 1), collaborative (Fig. 2) and fast (Fig. 3) process. Both employ an expansive toolbox of methodologies and ways of thinking (Fig. 4) toward reaching multiple solutions.

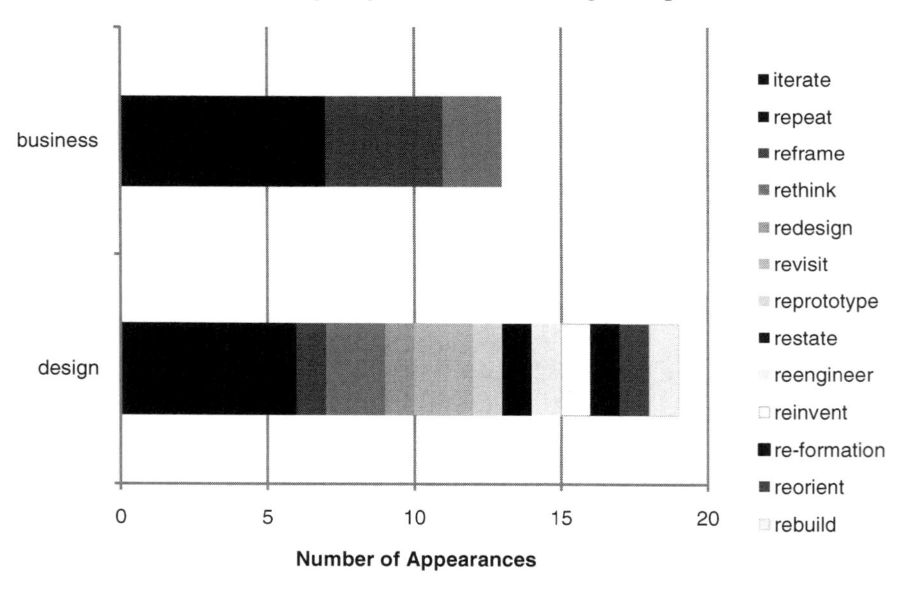

Fig. 1 Design thinking is a repetitive process

[1] Texts include but are not limited to: Leavy, "Design Thinking—a new mental model of value innovation", Strategy and Leadership, 2010. Wylant, "Design Thinking and the Experience of Innovation", Design Issues, 2008. Teal, "Developing a (Non-linear) Practice of Design Thinking", The International Journal of Art and Design Education, 2010. Ward, Runcie and Morris, "Embedding innovation: design thinking for small enterprises", Journal of Business Strategy, 2009. Brown, "Change by Design", Harper Business, 2009. Martin, "The Design of Business", Harvard Business School Press, 2009.

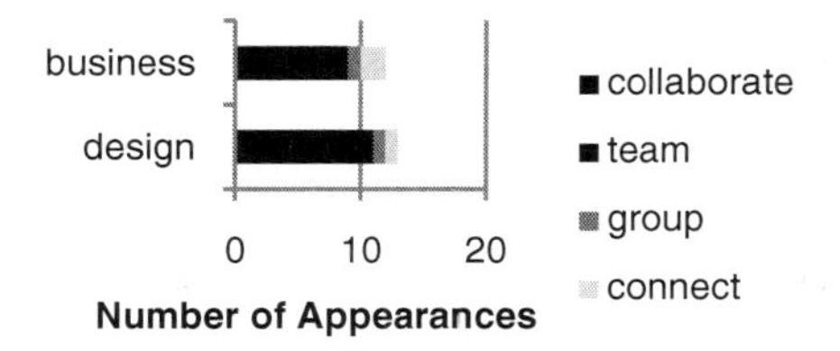

Fig. 2 Design thinking is a collaborative process

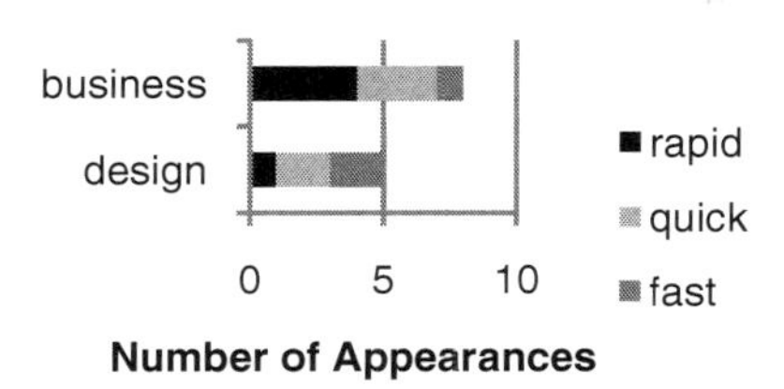

Fig. 3 Design thinking is a fast process

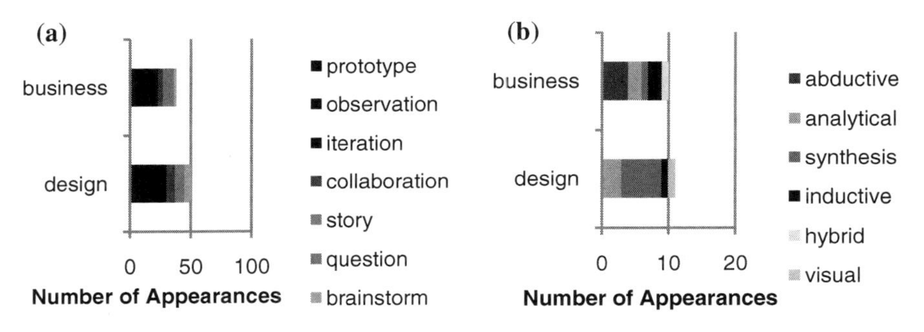

Fig. 4 Design thinking has multiple methodologies (a) and ways of thinking (b)

The similarities visible within the design thinking community not only help to bridge the gap between the design and business worlds but help give us a framework within which to identify differences as well.

3.2 Semiotic Analysis: Differences

Designers tend to look at design thinking as more of an experiential and immersive process (Figs. 5, 6 and 7) whereas business sees it as more of a knowledge based process (Fig. 8a). This difference in perspective has various practical implications. To designers, design thinking is more about what you do not know than what you do. It is a process guided by curiosity and asking questions which places a strong emphasis on decision making through experimenting and learning (Fig. 8b). The business community on the other hand favors a knowing and judging model

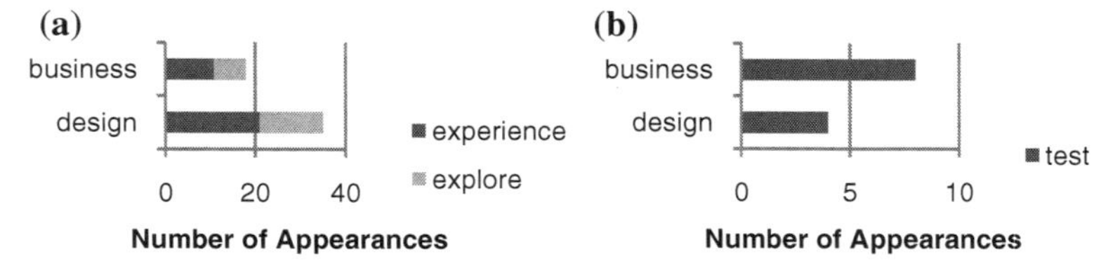

Fig. 5 Design emphasizes a more experiential process (**a**) whereas business emphasizes a process based on testing (**b**)

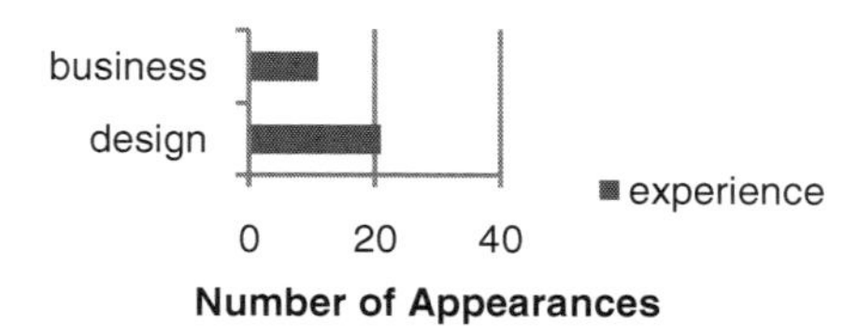

Fig. 6 Design places a stronger emphasis on experience

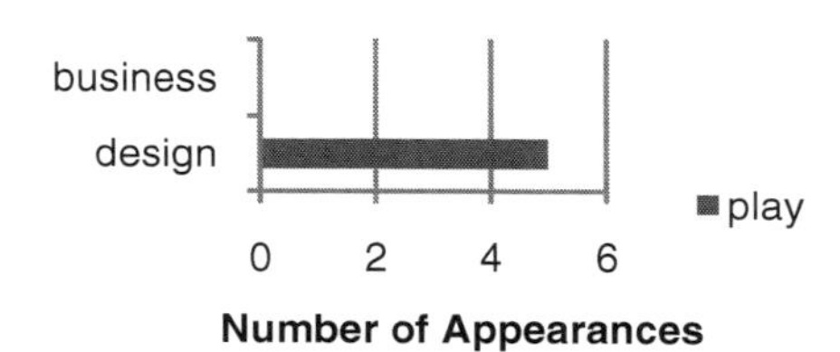

Fig. 7 Design emphasizes a playful process

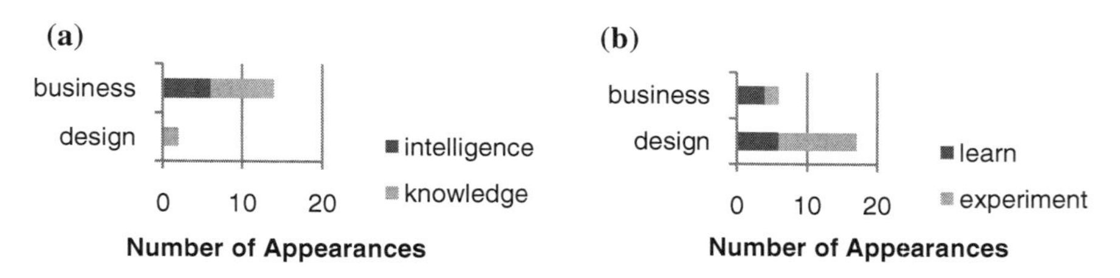

Fig. 8 Business emphasizes knowledge and intelligence (**a**) whereas design emphasizes a process based on learning and experimentation (**b**)

(Figs. 8a and 9). Design's leanings toward learning and experimentation explain its openness to failure as a learning experience and an important part of the design thinking process (Fig. 10).

An additional difference between the two communities is business's preference for verbal communication (Fig. 11a) as compared to design's focus on visual communication (Fig. 11b). This in mind, it is no surprise that the design community also puts a strong emphasis on complexity and patterns while the business

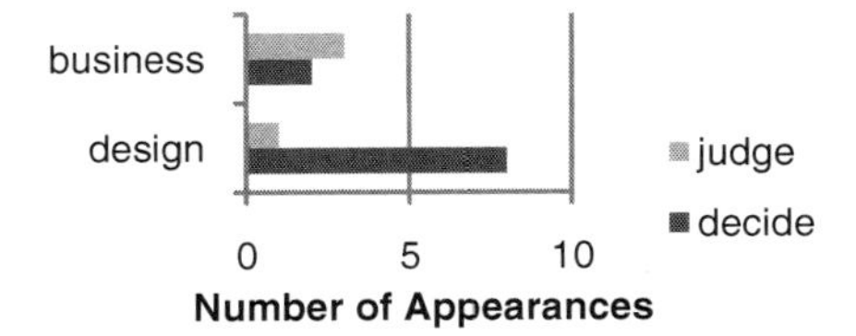

Fig. 9 Business emphasizes judging while design emphasizes deciding

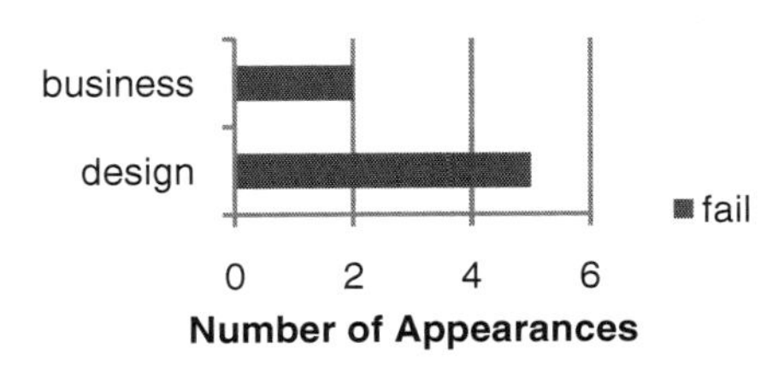

Fig. 10 Design puts a stronger emphasis on failure as part of the process

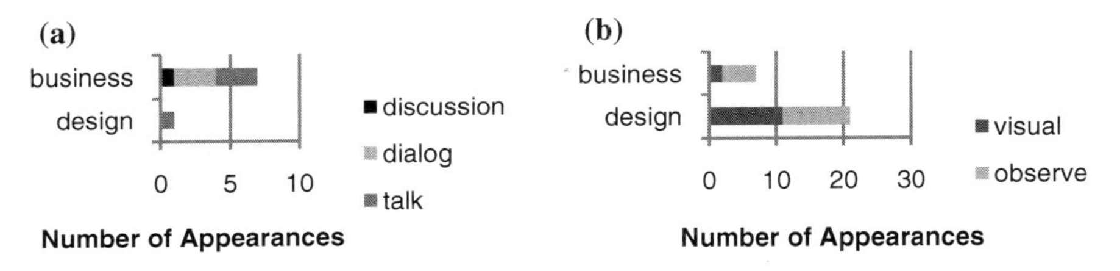

Fig. 11 Business emphasizes verbal communication (**a**) whereas design emphasizes visual communication (**b**)

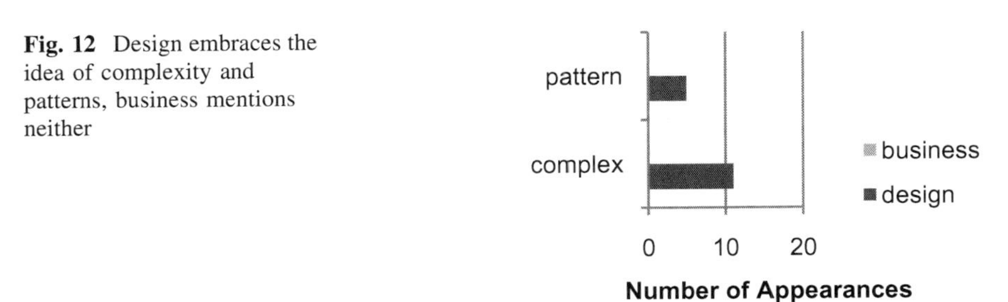

Fig. 12 Design embraces the idea of complexity and patterns, business mentions neither

community does not (Fig. 12). Another significant difference is the business community's emphasis on innovation. While both communities give equal weight to creativity and imagination, the term innovation appeared much more frequently in business texts (Fig. 13).

Many of the differences highlighted above are interconnected and follow certain patterns that align with their respective communities' various ideologies and philosophies.

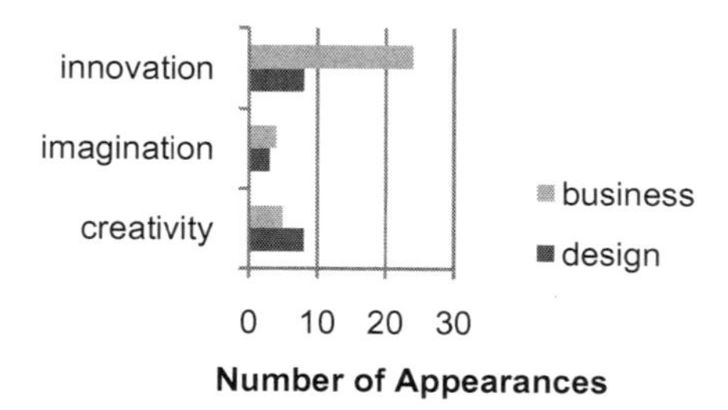

Fig. 13 Both categories put a similar emphasis on imagination and creativity but business emphasizes innovation

3.3 Factorial Analysis

For the most part the factorial analysis (Fig. 14) was not an independent data source but rather a tool used to interpret the results of the semiotic analysis. The results of the two forms of analysis were very different. This was because the factorial analysis answered a *yes* or *no* question: "was (a form of) the term present in the text at least once?" The semiotic analysis on the other hand examined the terms in relation to form and meaning. Terms could be counted multiple times in

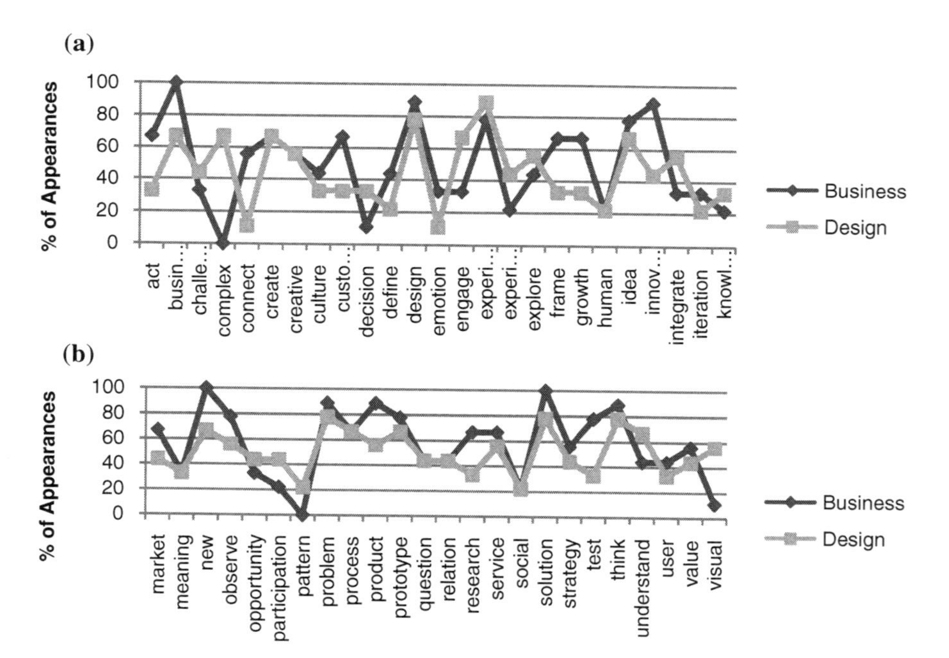

Fig. 14 a, b Factorial analysis shows the percentage of texts in which the 50 most prominent words from the semiotic analysis appeared

individual texts if they were used to describe different aspects of design thinking or if they were used in a different form (either grammatical or part of a larger term, e.g., product and product design).

It is interesting to note the prominence of the words 'design' and 'business' in both categories of texts (Fig. 14a) proving the interconnectedness of the two industries. Additionally, the factorial analysis highlights the importance of the problem solving process as part of the design thinking philosophy (Fig. 14b).

4 Discussion

Design thinking is comprised of many possible elements. Some of these elements are unique to a particular individual, studio or organization; others are consistent within a particular community (design and business being the ones examined here), a few appear to be universal. The semiotic and factorial analyses resulted in a variety of interesting findings that highlight various lexical patterns within the design thinking literature that provide us with insight into these different elements allowing us to pinpoint commonalities and differences between the two communities. These commonalities and differences are reflective of the philosophies of the individual communities as well as their relationship to one another.

4.1 Commonalities

The disciplines of design and business are undeniably interconnected. Business provides design with a platform for identifying problems and distributing the solutions, whereas design provides business with added value and a way to connect with users on an emotional level. In order to work together the two must have some common ground. Design thinking provides this common ground by acting as a vehicle for the creative process. It allows business people to be more imaginative and adventurous in their work and allows designers to implement their strategies in a more structured environment.

One of the reasons design thinking serves this purpose so well is because of its emphasis on choice. Design thinking does not force one single methodology or way of thinking on a person but rather provides a toolbox of various methodologies and ways of thinking (Fig. 4) along with a loose framework in which to implement the chosen tools.

While different communities may favor different tools (for example design favors observation whereas business favors questions) the framework enables the tools to be used in a similar manner and for similar goals. The basis for this framework is repetition (Fig. 1), collaboration (Fig. 2) and speed (Fig. 3). These three characteristics allow design thinkers from either discipline to churn out and test multiple solutions to any number of problems as quickly and inexpensively as possible.

Additionally, this framework allows for the design process to be more interactive giving people the opportunity to provide feedback and learn along the way [9].

4.2 Differences

It is when one starts to look at some of the underlying principles behind the different communities' use of design thinking that differences come out. Designers tend to view design thinking as a learning-based process (Fig. 7b) whereas the business community sees it as a knowledge-based process (Fig. 7a). This dichotomy results in designers having a much more immersive design thinking experience. For example, design texts tend to emphasize experience, exploration, experimentation, play and failure (Figs. 5a, 6, 7b, 8b and 10) and business texts emphasize a more passive approach using tools such as testing (Fig. 5b) and intelligence (Fig. 8a). This dichotomy can also be used to explain business' preference for judging (a more top-down, exclusionary action) over design's more active choice of deciding (Fig. 9).

> For designers, the process of building, prototyping and trying things is the decision making process. Instead of boiling down a problem to one large decision, designers make lots of little decisions, learning as they go. As they build and learn, something interesting happens: through the iterations, the best option often reveals itself and the other less-appropriate options fall by the wayside. [10, p. 36]

The debate between the learning-based model and knowledge-based model is most likely related to what Martin, Dean of the Rotman School of Management, refers to as the validity versus reliability debate. Designers tend to search out valid solutions. Valid solutions are solutions whose success cannot be proved at the moment, their success can only be predicted through experimentation and learning. Reliable solutions on the other hand can be proved using statistics, facts and past events. Traditionally, designers tend to search out valid solutions whereas business people tend to search out reliable ones [11]. The learning-based model arose out of the design community's search for valid solutions, whereas the knowledge-based model arose out of the business community's search for reliability.

Another difference between the two communities is the design community's emphasis on visual communication versus business' emphasis on verbal communication (Fig. 11). This could be because of design's emphasis on complexity (and in turn patterns) a theme that reoccurs in many design thinking texts but was not present in the business texts (Figs. 12 and 14). Visualizations can be used as an aid to understand complex problems and to search out patterns. Additionally, the interviews conducted revealed that many designers found verbal language to be an unclear communication method. One example is that many (even those who wrote extensively on the subject) stated in the interviews that they did not like the *term*

[2] Based on an interview with Nir in Moshav Bnei Ataroton July 18, 2011.

design thinking, a possible reason for this being that design has always been able to rely on the visual in order to communicate[2] and so verbal communication becomes less important.

Lastly, there is business' focus on innovation. While both communities put similar emphasis on creativity and imagination business places a much stronger emphasis on innovation (Fig. 13). While the two words are often used interchangeably there are differences. For example, according to Bob Hambly, creative director of Hambly and Wooly,[3] innovation is about moving forward and challenging what is out there while creativity can go in any direction (including examining the past). According to Yariv Sade of Igloo Design[4] innovation is external while creativity is internal. To be innovative one's idea or product has to be something new when compared to what is out there but to be creative one only has to do what is new for them. According to Mark Leung, associate director of Rotman Design Works[5] innovation is about doing something differently from the status quo; it is applied creativity and therefore all businesses are creative. While design texts gave the same prominence to both words, business texts favored the term innovation. This could be for two reasons. The first being that the recent fixation with innovation came about in the business world as a response to a need to adapt to the 21st century consumer experience [5] whereas in the design world innovation is less about a specific objective and more about something that just goes hand in hand with practice. Another reason for this could relate to the fact that innovation is itself seen as a buzzword, and that generally speaking business people are more comfortable with rhetoric and dialog (Fig. 11a).

5 Conclusion

Despite differences in how individuals define and describe design thinking there are definite linguistic patterns evident within the texts available from each community. These patterns can provide insight into how design thinking manifests in the real world, as well as reveal certain truths about the design and business communities. Additionally, the design thinking community as a whole does have many notable similarities when it comes to their vision of design thinking. These patterns are what will enable design thinking to become an effective communication tool among designers and business people, allowing design thinking to hold a strategic role in the future of design and business not only as individual disciplines, but as collaborative partners. This collaboration will elevate the role of designer to one that is more central in the business process as well as provide businesses with better product offerings and a more integrated business process.

---

[3] Based on an interview with Hambly on April 28, 2011 at Hambly and Wooly.

[4] Based on an interview with Sade on June 15, 2011 at the Technion.

[5] Based on an interview with Leung conducted on April 27, 2011 at Design Works.

Acknowledgments Thank you to Professor Ayala Cohen for her assistance with the factorial analysis and to all the individuals who took the time to be interviewed over the course of this study.

References

1. Simon HA (1969) The sciences of the artificial. MIT Press, Cambridge
2. McKim R (1973) Experiences in visual thinking. Brooks/Coles Publishing Co, Belmont
3. Rittel HWJ, Webber MM (1973) Dilemmas in a general theory of planning. Policy Sci 4: 155–169
4. Shedroff N (2010) An opportunity for shared strategy. Innov J 29:45–48
5. Kumar V (2009) A process for practicing design innovation. J Bus Strategy 30:91–100
6. Lockwood T (ed) (2010) Design thinking. Allworth Press, New York
7. Manning PK (1987) Semiotics and fieldwork. Qual Res Methods 7
8. Chandler D (2004) Semiotics: the basics. Routledge, New York
9. Kelley T (2001) Prototyping is the shorthand of innovation. Des Manage J 12:35–42
10. Raney C, Jacoby R (2010) Decisions by design: stop deciding and start designing. Rotman Mag 34–39
11. Martin R (2007) Design and business: why can't we be friends? J Bus Strategy 28:6–12

Advancing Design Research: A "Big-D" Design Perspective

Christopher L. Magee, Kristin L. Wood, Daniel D. Frey and Diana Moreno

Abstract Advances in design research representations and models are needed as the interfaces between disciplines in design become blurred and overlapping, and as design encompasses more and more complex systems. A conceptual framework known as "Big-D" Design, as coined by Singapore's newest national university (the Singapore University of Technology and Design or SUTD), may provide a meaningful and useful context for advancing design research. This paper is an initial examination of the implications for scientific design research on using this particular framework. As part of the analysis, the paper proposes a simplified decomposition of the broader concept in order to explore potential variation within this framework. It is found that many research objectives are better investigated when the broader design field is studied than in a singular category or domain of design. The paper concludes by recommending aggressive attempts to (1) arrive at a coherent set of terminology and research methodologies relative to design research that extend over at least all of technologically-enabled design and

C. L. Magee (✉)
SUTD-MIT International Design Center Cambridge, MA and Singapore MIT,
77 Massachusetts Avenue, Building E38-450, Cambridge, MA 02139-4307, USA
e-mail: cmagee@mit.edu

K. L. Wood · D. Moreno
SUTD-MIT International Design Center Cambridge, MA and Singapore SUTD,
20 Dover Drive, Dover 138682, Singapore
e-mail: kristinwood@sutd.edu.sg

D. Moreno
e-mail: Diana_moreno@sutd.edu.sg

D. D. Frey
SUTD-MIT International Design Center Cambridge, MA and Singapore MIT,
77 Massachusetts Avenue, Building 3-449D, Cambridge, MA 02139-4307, USA
e-mail: danfrey@mit.edu

A. Chakrabarti and R. V. Prakash (eds.), *ICoRD'13*, Lecture Notes in Mechanical Engineering, DOI: 10.1007/978-81-322-1050-4_4, © Springer India 2013

(2) perform epistemological and ontological studies of the relationship of engineering science and technologically-enabled design science as there is more overlap between them than is generally recognized.

Keywords Design research · Technologically-intensive design · Heuristics · Principles · Design theory

1 Introduction

A novel concept central to future innovation economies and the fields of engineering and architecture is "Big-D Design. As created by and used as a vision for Singapore's newest national university (the Singapore University of Technology and Design or SUTD), "***Big-D Design***" *includes all technologically-intensive design, from architectural design to product design, software design, and systems design. It is design through conception, development, prototyping, manufacturing, operation, maintenance, recycling, and reuse—the full value chain. It includes an understanding and integration of the liberal arts, humanities, and social sciences.* In short, Big-D encompasses the art and science of design; for more information about the university and its concepts, see references [1, 2].

The authors are all associated with the SUTD-MIT International Design Center which is metaphorically seen as the cardiovascular system of SUTD. Our research agenda aims to extend Big-D Design science and Big D Design practice through new methods, theories, principles, heuristics, pedagogy, technologies, processes, and the development of leaders in Design. This paper explores the implications on ours and any design research agenda of taking a Big-D Design perspective. Three elements of this Design perspective and the nature of the questions they lead to about design research are:

1. All technologically-enabled design: what are the advantages and disadvantages of committing to research across this breadth of fields?
2. Full Value Chain: What are the benefits and potential pitfalls of engaging the full value chain in Design research?
3. Art and Science of Design: Does design research entail both of these? Must it?

In order to adequately address these questions, we examine prior design research and establish criteria we are using in this examination. Our criteria are twofold: the first is to continue building a cumulative research enterprise around design so that future research builds upon a reliable base and has continually more impact on Big D Design research communities, and Big D Design education. The second criterion, which we consider equally important, is to impact design practice favorably, in new and exciting ways. Research results can impact practice favorably by development of new methods, theories, guidelines, heuristics and principles that when applied directly lead to superior results for practicing engineers and teams. Design research output can also favorably impact practice through results that point to superior education methods that can involve better basic knowledge structure to support design and better exposure to methods and experiences that are effective in design practice.

The next three sections of the paper examine, in order of the questions, each of the three aspects of the Design perspective introduced here. Each section considers the potential advantages and disadvantages based upon the criteria discussed in the preceding paragraph, relying upon examination of a wide range of design research and education literature. The final section (Sect. 5) draws together the separate elements in the other sections examining interactions among the three separate aspects of Big D Design. This section also discusses the relationship of "Design Science" with "Engineering Science." It is acknowledged that when exploring Big-D Design Science, there is much overlap with engineering science.

2 All Technologically: Enabled Design

The first aspect of Big D design is that it includes "All" technologically-enabled design; thus question 1 asks what are the advantages and disadvantages of committing to research across this breadth of fields. To address this question, it is necessary to define what is not in Big D Design following this definition and also about alternative ways to categorize the domains that are contained within Big D Design. This construct for design, although broad, does not envelope fields that are focused on aesthetics as the primary or exclusive criterion. It does not include design of visual art such as sculptures or paintings, or design of music or poetry, but does include technologically-enabled fields with a high aesthetic content such as architecture or product design. There are some fuzzy boundaries to any definition, but for our purposes debating specific cases is not important to our overall agenda. Design as used in the Big-D context is broad but does not include all activities that are legitimately called design.

In order to examine what might be gained or lost as one performs design research in the "Big D" perspective, it is useful to consider how one might establish categories included within Big D Design. Categories are desired that connect naturally with cognition and the cognitive science that pervades design as a process, methods, and science. Since "Big D" Design includes all engineering design and architecture, one approach is to think about design across all "typical" departments in an engineering school plus architecture. Such a listing is shown in the first column in Table 1. A second approach shown in the second column in Table 1 comes from a paper by Purao et al. [3] which reports on a workshop where a significant attempt was made to have presentations across a wide array of design domains. The domains listed are those describing the presentation disciplines at the workshop. We note only modest overlap between the lists in the first two columns of Table 1 but believe *both* are fully within "Big D" Design. The listing in the paper by Purao et al.—even though resulting from an attempt to encourage

Table 1 Categorizations within "Big D" design

Typical university departments	Disciplines listed in [1, 3]	Simplified categorization suggested here
Aerospace engineering architecture	Computer science	Software (algorithm and program) design
Biological engineering	Environmental design	Electromechanical-architectural artifacts and systems design
Chemical engineering	Human computer interaction	Socio-economic-technical systems design
	Informatics	Materials and molecular-level design
	Information sciences	
Civil engineering	Information studies	
	Information systems	
	Management science	
Computer (software) engineering and science	Production and management	
Electrical engineering	Software engineering	
Engineering mechanics		
Environmental engineering		
Industrial Engineering		
Materials engineering and science		
Mechanical engineering		
Nuclear engineering		
Systems and socio-technical engineering		

and obtain very broad design input—was organized by a MIS (Management Information Systems) group and many of the disciplines are from the area where management and engineering overlap—an area labeled systems socio-technical engineering in column 1. Interestingly, this field is one not stabilized within typical university engineering schools.

Although there is not uniformity in choices of departments or names within all universities, the listing in the first column of Table 1 is the most objectively defined partly because accrediting boards and fields of practice dictate some level of uniformity in names and content. This stability is an argument for exploring question 1 using this list. However, taxonomy criteria (and we are in fact discussing taxonomy for technologically-enabled design) more importantly express the desirability for internally homogenous categories and for the entire taxonomy to be collectively exhaustive and mutually exclusive.

In our attempt to answer question 1, we want to know if homogeneous, mutually exclusive categories are strong enough to support a set of homogeneous principles or heuristics about design. In reality, categorization attempts rarely arrive at homogeneous, mutually exclusive categories in a collectively exhaustive set, but it is our judgment that the first two columns of Table 1 fail badly enough to make further analysis using either approach potentially meaningless. Column 2 is clearly not collectively exhaustive based upon cursory analysis of design research in an engineering school within a typical university. Because specific university engineering departments engage in a broad variety of types of design and typically have overlap in each type, the organizing approach in column 1 misses badly on mutual exclusivity and internal homogeneity. Consider, for example, software engineering that is carried out in nuclear engineering, aerospace engineering, civil engineering and in other places beyond computer engineering or science in a typical engineering school. Similarly diverse placement of materials invention, systems engineering, structural engineering, fluid dynamics, and others indicate that to consider design in any one engineering field could verge on equivalence to studying it across all of "Big D" Design. Thus decomposition to categories at the university department level—while useful for other purposes—does not seem useful to analysis of design research. Perhaps a consolidation or purification of these fields can yield categories more useful to our needs.

Thus, a further attempt to develop categories within technologically-intensive design is undertaken. Technologically-intensive design is a very broad term covering many types and types of design output. To name just a few specific examples, the act and output of such design includes halogen light bulbs, a personal water purification device for developing countries, LED lighting, improved supercapacitors, nano-materials for water purification, a new soccer robot, a new military aircraft, software for controlling air flow in a large building, a new large–scale building, the Internet, and the road system in a large city. Depending upon the specifics of the typology one might think about, it might be possible to define hundreds if not thousands of technologically-intensive design domains. For example, the US patent system has more than 400 classes of patents in its highest classification category and more than 200,000 in its most granular categorization. One must recognize that each of these hundreds and even thousands of "domains" in fact has unique characteristics that might affect how design is performed. Specifying these characteristics (or especially trying to teach or carry out research in a cohesive fashion) for hundreds or even thousands of design fields is not especially feasible and certainly not very useful.

For the purpose of condensation for this paper, our reading of the design literature—particularly the references given in the next subsections, our experience with specific examples of design we have pursued, and our discussions with a variety of technologically-intensive designers, is synthesized to suggest four prototypical classes. These categories are our attempt to capture the breadth of the field, while designating classes likely to contain similar design fundamentals and methods. In other words, it is our judgment that these classes represent some of the most important differences likely to have effect in terms of advancing research fields and performing design. The third column of Table 1 gives the following names to these four classes:

- Software (algorithm and program) design;
- Electromechanical-architectural artifacts and systems design;

- Socio-economic-technical system design; and
- Materials and molecular-level design.

We now briefly discuss each of these classes in order to summarize what the literature analysis suggests is homogeneous in each category. The brief descriptions also are meant to support our contention that these four classes are to a large degree mutually exclusive. It is also clear that no simple set represents a perfect decomposition of “Big D” Design, but these four are more useful to us than any identified alternative in considering question 1. Since we are attempting to include all technologically-intensive design within the four categories, we are using the terms more broadly than may be customary for most readers.

2.1 Software Algorithms and Program Design

Software design relates to the digital and not the material world, where the output of this category of design are programs, or software systems, that accomplish many different functions and have a range of sizes (usually characterized by lines of code even as all recognize the imperfections of the metric). In cases where control software is highly integrated with physical artifacts, there exists a clear connection with our next class (electromechanical-architectural artifact and system design). For very large scale software systems where the software, hardware and the users are tightly coupled (for example, an air traffic control system), we consider such design problems to be contained in our third category (socio-economic-technical system design). Thus, the category we discuss here is for relatively pure software but the descriptions connect, interact, and apply to software subsystems in our other categories.

Software design is relatively new as a practice domain but nonetheless has received a large amount of attention academically [4–10].[1] Pressman [4] has summarized the “evolution of software design” as an ongoing process that has spanned four (and now more) decades which early on concentrated on modular programs and methods for refining software structures in a top-down manner. Later work proposed methods for translation of data flow or data structure into a design definition. In the 90s (and beyond) emphasis was on an object-oriented approach to design derivation. Software architecture and design patterns have also recently received emphasis in software design. Abstraction, complexity and re-use are also fundamental concerns in software design whereas the basic knowledge needed by designers in this category centers on discrete mathematics. Representation and possibly cognitive differences between this category of design and others have not been demonstrated but would be interesting to pursue.

---

[1] These references are only a modest fraction of the books in this general area.

2.2 *Electromechanical-Architectural Artifacts and Systems Design*

Electromechanical-architectural artifact and system design produces output that is generally the most visible and tangible of our four categories. As used in this paper, it includes almost all of what are commonly called products (e.g., automobiles, home appliances and furnishing, PCs, cell phones, cameras, etc.) and extends in scale and function beyond what is usually referred to as "products" to include boats, air conditioning systems, elevators, cranes, houses, buildings, locomotives, etc. In our definition, even quite large artifacts such as airplanes, electric power generation turbines and plants, aircraft carriers and large buildings are included in this category. When these large scale systems include a large social, economic and human-enterprise component, they can be categorized within the third category (socio-economic-technical systems).

In our definition, human-designed physical systems that process energy are also classified as electromechanical artifacts. Thus the output of this category includes much of the human-made physical world. Possibly because of the visibility and prevalence of its output, much popular and academic thinking equates this category with the totality of what is meant by technologically-intensive design. However, Design, as defined within the Big-D Design concept and at SUTD, recognizes a much broader domain of technologically-intensive design, so we do not restrict "Big D" Design to this category. Two important sub-fields in the electromechanical-architectural design field tend to, in most instances, even more narrowly define design, using the term to focus on the actual kind of design they do: those sub-fields are industrial design and architecture. These fields which are leading areas in key sub-fields of electromechanical-architectural artifacts and systems—for example, aesthetically and spatial sensitive electromechanical-architectural systems—deserve special attention.

Electromechanical-architectural artifacts and system design has—not surprisingly—resulted in a number of textbooks that are used in universities and by practitioners. References [11–25] give a small sample of the many diverse published books that treat this category of Big-D Design. Given the physical nature of these systems, consideration of space (geometry) is fundamental to electromechanical-architectural design. However, electromechanical-architectural design goes well beyond space considerations to include energy and information feedback. Due to the wide variety of designed objects, the fundamental topics of interest include function, materials, architecture and flexibility. The basic knowledge that underlies electromechanical-architectural design is centered, in part, on physics and mathematics. Practical knowledge in this domain often includes visual representation from sketching to complex 3D geometric representation systems, it also often includes knowledge of fabrication, materials and manufacturing of discrete products and it often includes deep knowledge of systems dynamics, modeling, making, and testing.

2.3 Socio-Economic-Technical System Design

Of the four categories we define for this study, historically, recognition of the concept of socio-economic-technical system design occurred latest. Although interest in large-scale technical systems with major social and economic impact has existed for a few decades [26, 27], it is only more recently that such a category was recognized as critical in the world of design and needing to be addressed from an engineering/technical perspective [28–34]. The boundary between socio-economic-technical systems with both large-scale electromechanical-architectural systems and large scale software systems is the inclusion within the design problem of complex social elements. At times, technical designers leave these social aspects to others, such as those from management or policy fields. Only if such problems are considered as part of the design do we consider the example to be in the socio-economic-technical design category. Here are two specific examples which might shed light on our use of the term: (1) Some might consider design of an air-traffic control system as only concerned with the radar sensing system and the software; (2) The design of a corporate control and improvement system such as the Toyota Production System (TPS) has been considered by some to only include the protocols, plant layouts and technical heuristics. In our use of the term socio-economic-technical design, however, these two examples also include: (1) the personnel, organizational and communication problems (pilot to controller, controller to controller, pilot to pilot, controller to supervisor etc.) and (2) the problem-solving approach, the redesigned role of management, cooperative teams and personnel incentives. It is the nature of socio-economic-technical systems [28–30] that if the complete design effort is constrained within the purely technical domain, the system will be much less effective than it would otherwise be.

Socio-economic-technical system design thus has prominently among its concerns considerations of stakeholders, decision processes, protocols, and standards. Because of their large scale and typically societal importance, architecture, flexibility, sophisticated design processes such as systems engineering and re-use are also top concerns in design of such systems. Representation of various types including process flow, as well as sophisticated programs for requirements and stakeholders are generally associated with this category of design. Of the four categories we have proposed, socio-economic-technical system designers have the most need for fundamental understanding of operations research and social science approaches and theories.

2.4 Materials and Molecular Level Design

Even though a relatively large fraction of technological progress [35–37] is due to design (invention and improvement) of materials and fabrication processes, there has been relatively little attention paid to materials design research and theory as a subject of enquiry. This lack of attention occurs despite (or perhaps because) materials and molecular level design predates even engineering and science as we know them. There have been a few papers describing the expanding knowledge that underlies particularly exciting new materials [38–40], but only Olson's contribution [40] contains significant attention to materials design in a broader sense. In many design textbooks [11–25], design *of* materials is not covered. In a few of these books [11, 14], design *with* materials is discussed including the introduction of Ashby diagrams [41] that systematize materials choice in a variety of design problems. However, choosing the best available material for a given application is ***not*** the focus of what we mean here by materials design. Instead materials design is the process of changing fundamental materials, processes, and processing parameters to create novel and useful materials. Examples of these include new nano-materials processing techniques for Li-ion batteries, vapor deposition of low band-gap semiconductors on Si for solar photovoltaic improvement, new thermoforming techniques for polymeric materials, and literally many hundreds of other specific novel useful materials and processes documented in the patent literature each year. In the solar PV field alone (about ½ of one of the 400 categories in the US patent database), there are about 75 "materials design" patents per year [42].

For consistency, we do not consider design of new materials systems such as large scale materials manufacturing systems or photovoltaic arrays to be materials design, but instead categorize these in either socio-economic-technical or electromechanical-architectural system design. Thus, our definition of materials design positions itself at the relatively small end of a dimensional scale. Perhaps most importantly, materials design always is intimately involved with processing (fabrication of the material). Olsen [40] and others [35] are clear that when materials designers undertake their creative steps, processing can come first: concurrent consideration of making and creating is not a new procedure for materials and molecular level designers. Materials are used as important elements in other artifacts and systems so materials design often aims at improving properties that are known to be important rather than directly aiming to improve an end user function. The fundamental knowledge important to materials and molecular design includes physics, biology, and chemistry at multiple scales; important practical knowledge includes deep and broad knowledge of material processing approaches and understanding of functional requirements that link to properties of various kinds.

2.5 Design Research from Narrower or Wider Perspectives

The fundamental knowledge and approaches used by the four categories of technologically-intensive designers have clear differences even in the brief discussions just presented. In addition, it seems quite reasonable to expect some cognitive processing differences even though this subject has not yet been researched. Thus, there is significant and strong rationale for conducting much design research within such domains and in even finer categories where specific methods and approaches might have value. The benefits from a narrower focus can be consideration of specific important problems (for example, flexibility—see [29, 44, 45] or very specific design methods (for example, objects that transform as part of their function—see [45]). Is there any evidence for value in research from broader perspectives? In fact, there is much work that has produced valuable output while taking a very broad view of design. Indeed, two of the most cited and most important contributors to a cumulative design research agenda are Simon [46] and Schön [47]. Both of these "founding fathers" of design research considered design quite broadly. Based upon their work, the benefits from a broad agenda are deeper insights, improved generalizability and improved capacity for differentiating fundamental from contingent aspects of design.

We also used two other approaches for input to answering the first of our questions. The first additional approach was to review 56 design papers relative to differences in "Type of Theory" from publications in different design domains. The theory typology (or taxonomy) that we followed is from Gregor [48, 49] who considers Information Systems design but argues for looking at design broadly. Table 2 shows Gregor's taxonomy and Table 3 shows the distribution of theory types as a function of papers that are predominantly in the differing design domains shown. Class IV of Gregor's taxonomy is the theory type that is most consistent with the establishment of a cumulative research agenda for design. It is important that our classification of the reviewed papers shows a significant fraction in this theory type and that those papers appear in all the different design categories. Further research will extend this analysis and seek causality connections and implications on cumulative Design theory.

Table 2 A taxonomy of theory in information systems research after Gregor [45]

Theory type	Distinguishing attributes
1. Analysis	Concerns what is. The theory does not extend beyond analysis and description. No casual relationships among phenomena are specified and no predictions are made.
2. Explanation	Concerns what is, how, why, when, and where. The theory provides explanations but does not aim to predict with any precision. There are no testable propositions.
3. Prediction	Concerns what is and what will be. The theory provides predictions and has testable propositions but does not have well-developed justificatory causal explanations.
4. Explanation and prediction	Concerns what is, how, why, when, where and what will be. Provides predictions and has both testable propositions and causal explanations
5. Design and action	Concerns what and how to do something. The theory gives explicit prescriptions (e.g., methods, techniques, principles of form and function) for constructing an systems, artifact (product), or process.

Table 3 Theory type distribution of analyzed papers (references are papers analyzed)

Theory type	Materials and molecular-level design [38–40, 50–55]	Electromechanical-architectural artifacts and systems design [43–45, 56–85]	Software (algorithms and program) design [86–92]	Socio-economic-technical systems design [28, 37, 92–96]	Total
Analyze	1	8	2	2	13
Explain	0	4	1	1	6
Predict	0	2	0	0	2
Explain, predict	6	10	2	4	24
Action, design	2	9	2	0	11

In addition to examining theory type distributions, we also briefly examined design principles that have resulted from design research. In the spirit of design science, much research and writing on design attempts to identify principles that can be used beyond single cases. In some instances, these are called heuristics or guidelines [94] and axioms [95]. In other cases [96, 97], researchers attempt to describe overall systems of interlinked principles for invention (such systems are of potential relevance here since the most novel design outputs are inventions).

Much of the work on design principles and heuristics has been carried out within a particular design context. For two examples, we consider the 180 plus heuristics given in Rechtin and Maier's book [97] and the 201 principles discussed by Davis [101]. In the former case, the principles are clearly framed in terms of design of classes of large-scale complex technical (and socio-economic-technical) systems, while in the latter case the principles are intended to guide software development. Analyzing these carefully, one can identify a number that have wider applicability but some—not surprisingly—are clearly not relevant in other domains. Specific examples from each study—two that have potentially general interest across domains (G) and one too narrow to be general (S) are:

- Rechtin and Maier (G): The first line of defense against complexity is simplicity of design;
- Rechtin and Maier (G): You can't avoid **redesign**. It's a natural part of design;
- Rechtin and Maier (S): If social cooperation is required, the **way** in which a system is **implemented** and introduced must be an **integral part** of its architecture;
- Davis (G): The design process should not suffer from "tunnel vision;"
- Davis (G): The design should be structured to degrade gently, even when aberrant data, events or operating conditions are encountered; and
- Davis (S): The design should "minimize the intellectual distance" between the software and the problem as it exists in the real world.

Similar to these examples, most principles of design are framed within a limited context and are often judged to be useful and instructive within that context.

Most research papers published—dissimilar to the textbooks just discussed—specify principles only for the intended problem and domain (examples are [42–45]). There are clear overlaps with principles between the two texts just reviewed sometimes with quite similar and sometimes dissimilar terminology (decomposition, integration, function and customer concerns are obvious ones that arise). Thus, it is worth exploring if a good starting point to examine design principles from a Big-D Design perspective already exists. As far as the authors are aware, only two attempts have been made to define general design principles and these will be considered next.

The first is the work done by Nam Suh and described in his book *The Principles of Design* [98]. The book–as opposed to the references noted in the preceding paragraphs–does not list a large number of principles or heuristics; instead it focuses on a very small number of what the book terms axioms. In fact, the two key "axioms" are the independence axiom (each functional requirement should be independent of other functional requirements) and the information axiom (among the designs that satisfy the independence axiom, the design that has the smallest information content is the best design). Suh's work in this book and other writings [67] uses these two axioms to "derive" larger numbers of theories and corollaries. On one hand, the independence concept is fairly widely applicable to thinking about designs across our full range of design domains. On the other hand—despite the terminology—the basic axioms are not as fundamental as this mathematical terminology implies. Indeed, while independence has a number of advantages, many designs that do *not* follow it are superior to alternatives that do. In this sense, it is much like the other "principles and heuristics" that have been postulated and is not in any sense truly axiomatic. The derived theories and corollaries are similar principles that can be seen as implications of the two major principles (independence and information). The strength of "axiomatic" design is that the principles apparently have wider application than others. In addition, there have been a number of conferences and workshops held on axiomatic design and some use in industry; however, at the present time this is not a fully developed set of principles for use across all Design.

The second effort that apparently attempts to develop generally applicable design principles is the work initiated by Altshuller [100, 101] in the 1940s and still actively pursued today. This work, known both by its Russian acronym (TRIZ) and by English terminology (Theory of Inventive Problem Solving or TIPS), has its empirical basis in study and classification of patents. Four different aspects of TRIZ include:

1. TRIZ identifies eight "laws" of technical systems evolution which are useful in predicting the nature of desired future design changes;
2. TRIZ identifies thousands of "effects" that are characterized as domain independent;
3. TRIZ identifies 40 design "principles" for resolving contradictions (TRIZ hypothesizes that contradictions in existing solutions are the major way to specify inventive opportunities for the future);
4. TRIZ identifies ~75 "standard solutions" that deal with identified problems.

The translation of TRIZ to "Big D" Design is challenging because the TRIZ literature does not discuss the breadth of applicability and tends to not recognize what aspects of "technologically-intensive design" that it may be neglecting. Moreover, most of the examples shown in the literature are from the electromechanical-architectural design field which may be a result of the background of practitioners and supporters of TRIZ.

The TRIZ laws of evolution are largely descriptive and some may seem difficult to make operational. For example, evolutionary law number 2 ("increasing ideality") simply says that output per resource increases over time. This is better stated by the exponential improvements seen in various output per resource as first documented by Moore [102] and now known to be much more general [103, 104]. Nonetheless, many of the design principles appear quite general and can be imagined to apply across all "Big D" Design domains. For example, principle number 13 "the other way around" suggests the powerful heuristic to examine the problem in a reverse (or with the inside out or in different temporal order or). However, many principles appear to be more limited in their application across domains (examples include #7 "Nesting", #8 "counterweight", #18 "Mechanical vibration", #28 "Replacement of a mechanical system", #29 "pneumatic or hydraulic construction", #32 "Changing the Color", #35 "transformation of the chemical or physical states of an object", #37 "Thermal expansion"). Although these apparent limitations may relate to terminology and translation from the theory's source language, research and advancement of TRIZ are needed to understand this system's application for all technologically-intensive design.

Recognizing exceptions such as the relatively general #13, neither the TRIZ principles nor the solutions appear to have direct application to software or socio-technical design—perhaps because of the scarcity of such solutions in the patent database that underlies the approach. It is also not clear how well the approach covers materials design despite its prevalence in the patent database. The principles with clear materials content are about materials change or substitution, not about inventing new materials (as examples, #30 "Flexible membranes or thin films" and #31 "Use of porous materials"). Thus, despite some uptake in practice and ongoing documented work [105], TRIZ is also not a fully developed set of Big-D Design principles.

Overall, based upon this preliminary analysis, it appears that sets of broadly applicable design principles are potentially derivable which gives tentative support for a positive answer to question 1. However, the current general approaches do not seem adequate. From the commonality seen in the lists examined, one infers that by some work an overall listing might be developed giving principles in an organized framework but doing this (or even proving its value) will require significant additional work.

3 The Full Value Chain

Question 2 in the first section asks: What are the benefits and potential harm of engaging the full value chain in design research? There are clear practice benefits from considering the full value chain in design as the extensive practice-oriented work done on concurrent engineering signals. There are also clear educational benefits both from a leadership education and understanding design in context viewpoint. Thus, from a university such as SUTD, there is great value in defining design as broadly across the value chain as it does. However, from a research perspective, there may be only a few research objectives that benefit from the wider lens—design for sustainability, value, manufacturability [106, 107] and other DFX areas are examples. Since the full value chain differs in the categories we consider (software does not have physical facilities or tools, materials processing is mostly continuous vs. the discrete product or system manufacturing in the other categories, the nature of customers, clients and stakeholders are different), Design for manufacturability research naturally occurs in narrower domains than ***all*** technologically-intensive Design. Based on these examples, care must be taken in understanding how to develop and engage in design research from the broader Big-D context in regard to the value chain.

4 Art and Science of Design

The question of interest in this section is whether design research must involve both the art and science of design. Our criteria for assessing design research state that such research must impact practice in order to be of value. Since the practice of design is essentially about creating something that has not previously existed, an irreducible element of art is involved in the practice of all technologically-intensive design. This conclusion combined with our criterion for research value and the fact that research is the process for developing new science dictates that all design research includes both the art and science of design.

While almost no-one would disagree with design practice having at least some artistic aspect, there are some [10] who object to a Science of Design (thereby implicitly or explicitly arguing that design research is not viable). This position seems indefensible given the progress that has been made in design research. In our study of design principles (Sect. 2), we find some principles that apply quite widely (modularity or independence of function) and much opportunity exists to explore others. Moreover, there is much more understanding of the importance of expertise [108] than there was when the cumulative design research agenda was initiated almost 50 years ago. Similarly, the importance of analogical transfer in design has been much more strongly established [109, 110] including some work [111] that points towards the best "knowledge structure" for enabling this process.

5 Concluding Remarks

Although we have chosen to discuss the three elements of "Big-D" Design separately (1-all domains of technologically-intensive design, 2-full value chain and 3-art/science combination), there are clear and important interactions among these dimensions. One example of the interconnectedness of these elements is that when research is performed that combines the art and science of design, valuable work has been done that examines design in essentially all domains [46, 47] as well as by looking at more specific problems within a domain [43–45]. A second example of the interactions among the elements is that when research is carried out on the full design value chain, more practical (or art content) is introduced as well as more scientific content [106]. A third example—among many that can be noted—is that as mentioned in Sect. 3, the full value chain has very different content in the different domains that we have described.

Our consideration of the impact of taking a "Big-D" perspective in design research has in all cases shown potential value for broader viewpoints while clearly avoiding any requirement to do so. A 2008 paper by Kuechler and Vaishnavi [93], that argues for broadening the scope of Information Systems Design Research (ISDR), criticizes ISDR for missing important contributions from the "designerly way of knowing" schools [112] and that the ISDR literature contains little in citations to design work outside ISDR. This is not apparently so in all design research domains, but a tendency to fragment might be working to overcome the early start by Simon and others in a broader way. In addition, there are valuable results in the literature that come from considering design beyond technologically-intensive domains. In regard to combining art (practice) and science (research), we have already argued that this is a natural outcome of carrying out research with one objective being to impact the practice of design favorably. However, we do not believe that all design research must involve designing something new as this would amount to the methodological straight-jacket (elimination of valuable research projects) noted by Purao et al. [3]. Research on the art of design can uncover theory that is at least partly scientific, but this can be accomplished by a variety of methods beyond designing something new—for example by systematic study of much design output (empirical studies) [43, 113, 114] or by systematizing observed designer methods [59, 81].

Arguing as we have for a broader (technologically—enabled) perspective for much design research introduces two issues that can limit the value of the work. The first issue is one articulated well in Purao et al. [3] after participating in presentations and extensive discussion among the fields of design shown in the second column of Table 1; one participant said:

> The lack of a common language constitutes a danger to the nascent design sciences. The danger is that our joint efforts will dissolve into incoherence, as exemplified by the myth of the ill-fated Tower of Babel.

Analyzing a wide variety of literature from across design research domains reinforces this point. As one example, many in software design consider design only the creative core of the process so design as used by them does not include specifying, coding or testing; whereas in most electromechanical-architectural design literature, design includes specifying and testing and often manufacturing. Multiplying this example by the many other words that are used quite differently shows that the Tower of Babel danger is real and present (even within domains there is surprising variety in terminology). Thus, one necessary step in pursuing a broader and effective design research agenda is a serious attempt to arrive at a more coherent terminology.

A second major issue in pursuing a research agenda across all technologically-intensive design is the epistemological relationship of such design research to "Engineering Science"—the reigning academic standard in engineering schools worldwide. There is extensive discussion in the design science literature about the epistemological relationship of design to natural science, and there is significant discussion of its relationship to the social sciences. However, there is almost none discussing the relationship of engineering science with technologically-intensive design science. This silence is almost surely related to the fact that the epistemological basis of engineering science has not been considered very deeply. In fact, the arguably best and perhaps only serious consideration of engineering science—Vincenti's 1990 book "What Engineers Know and How They Know it" [114]—does not use the term engineering science despite discussing knowledge that most engineering scientists would consider appropriate to the term. Most interestingly, the major conclusion by Vincenti appears to be that the difference in the science that engineers do compared with natural science, is that "[engineering] science" is fundamentally oriented to make the findings of natural science useful in *design*. Thus, one can probably consider "engineering science" and "design science" intertwined and one possibly a sub-set of the other. An aggressive attempt to clarify this relationship would have great value in setting an agenda for pursuing design research—particularly over the broad spectrum of "all technologically-enabled design."

As a conclusion to this paper, it is clear that we have only examined a small fraction of the issues and foundations needed to create a Big-D perspective of Design research. At the core of our analysis is an understanding of technologically-intensive design as categories, as the study, identification, formalism, and use of design principles and heuristics, as the full value chain, and inclusive of art and science. While the supporting literature of this paper generally supports this view, significantly more analysis is needed on this literature, in addition to integration with design research methodologies and other segments of the design research literature, including [115–124] and beyond.

Acknowledgments The authors thank Professor Robert W. Weisberg for many helpful comments on an earlier and partial draft. This work is supported by the SUTD-MIT International Design Center.

References

1. Wood KL, Rajesh Elara M, Kaijima S, Dritsas S, Frey D, White, CK, Crawford RH, Moreno D, Pey K-L (2012) A symphony of designiettes—exploring the boundaries of design thinking in engineering education. ASEE annual conference, San Antonio
2. Magee CL, Leong PK, Jin C, Luo J, Frey DD (2012) Beyond R&D: what design adds to a modern research university. Int J Eng Educ 28:397–406
3. Purao S, Baldwin CY, Hevner A, Storey V, Pries-Heje J, Smith B (2008) The sciences of design: observations on an emerging field. Commun Assoc Inf Syst 3
4. Pressman RS (2001) Software engineering : a practitioners approach, 5th edn
5. Somerville I (2011) Software engineering, 9th edn
6. McConnell S (2004) Code complete: a practical handbook of software construction
7. Pilone D (2008) Head first software development
8. Braude EJ (2010) Software engineering: modern approaches
9. van Vliet H (2008) Software engineering: principles and practice
10. Brooks FP (2010) The design of design: essays from a computer scientist
11. Dym CL, Little P (2009) Engineering design: a project-based introduction, 3rd edn
12. Ulrich KT, Eppinger ST (1995) Product design and development
13. Otto KN, Wood KL (2001) Product design: techniques in reverse engineering and new product development. Prentice-Hall, Englewood Cliffs
14. Dieter GE, Schmidt LC (2008) Engineering design
15. Anderson J (2010) Basics architecture: architecture design
16. Legendre GL (2011) Mathematics of space: architectural design
17. Norman DA (2002) The design of everyday things
18. Lidwell W (2010) Universal principles of design, revised and updated: 125 ways to enhance usability, influence perception, increase appeal, make better design decisions, and teach through design
19. Cross N (2008) Engineering design methods: strategies for product design, 4th edn. Wiley, Chichester
20. Ullman D (2009) The mechanical design process, 4th edn. McGraw-Hill, New York
21. Venturi R (1966) Complexity and contradiction in architecture. The Museum of Modern Art
22. Thackara J (2005) In the bubble: designing in a complex world. MIT Press, Cambridge
23. Brawne M (1992) From idea to building: issues in architecture. Butterworth Architecture
24. Addis W (2007) Building: 3,000 years of design engineering and construction. Phaidon, London
25. McDonough W, Braungart M (2002) Cradle to cradle: remaking the way we make things. North Point Press, New York
26. Hughes TP (1983) Networks of power: electrification in western society, 1880–1930
27. Hughes TP (1998) Rescuing prometheus
28. Maier MW (1999) Architecting principles for systems-of-systems. Syst Eng
29. de Neufville RA, Scholtes S (2011) Flexibility in engineering design
30. de Weck OL, Roos D, Magee CL (2011) Engineering systems, meeting human needs in a complex technological world (Chap. 6)
31. Simchi-Levi DS (2010) Operations rules: defining value through flexible operation
32. Hopp WJ, Spearman ML (2007) Factory physics, 3rd edn
33. Boland R, Callopy F (2004) Managing as designing
34. Boland R, Callopy F, Lyytinen K, Yoo Y (2008) Managing as designing: lesssons for organization leaders from the design practice of frank gehry. Desig Issues 24(1):10–25
35. National Research Council study (1989) Materials science and engineering for the 1990s: maintaining competitiveness in the age of materials. In: Chaudhari P, Flemings MC (eds) National Academy Press, ISBN: 0-309-57374-2
36. Magee CL (2010) The role of materials innovation in overall technological development. JOM
37. Magee CL (2012) Toward quantification of the role of materials innovation in overall technological development. Complexity, 18:10–25

38. Langer R, Tirrell DA (2004) Designing materials for biology and medicine. Nature 428:487–491
39. Ortiz C, Boyce MC (2008) Bio-inspired structural materials. Science 319:1053–1054
40. Olsen GB (2000) Designing a new material world. Science 288:993–998
41. Ashby MF (1999) Materials selection in mechanical design, 2nd edn
42. Benson CL, Magee CL (2012) A framework for analyzing the underlying inventions that drive technical improvements in a specific technological field. Eng Manage Res 1:2–15
43. Keese DA, Tilstra AH, Seepersad CC, Wood KL (2007) Empirically-derived principles for designing products with flexibility for future evolution. ASME international design technical conferences
44. Tilstra AH, Backlund PB, Seepersad CC, Wood KL (2008) Industrial case studies in product flexibility for future evolution: an application and evaluation of design guidelines. ASME international design technical conferences DETC2008-49370, ASME
45. Weaver J, Wood KL, Crawford RL, Jensen D (2010) Transformation design theory: a meta-analogical framework. J Comput Inf Sci Eng 10:013012-1–013012-11
46. Simon HA (1996) The sciences of the artificial, 3rd edn. MIT Press, Cambridge
47. Schön DA (1983) The reflective practitioner: how professionals think in action. Basic Books, New York
48. Gregor S (2006) The nature of theory in information systems. MIS Q 30:611–642
49. Gregor S, Jones D (2007) The anatomy of a design theory. J Assoc Inf Syst 5:313–335
50. Mooney DJ, Baldwin DF, Suh NP, Vacanti JP, Langer R (1996) Novel approach to fabricate porous sponges of poly(D, L-Lactic-co-glycolic acid) without the use of organic solvents. Biomaterials 17:1417–1422
51. Bruet BF, Song J, Boyce MC, Ortiz C (2008) Materials design principles of ancient fish armor. Nat Mater 7:748–756
52. Suresh S (2001) Graded materials for resistance contact deformation and damage. Science 292:2447–2451
53. Gao H, Ji B, Jager IL, Arzt E, Fratzl P (2003) Materials become insensitive to flaws at nanoscale: lessons from nature. Proc Nat Acad Sci (US) 100:5597–5600
54. Kuehmann CJ, Olsen GB (2011) ICME: success stories and cultural barriers. In: Arnold S, Wong T (eds) Integrated computational materials engineering, ASM
55. Han L, Grodzinsky AJ, Ortiz C (2011) Nanomechanics of the cartilage extracellular matrix. Annu Rev Mater Res 41:133–168
56. Gunther J, Ehrlenspeil J (1999) Comparing designers from practice and designers with systematic design education. Des Stud 20:439–451
57. Paramasivam V, Senthil V (2009) Analysis and evaluation of product design through design aspects using digraph and matrix approach. Int J Interact Des Manuf 3:13–33
58. Sorenson CG, Jorgenson RN, Maagaard J, Bertelsen KK, Dalgaard L, Norremark M (2010) Conceptual and user-centric design guidelines for a plant nursing robot. Biosyst Eng 105:119–129
59. Yilmaz S, Seiffert CM (2011) Creativity through design heuristics: a case study of expert product design. Des Stud 32:384–415
60. Wu JC, Shih MH, Lin YY, Shen YC (2005) Design guidelines for tuned liquid column damper for structures responding to wind. Eng Struct 27:1893–1905
61. Galle P (1996) Design rationalization and the logic of design. Des Stud 17:253–275
62. Joseph S (1996) Design systems and paradigms. Des Stud 17:227–239
63. Berends J, Reymen I, Stultiens RGL, Peutz M (2011) External designers in product design processes of small manufacturing firms. Des Stud 32:86–108
64. Dorst K, Veermas PE (2005) John Gero's function-structure-behavior model of designing: a critical analysis. Res Eng Des 16:17–26
65. Magee CL, Thornton PH (1978) Design considerations in energy absorption by structural collapse. Soc Automot Eng Trans SAE 780434
66. Matthews PC, Blessing LTM, Wallace KM (2002) The introduction of a design heuristics extraction method. Adv Eng Inform 16:3–19

67. Suh NP (1998) Axiomatic system design. Res Eng Des 10:189–209
68. Wood KL, Jensen D, Singh V (2009) Innovations in design through transformation: a fundamental study of tRaNsFoRmAtIoN principles. ASME J Mech Des 131:8
69. Frey DD, Herder PM, Wjnia Y, Subrahmaniam E, Katsikopoulos K, Clausing DP (2009) The pugh controlled convergence method: model-based evaluation and implications for design theory. Res Eng Des 20:41–58
70. Frey DD, Herder PM, Wjnia Y, Subrahmaniam E, Katsikopoulos K, de Neufville RA, Clausing DP (2010) Reply: the role of mathematical theory and empirical evidence. Res Eng Des 21:341–344
71. Hirtz J, McAdams DA, Sykman S, Wood KL (2002) A functional basis for engineering design: reconciling and evolving previous efforts. NIST technical note 1447
72. Stone RB, Wood KL, Crawford RH (2000) A heuristic method for identifying modules for product architectures. Des Stud 21:5–31
73. Rajan PKP, van Wie M, Campbell MI, Wood KL, Otto KN (2005) An empirical foundation for product flexibility. Des Stud
74. Hey J, Linsey J, Agogino AM, Wood KL (2008) Analogies and metaphors in creative design. Int J Eng Educ 24:283–294
75. Qureshi A, Murphy JT, Kuchinsky B, Seepersad CC, Wood KL, Jensen DD (2006) DETC 2006-99583
76. Weaver JM, Kuhr R, Wang D, Crawford RH, Wood KL, Jensen D, Linsey JD (2009) Increasing innovation in multi-function systems: evaluation and experimentation of two ideation methods for design. DETC 2009-86526
77. Rajan PKP, van Wie M, Campbell MI, Otto KN, Wood KL (2003) Design for flexibility—measures and guidelines. ICED03
78. Moe RE, Jensen DD, Wood KL (2004) Prototype partitioning based upon requirement flexibility. DETC 2004-57221
79. Weaver J, Wood KL, Jensen D (2008) Transformation facilitators: a quantitative analysis of reconfigurable products and their characteristics. DETC2008-49891
80. Singh V, Skiles SM, Krager JE, Wood KL, Jensen D, Sierokowski R (2009) Innovations in design through transformation: a fundamental study of transformation principles. J Mech Des 131:081010-1–081010-18
81. Chrysikou EG, Weisberg RW (2005) Following the wrong footsteps: fixation effects of pictorial examples in a design problem-solving task. J Exp Psychol: Learn Mem Cogn 31:1134–1148
82. Weisberg RW (2009) On 'out-of-the-box' thinking in creativity. In: Wood K, Markman A (eds) Tools for innovation, pp 23–47
83. Rowe PG (1987) Design thinking
84. Luo J, Olechowski AO, Magee CL (2012) Technologically-based design as a strategy for sustained economic growth. Technovation, to appear
85. Magee CL, Frey DD (2006) Experimentation in engineering design: linking a student design exercise to new results from cognitive psychology. Int J Eng Educ 22(3):85–103
86. Gill GR, Hevner AR (2010) A fitness-utility model for design science research
87. MacCormack AD (2001) Product development practices that work: how internet companies build software. Sloan Manag Rev 42:75–84
88. Kim H (2010) Effective organization of design guidelines reflecting designer's design strategies. In J Indus Erg 40:669–688
89. Hevner AR, Ram S, March ST, Park J (2004) Design science in information systems research. MIS Q 28:75–105
90. MacCormack AD, Rusnak R, Baldwin CA (2006) Exploring the structure of complex software designs: an empirical study of open-source and proprietary code. Manag Sci 52:1015–1030
91. Venables JR (2010) Design research post Hevner et al: criteria, standards, guidelines and expectations. In: DESRIST proceedings
92. Shaw M, Garlan D (1996) Software architecture: an emerging discipline

93. Kuechler W, Vaishnavi V (2008) The emergence of design research in information systems in North America. J Des Res 7:1–16
94. Poole S, Simon M (2007) Technological trends, product design and the environment. Des Stud 18:237–248
95. Jarvinen P (2007) On reviewing of results in design research. In: ECIS, Proceedings of the fifteenth European conference on information systems, pp 1388–1397
96. Indulska M, Recker JC (2008) Design science in IS research : a literature analysis. In: 4th biennial ANU workshop on information systems foundations, 2–3 Oct 2008
97. Rechtin E, Maier MW (2009) The art of system architecting, 3rd edn
98. Suh NP (1990) Principles of design
99. Altshuller G (1984) Creativity as an exact science
100. Sickafus E (1997) Unified structured inventive thinking: how to invent
101. Davis AM (1995) 201 principles of software development
102. Moore GE (1965) Cramming more components onto integrated circuits. Electron Mag 8:38
103. Koh H, Magee CL (2006) A functional approach for studying technological progress: application to information technology. Technol Forecast Soc Chang 73:1061–1083
104. Koh H, Magee CL (2008) A functional approach for studying technological progress: extension to energy technology. Technol Forecast Soc Chang 75:735–758
105. The TRIZ Journal is published regularly, see http://www.triz-journal.com/
106. Boothroyd G (2005) Assembly automation and product design, 2nd edn
107. Boothroyd G, Dewhurst P, Knight W (2002) Product design for manufacture and assembly, 2nd edn
108. Weisberg RW (2006) Creativity: understanding innovation in problem solving, science, invention, and the arts. Wiley, New York
109. Christensen CB, Schunn CD (2007) The relationship of analogical distance to analogical function and pre-inventive structure: the case of engineering design. Mem Cogn 35:29–38
110. Markman AB, Wood KL, Linsey JS, Murphy JT, Laux J (2009) Supporting innovation by promoting analogical reasoning. In: Markman AB, Wood KL (eds) Tools for innovation. Oxford University Press, New York, pp 85–103
111. Linsey JS, Wood KL, Markman AB (2008) Modality and representation in analogy. Artif Intell Eng Des Anal Manuf 22:85–100
112. Cross N (1984) Developments in design methodology
113. Baldwin CY, Clark KB (2006) Between 'knowledge' and 'the economy': notes on the scientific study of designs. In: Kahin B, Foray D (eds) Advancing knowledge and the knowledge economy. MIT Press, Cambridge
114. Vincenti W (1990) What engineers know and how they know it: analytical studies from aeronautical history
115. Friedman K (2003) Theory construction in design research: criteria: approaches, and methods. Des Stud 24:507–522
116. Collins A, Josepjh D, Bielaczyc K (2004) Design research: theoretical and methodological issues. J Learn Sci 13(1):15–42
117. Horvath I (2004) A treatise on order in engineering design research. Res Eng Des 15:155–181
118. Eder WE (2011) Engineering design science and theory of technical systems: legacy of vladimir hubka. J Eng Des 22(5):361–385
119. Ball P (2001) Life's lessons in design. Nature 409:413–416
120. Dorst K (2008) Design research: a revolution-waiting-to-happen. Des Stud 29:4–11
121. Galle P (2008) Candidate worldviews for design theory. Des Stud 29:267–303
122. Farrell R, Hooker C (2012) The Simon-Kroes model of technical artifacts and the distinction between science and design. Des Stud 33:480–495
123. Reich Y (2010) The redesign of research in engineering design. Res Eng Des 21:65–68
124. Andreasen MM (2011) 45 years with design methodology. J Eng Des 22(5):293–332

Proposal of Quality Function Deployment Based on Multispace Design Model and its Application

Takeo Kato and Yoshiyuki Matsuoka

Abstract Due to the specialization and professionalization of the design work, sharing the product information between the product development members has been important in the product development process. Quality Function Deployment is one of the effective methods that enables the development members to share the information of the product using the quality matrices that describes the relationship between design elements needed to be considered. This paper improves the quality matrices by introducing the multispace design model and the Interpretive structural modeling. The proposed quality matrices are applied to a disc brake design problem, and their applicability is confirmed.

Keywords Design methodology · QFD · ISM · DSM

1 Introduction

Functions and mechanisms of products have been diversified and complicated recently. Therefore, design work has been specialized and professionalized [1]. In the situation, the members of the product development should share the product information, including the concept and knowledge. However, the information tends to be left in their mind (i.e. not to be transmitted to others), and this causes quality

T. Kato (✉)
Department of Mechanical Engineering, Tokai University, Hiratsuka, Japan
e-mail: t.kato@tokai-u.jp

Y. Matsuoka
Department of Mechanical Engineering, Keio University, Yokohama, Japan
e-mail: matsuoka@mech.keio.ac.jp

A. Chakrabarti and R. V. Prakash (eds.), *ICoRD'13*, Lecture Notes in Mechanical Engineering, DOI: 10.1007/978-81-322-1050-4_5,

issues. Quality Function Deployment (QFD) is one of the effective methods to solve these problems and used in many company throughout the world [2]. QFD is suitable for the members to share the information using the quality matrices as shown in Fig. 1 [3]. Quality matrices are composed by the deployment charts including allied design elements needed to be considered in the design process and the relationship matrices which represent the relationship between design elements in different deployment charts. Using the quality matrices, design elements of customer demands can be translated into that of engineering tasks (engineering characteristics, product's function, parts and etc.). This enables the product development members (product planners, designers, manufacturing staff and etc.) to share the product's information and assist to implement the ideal product design free of the quality issues. However, the quality matrices have some problems as follows:

1. Applying to a whole new product development is difficult because they are assumed to be applied to the design for improving existing product [3];
2. Sharing the product's information between the development members is counteracted because the quality matrices are different depend on the design process. For example, the relationship matrix between engineering characteristics and parts is only used by the engineering designers or manufacturing staff in the detail design process [4];
3. The circumstance of the developing product is not clarified in the detail design process (i.e., the design elements meeting the circumstance cannot be extracted);
4. The relationship between design elements in the same deployment chart cannot be identified because the relationship matrices represent only the relationship between design elements in different deployment chart. However, the relationship of design elements in the deployment chart of engineering characteristics is considered in some conventional study [5].

This study proposes new QFD including both the Multispace design model (MDM) and the Interpretive Structural Modeling (ISM) method to overcome the above problems. Section 2 describes a brief description of the MDM and the

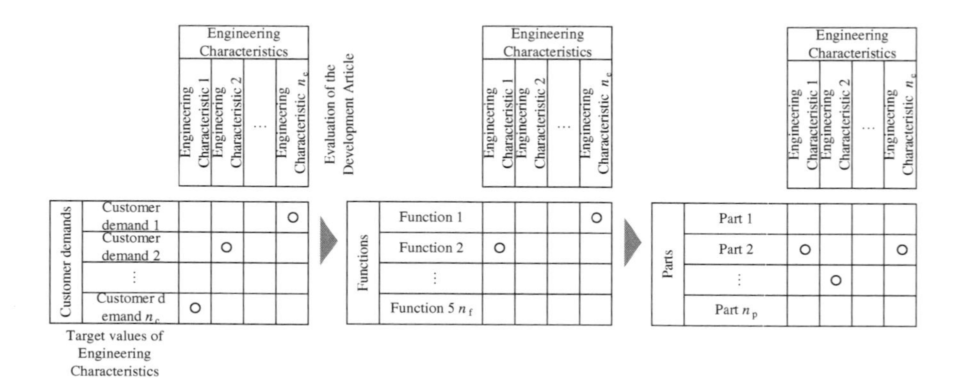

Fig. 1 Conceptual diagram of quality matrices used in QFD

introduction of the MDM into the QFD. Section 3 describes the overview of the ISM method and the introduction of the ISM into the QFD. Section 4 illustrates an application of the proposed QFD to a disc brake, while Sect. 5 provides conclusions and the future research direction.

2 Introduction of MDM into QFD

The MDM aims to comprehensively deal with design and is comprised of the thinking space and knowledge space (Fig. 2) [6]. The thinking space includes a reasoning model for four types of spaces and inter-spaces: value space, meaning space, state space, and attribute space. These space are defined as follows:

1. The value space is a set of the value elements. The value elements are psychological elements relating to values that the user thinks about products. For example, functional value, social value, and so forth;
2. The meaning space is a set of the meaning elements. The meaning elements are psychological elements relating to meanings that the user thinks about products. For example, function, image, and so forth;
3. The state space is a set of supposed circumstance and the state elements. The circumstance is physical environment for which products are used. For example, time, external force, users physique and so forth. The state elements

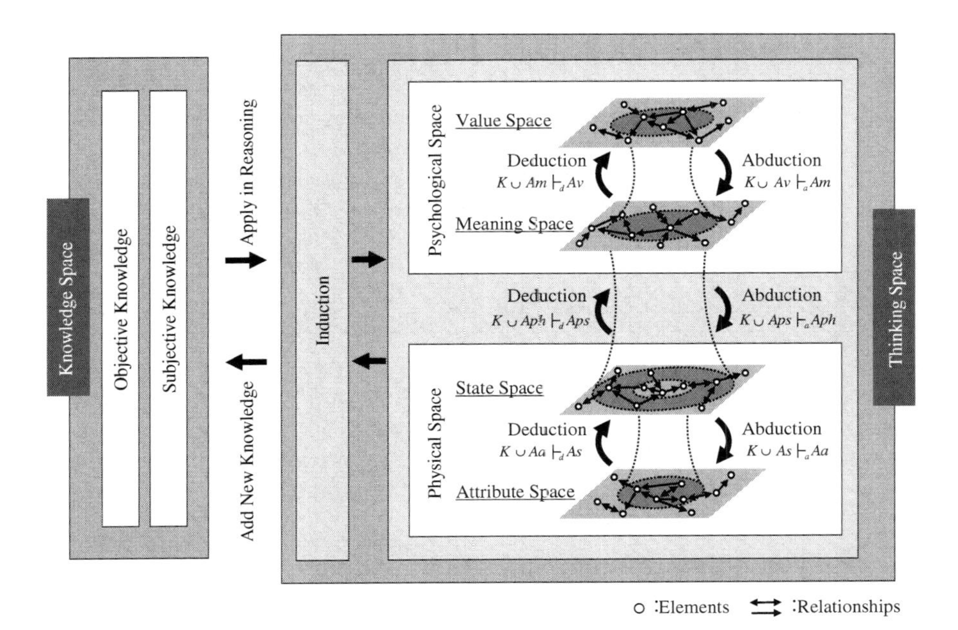

Fig. 2 Conceptual diagram of multispace design model

are physical quantity generated when products are in the circumstance. For example, stress, acceleration produced when an external force acts on products and so forth;
4. The attribute space is a set of the attribute elements. The attribute elements are geometrical and physical property of the products. For example, dimension and material like being shown in the technical drawing and so forth.

Value space and meaning space composes the psychological space, whereas state space and attribute space composes the physical space. Knowledge space is comprised of objective knowledge and subjective knowledge. The objective knowledge holds generalities such as theories and methodologies, including physical laws in natural science, social sciences, and humanities. In contrast, the subjective knowledge contains specialties that depend on individual contexts. The knowledge space is the basis for identifying intra-space and inter-space models in the thinking space.

One of the features of the MDM is to describe whole design process (from the early to late process). As a feature of design process, in the early process of design, the designers focus both on the psychological and physical design elements, and search design solution candidates by considering the relationship between the elements through trial and error (bidirectional design). On the other hand, in the late process of design, the designers derive a unique design solution by optimizing both psychological and physical design elements systematically (unidirectional design). The features of both design processes can be properly described by the four space of the MDM (value, meaning, state, and attribute space). Conventional QFD seems to assume the late design process because of its procedure in which the designers deploy the design elements from the customer demands (psychological design element) to the product's parts (physical design element) systematically as shown in Fig. 1. This causes the difficulty of applying QFD to a whole new product development, which needs the bidirectional design in the early process of design, described in Sect. 1. Therefore, to overcome the problem, the concept of the MDM (four space) is introduced into the QFD. Specifically, the four space deployment charts and three relationship matrices are put into the quality matrices in order to bidirectional design in the early process of design (Fig. 3). The quality matrices can also be used in the late process of design. This means that the common quality matrices can be used throughout the whole process of design. Hence, this also overcome the problem regarding the sharing of the product's information between the development members described in Sect. 1.

Another feature of the MDM is a clear definition of the product circumstance. The MDM describes that the function of the product is generated not only based on the product physical characteristics but also on the circumstance of the product. Unless the circumstance is adequately considered or clarified, the following problems have a potential to emerge:

1. The variance of the circumstance worsens the product's objective characteristic;
2. In the redesign (improvement design) after the development, the product information during the development cannot be utilized.

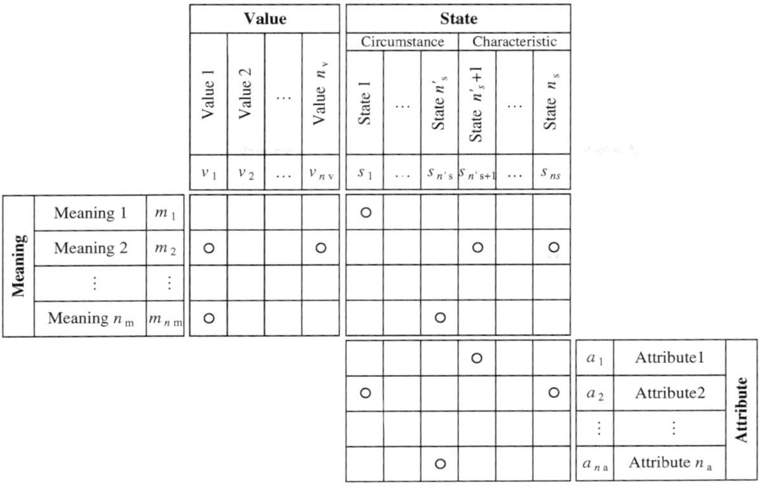

		Value				State							
						Circumstance			Characteristic				
		Value 1	Value 2	…	Value n_v	State 1	…	State n'_s	State n'_s+1	…	State n_s		
		v_1	v_2	…	v_{nv}	s_1	…	$s_{n's}$	$s_{n's+1}$	…	s_{ns}		
Meaning 1	m_1					O							
Meaning 2	m_2	O			O				O		O		
⋮	⋮												
Meaning n_m	m_{nm}	O						O					
									O			a_1	Attribute1
						O					O	a_2	Attribute2
												⋮	⋮
								O				a_{na}	Attribute n_a

(Row group labels: Meaning; Attribute)

Fig. 3 Proposed QFD include the concept of multispace design model

Therefore, the MDM divides the physical elements into state elements affected to the circumstance and attribute elements and clearly categorizes the circumstance elements in the state space. In the conventional QFD, the information about the circumstance does not tend to be carried to the engineering designers or manufacturing staff in the late process of design because the circumstance descriptions are in the customer demands or product's function in most cases. This study creates a category of the circumstance in the state deployment chart based on the concept of the MDM (Fig. 3) and clarifies the design elements of the current circumstance to overcome the problem as mentioned in Sect. 1.

3 Introduction of ISM into QFD

The ISM method is one of the design methods to visually express the complex relationship between design elements by using matrix operation [7, 8]. In the ISM method, the direct affective matrix **X** (Fig. 4a), which expresses the relationship between design elements, is firstly constructed as following equation:

$$X = \begin{pmatrix} X_{11} & \cdots & X_{1j} & \cdots & X_{1n} \\ \vdots & & & & \\ X_{i1} & & \ddots & & \vdots \\ \vdots & & & & \\ X_{n1} & & \cdots & & X_{nn} \end{pmatrix}, \tag{1}$$

where, n is the number of design elements and X_{ij} are calculated as:

$$X_{ij}\begin{cases} 1 & \textit{if ith element relates to jth element} \\ 0 & \textit{else} \end{cases} \quad \begin{pmatrix} i = 1, 2, \ldots, n \\ j = 1, 2, \ldots, n \end{pmatrix}. \tag{2}$$

Secondly, the reachable matrix $\mathbf{M}_{\mathrm{R}}$ (Fig. 4b) is derived using the matrix $\mathbf{M} = \mathbf{X} + \mathbf{I}$, where $\mathbf{I}$ is a unit matrix, as shown in the following equation:

$$M_R = M^r \qquad \left(M^r = M^{r-1}\right) \tag{3}$$

Finally, the reachable matrix MR is transformed into the skeleton matrix M (Fig. 4c) and the structural model (Fig. 4d) is constructed based on the relationship in the matrix. Where, the skeleton matrix can represent the relationship of the reachable matrix using minimum relationships [8]. This paper omits the detail calculation of the skeleton matrix.

This study introduces the correlation matrix to each of the four space deployment charts in order to overcome the problem of unclear relationship between them described in Sect. 1. The correlation matrix is described as Fig. 5. In the matrices, the unidirectional relations (i.e. element "A" causes "B" but "B" does not causes "A") are described as arrows, whereas, the bidirectional relations (i.e. element "A" causes "B" and "B" also causes "A") are described as "○". In Fig. 5, design element 1 (d_1) affects both d2 and dn and is affected by d_2. Additionally, this study introduces the ISM method to figure out the relationships between design elements in each deployment chart. This paper describes an introduction of the ISM method to

(a)

	x_1	x_2	x_3	x_4	x_5	x_6	x_7	x_8
x_1	0	0	0	0	1	0	1	0
x_2	0	0	1	0	1	0	0	0
x_3	0	1	0	0	0	1	0	1
x_4	0	0	0	0	1	0	1	0
x_5	0	1	1	0	0	0	0	0
x_6	0	0	0	0	0	0	0	0
x_7	0	0	1	0	0	0	0	0
x_8	0	0	0	0	0	1	0	0

(b)

	x_1	x_2	x_3	x_4	x_5	x_6	x_7	x_8
x_1	1	1	1	0	1	1	1	1
x_2	0	1	1	0	1	1	0	1
x_3	0	1	1	0	1	1	0	1
x_4	0	1	1	1	1	1	1	1
x_5	0	1	1	0	1	1	0	1
x_6	0	0	0	0	0	1	0	0
x_7	0	1	1	0	1	1	1	1
x_8	0	0	0	0	0	1	0	1

(c)

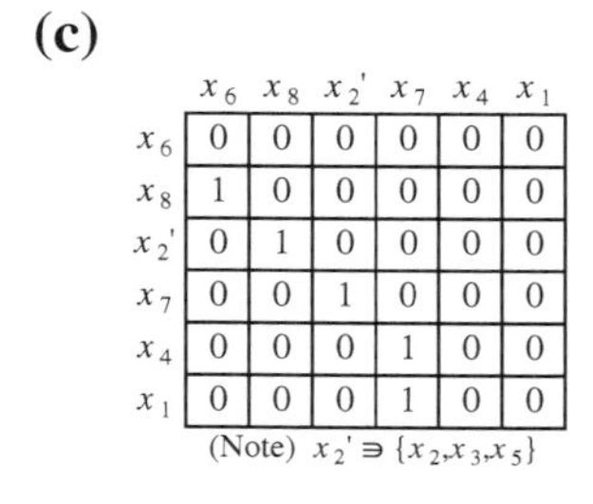

	x_6	x_8	x_2'	x_7	x_4	x_1
x_6	0	0	0	0	0	0
x_8	1	0	0	0	0	0
x_2'	0	1	0	0	0	0
x_7	0	0	1	0	0	0
x_4	0	0	0	1	0	0
x_1	0	0	0	1	0	0

(Note) $x_2' \ni \{x_2, x_3, x_5\}$

(d)

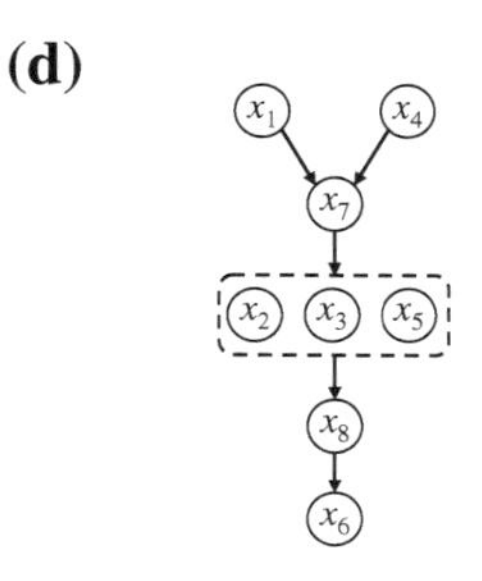

Fig. 4 Conceptual diagram of ISM

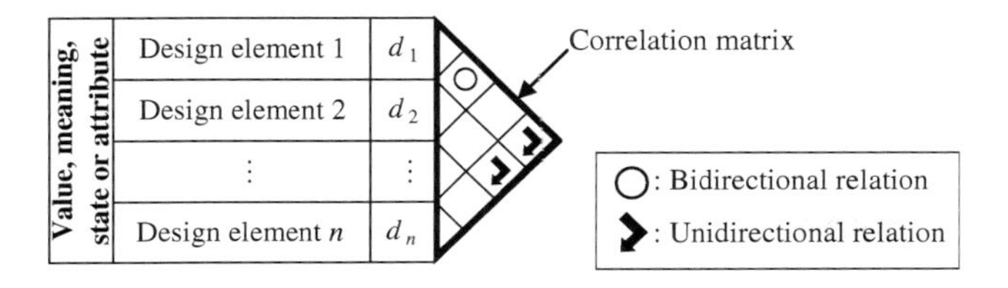

Fig. 5 Correlation matrix

the attribute (product's parts) design elements, whose relationships are complicated due to the greatest element number, in order to construct the parts design procedure. The procedure is constructed by not only their relationships but also the state (engineering characteristic) elements relationships. For example, the parts elements, which are not related to each other but related to the same engineering characteristics, assumed to have relationship and should be designed concurrently. In this study, the sum of both direct affective matrices **X** derived from the attribute correlation matrix (Eq. 1) and **X**′ derived from the engineering state correlation matrix are used to construct the parts design procedure.

$$X + X' \quad \left(X' = \begin{pmatrix} X'_{11} & \cdots & X'_{1j} & \cdots & X'_{1n_p} \\ \vdots & & & & \\ X'_{i1} & & \ddots & & \vdots \\ \vdots & & & & \\ X'_{n_p1} & & \cdots & & X'_{n_pn_p} \end{pmatrix} \right), \tag{4}$$

where n_a is the number of the attribute elements, and X'_{ij} are calculated as:

$$X'_{ij} \begin{cases} 1 & \textit{if both } a_i \textit{ and } a_j \textit{ relate to } s_k \\ 1 & \textit{if } s_k \textit{ (relating } a_i) \textit{ relates to } s_j \textit{ (relating } s_j) \\ 0 & \textit{else} \end{cases} \quad \begin{pmatrix} i = 1, 2, \ldots, n_a, \\ j = 1, 2, \ldots, n_a, \\ k = 1, 2, \ldots, n_a, \\ l = 1, 2, \ldots, n_a \end{pmatrix}, \tag{5}$$

where a_i is ith attribute element and s_k is kth state element.

4 Procedure of proposed QFD

The proposed quality matrices including the deployment charts of the four space and their correlation matrices are shown in Fig. 6. The procedure of the proposed quality matrices is as follows. First, the design elements (including value, meaning, state, and attribute elements) are extracted (Step 1 in Fig. 6). Second, the relationship matrices between them are developed (Step 2). Third, the correlation

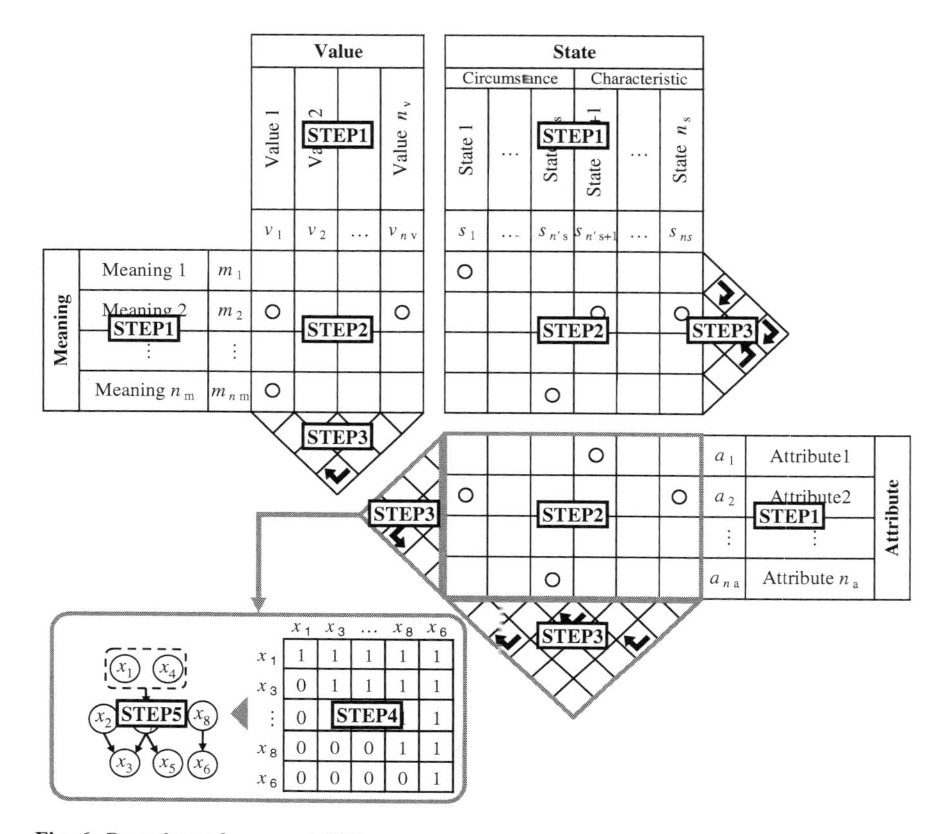

Fig. 6 Procedure of proposed QFD

matrices of each space is constructed (Step 3). Where, Step 1–3 are repeated until the product development members are satisfied. Fourth, the state and attribute correlation matrices are transformed into the direct affective matrices using Eqs. 1 and 5 (Step 4). Where, the value and meaning correlation matrices can be transformed if the members need. Finally, the structural models of them are constructed by the ISM method using the direct affective matrices (Step 5). Based on the structural models, the designers can proceed the parts (attribute) design without design change which is caused by the inadequate design procedure.

5 Illustrative Example

The proposed quality matrices were applied to a design problem of a disc brake to confirm their effectiveness. The common disc brakes generate brake torque by pushing the disc rotor (armature) to the brake pad (friction material) due to the spring force. Figure 7 shows a conceptual diagram of the disc brakes. In this figure, the coil springs located between the coil case and armature push the armature to the brake pad when braking. Whereas, when releasing the brake, the armature is attracted to the coil case by electromagnetic force. To sense that the brake is braking or released, the brake switch is installed on the coil case and flipped by the striker bolt set on the armature. In the disc brake design, the designers should consider a lot of design elements (e.g., the spring characteristics related to the brake torque, the coil characteristics to specify the electromagnetic force, the armature stroke which concerns both drive noise and brake switch characteristics) to realize the ideal brake characteristic (e.g., high brake torque, low drive noise, no brake switch glitch). However, there are trade-off relationship between the characteristics. Hence, there is a high possibilities that design change caused by inadequate design process is occurred.

Figures 8 describe the proposed quality matrices and structural model of the disc brake. These figures shows the followings:

1. The quality matrices composed by the four space deployment charts can describe both the design elements considered in the early process of design (e.g. safety and comfort) and that in the late process (e.g. brake torque and coil material). This enables designers to implement bidirectional design for a whole new product development. Additionally, the matrices can be used through the whole design process, and therefore promote the sharing of the product's information between the development members.
2. The structural model of the attribute elements simply describes the relationship between them and contributes the construction of the ideal product design process free of the design change. Additionally, the structural model of the state elements describe the circumstance elements, and the designers can easily extract and manage the elements related to the circumstance in redesign (improvement design).

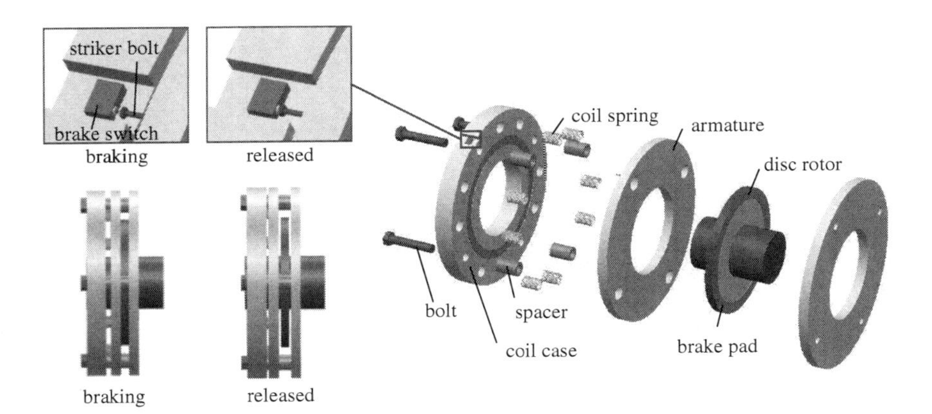

Fig. 7 Conceptual diagram of disc brake

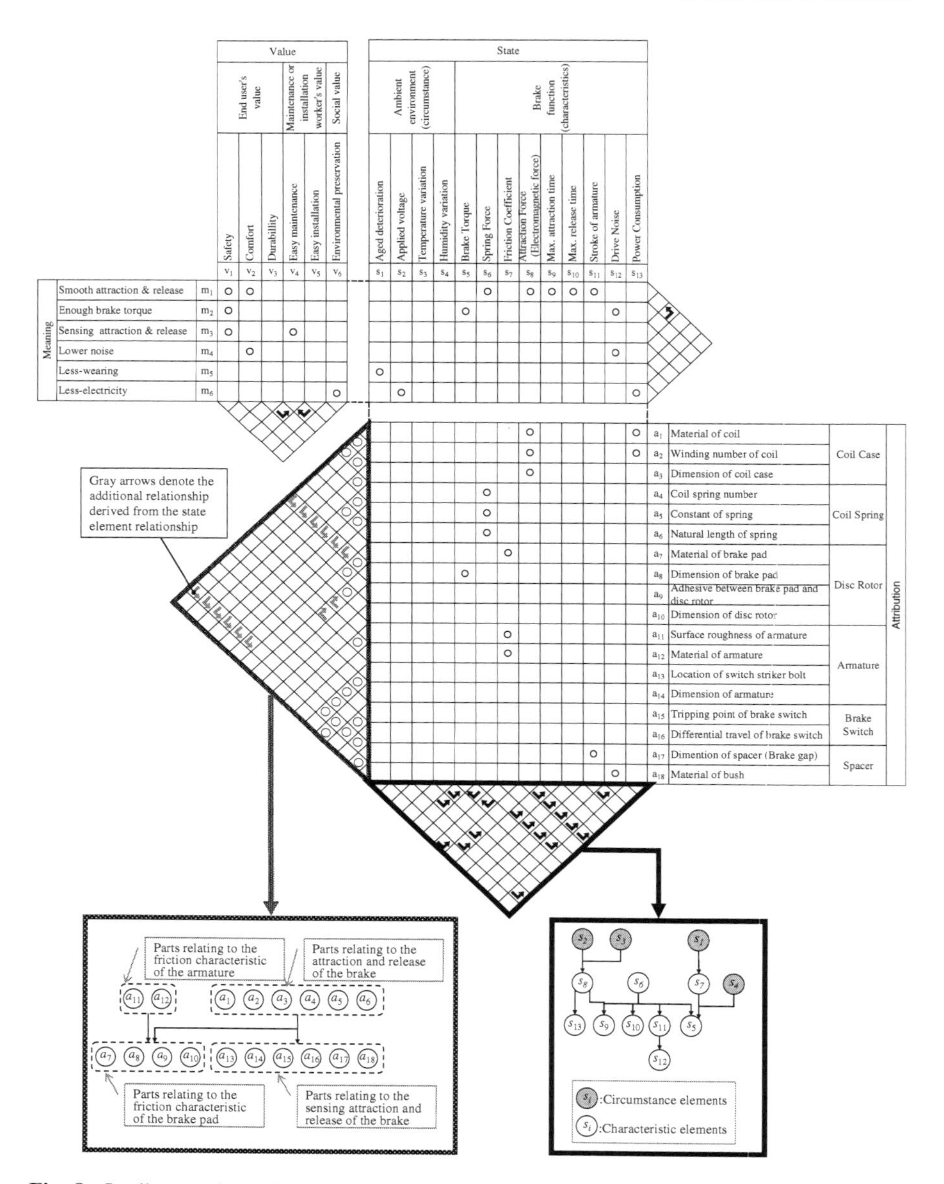

Fig. 8 Quality matrices of disc brake

6 Conclusion

This study introduced the Multispace Design Model and Interpretive Structural Modeling into the Quality Function Deployment (QFD). This introduction was expected to have the following effects:

1. Promoting the bidirectional design which enables the QFD to be applied to the whole new product development;
2. Increase the sharing information of the product between the development members;
3. Easier extracting and managing the elements related to the circumstance of the design object;
4. Assisting the construction of the ideal product design process free of the design change.

Additionally, the proposed QFD was applied to a design problem of the disc brake, and its applicability was confirmed. Future work should implement the followings:

1. Application of both the proposed and the conventional QFD to the same product development in order to confirm the effectiveness of the proposed one;
2. Many design application, including novel design, redesign, and improvement design, in order to confirm the versatility;
3. Designer survey on user friendliness.

Acknowledgments This work was supported by the Japan Society for the Promotion of Science, Grant-in-Aid for Scientific Research (C) (23611037).

References

1. Matsuoka Y (2010) Design science. Maruzen
2. Poel IVD (2007) Methodological problems in QFD and directions for future development. Res Eng Design 18:21–36
3. Akao Y (1990) Quality function deployment. Integrating customer requirements into product design. Productivity Press, Boca Raton
4. Akao Y (1989) Result of Questionnaire for application status on quality deployment—report of quality deployment research section. Qual J JSQC 19(1):35–44
5. Bode J, Fung RYK (1998) Cost engineering with quality function deployment. Comput Ind Eng 35:587–590
6. Matsuoka Y (2010) Multispace design model as framework for design science towards integration of design. In: Proceedings of international conference on design engineering and science 2010 (ICDES2010), Tokyo
7. Olsen SA (1982) Group planning and problem-solving methods in engineering. Wiley, London
8. Warfield JN (1976) Societal systems: planning, policy and complexity. Wiley, London

Exploring a Multi-Meeting Engineering Design Project

John S. Gero, Jiang Hao and Sonia da Silva Vieira

Abstract This paper reports a case study of a multi-meeting engineering design project lasting 5 months, unlike most design studies that focus on a single meeting. The project involved an engineering consultancy for the design of a robot controller. The design team consisted of engineers with different backgrounds. Eight sequential design meetings were studied using protocol analysis. The video recordings of these meetings were transcribed and then segmented and coded using an ontologically-based coding scheme. The analysis of these meetings focused on differences in the distributions of design issues and syntactic design processes between adjacent meetings. Statistically significant differences between some adjacent meetings were observed, which implies changes in design behavior between those meetings.

Keywords Multi-meeting project · Protocol analysis · FBS ontology

J. S. Gero (✉)
Krasnow Institute for Advanced Study, George Mason University, Fairfax, USA
e-mail: john@johngero.com

J. S. Gero
University of North Carolina, Charlotte, USA

J. Hao
Division of Industrial Design, National University of Singapore, Kent Ridge, Singapore
e-mail: didjh@nus.edu.sg

S. da Silva Vieira
IDMEC, Faculty of Engineering, University of Porto, Porto, Portugal
e-mail: sonia.vieira@fe.up.pt

A. Chakrabarti and R. V. Prakash (eds.), *ICoRD'13*, Lecture Notes in Mechanical Engineering, DOI: 10.1007/978-81-322-1050-4_6,

1 Introduction

The overwhelming majority of studies into engineering design, whether they are conducted in the laboratory or in the office, are of single design sessions or meetings. There are a few studies of engineering design that involve two meetings (e.g., [1]). In practice most design projects are spread out over time and involve the design team in multiple meetings. It is important to study such multi-meeting projects both to determine differences in design behavior between multi-meeting and single meeting design projects and differences in design behavior between several meetings of one project. This paper reports on the results of comparing adjacent meetings of a case study of a multi-meeting engineering design project.

Design meetings are considered as a sampling technique for investigating a lengthy design projects [1]. In the field of design cognition research, protocol analysis has been identified as the dominant methodology, aiming to explore cognitive processes underlying designers' behaviors [2–4]. This is a resource intensive methodology with a high ratio of analysis time to observation time; the observed design activities were usually of a limited duration, ranging from a few minutes to 1 or 2 h [5]. It is thus inappropriate to directly apply the existing methods focusing on the observation scale of minutes or hours on a lengthy project lasting a few weeks or months. Some adoptions were made by reducing the resolution of observation, e.g., omitting the minute-by-minute details, to track a longer design project [6–8]. This kind of approach fails to capture the transient cognitive events and interactions/transitions between thoughts.

This paper uses eight sequential meetings to represent an engineering robotic controller design project lasting 5 months. The remainder of the paper commences by describing the engineering design project and providing an overview of the eight meetings that made up the design sessions of the project with a qualitative description of the activities in each meeting. This is followed by an outline of the protocol analysis method used to produce the base data, which is in the form of a sequence of design issues and design processes for each of the eight meetings. These design issues and design processes are derived from an analysis of the transcriptions of the meeting videos using the Function-Behavior-Structure (FBS) ontologically-based segmentation and coding approach. The sequences of design issues and design processes are then analyzed as statistical distributions and comparisons between adjacent meetings are made.

2 An Engineering Design Project

In this engineering design case study, eight sequential meetings took place during a period of 5 months for the design of a robot controller. This project was the subject of a research project developed in collaboration with a design team of engineers with different backgrounds in mechatronics, namely: software/hardware,

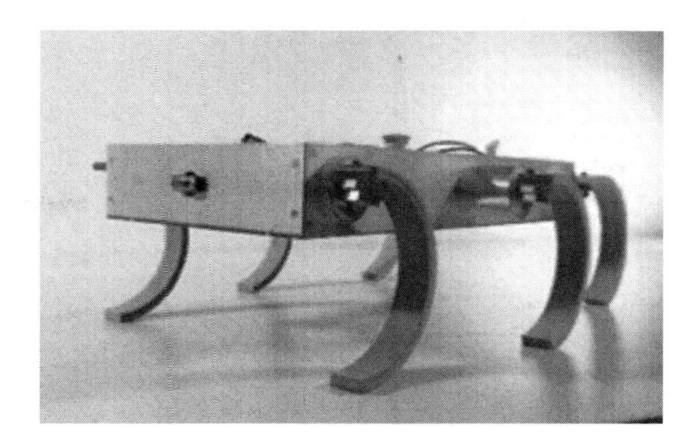

Fig. 1 Prototype of the robot hexapod

control, aerospace and electronics engineering. The design was based on a previous similar robotic controller, nevertheless the team faced several unexpected situations and challenges. The robot prototype is illustrated in Fig. 1.

Table 1 provides an overview of the meetings, their lengths, topics, team members' attendance and qualitative division of the eight meetings into two fundamental stages. Each of these meetings lasted approximately 1 h. In the first month, three meetings were dedicated to analyzing and clarifying specifications and production planning. In the second month a fourth meeting initiated the testing and detailing tasks that lasted until the end of the observation period. The three meetings in the last 2 months focused more on evaluations of problems, detailing and testing. Issues of specification analysis, connection systems, power supply, costs, and identification and analysis of problems were mostly discussed in these meetings.

3 Ontologically-Based Protocol Analysis

Each these eight meetings was videotaped, the utterances in them were transcribed and the transcriptions were then converted into a sequence of design issues using a principled coding scheme developed from the Function-Behavior-Structure (FBS) ontology [9, 10]. The FBS ontology models designing by three classes of ontological variables: function, behavior, and structure. The function (F) of a designed object is defined as its teleology, the behavior (B) of that object is either derived (Bs) or expected (Be) from the structure, where structure (S) represents the components of an object and their compositional relationships. These ontological classes are augmented by requirements (R) that come from outside the designer and description (D) that is the document of any aspect of designing, Fig. 2.

The FBS ontologically-based coding scheme consists of these six codes, each represents a particular aspect of design cognition. Application of this coding scheme can segment and encode the meeting videos (i.e., design conversations and gestures, etc.) into a sequence of design issues denoted with semantic symbol, i.e.,

Table 1 Overview of meetings during the design of a robot control

Month		1st			2nd	3rd	4th		5th
Meeting		1	2	3	4	5	6	7	8
Duration		1 h 06 min	1 h 03 min	1 h 08 min	52 min	34 min	52 min	1 h 01 min	58 min
Topic		Detailed discussion about the specifications and solutions	Planning, outsourcing and power supply	Discussion about the internal communication of the robot	Assembly of sub-parts, connections, testing and details	Discussion of software, defining tests and connections	Identification of problems, connections and detail	Complete assembly, Identification of problems, detailing and testing	Identification of problems, detailing and testing
Stage		Clarifying specifications, production planning				Concept generation, evaluations of problems, detailing and testing			
Team member	Leading researcher	√	√	√	√	√	√	√	√
	Electronics Engineer	√	√	√	√	√	√	√	√
	Software Engineer	√	√	√		√	√	√	√
	Technician	√	√		√	√	√	√	√

LR: Leading Researcher, **EE**: Electronics Engineer, **SE**: Software Engineer, **TC**: Technician

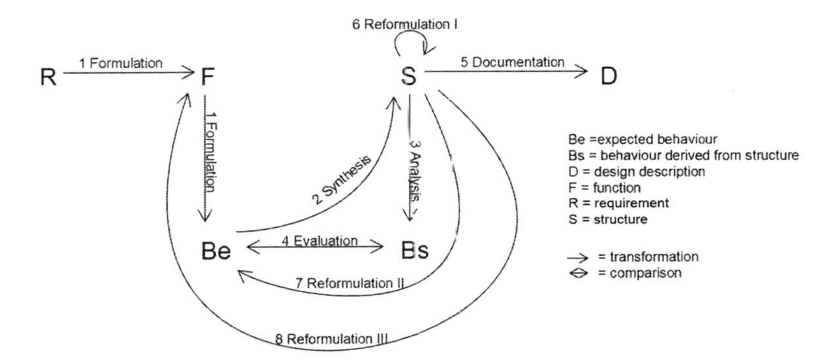

Fig. 2 The FBS ontology (after [10])

the FBS codes. The transitions between adjacent design issues were then defined as eight types of syntactic design processes, as numbered in Fig. 2 [9, 10].

The transformed data of these eight design meetings, namely eight sets of design issues and syntactic design processes, became the foundational data for subsequent analyses. Design cognition is multifaceted. Design issues and syntactic design processes measure two orthogonal dimensions of design cognition, respectively responding to the content-oriented and process-oriented analyses of design cognition.

4 Method of Analysis

Each design meeting's frequency distributions of design issues and syntactic design processes summarize the overall characteristics of the design cognition manifested in that meeting. The cognitive differences between two meetings can be examined by Pearson's Chi square test for independence. When a statistical significance is identified ($p < 0.05$), Cramer's V coefficient is calculated as the effect size to describe the relative strength of the difference between two meetings' issue/process distributions. The possible value of Cramer's V varies from 0 to 1. This study used the value of 0.15 as the threshold to indicate a substantive difference [11, 12].

The cross tabulations (here referred to as cross tabs) are then used as a *post hoc* test to further investigate which specific design issue(s) or syntactic process(es) contributes to the overall cognitive differences between two meetings. Adjusted residuals in a cross tab provide an estimation of the differences between observed and expected values (by assuming the distributions under comparisons are identical to each other). The design issues/processes with a high absolute value of adjusted residuals ($\geq$2) indicate that designers are more engaged in those aspects of design cognition in the meeting corresponding to the positive cells, than the other one.

5 Results

5.1 Coding Results

This paper presents preliminary coding results carried out by a single coder. The frequencies of design issues and syntactic design processes were normalized by converting them into percentages; this eliminates the different lengths of the design meetings and the subsequent different number of segments in each meeting. The design issues and syntactic design processes for all the eight design meetings are aggregated and are plotted in Fig. 3 along with the standard deviations. The means for the eight sessions shown in Fig. 3 provide an overall indication of the design cognition of the entire design activity while the standard deviations provide an indication of the variability across the meetings.

The distributions of design issues, Fig. 3(a), indicate that, in each meeting, the majority of design issues were structure and behavior from structure. These two solution-related issues represented about 85 % of total issues. The requirement issues, i.e., input from outside of the design teams, on the other hand, were negligible in this project, only occupying 0.23 % of the total issues. The requirement issue was thus excluded in the following Chi square analysis of design issue distributions.

The most frequent syntactic design processes, shown in Fig. 3(b), were associated with reasoning about the solution space, namely the processes of reformulation I ($M = 35.59$, $SD = 9.91$), analysis ($M = 32.59$, $SD = 4.96$) and evaluation ($M = 16.65$, $SD = 9.54$). Three problem-related processes, i.e., formulation, reformulation II (of expected behaviors) and reformulation III (of

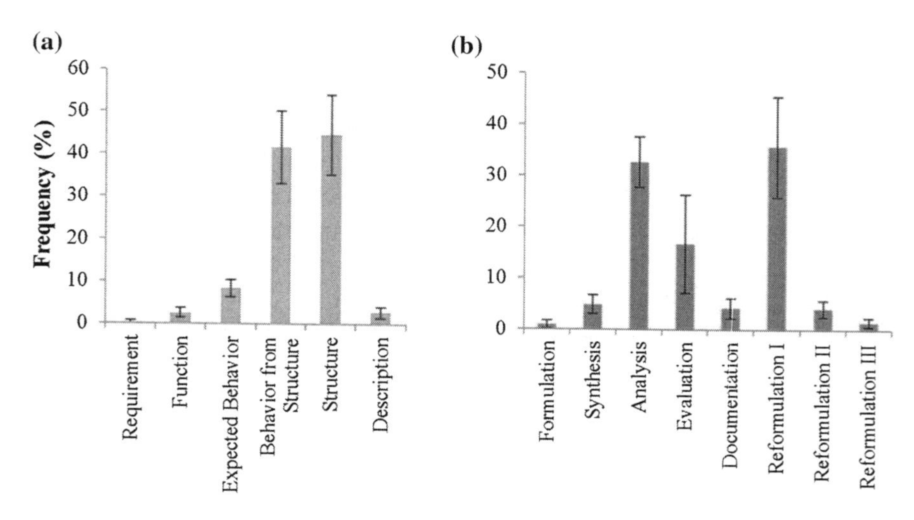

Fig. 3 Frequency distribution of (**a**) design issues and (**b**) syntactic design processes

functions), consumed less cognitive effort; each was less than 5 % of the total processes. They were thus combined as a single category in the Chi square analysis of syntactic process distributions.

5.2 Cognitive Shifts Between Two Adjacent Meetings

The analysis of cognitive progress during the 5 month observation was undertaken by comparing the cognitive changes between two adjacent meetings. The Chi square test results are summarized in Table 2. Two significant cognitive shifts were identified in adjacent meetings: from Meeting 3 to Meeting 4; and from Meeting 6 to Meeting 7.

5.2.1 Cognitive Transition Between Meetings 3 and 4

The comparisons of specific design issues and syntactic design processes between Meetings 3 and 4 are presented in two cross tabs, Table 3. Meeting 3 was more engaged in the generative aspect of design cognition, indicated by significantly higher percentages of structure issue and the syntactic design process of reformulation of structure (reformulation I). Meeting 4 then shifted to behavioral aspect of design cognition, indicated by higher percentages of expected behavior and behavior from structure issues, as well as the syntactic design process of evaluation.

Table 2 Comparisons of issue/process distributions of adjacent meetings

Comparison	Distr. of issue/process	*df*	Chi square statistics	*p* value	Cramer's V
Meeting 1 vs 2	Issue	4	5.162	0.271	0.092
	Process	5	5.835	0.323	0.123
Meeting 2 vs 3	Issue	4	12.902	0.012*	0.134
	Process	5	5.704	0.336	0.112
Meeting 3 vs 4	Issue	4	19.712	0.001**	0.169
	Process	5	27.161	0.000***	0.263
Meeting 4 vs 5	Issue	4	7.631	0.106	0.124
	Process	5	11.739	0.039*	0.207
Meeting 5 vs 6	Issue	4	1.907	0.753	0.065
	Process	5	0.973	0.965	0.060
Meeting 6 vs 7	Issue	4	20.423	0.000***	0.173
	Process	5	30.06	0.000***	0.282
Meeting 7 vs 8	Issue	4	9.522	0.049*	0.113
	Process	5	12.429	0.029*	0.187

*$p < 0.05$, **$p < 0.01$, ***$p < 0.001$

Table 3 Comparison between meetings 3 and 4, significant differences are highlighted in bold

(a) Comparison of the design issue distributions

Design issue	Meeting Number	3	4
Function	Count	17	10
	% within the meeting	5.0	2.8
	Adjusted residual	1.5	−1.5
Expected behavior	Count	25	46
	% within the meeting	7.4	13.1
	Adjusted residual	−2.5	**2.5**
Behavior from structure	Count	123	163
	% within the meeting	36.5	46.4
	Adjusted residual	−2.6	**2.6**
Structure	Count	166	124
	% within the meeting	49.3	35.3
	Adjusted residual	**3.7**	−3.7
Design description	Count	6	8
	% within the meeting	1.8	2.3
	Adjusted residual	−0.5	0.5
Total	Count	337	351
	% within the meeting	100.0	100.0

(b) Comparison of the syntactic process distributions

Syntactic design process	Meeting#	3	4
(Re-)formulation of function and expected behavior	Count	17	10
	% within the meeting	8.5	5.2
	Adjusted residual	1.3	−1.3
Synthesis	Count	9	12
	% within the meeting	4.5	6.3
	Adjusted residual	−0.8	0.8
Analysis	Count	61	55
	% within the meeting	30.3	28.6
	Adjusted residual	0.4	−0.4
Evaluation	Count	21	56
	% within the meeting	10.4	29.2
	Adjusted residual	−4.7	**4.7**
Documentation	Count	6	7
	% within the meeting	3.0	3.6
	Adjusted residual	−0.4	0.4
Reformulation I (of structure)	Count	87	52
	% within the meeting	43.3	27.1
	Adjusted residual	**3.4**	−3.4
Total	Count	201	192
	% within the meeting	100.0	100.0

Table 4 Comparison between meetings 6 and 7, significant differences are highlighted in bold

(a) comparison of the design issue distributions

Design issue	Meeting#	6	7
Function	Count	6	9
	% within the meeting	2.0	2.4
	Adjusted residual	−0.3	0.3
Expected behavior	Count	23	31
	% within the meeting	7.7	8.1
	Adjusted residual	−0.2	0.2
Behavior from structure	Count	111	202
	% within the meeting	37.0	53.0
	Adjusted residual	−4.2	**4.2**
Structure	Count	155	134
	% within the meeting	51.7	35.2
	Adjusted residual	**4.3**	−4.3
Design description	Count	5	5
	% within the meeting	1.7	1.3
	Adjusted residual	0.4	−0.4
Total	Count	300	381
	% within the meeting	100.0	100.0

(b) Comparison of the syntactic process distributions

Syntactic design process	Meeting#	6	7
(Re-)formulation of function and expected behavior	Count	11	6
	% within the meeting	5.8	3.2
	Adjusted residual	1.2	−1.2
Synthesis	Count	8	5
	% within the meeting	4.2	2.6
	Adjusted residual	0.8	−0.8
Analysis	Count	58	83
	% within the meeting	30.7	43.9
	Adjusted residual	−2.7	**2.7**
Evaluation	Count	25	51
	% within the meeting	13.2	27.0
	Adjusted residual	−3.3	**3.3**
Documentation	Count	5	4
	% within the meeting	2.6	2.1
	Adjusted residual	0.3	−0.3
Reformulation I (of structure)	Count	82	40
	% within the meeting	43.4	21.2
	Adjusted residual	**4.6**	−4.6
Total	Count	189	189
	% Within the meeting	100.0	100.0

5.2.2 Cognitive Transition Between Meetings 6 and 7

The cognitive shift between Meetings 6 and 7 is articulated in Table 4. Resembling the previous cognitive change between Meetings 3 and 4, the latter meeting shifted from an emphasis on generative aspect of design cognition (indicated by higher percentages of structure issue and the process of reformulation I) to engage more in the evaluative aspect of design cognition (indicated by higher percentages of behavior from structure issue and the processes of analysis and evaluation).

6 Discussion

The engineers in this design project had previous experience in designing robot controllers. Many robotic components, such as the microcontroller, and battery were continuously discussed from the first meeting on. This may explain the descriptive statistics result that solution-related design issues and solution-related syntactic design processes constituted the majority of design reasoning in all the eight meetings. The two significant cognitive changes identified in this case study were shifting from a relative focus on the solution generation to an increased focus on the analysis and evaluation of the proposed solutions.

The quantitative comparisons were then triangulated with qualitative assessments of the individual meetings. Meeting 3 focused on the discussion of structure components introduced in Meeting 1. Similar to Meeting 7, Meeting 3 did not continue the topics raised in Meeting 2. The control aspects of the robot are introduced in the next meeting. Meeting 4 seemed to be a "bridge meeting," discussing some design considerations more in depth, attempting to make connections to other considerations. This may explain that, in this meeting, the cognitive effort spent on reasoning about structure decreased, while the meeting was more focused on the expected consequence of solutions.

Meeting 6 was mainly targeted at a particular technical problem "how to solve the overheating of the board." A number of alternative solutions were proposed accordingly. Due to the focus on this topic, the percentages of the structure issue and the syntactic process of reformulation I increased in this meeting compared to the previous meetings.

There was a topic shift between Meetings 6 and 7. The latter meeting did not continue the topics raised in Meeting 6. Rather, it reactivated the topics introduced in Meetings 1 and 4, such as CPU and batteries, during the testing process. Behavioral and evaluative aspects of design cognition thus became the focus of this meeting. The cognitive shifts between generative and evaluative modes of designing also indicate the iterative nature of engineering design activities. Later studies will present detailed analyses of design cognition during the critical situations leading to design decisions.

7 Conclusion

This paper presents a preliminary analysis of a multi-meeting engineering design project lasting 5 months during which there were eight design meetings. Design projects in practice regularly involve multiple meetings, and it is important in the development of the understanding of designing that such multiple meeting design projects be studied and comparisons made with single meeting design projects to determine differences. The increased scale of observation, compared to a single design session of 1 or 2 h in most design protocol studies, provides a more nuanced understanding of designing as can be seen in the statistically significant differences found between a number of the design meetings.

When the eight meetings are aggregated into a single set of measurements of design issues and syntactic design processes, Fig. 3, the design issues and syntactic design processes distributions follow the general behavior observed in the single engineering design meetings/sessions used in studies of designing [13, 14] masking any detailed behavioral differences that occur over time. This points to the need for a more detailed study of multi-meeting designing.

Multiple meetings with time gaps provide opportunities for incubation that are not directly available in single meeting design sessions [15, 16]. Incubation plays a role in all areas of human cognition but insufficient is known about the design cognition of incubation. Studies of multiple meetings of professional designers in practice are an alternate to laboratory studies of the cognition of incubation. They may provide the basis of insight into incubation in designing [17].

This paper has demonstrated that it is feasible to carry out a design cognition study of a multi-meeting engineering design project in such a manner that the results are commensurable with such studies of single design meetings. Multiple meetings are the norm in professional engineering design practice. Studying them is critical to the development of our understanding of engineering design. However, it may be that multiple meetings do not exhibit design behaviors that differ from single design meetings but this needs to be tested empirically. If that hypothesis is shown to be supported by the empirical evidence then design scientists need only study individual design meetings.

Later papers will present detailed comparisons of design cognition derived from multi-meetings with the behavior observed in single meeting design sessions. The specific findings in this paper are based on a preliminary coding of the meetings and need to be confirmed.

Acknowledgments This research is supported in part by grants from the US National Science Foundation Grant (Nos. IIS-1002079 and SBE-0915482). The authors gratefully acknowledge the Delft Center for Systems and Control for practical support of this research.

References

1. McDonnell J, Lloyd P (eds) (2009) About: designing: analysing design meetings. CRC Press, Boca Raton
2. Cross N, Christiaans H, Dorst K (1996) Analysing design activity. Wiley, Chichester
3. Gero JS (2010) Generalizing design cognition research. In: Dorst K et al (eds) DTRS 8: interpreting design thinking. University of Technology Sydney, Sydney, pp 187–198
4. Jiang H, Yen CC (2009) Protocol analysis in design research: a review. In: Proceedings of International Association of Societies of Design Research (IASDR) 2009 Conference, Seoul, Korea
5. Bainbridge L, Sanderson PM (2005) Verbal protocol analysis. In: Wilson JR, Corlett EN (eds) Evaluation of human work: a practical ergonomics methodology, 3rd edn. Taylor and Francis, Florida, pp 159–184
6. Waldron MB, Waldron KJ (1988) A time sequence study of a complex mechanical system design. Des Stud 9:95–106
7. Rowe PG (1987) Design thinking. MIT Press, Cambridge
8. Rowe PG (1982) A priori knowledge and heuristic reasoning in architectural design. J Architectural Educ 36:18–23
9. Gero JS (1990) Design prototypes: a knowledge representation schema for design. AI Mag 11:26–36
10. Gero JS, Kannengiesser U (2004) The situated function-behaviour-structure framework. Des Stud 25:373–391
11. Crewson P (2011) Applied statistics handbook (version 1.2). Available: http://www.acastat.com/statbook.htm
12. Dancey CP, Reidy J (2011) Statistics without maths for psychology, 5th edn. Prentice Hall/Pearson, Harlow
13. Kan JWT, Gero JS (2007) Using the FBS ontology to capture semantic design information. In: McDonnell J, Lloyd P (eds) DTRS7. University of the Arts, London, pp 155–165
14. Williams CB, Gero JS, Lee Y, Paretti M (2011) Exploring the effect of design education on the design cognition of mechanical engineering students. ASME IDETC2011 DETC2011-48357
15. Smith SM, Dodds RA (1999) Incubation. In: Runco MA, Pritzker SR (eds) Encyclopedia of creativity, vol 2. Associated Press, San Diego, pp 39–44
16. Smith S, Blankenship S (1991) Incubation and the persistence of fixation in problem solving. Am J Psychol 104:61–87
17. Gero JS (2011) Fixation and commitment while designing and its measurement. J Creative Behav 45:108–115

Integrating Different Functional Modeling Perspectives

Boris Eisenbart, Ahmed Qureshi, Kilian Gericke and Luciënne Blessing

Abstract The paper proposes a modular functional modeling framework, which aims at integrating the different functional modeling perspectives, relevant to different disciplines. The results of two extensive literature studies on diverse functional modeling approaches proposed in a variety of disciplines are consolidated. These studies identified specific needs for an integrated functional modeling approach to support interdisciplinary conceptual design. The presented framework aims at fulfilling these needs. It consists of a variety of associated views, represented through different matrices. This matrix-based representation facilitates the analysis of different functional modeling perspectives and their interdependencies. Finally, the implications of the presented approach are discussed.

Keywords Functional modeling · Functional modeling perspectives · Modeling framework · Interdisciplinary design

B. Eisenbart (✉) · A. Qureshi · K. Gericke · L. Blessing
Engineering Design and Methodology Group, University of Luxembourg, Luxembourg
e-mail: boris.eisenbart@uni.lu

A. Qureshi
e-mail: ahmed.qureshi@uni.lu

K. Gericke
e-mail: kilian.gericke@uni.lu

L. Blessing
e-mail: lucienne.blessing@uni.lu

A. Chakrabarti and R. V. Prakash (eds.), *ICoRD'13*, Lecture Notes in Mechanical Engineering, DOI: 10.1007/978-81-322-1050-4_7,

1 Introduction

Functional modeling is proposed in systematic design approaches across disciplines. It is intended to support early concept development for a technical system, i.e. the transition from a design problem to an early solution concept. Functional modeling results in a first abstract representation of the technical system under development. The term "technical system" encompasses both technical products as well as product/service-systems (PSS) in this paper.

Across and within different disciplines a large variety of function models is proposed and a common approach to functional modeling can hardly be found [1–4]. As a consequence, diverse ways of representing functions are competing when designers from different disciplines collaborate, potentially hindering the exchange expertise [1, 5]. Approaches to bridge the existing diversity, so far, have not been successful [2, 6].

This paper consolidates the results of two extensive literature studies on the diverse functional modeling approaches, proposed in a variety of disciplines [3, 4]. The considered studies identified specific needs for an integrated functional modeling approach linking the different functional modeling perspectives, which are prominently addressed in the proposed function models from different disciplines. This paper presents the concept of an integrated functional modeling approach, which aims at satisfying the identified needs.

2 Towards Integrated Functional Modeling

Eisenbart et al. [3] analyzed function models proposed in mechanical engineering design, electrical engineering design, software development, mechatronic system development, service development and PSS design. The particular content addressed by individual function models is linked to different functional modeling perspectives. Seven central perspectives have been identified, which are described in Table 1.

None of the reviewed function models from the different disciplines addresses all identified functional modeling perspectives [3]. In each considered discipline a different set of functional modeling perspectives is prominently addressed. However, Eisenbart et al. [4] identified the *transformation process perspective* to be prominently addressed in functional modeling approaches proposed across all reviewed disciplines. It may, hence, serve as a common basis in an integrated functional modeling approach. Based on the two literature studies by Eisenbart et al.[1] specific needs for such a modeling approach can be formulated. Accordingly, an adequate integrated functional modeling approach should:

[1] Eisenbart et al. considered 70 function models (54 original models plus variants proposed by different authors) and 41 systematic design approaches. The respective references may be taken from [3, 4].

Table 1 Functional modeling perspectives, after [4]

States	Representation of the states a system can be in, or of the states of operands before (input) and after (output) a transformation process. Operands are typically specifications of energy, material, and information
Effects	Representation of the required physiochemical effects, which have to be provided to enable, respectively support, the transformation process(es) changing one state into another state
Trans-formation processes	Representation of the processes executed by stakeholders or technical systems, which (from the designers' perspective) are part of the technical system under development in order to change the state of the system or of operands. *Technical processes* are transformation processes related to technical systems, while *human processes* are related to stakeholders (thus, including service activities)
Interaction processes	Representation of interaction processes of stakeholders or of other technical systems, which (from the designers' perspective) are *not* part of a system, with stakeholders or technical systems, which *are* part of the system under consideration
Use case	Representation of different cases of applying the technical system. This is typically associated to the interaction of stakeholders or another technical system with the technical system under development, which triggers, respectively requires subsequent processes to take place
Technical system allocation	Representation of the role of a technical system, which is supposed to perform or enable a (sub-) set of required *effects* or *processes*, either as part of the technical system under consideration or by interacting with it
Stakeholder allocation	Representation of the roles of different stakeholders, which may be users benefitting from a system or operators contributing to the system, e.g. through executing required processes or providing resources, etc

- …link the identified functional modeling perspectives, in order to relate between information, which is relevant to the designers from the different disciplines.
- …enable flexibly switching between considered functional modeling perspectives, in order to facilitate adaption of the modeling approach to different design approaches.
- …provide a condense and clearly structured representation, in order to ease comprehension of the modeled functions among collaborating designers. Often multiple complementary models are proposed in a functional modeling approach. Comprehensively capturing information distributed across different models can be a difficult cognitive task. However, *one* model covering a large number of functional modeling perspectives may quickly become confusingly packed with information.
- …facilitate linking functions in different ways, in order to be adaptable to discipline-specific representations. Depending on the particular discipline, functions may essentially be linked related to *time* (particularly prominent in software and service development), *input/output relations* (particularly prominent in mechanical engineering design) or *hierarchy*.

- ...address impacts from, respectively on the environment, in order to facilitate finding viable solution concepts [7]. Only few authors explicitly consider the environment within functional modeling (e.g. [8, 9]). However, impacts from the environment on a technical system may impair function fulfillment. In turn, impacts from a technical system on the environment may be critical to e.g. safety requirements or environmental legislation.

Beyond these needs, additional options are discussed by Eisenbart et al. [3], which may considerably support the reasoning about functions within system conceptualization; such as considering function-sharing, the inclusion of quantities, as well as a stronger link between the functional model of a system and its structure.

3 Integrated Functional Modeling Framework

In order to meet the needs discussed in the previous section, this paper proposes the integrated functional modeling framework (IFM framework), which aims at supporting integrated modeling of the identified functional modeling perspectives.

3.1 Development of the IFM Framework

In the development of the functional modeling framework, different alternatives have been generated. Firstly, an attempt was made to adapt existing functional modeling approaches to satisfy the specific needs discussed above. For this, several approaches have been selected, which already cover a large variety of the different functional modeling perspectives. Each generated alternative has been applied for re-modeling examples of existing function models from the literature as well as an example from industry. The generated models and approaches have been comparatively evaluated.

From the authors' point view, merely expanding existing approaches has not resulted in suitable integrated modeling approaches: The respective models frequently seemed overburdened with the represented information and thus quickly became very difficult to comprehend. Often, the dependencies between the different functional modeling perspectives in relation to the central *transformation process perspective* could not adequately be addressed. Also, the link between individual functional modeling perspectives often became fuzzy, with the result that individual perspectives could hardly be reasoned upon disconnected from others.

Existing function models typically use blocks or circles for depicting transformation processes, states, effects, etc. These elements are typically arranged *circular* or in *vertical/horizontal* flows. The functional modeling framework presented in the following, instead, uses a modular, *matrix-based* representation, which allows modeling and retrieving information more clearly. The developed approach is related to the concept of multi-domain matrices (MDM) proposed by Lindemann and Maurer [10]. MDM map different design information, in order to facilitate analysis and representation of interdependencies. Similarly, the IFM framework aims at clearly representing information associated to the individual modeling perspectives and their dependencies.

3.2 Outline of the IFM Framework

The IFM framework consists of associated modular matrices representing different views onto the functions of a system under development. The central view (*process flow view*) addresses the *transformation process perspective*, which is prominent in functional modeling approaches across disciplines. Associated views use matrices to represent information about the different entities and their interdependencies in the modeling framework. The entities are directly linked to the specific functional modeling perspectives discussed above (see Table 2).

The framework of modular, adjacent views provides a clearly structured representation and allows taking different views on the functions of a technical system. This modular structure allows addition or omission of views related to the specific needs of the involved designers. The following sub-section describes the entities and their relations, which form the basis of the IFM framework. Section 3.2.2 describes the associated views, which form the representation of the IFM framework.

3.2.1 Entities and their Relations

The class diagram in Fig. 1 represents the relations between individual entities in the developed modeling framework. A technical system under development may support one or more use cases. Each use case may be decomposed into sub-use cases. Use cases may have dependencies among each other that may be bound by

Table 2 Entities in the IFM framework and addressed functional modeling perspectives

Entity	Addressed functional modeling perspective
Use case	Use case perspective
State	States (operands and system)
Process	Transformation process and interaction process perspectives
Effect	Effect perspective
Actor	Stakeholder and technical system allocation perspective; system state perspective
Operand	Operand state perspective

specific constraints (mutually exclusive, mutually inclusive etc.,). For all other situations, in Fig. 1, the dependencies shown will be used to depict the similar constraints.

A use case may have one or more transformation process associated to it. There may be dependencies between individual transformation processes, which may or may not be also composed of sub-processes. A transformation process results in the transformation of one or more operand and/or actor from a given state into another. Such state transformations are enabled, respectively supported by effects, which are provided by actors. Actors, by providing the necessary effects, act as operators in transformation processes. Actor is a super class which contains the sub classes of stakeholder, technical (sub-) system, and environment. The actor sub-class of stakeholder comprises (groups of) animate beings affected by or affecting the technical system under consideration (including any related services). The actor sub-class of technical (sub-) system encompasses technical systems which are sub-systems to the technical system under development. It can also be composed of more technical (sub-) systems. Actors also may have dependencies among each other. Environment includes all active and passive parts of nature in general surrounding the system under development.

3.2.2 Associated Views

The different views are strongly linked to each other through the adjacent placement and the respectively shared header rows and header columns in the specific matrices forming the individual views (see Figs. 2, 3, 4, 5). The aim behind this specific set-up is to interlink all the different functional modeling perspectives (i.e. the corresponding views), prominent in the different disciplines, via the *transformation process perspective* (i.e. the central *process flow view*), which is

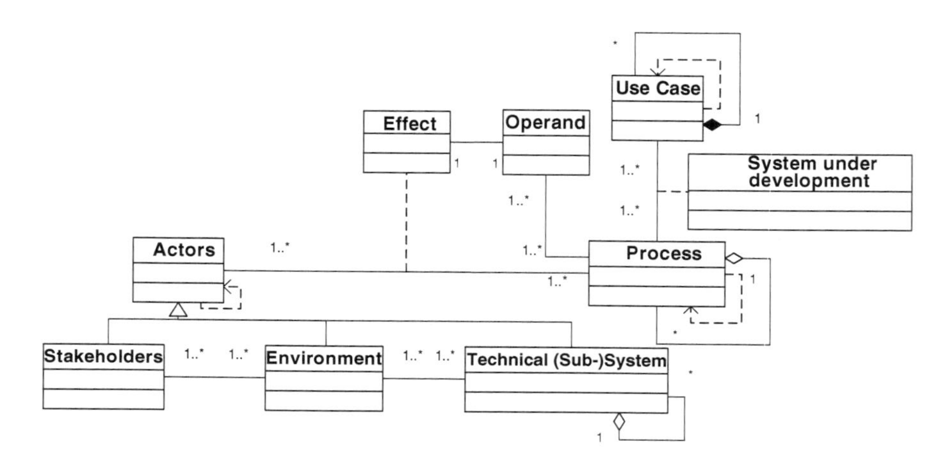

Fig. 1 Class diagram of the developed functional modeling framework

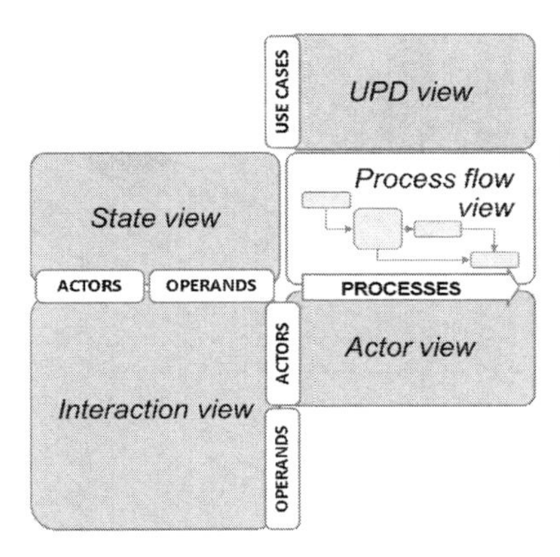

Fig. 2 Adjacent views in the IFM framework

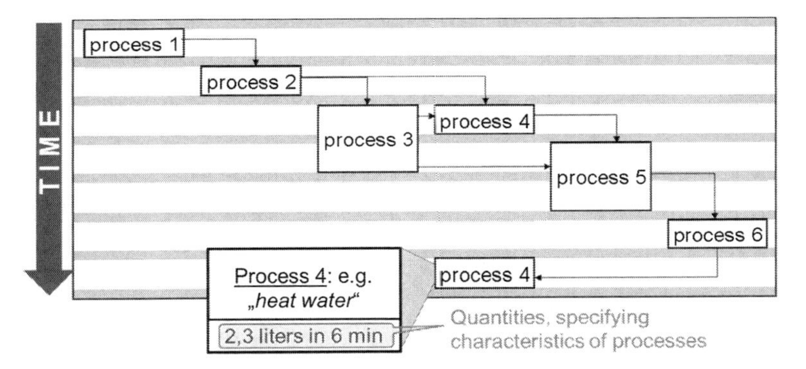

Fig. 3 Process flow view

prominent across disciplines. The individual views and how they link to each other are described in the following.

The *Process flow view* visualizes the flow of processes related to a specific use case. In the view individual processes are represented as chronologically numbered blocks. In the vertical direction, the process flow related to time is visualized. The flow qualitatively illustrates whether individual processes are expected to be carried out sequentially, in parallel or to be overlapping with other processes. The process blocks are furthermore spread horizontally from left to right, so as to enable a direct link to the *actor view* matrix, which is described further down. As an option, quantities related to individual processes can be included to specify processes further, as illustrated in Fig. 3.

The *Effect view* represents the effects, which enable individual transformation processes and are provided by actors. For each process block in the *process flow view*, a separate *effect view* may be created. Similar to the *process flow view*,

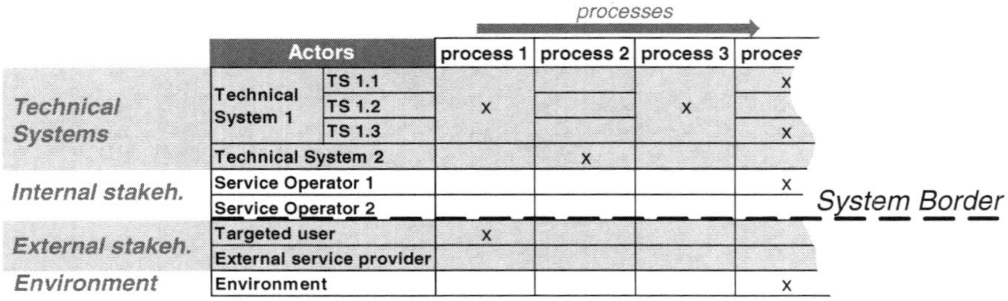

processes →

	Actors		process 1	process 2	process 3	proces
Technical Systems	Technical System 1	TS 1.1				x
		TS 1.2	x		x	
		TS 1.3				x
	Technical System 2			x		
Internal stakeh.	Service Operator 1					x
	Service Operator 2					
System Border						
External stakeh.	Targeted user		x			
	External service provider					
Environment	Environment					x

Fig. 4 Actor view matrix

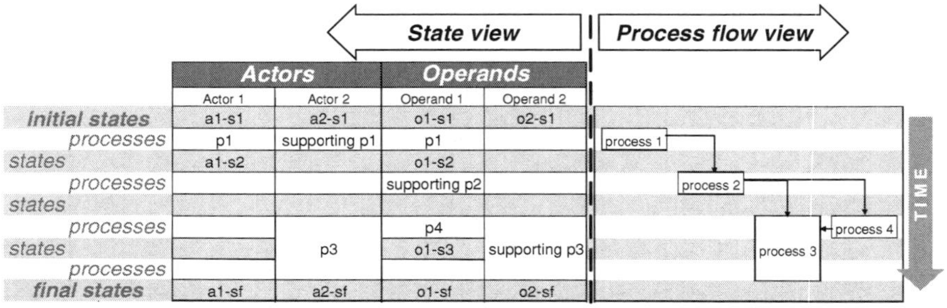

State view | Process flow view

	Actors		Operands	
	Actor 1	Actor 2	Operand 1	Operand 2
initial states	a1-s1	a2-s1	o1-s1	o2-s1
processes	p1	supporting p1	p1	
states	a1-s2		o1-s2	
processes			supporting p2	
states				
processes			p4	
states		p3	o1-s3	supporting p3
processes				
final states	a1-sf	a2-sf	o1-sf	o2-sf

process 1
process 2
process 3
process 4
TIME

Fig. 5 System states view associated to process flow view

effects can be modeled related to time or flow of operands. Hierarchical trees or alternative models (similar to [7, 8]) may also be applied.

The *Use case/process dependency* (*UPD*) *view* indicates the involvement of individual processes within different use cases. The individual use cases are listed in the header column. The matrix is directly linked to the *process flow view*. The individual—strictly horizontally ordered—process blocks build up the header row for the *UPD view* (see Fig. 2). Dependencies between use cases and processes could affect their operability. For instance, the processes of "heating water" in one use case and the process of "cooling water" in another use case should not be executed in parallel for the same water sample; hence, neither should the respective use cases.

The *Actor view* indicates the involvement of specific actors in the realization of transformation processes. Transformation processes are spread in the header row, associated to the process flow view (see Fig. 2). Within the matrix, involvement may initially be indicated with an "x". As more information becomes available in the design process, the particular role of actors (e.g. as either "affecting" or "being affected" by a process) can be more concretely specified.

The *actor view* allows differentiating actors according to whether they—from the designers' point of view—are part of the system under development (e.g. service operators as part of a PSS) or not (e.g. the targeted users or external service providers). This differentiation is particularly important in PSS design [11]. Through this differentiated allocation, individual processes are separated between *transformation processes* (enabled by actors which are part of the system) and *interaction processes* (enabled by actors which are not part of the system).

The *States view* represents the specific states of operands and agents as well as the state changes caused by individual processes. The *system states view* consists of the *actor state matrix* and the *operand state matrix*, and is a modular addition to the *process flow view* (see Figs. 2 and 5). The adjacent placement of the *system state* and *process flow views*, as shown in Fig. 5, allows the development of the views and the verification of their consistency. Considering the required changes from initial to final states of operands and actors facilitates the development of the *process flow view* and vice-versa. The system state view also allows the indication of operands supporting a transformation process without changing their states.

Figure 5 illustrates, e.g., that a process "heat water" (process 1) is linked to a change of the technical system (actor 1) from switched-off (a1-s1) to switched-on (a1-s2), supported by an operator (actor 2), as well as a change of the state of water (operand 1) from liquid (state o1-s1) to steam (new state o1-s2). The steam may then be used to drive a turbine (process 2), which may give rise to changes to other operands and/or actors. During this process, the state of the water (i.e. steam) is not changing, but supporting process 2.

The *interaction view* depicts the specific interactions between actors and operands, as well as among each other, in the realization of processes. The view uses operands and actors as both heading column and heading row, as illustrated in Figs. 2 and 6. The specification of the interaction between actors or operands includes the number of the respective process (to provide clarity, as numerous interactions may occur related to different processes) and a short statement specifying the interaction. Analysis of interferences between actors and/or operands may highlight problems with function fulfillment. Also, information about how actors and operands may impact on each other facilitates the design of the interfaces between them accordingly in later design phases.

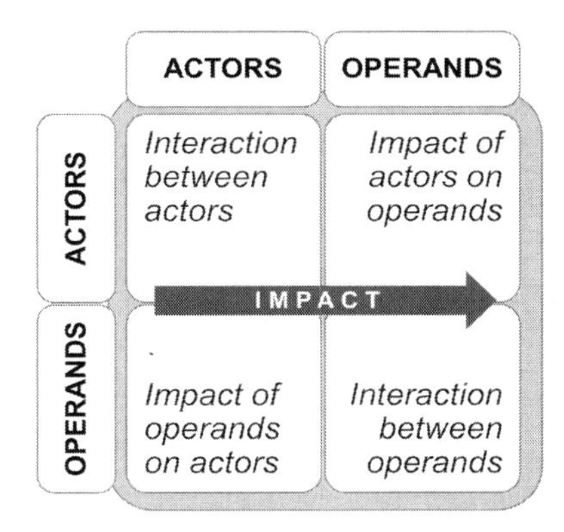

Fig. 6 Concept of interaction view matrix

To give an example, a stakeholder (actor) may have an impact on a technical system (actor) by pushing a button. Similarly, hot water (operand) may impact on a technical system (actor) through transmitting heat or vise-versa. Finally, operands may also impact on each other, as for instance, cold water (operand) may be used to cool hydraulic fluids (operand) and vise-versa. In case, the specifics of an interaction cannot be specified at an early point in the modeling process, the respective cell may initially be marked with an 'x'. Optionally, information about how the interaction is embodied may be included; such as 'mechanical contact' between the stakeholder's finger and the button being pushed.

3.3 Application

The presented framework may be applied in different ways, i.e. depending on the specific approach taken by designers, alternative entry points and sequences of steps may be applied. One potential sequence of modeling activities for an original design project is described in the following. Starting point may be a comprehensive requirements specification (or similar).

- Step 1—*Use Case definition* includes the consolidation of the different use cases (and their sub-use cases, if applicable) the system under development is expected to support in the different phases of its life-cycle. The use cases are represented in the respective column in the *UPD view*.
- Step 2—*Process flow modeling* involves modeling separate flows of required transformation and interaction processes related to each (sub-) use case. A multitude of alternative process flows may fulfill a use case. As described above, modeling and selecting an alternative process flow may be facilitated through considering the required state changes of (supporting) operands in the *operand states matrix* (as part of the *state view*). While modeling the process flows, the involvement of processes in multiple use cases (represented in the *UPD view*) needs to be considered.
- Step 3—*Effect modeling* involves modeling the required effects related to the specific process flows. Considering the basic required effects enabling transformation processes may considerably support the allocation of actors in the following step.
- Step 4—*Actor allocation* includes allocation of the actors, which are involved in the individual processes, either as affecting or being affected through the delivered effects. Actor allocation may be supported through applying the function-means pattern, morphological charts or similar approaches. Carefully considering re-use of allocated actors in different processes and use cases may facilitate function sharing.
- Step 5—*State modeling* includes modeling the state changes of allocated actors in the actor state matrix (as part of the *state view*) related to the chosen process flows.

- Step 6—*Interaction specification* involves analyzing and detailing the specific interactions (i.e. the bilateral impacts) among actors, among operands, and between actors and operands in the realization of processes.

There can be iterations within and between individual steps. For instance, depending on the specific choice of realizing actors, the chosen process flows may have to change, requiring iterations between steps 2–4. *Actor allocation* essentially marks the transition from the problem to the solution. However, the final set of *process flow view* and *actor view* merely represents one potential concept out of large number of variants.

Modeling starts on a high level of abstraction defining the use cases, associated processes etc. On the next level of detail, individual process blocks may then be regarded as use cases comprised of sub-processes. These are enabled by technical (sub-) systems (which may again be comprised of general function carriers or "organs" [8], which are gradually concretized) including any related service operator etc. Thus, the framework allows modeling the functions and actors of a system under development from very abstract to very detailed and concrete.

4 Discussion and Conclusion

Functional modeling is proposed across disciplines to support early concept development. The different functional modeling perspectives prominently addressed in the different disciplines need to be integrated in order to support interdisciplinary functional modeling. In this paper, an integrated modeling framework has been proposed, which aims at linking the different functional modeling perspectives. The proposed framework uses interlinked modular matrices, representing different views on the functions of a system under development. The different views represent individual functional modeling perspectives and/or dependencies between them. It is expected to provide designers from different disciplines with a valuable approach to functional modeling, as it.

- uses an established matrix-based approach for analyzing and representing interdependencies between functional modeling perspectives, similar to MDM;
- considers all identified functional modeling perspectives and their interdependencies;
- is expected to ease communication across disciplines, as the different views are linked via a central view, which is commonly prominent across disciplines;
- is modular, which enables addition or omission of views and related modeling activities depending on whether these are needed in a specific design context;
- allows using different views separately; the designers may flexibly switch between considered views, which allows focusing on specific functional modeling perspectives;

- the strong links between the modular matrices representing the different views provide a clearly structured representation supporting comprehension of complex systems;
- allows embedding existing (discipline-specific) function models[2];
- is open for existing functional taxonomies to be embedded;
- can address the functions of a technical system on different level of detail/ abstraction;
- is expected to be easily transferrable into a software tool;
- finally, integrated the consideration of the environment.

In summary, the proposed functional modeling framework aims at fulfilling the formulated needs and—through its specific structure—is expected to support the exchange of discipline-specific expertise during system conceptualization. The explicit inclusion of the specific interactions between individual actors is further expected to provide links to models used in subsequent design phases (e.g. system structure, interface matrix, etc.,). Apart from the presented views, the framework may be further expanded. For instance, additional views may address the dependencies among different states (for both operands and actors), different use cases, different processes (across use cases) etc.

Future research will address the practical application of the developed framework by designers in industry. That will include workshops, wherein practical designers from different disciplines apply the developed framework in conceptual design of mechatronic systems and PSS. It will be of particular interest which specific functional modeling perspectives are most relevant to designers from different disciplines and how designers reason between the different proposed views in different design contexts. The gained insights and feedback from the designers will be used to develop the framework further, in order to improve its applicability in different design contexts.

Acknowledgments The authors would like to thank the Fonds Nationale de la Recherche Luxembourg for funding this research as well as Prof. Mogens Myrup Andreasen (DTU Copenhagen), Prof. Pierre Kelsen (Luxembourg University), and Hubert Moser (LuxSpace) for valuable feedback on earlier versions of the presented functional modeling framework.

References

1. Erden M, Komoto H, van Beek TJ, D'Amelio V, Echavarria E, Tomiyama T (2008) A review of function modeling—approaches and applications. Artif Intell Eng Des Anal Manuf 22:147–169

[2] This has only been briefly discussed related to the description of the *effect view*. However, numerous other cases wherein alternative existing models may be applied (either complementing or in exchange for matrices associated to a specific view) can be thought of.

2. Vermaas P (2011) Accepting ambiguity of engineering functional descriptions. Proceedings of 18th international conference on engineering design, ICED, pp 98–107
3. Eisenbart B, Blessing LTM, Gericke K (2012) Functional modeling perspectives across disciplines—a literature review. Proceedings of 12th international design conference, design
4. Eisenbart B, Gericke K, Blessing L (2012) A shared basis for functional modeling. Proceedings of 9th Nord design conference
5. Müller P, Schmidt-Kretschmer M, Blessing L (2007) Function allocation in product-service-systems—are there analogies between PSS and mechatronics? AEDS Workshop
6. Carrara M, Garbacz P, Vermaas P (2011) If engineering function is a family resemblance concept. Appl Ontol 6:141–163
7. Pahl G, Beitz WFJ, Grote K-H (2008) Engineering design—a systematic approach. Springer, Berlin
8. Hubka V (1984) Theorie technischer systeme—grundlagen einer wissenschaftlichen konstruktionslehre. Springer
9. Watanabe K, Mikoshiba S, Tateyama T, Shimomura Y, Kimita K (2011) Service design methodology for cooperative services. Proceedings of the ASME IDETC/CIE
10. Lindemann U, Maurer M (2007) Facing multi-domain complexity in product development. In: Krause FL (ed) The future of product development, Springer, Berlin
11. Tan AR (2010) Service-oriented product development strategies

Part II
Design Creativity, Synthesis, Evaluation and Optimization

Information Entropy in the Design Process

Petter Krus

Abstract In this paper the design process is viewed as a process of increasing the information of the product/system. Therefore, it is natural to investigate the design process from an information theoretical point of view. The design information entropy is introduced as a state that reflects both complexity and refinement, and it is argued that it can be useful as some measure of design effort and design quality. The concept of design information entropy also provides a sound base for defining creativity as the process of selecting areas for expanding the design space in useful direction, "to think outside the box", while the automated activity of design optimization is focused, so far, on concept refinement, within a confined design space. In this paper the theory is illustrated on the conceptual design of an unmanned aircraft, going through concept generation, concept selection, and parameter optimization.

Keywords Information entropy · Design complexity · Product platform

1 Introduction

During the process of design, information is gradually increased as the design progress, and the uncertainty of the design is reduced. Every design decision reduces the uncertainty of the design, as well as parameter calculations do. It could be argued that this is the central aspect of design. Design in general is about

P. Krus (✉)
Department of Management and Engineering, Division of Fluid and Mechatronic Systems, Linköping University, Linköping, Sweden
e-mail: petter.krus@liu.se

A. Chakrabarti and R. V. Prakash (eds.), *ICoRD'13*, Lecture Notes in Mechanical Engineering, DOI: 10.1007/978-81-322-1050-4_8, © Springer India 2013

increasing the information of the product/system. This can be viewed as a learning process, as described in [1]. Therefore, design theory should really be a theory of design information. Although there is a rich literature regarding design process, there has been little or no effort to describe the generation of information in quantitative terms.

The classical information theory of communication was founded in 1948 by C.E. Shannon with his paper "A Mathematical Theory of Communication" [2]. Subsequently it has been recognized that information is a key property of design, and for describing and analyzing the design process. The notion of information theory in design has been introduced by several authors. Notables are Suh [1, 3], Kahn and Angeles [4] and Frey and Jahangir [5]. The two first are discussed later in the text. Frey and Jahangir deals with the transformation of information content in design parameters to information content in the functional characteristics, so is Bras and Mistree [6] where robustness is defined from maximizing the signal to noise ratio, which also deals with the relation between design parameters and functional characteristics. Information theory was also used to analyze design optimization in Krus and Andersson [7]. Information theory has also been used to define complexity in software, and notably by Bansiya et al. [8] to describe complexity in object oriented systems, which is very close to general design.

The notion of information is used in the second axiom of Suh's Axiomatic design, which states that the information content in a design should be minimized [9].

2 The Characteristics of Design Information Entropy

2.1 Design Information

In a product, the design information x is transformed into functional attributes y

$$\mathbf{y} = \mathbf{f}(\mathbf{x}) \tag{1}$$

The design information could be the bit string in a CAD-file, and/or the set of parameters in a design that at some stage are free. It could also be the genome in an organism. The design information is the code of the design space D; every bit-combination represents a unique design in the design space. At this stage it is not necessary to separate design parameters from system architecture, it is all included in x. If only little information is present, only parts of the design space can be excluded, the design can be any of several unique designs. This introduces a presence of uncertainty, since the design is not precisely known. An important aspect of design information is that it can only be defined relative to a design space. The design space need not, however, be static, but can be expanded if found necessary. This is also the case in the genome in biological systems, where, different organism has different sizes of the genome. The definition of design information used here is therefore:

> Design information is the information needed to define a design, relative to a design space, to within a certain precision.

As a consequence, information is the inverse of uncertainty since lower precision results in less design information needed to describe a design.

2.2 Design Space

In order to generate a concept, a design space has to be established first. The design space contains all the possible designs. A Lego™ set is an example of a design space. A large number (although finite) number of designs can be build from a particular set. Another design space is represented by all the different Lego pieces. Different finite design spaces are then represented by the different number of pieces allowed in the design i.e. 1, 2... n pieces.

Figure 1 The design space of a set of Lego bricks represents all (discrete) combinations of arranging these bricks, n_{Dstate}. With a set of only two bricks with four knobs on each there are 51 discrete possible arrangements (two of these represents picking only one brick and one state is to pick no one).

The 51 different configuration (states) means that the amount of information needed to specify a particular design is:

$$I_x = \log_2 n_{Dstate} = \log_2 51 = 5.7 \text{ bits} \tag{2}$$

Another example is the design space provided by all standard components, or a product platform i.e. a car platform that is used to generate different cars in a product family. Design space generation is also made in parameterization of models such as CAD models or simulation models. By coupling parameters to each other to reflect different constraints in the design, a smaller more efficient design space can be produced, where waste in the form of unfeasible designs is

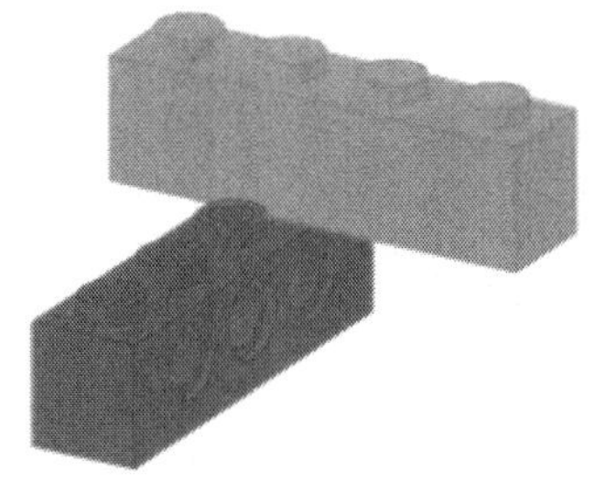

Fig. 1 Design in a design space of two Lego bricks

minimized. This means that less information is needed to arrive at a particular design from the design space.

2.3 Design Information Entropy

The definition of Information entropy for the discrete case is defined by Shannon [2]

$$H_d = \sum_{i=1}^{n} p_i \log_2 p_i \tag{3}$$

Here the system can be in n different states with probabilities p_i for each of them.

A more general definition than the information entropy for the discrete case is the differential information entropy for continuous signals, defined by Shannon [2] as:

$$H_c = -\int_{-\infty}^{\infty} p(x) \log_2(p(x))dx \tag{4}$$

This gives a measure of the average information content of a variable x. Here $p(x)$ is the probability density function. One problem with this expression is that it does not make sense unless x is dimensionless, since the probability density function has the unit of the inverse of x. If not, the differential entropy of the probability density function $p(x)$ needs to be related to another distribution $m(x)$. The result is called the Kullback-Leibler divergence [10] from the distribution $m(x)$. This is the relative entropy, and it is defined as:

$$H_{rel} = \int_{-\infty}^{\infty} p(x) \log_2\left(\frac{p(x)}{m(x)}\right)dx \tag{5}$$

This is the difference in entropy between having information that a random variable is within $m(x)$, and knowing that it is within the distribution $p(x)$. Furthermore, it represents a measure of information in bits. It can also be generalized to any dimensionality.

$$H_{rel} = \int_{-\infty}^{\infty} \cdots \int_{-\infty}^{\infty} p(x_1 \ldots x_n) \log_2\left(\frac{p(x_1 \ldots x_n)}{m(x_1 \ldots x_n)}\right) dx_1 \ldots dx_n \tag{6}$$

A rectangular distribution of $m(x)$ in the bounded interval $x \in [x_{\min}, x_{\max}]$, with $x_R = x_{\max} - x_{\min}$, would mean that the distribution $m(x)$ of the *design space* is a space of equal possibilities, where no particular region can be considered more likely than another a priori. Other distributions can also be considered but they can always be mapped on a rectangular distribution by transforming the design space, which can be very useful (this also includes infinite distributions), for i.e. design optimization. Equation (5) can then be rewritten as:

$$I_x = H_{rel}(x) = \int_{x_{\min}}^{x_{\max}} p(x)\log_2(p(x)x_R)dx \tag{7}$$

The letter I is here used here to indicate relative information entropy related to a rectangular distribution, and it has the unit *bits*. For the multidimensional case it becomes:

$$I_x = \int_{x_{1,\min}}^{x_{1,\max}} \dots \int_{x_{n,\min}}^{x_{n,\max}} p(x_1 \dots x_n)\log_2(p(x_1, \dots x_n)x_{R1} \dots x_{Rn})dx_1 \dots dx_n \tag{8}$$

This can also be written in a more compact form as:

$$I_x = \int_D p(\mathbf{x})\log_2(p(\mathbf{x})S)d\mathbf{x} \tag{9}$$

where D is the design space. I_x is defined as the *design information entropy* where the design $\mathbf{x}$ is defined within the design space D. S is the size of the design space and is defined as:

$$S = \int_D \mathbf{x}d\mathbf{x} \tag{10}$$

If the range of one variable is divided into equal parts Δx that have the same probability, the probability density distribution will be:

$$\begin{aligned} p(x) &= \frac{x_R}{\Delta x} : x \in \left[x_{0,\min}, x_{0,\max}\right] \\ p(x) &= 0 : x \notin \left[x_{0,\min}, x_{0,\max}\right] \end{aligned} \tag{11}$$

where:

$$x_R = x_{0,\max} - x_{0,\min} \tag{12}$$

This yields the information content (in bits) for that variable as:

$$\begin{aligned} I_x &= \int_{-x_{\min}}^{x_{\max}} \frac{x_R}{\Delta x}\log_2\left(\frac{x_R}{\Delta x}\right)dx \\ &= \log_2\frac{x_R}{\Delta x} = \log_2\frac{1}{\delta_x} = -\log_2\delta_x \end{aligned} \tag{13}$$

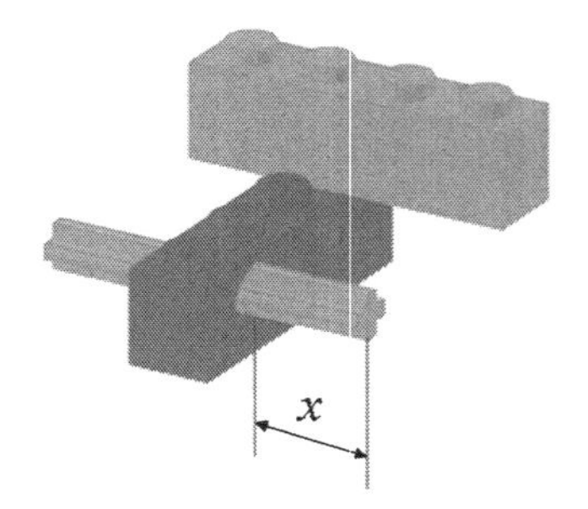

Fig. 2 Design with both discrete and continuous variables

where Δx is the uncertainty of the variable, and x_R its design range. δ_x is introduced as the relative uncertainty in parameters. The same expression holds if the probability distribution is normally distributed. In that cases:

$$\delta_x = \frac{2\sigma_x}{x_R} \tag{14}$$

Here σ_x is the standard deviation in x. In the following text it is assumed that the uncertainty can be described by δ_x. If the lego^{TM} example is expanded with an axis, the position of the inserted axis represents a continuous variables x. The information entropy associated with that, is dependent on the accuracy Δx with which it is specified, and the number of discrete positions (three) where it can be placed Fig. 2.

$$I_x = \log_2 n'_{Dstates} + \log_2 \frac{x_R}{\Delta x} \tag{15}$$

The axis can be in three positions (adding three discrete states) and if the position of the axis within one hole is specified within 10 % the total information entropy is:

$$I_x = \log_2(51 + 3) + \log_2 \frac{1}{0.1} = 8.2 \text{ bits} \tag{16}$$

The concept of design information entropy hence provides a framework for defining design information in very general terms. It is the information that causes the uncertainty of the design to be shrunk from an initial state high uncertainty, to another state with less uncertainty.

3 Design Information Entropy in the Design Process

3.1 Design Space Generation

In information theoretical terms the design space corresponds to the reference distribution $m(x)$ of the Kullback-Leibler divergence, or in this case the design

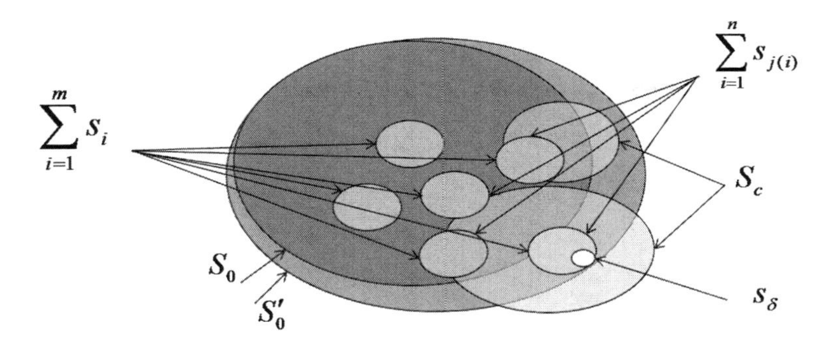

Fig. 3 Subspaces of the design space

space D_0, with the size S_0, against which the design information entropy is defined. Within the design space concepts s are generated. In general only part of the design space falls within the constraints of the design D_c, with the size S_c, finally the refined final design represents only a small fraction of a concept Fig. 3 s_δ.

3.2 Concept Generation

In concept generation, the design space is limited by selecting n_O subsets of possible concepts for further analysis. This represents an increase in information. If the design space distribution $m(x)$ is rectangular and uniform over the design space D_0 and each concept i occupies the region with the size s_i the information generated in this phase is:

$$I_I = -\log_2 \frac{\sum_{i=1}^{n_0} s_i}{S_0} \tag{17}$$

In general the amount of information generated in this stage is quite large. It involves selecting a few concepts from a highly dimensional multi modal design space.

3.3 Concept Screening

In this phase concepts that have no or little possibility of being successful are eliminated. If the concepts are reduced from n_0 to n concepts in this phase, the information added is:

$$I_{II} = -\log_2 \frac{\sum_{i=1}^{n} s_{j(i)}}{\sum_{i=1}^{n_0} s_i} \tag{18}$$

For the special case that all regions are of the same size it becomes:

$$I_{II} = -\log_2 \frac{n}{n_0} \tag{19}$$

3.4 Aircraft Design Example

Aircraft design is a good example of a complex product that involves all the phases illustrated in Fig. 6.

Design space generation constitute the process of collecting possible elements needed for the design in response to functional requirements. Looking at existing design there is a variety in concepts that can be dissected into components to recreate a design space from where they all can be derived. From the example in Fig. 4, there are various arrangements for wing and tail arrangements, engine location, etc.

There is a tail at the end of the fuselage, or at a twin boom arrangements, there are inverted butterfly tail and conventional tail. The engines can be mounted front or rear. The horizontal stabilizer can be mounted aft or forward (canard configuration) and

With these as example a design space can be defined some design elements can be identified.

- The horizontal stabilization front, aft or integrated in the main wing. In the aft configuration it can also be integrated with the vertical stabilizer in a butterfly tail.
- Vertical stabilization can be central or at wing tip, or integrated with horizontal configuration. It can be upwards or downwards.
- The tail arrangement can be located on a single fuselage or on twin boom arrangements.

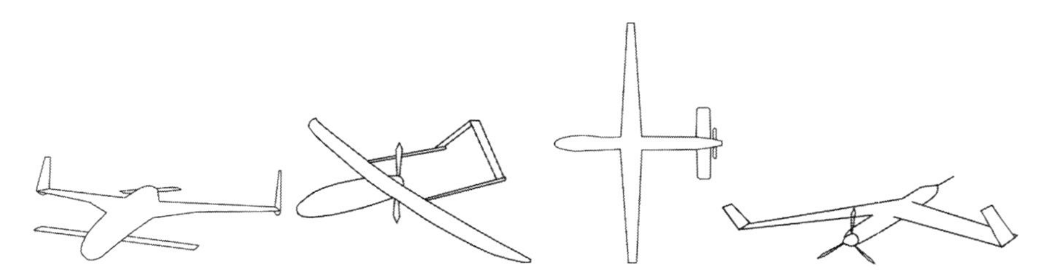

Fig. 4 Some UAV configurations

Table 1 Morphological matrix for aircraft

Design elements	Alternative solutions			
Horizontal stabilization	Front (canard)	Aft	Aft fin integrated	Wing integrated
Vertical stabilization	Central	Wing tip	Integrated	Upper
Tail mount	Single fuselage	Twin Boom		
Propulsion	Tractor	Pusher		

- Although these example all have pusher prop (in order to have a clear front view for sensors), a prop in the front is of course also possible. The fuselage could be a single or with a twin boom after section.

One popular tool for concept generation is the morphological matrix (or table). It was introduced in [6]. Here a table is set up where, for each function, a list of alternative solution elements is presented. A specific concept is obtained by selecting one solution element for each function. The morphological matrix represents a tool to display a design space of possibilities. Table 1 shows a morphological matrix for aircraft configuration.

The total number of possibilities, n_s, is in the general case:

$$n_s = \prod_{i=1}^{n_f} n_{m,i} \tag{20}$$

where n_s is the number of functions. In the example this becomes functions:

$$n_s = 4 \times 5 \times 2 \times 2 = 80 \tag{21}$$

This represents information entropy of

$$I_w = I_c = -\log_2 \frac{s_c}{s} = -\log_2 \frac{1}{80} = 6.32 \text{ bit} \tag{22}$$

That means that selecting one of the configurations in the design space represents 6.32 bits of information. An interesting property of the information entropy is that it is roughly proportional to the number of design elements.

3.5 Concept Optimization and Selection

For making concept selection, it is really necessary to do an optimization of each concept left, to investigate its properties at the optimal parameter set. The optimization process then represents the contraction of the domain for each concept down to a domain of specified tolerance s_δ. The increase in information for optimizing all concepts and selecting one concept, k, is represented by:

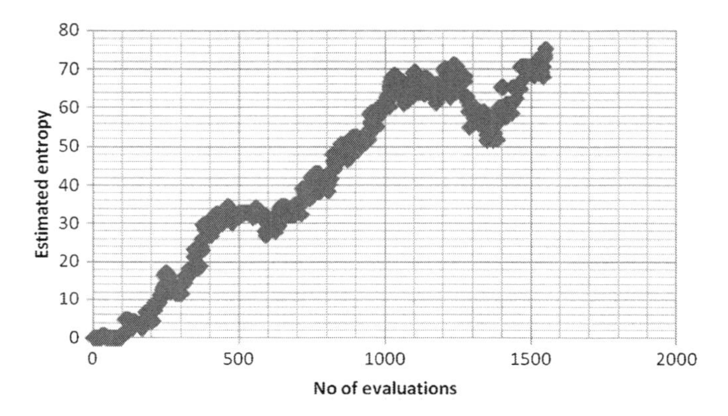

Fig. 5 Accumulation of information as a function of number of objective function evaluations

$$I_{III} = -\log_2 \frac{s_{\delta k}}{\sum_{i=1}^{n} s_{j(i)}} \tag{23}$$

An initial optimization of aircraft concept can be done using handbook formula based on physics and statistics from existing designs as in [11], especially if the concept is similar to an existing product.

For the aircraft example typical design parameters would be; wing span, root cord, tapering, thickness, and sweep, structural weight, fuel weight, engine size, wing position, span of horizontal tail, cruise speed.

Other parameters could be established using simple design rules, e.g. the vertical fin. Here the Complex-RF method [7] is used for the optimization. It is a method that is interesting because it is easy to estimate the information entropy as the optimization progress as a function of the spread of parameter set, as the optimization start with a spread the same size as the original design space (given by the constraints of the parameters). In this example 11 parameters was used as the accumulation of information entropy for one particular concept is shown in Fig. 5. The information entropy is estimated as:

$$\hat{I}_x = -n \log_2\left(\max\left(\delta_{x,i}\right)\right) \tag{24}$$

where $\delta_{x,i}$ is the relative uncertainty in the ith parameter. It is defined as:

$$\delta_{x,i} = \frac{x_{i,\max} - x_{i,\min}}{x_{0,i,\max} - x_{0,i,\min}} \tag{25}$$

Here the denominator represents the original design space for the optimization. At about 1,200 evaluations there is a reduction in estimated entropy as the solution moves away from a false optimum. In the end the entropy has increased about 75 bits.

4 Discussion

The design information entropy should be seen as a measure of the design space that has been under consideration during the design process. As such it provides some measure of the effort that has been going into the design.

To be effective it is desirable to have a small design space, but that still contain sufficiently good designs. A hallmark of a good design space is therefore that it is easy to assemble viable designs from a limited set of design elements, where there are ready to use sub systems and components that can be combined into new products e.g. like in a Lego set, or a good product platform. It can also be applied to parameterization of a design. In a perfect parameterization, all parameter combinations should yield viable designs, or at least possible geometries. In fact the fraction of viable parameter combinations can be seen as a quality of a parameterization [9]. This means that a smaller design space needs to be searched during design optimization.

The concept of design information entropy also provides a sound base for defining creativity as the process of selecting areas for expanding the design space, "to think outside the box", it can also be the process of navigation in highly dimensional multimodal design spaces for concept generation, which is simplified by a limited design space, with few unviable designs.

5 Conclusions

A formal theory of design should be based on the generation and transformation of information during the design process, and information theory provides a set of tools that can be used in this context. In this paper it is demonstrated that introducing design information entropy as a state, can be used for quantitative description for various aspects in the design process, both regarding structural information regarding architecture and connectivity, as well as for parameter values, both discrete and continuous. It is consistent with the view that the design process is a learning process i.e. where information is gained as uncertainty is reduced.

It is also clear that the design of the design space as such, is critical to promote creativity by making viable design alternatives clear to the designer, not obstructed among noise of a sea of unviable designs.

References

1. Ullman DG (1992) The mechanical design process. McGraw-Hill Book Co, Singapore. ISBN 0-07-065739-4
2. Shannon D (1948) A mathematical theory of communication. Bell Syst Tech J 27:379

3. Suh NP (2001) Axiomatic design: advances and applications. Oxford University Press, USA, ISBN-0-19-513466-5
4. Khan WA, Angeles J (2007) The role of entropy in design theory and methodology. In: Proceedings of CDEN/C2E2 2007 conference, Winnipeg, Alberta, Canada
5. Frey D, Jahangir E (1999) Differential entropy as a measure of information content in axiomatic design. In: Proceedings of the 1999 ASME design engineering technical conference, Las Vegas, USA
6. Bras B, Mistree F (1995) Compromise design decision support problem for axiomatic and robust design. J Mech des Trans ASME 117
7. Krus P, Andersson J (2004) An information theoretical perspective on design optimization. ASME, DETC, Salt Lake City, USA
8. Bansiya J, Davies C, Etzkorn L (1999) An entropy-based complexity measure for object oriented design. Theor Pract Object Syst 5(2):111–118
9. Amadori K, Lundström D, Krus P Automated design and fabrication of micro-air vehicles. Accepted for publication on journal of aerospace engineering, proceedings of the institution of mechanical engineers part G [PIG], doi:10.1177/0954410011419612
10. Kullback S, Leibler RA (1951) On information and sufficiency. Ann Math Stat 22:79–86
11. Torenbeek E (1982) Synthesis of subsonic airplane design. Kluwer Academic Publishers, ISBN 90-247-2724-3
12. Zwicky F (1948) The morphological method of analysis and construction. Courant, Anniversary volume, Intersciences Publisher, New York, pp 461–470
13. Krus P, Jansson A, Palmberg JO (1993) Optimization using simulation for aircraft hydraulic system design. In: Proceedings of IMECH international conference on aircraft hydraulics and systems, London
14. Rosenbrock HH (1960) An automatic method for finding the greatest or least value of a function. Comput J 3:175–184

Mitigation of Design Fixation in Engineering Idea Generation: A Study on the Role of Defixation Instructions

Vimal Viswanathan and Julie Linsey

Abstract Design fixation is considered to be a major factor influencing engineering idea generation. When fixated, designers unknowingly replicate the features from their own initial ideas or presented examples. The study reported in this paper investigates the effects of warnings about the undesirable features on design fixation. The authors hypothesize that if designers are given warnings about the undesirable example features along with the reasons for those warnings, fixation to those features can be mitigated. In order to investigate this hypothesis, a controlled experiment is conducted with novice designers. The participants are randomly assigned to one of the three experiment groups: a Control, Fixation or Defixation. Participants in all the groups generate ideas for the same design problem. It is observed that even when the warnings are present, designers replicate the flawed features in their ideas. Further, this paper compares said result with the findings in the existing literature.

Keywords Concept generation · Defixation instructions · Design fixation

1 Introduction

In the current competitive economy, introduction of novel products and services to the market is necessary for any industry to exist and being profitable. Engineering design attempts to satisfy the needs in the market through effective generation,

---

V. Viswanathan · J. Linsey (✉)
School of Mechanical Engineering, Georgia Institute of Technology,
Atlanta, GA 30332, USA
e-mail: julie.linsey@me.gatech.edu

V. Viswanathan
e-mail: v.viswanathan@gatech.edu

A. Chakrabarti and R. V. Prakash (eds.), *ICoRD'13*, Lecture Notes in Mechanical Engineering, DOI: 10.1007/978-81-322-1050-4_9,

development and implementation of novel and creative ideas. In this process, early concept generation plays a very vital role. For the success of any engineering design effort, generation of novel and creative ideas at this stage is highly crucial.

While considering creativity and innovation in early concept generation, it is essential to consider design fixation. According to Jansson and Smith, design fixation is the adherence of a designer to his or her initial ideas or features of presented examples [1]. When design fixation is present, it limits the solution space where the designers search for their ideas. This restricts the generation of novel ideas to a great extent. In most of the cases, design fixation causes serious threat to the early concept generation stage.

In recent times, many researchers have investigated the issue of design fixation and potential ways of mitigating fixation [2–6]. In this paper, the authors investigate the role of defixating instructions in the forms of warnings in the mitigation of design fixation to a flawed example. Building on the prior work [2, 5], the authors hypothesize that with the help of proper warnings about the undesirable features in a fixating example, design fixation can be mitigated. This paper outlines a controlled between-subject experiment conducted to investigate this hypothesis. The obtained results provide mixed support to the hypothesis. Further sections in this paper summarize the relevant literature review, method followed and a discussion of the results obtained.

2 Background

The first set of experiments that revealed design fixation effects in engineering design were performed by Jansson and Smith [1]. They found that when an example was present, both expert and novice designers replicated the features of that example in their solutions. After the publication of said study, many other researchers successfully demonstrated fixation effects of pictorial examples in concept generation. In a replication of Jansson and Smith's study, Purcell and Gero [7] found that design fixation effects varied across engineering disciplines. They observed that industrial design students fixated less compared to mechanical engineering students. More recent research showed that more realistic representations like photographs [8] and physical prototypes [9, 10] could also lead to design fixation.

Many researchers have investigated the possible ways to mitigate design fixation. Incubation and provocative stimuli are two potential candidates for this purpose. In incubation, fixated designers set their design problem aside temporarily. Many times, they can generate innovative ideas once they return their attention to the problem [11–13]. In the cases of provocative stimuli, some external stimuli provide a change of reference to the designers, in that process breaking their fixation [14, 15]. Linsey et al. [3] demonstrate that design fixation in faculty can be mitigated to a significant extent with the help of defixation materials that include alternate representations of the design problem. These defixation materials include a list of potential analogies, some back-of-the-envelope calculations and a direct list of energy sources that the designers can use. However, a more recent study shows that the defixating effect of these materials depends on the level of expertise of the designers [6].

In a very recent study, Youmans shows that when the example is presented in the form of a physical model and when the designers are allowed to test that model, they do not copy the example features in their designs [4]. In a very similar study, Viswanathan et al. [5] show that novice designers building and testing the physical models of their ideas, identify the problems with fixation and mitigate those gradually. Providing a physical example instead of a pictorial one also causes an improvement in the quantity of non-redundant ideas generated [10].

A very interesting study on mitigation of design fixation is conducted by Chrysikou and Weisberg [2]. They replicate Jansson and Smith's experiments [1] in extremely similar conditions. They observe that when the designers are warned about the flaws in an example, they tend to fixate less to the features of the same. These warnings also include explanations about why the features are flawed. Based on this result, Viswanathan et al. [5] provide warnings to novice designers about the flaws in a poor example, without specifying why those features are flawed. Interestingly, it is observed that the designers fixate more when they are given such warnings. One potential explanation is the curiosity of designers about the flawed features, due to lack of explanation about the flaws. Building upon said studies, this paper extends the investigations on the potential of proper warnings in mitigating design fixation to a more complicated design task. The hypotheses investigated further in this paper are the following:

Fixation Hypothesis: Designers fixate to the features of the flawed example presented to them.

Defixation Hypothesis: The fixation of designers to a flawed example can be mitigated with the help of warnings about the undesirable example features along with the reasons for those warnings.

3 Method

In order to evaluate the hypotheses presented above, a between-subject controlled experiment was conducted. This experiment was designed based on the some prior studies on design fixation [3, 6]. In the experiment, the participants were randomly assigned to one of the three experiment groups: Control, Fixation or Defixation. All the participants generated ideas to solve a design problem. The materials provided to the participants varied across the conditions. The occurrence of fixating features across the conditions was analyzed to infer the effects of fixating and defixating materials provided. The following subsections detail of the design problem, experiment conditions, procedure followed and the metrics used for evaluation of the data.

3.1 Design Problem

All participants in the experiment solved a "peanut sheller" design problem. This problem was successfully used in many prior studies dealing with various aspects of design cognition [3, 6, 16, 17]. This design problem instructed participants to generate ideas for a device that can quickly and efficiently shell a large quantity of peanuts without using electricity. The device was expected to be used by the farmers in West African countries. The device was expected to control the extent of damage to the peanuts while shelling them. This problem presented the challenges of a real-life design to the participants. None of the participants were familiar with this design problem before the experiment, but they all had experienced the routine task of shelling peanuts.

3.2 Experiment Groups

The participants were randomly assigned to one of the three experiment groups: Control, Fixation or Defixation. The details of each group are described below:

3.2.1 Control Group

The Control Group generated ideas to the above mentioned design problem without the help of any additional materials. They received a design problem statement along with the instructions to record their ideas. They were instructed to sketch and label their concepts along with one or two sentences describing the working of the concept on plain sheets of paper.

3.2.2 Fixation Group

This group received the same design problem and instructions as the Control Group along with the sketch of an example solution to the design problem. The sketch of the example provided to the participants is shown in Fig. 1. This example was developed originally by Linsey et al. [3, 16] and consisted of features that commonly appeared in participant solutions. These common solution features had higher potential to fixate designers [18, 19]. This example featured a gasoline (gas)-powered press that crushed the raw peanuts to shell them. The raw peanuts were imported to the system by a hopper and guided to the press by a conveyor-inclined surface combination. After shelling, the shells and peanuts fell into a collection bin at the bottom of the press. This system possessed certain shortcomings. A gas-powered system was too complicated and un-economical for manufacture and use in a less industrialized economy. With a gas-powered press, it was very difficult to control the damage to peanuts. Also, after shelling, this system did not necessarily separate the shells from the peanuts. Though these shortcomings were not stated explicitly, all the participants had sufficient mechanical engineering background to infer these.

3.2.3 Defixation Group

The defixation group received the same design problem and instructions along with the sketch of example solution shown in Fig. 1. In addition to those, the participants also received warnings about the flawed features of the example. Specifically, the following sentences were provided to the participants along with the example sketch: "Note that this system is NOT a good solution for peanut shelling. The gas press is costly and complex for usage in rural parts of African countries. This system does not have any control on the damage to peanuts from the pressure applied. Also, the system cracks the shell, but does NOT necessarily remove the shell from the nut." These warnings specifically warned the participants about the flawed features in the example along with the reasoning for those features to be flawed. According to the Defixation Hypothesis, these warnings were expected to help the participants in the mitigation of their fixation to the example.

3.3 Participants

All the participants in this study were senior undergraduate students attending a capstone design course at Mechanical Engineering Department of Texas A&M University. A total of 33 students (9 female) volunteered for this experiment and they were equally distributed across the three groups. One to four students par-

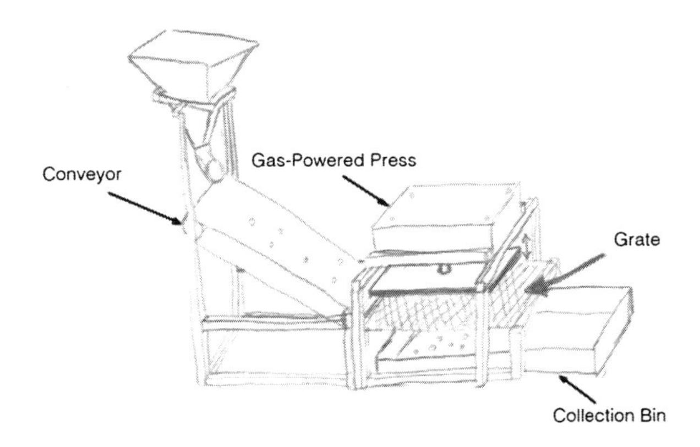

Fig. 1 Sketch of the example solution provided to the participants

ticipated in the experiment at a time, with their work spaces separated with dividers. The participants received extra credit in their design class or monetary compensation for their participation. In order to encourage generation of many ideas, a prize was announced for the person with greatest number of concepts. However, in order to ease the logistics, this prize was given to all the participants at the end of their experiment. None of the participants possessed significant design experiences.

3.4 Procedure

As the participants entered the experiment room, they were guided to randomly assigned work spaces. When the experiment began, they received the design problem statement and instructions along with the additional materials (example sketch or example with warnings about the flawed features), depending on their experiment group. They were given 5 min to read and understand the problem and instructions. Then they were instructed to generate as many concepts as possible within the available time (45 min) for idea generation. Multiple colors of pens were used to track the time interval in which a concept was generated. The pens of the participants were exchanged at 5, 10 and then every 10 min. At the end of idea generation, the participants were asked to mark any analogies they used to come up with their concepts. Finally, the participants were asked about their prior exposure to the peanut sheller design problem or its solutions.

3.5 Metrics for Evaluation

According to the hypotheses presented, the example sketch was supposed to fixate designers to its own features and the warnings about the flawed features in the example were expected to mitigate this fixation to some extent. For evaluating these arguments, design fixation at two different levels were measured: fixation to the example as a whole and fixation to the specific flawed features that the participants were warned about. The fixation to the whole example was measured using two different metrics: Quantity of non-redundant ideas and the percentage of example ideas used. The fixation to said flawed features was measured using the percentage of concepts with each flawed feature.

Quantity of non-redundant ideas was based on the metric originally proposed by Shah et al. [20] and further developed by Linsey et al. [16]. For the purpose of this study, an idea was defined as the one that solved one or more functions in the functional basis [21]. A number of ideas constituted a concept, which was defined as a solution to the problem in hand. Each concept generated by the participants was broken down to a number of ideas with the help of the functional basis. For calculating the quantity, repeated ideas were counted only once. For the Fixation and Defixation groups, the ideas presented by the example were counted as redundant and hence were not included in the calculation of the quantity. The number of example features used by a participant was normalized by the total number of ideas in the example to obtain percentage of example ideas. This metric provided a measure of the extent of fixation to example concept. These two metrics were found to be reliable with high inter-rater agreements in the prior studies by the authors [6, 10].

As specified by the warnings provided to the Defixation Group, the example concept contained three flaws: the use of gas as power source, lack of control to the damage of peanuts and the failure to separate the shelled peanuts from the shells. In order to measure the fixation to these flaws, the occurrences of these flaws in participants' concepts were identified. The number of occurrences of these flaws in a participant's concepts was normalized by the total number of concepts generated by that participant to obtain the percentage of concepts with flawed features.

4 Results and Discussion

This study aims to understand the usefulness of warnings about flawed features in mitigation of design fixation to a flawed example. As stated in the previous section, the fixation is studied in two different levels: fixation to the example as a whole and fixation to the flawed ideas that the participants are warned about. The following subsections provide the details of the results obtained for each of these categories.

4.1 Fixation to the Whole Example

4.1.1 Quantity of Non-Redundant Ideas

Variation of quantity across the experiment groups reveals interesting trends. Figure 2 shows the variation of mean quantity across the experiment groups. As evident from the figure, the Fixation Group produced a lower quantity of non-redundant ideas, demonstrating that the participants in that group are fixated to the presented example. At the same time, the Defixation Group produce a higher quantity compared to the Fixation Group, showing evidence of defixation. Overall, this metric provides support to the argument that designers can be defixated with the help of suitable warnings about the fixating features.

A one-way ANOVA with a priori contrasts [22] is performed to analyze the data statistically. The quantity of non-redundant ideas is not normally distributed, but it is homogeneous in variance. As the sample size is large, ANOVA is robust to the violation of normality. The results show that the quantity varies significantly across the experiment groups ($F = 5.13$, $p < 0.01$). Further, the results from a priori comparisons show that the Fixation Group varies significantly in quantity

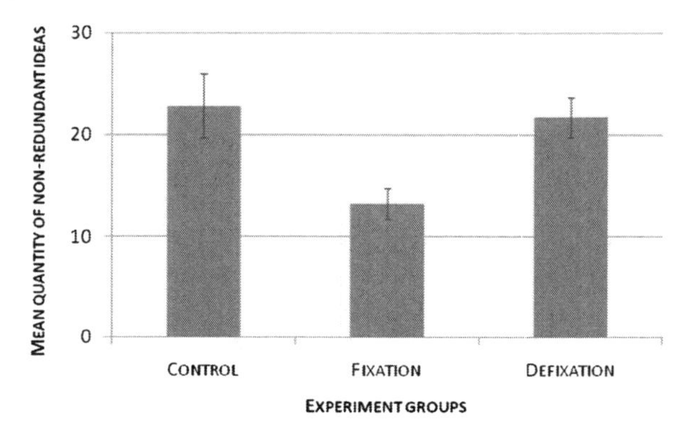

Fig. 2 Variation of mean quantity of non-redundant ideas across the experiment groups. The *error bars* show (±) 1 standard error of the mean

from the Control (F = 8.56, $p < 0.01$) and Defixation (F = 6.72, $p < 0.02$) groups. At the same time, there is no significant difference between the Control and Defixation groups.

These results provide support to both Fixation and Defixation Hypothesis. The significantly lower quantity in the Fixation Group compared to the Control Group provides evidence of design fixation. This indicates that the participants in the Fixation Group are replicating more example ideas in their concepts thereby reducing their overall quantity of non-redundant ideas. At the same time, providing warnings about fixating features does help them in improving their quantity in the Defixation Group, showing evidence for mitigation of design fixation.

4.1.2 Percentage of Example Ideas Used

This metric measures the extent of participants' fixation to the example provided. The results demonstrate interesting trends in the data. Figure 3 shows the variation of mean percentage of example ideas used by the participants in their concepts.

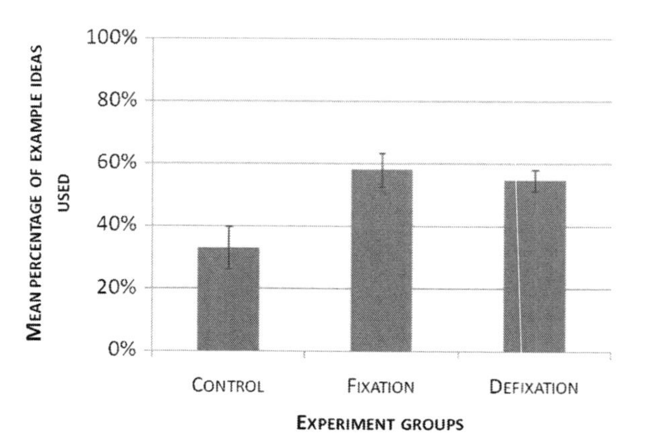

Fig. 3 Variation of mean percentage of example ideas used by the participants in their concepts. The *error bars* show (±) 1 standard error of the mean

As evident from Fig. 3, both the Fixation and Defixation groups copied the example ideas to the same extent in their concepts.

A one-way ANOVA with a priori contrasts shows evidence of significant difference of this metric across the experiment groups ($F = 6.29$, $p < 0.01$). These data also violates normality of distribution, but are homogeneous in variance, making ANOVA robust to the violation of normality. The a priori contrasts show that both the Fixation and Defixation groups differ significantly from the Control Group in percentage of example ideas used (Fixation: $F = 10.69$, $p < 0.01$; Defixation: $F = 8.00$, $p < 0.01$). Meanwhile, the Fixation and Defixation groups do not differ significantly from each other, statistically.

These results support the Fixation Hypothesis, but do not provide evidence supporting the Defixation Hypothesis. Evidently, when participants are exposed to the example solution, they replicate a higher percentage of ideas from the example in their concepts. As observed from Fig. 3, the Control Group also generates some ideas from the example. This is expected, as the example concept consists of most common ideas generated by participants in the previous experiments. At the same time, the Fixation and Defixation groups use the example ideas in significantly more times compared to the Control Group, showing design fixation. According to this metric, providing warnings about the flaws in the example concept does not help participants in reducing their fixation to that overall example.

4.1.3 Discussion: Fixation to the Whole Example

Both quantity and percentage of example ideas metrics provide strong support to the Fixation Hypothesis. They show that in the presence of an example, designers fixate to the features of that example. This result is consistent with many prior studies on design fixation [1, 3, 6–8, 10]. At the same time, these two metrics provide conflicting recommendations about the influence of warnings about flawed features on design fixation. The presence of warnings about the flawed features does cause an increase in the quantity of non-redundant ideas, but the participants still fixate to the same percentage of example ideas as the Fixation Group. It may be possible that the warnings lead participants deliberately generate non-redundant ideas for the flawed features leading them to higher quantity. At the same time, they may be still fixating to the example features that they are not warned about. In order to get a more complete picture, it is necessary to separately study the fixation to the individual flawed features. The following subsection deals with this issue.

4.2 Fixation to the Flawed Features of the Example

As described in the previous section, it is interesting to see if the warnings have any effect on the fixation to the flawed example features. Figure 4 shows the variation of the mean percentage of concepts that use the flawed example features in them. It is observed that in two cases (usage of gas as power source and separation of shells from the shelled peanuts), the warnings have some influence the extent of design fixation. At the same time, in the case of controlling the damage to the peanuts, there is no difference across the three experiment groups.

One-way ANOVA with a priori contrasts is performed on all these three metrics for the statistical analysis. The results show that for none of the metrics, the variation across the experiment groups is significant. Similarly, none of the a priori pair-wise comparisons provide significant differences across the pairs of groups.

Though these data do not provide any statistically significant comparisons, these do reveal interesting trends in the data. The Fixation Group generated more percentage of concepts with gas as power source, showing evidence of design fixation. At the same time, the Defixation Group generated lower mean percentage of concepts using gas as power source, showing evidence of mitigation of this fixation. At the same time, neither the presence of the example, nor the warnings influence the percentage of concepts that do not control damage to the peanuts. In all the three groups, only less than 30 % of concepts (approximately) provide explicit means to control the damage to the peanuts. One possibility is that during idea generation, the participants focus more on the novel ideas to shell the peanuts, ignoring the constraint about the extent of damage to peanuts, as that constraint limits their ideas. In the Defixation Group, the lack of control to the damage to the peanuts is presented as a design flaw. Even the knowledge about that design flaw does not help participants in eliminating that flaw from their designs.

The variation of percentage of concepts that separate shells from peanuts shows another interesting trend. As evident from Fig. 4, when the participants are presented with the warning that a concept that does not separate shells from peanuts is a flawed one, they generate more concepts that complete the said function. The lack of variation of this metric across the Control and Fixation groups suggests that the participants do not focus on this function, when they are not told explicitly about it. In other words, the participants do not perceive the need of separating the shells from the peanuts when they are not told that it is an important part of the design.

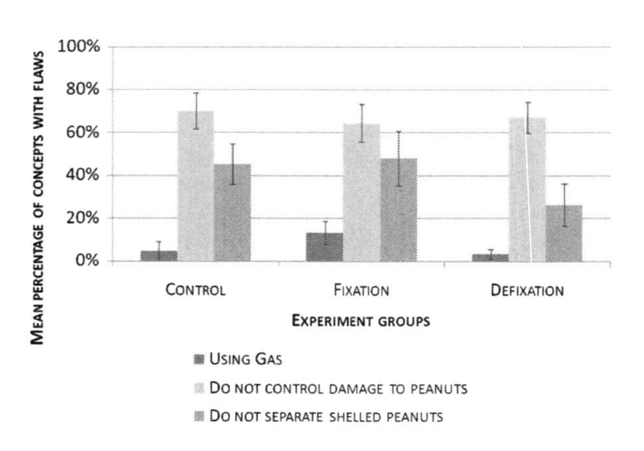

Fig. 4 Variation of the percentage of concepts with flawed example features across the experiment groups. The *error bars* show (±) 1 standard error of the mean

Overall, the results provide mixed support to the defixation effects of warnings about flawed features. Similar to the study by Chrysikou and Weisberg [2], the participants in the current study are provided with the warnings about the flawed features of the example and the reason for considering those as flaws. However, unlike [2], the warnings do not show any significant mitigation effect. The complexity of the design problem can be a potential factor affecting the usefulness of such defixation instructions. The design problem in the current study is significantly complex compared to that from [2] and consists of more functions to be solved by the designers' concepts. Psychological inertia [23] may be another contributing factor, because of which the designers are reluctant to change their way of solving complex problems, even in the presence of warnings. In summary, the effectiveness of such defixation tactics may vary across the levels of design problem complexity.

5 Conclusions

Design fixation is considered to be a major challenge in engineering idea generation. Building upon prior studies on mitigation of design fixation, this paper hypothesizes that the fixation to a flawed example can be mitigated with a help of a set of defixating instructions in the form of warnings about the flawed features of the example along with the reasons for those warnings. The results provide mixed support to the hypothesis. The warnings cause an increase in the quantity of non-redundant ideas; however, do not cause any reduction in the fixation to the example features. An investigation of the appearance of flawed features in the concepts generated by the participants further shows that the effect of warnings is very limited on the fixation to those specific features. These results contradict the prior findings in the literature. At the same time, this study uses a significantly more complex design problem compared to the prior studies. Overall, it can be argued that the defixation effects of warnings about flawed features in an example depend on the level of complexity of the design problem.

Acknowledgments Partial support for this work was provided by the National Science Foundation Award No. CMMI-1000954. Any opinions, findings and conclusions or recommendations expressed in this material are those of the authors and do not necessarily reflect the views of the National Science Foundation.

References

1. Jansson D, Smith S (1991) Design fixation. Des Stud 12(1):3–11
2. Chrysikou EG, Weisberg RW (2005) Following the wrong footsteps: fixation effects of pictorial examples in a design problem-solving task. J Exp Psychol Learn Mem Cogn 31(5):1134

3. Linsey J, Tseng I, Fu K, Cagan J, Wood K, Schunn C (2010) A study of design fixation, its mitigation and perception in engineering design faculty. ASME Trans J Mech Des 132: 041003
4. Youmans RJ (2011) The effects of physical prototyping and group work on the reduction of design fixation. Des Stud 32(2):115–138
5. Viswanathan V, Esposito N, Linsey J (2012) Training tomorrow's designers: a study on design fixation. ASEE Annual Conference, San Antonio, TX
6. Viswanathan VK, Linsey JS (2012) A study on the role of expertise in design fixation and its mitigation. ASME International Design Engineering Technical Conferences, Chicago. IL
7. Purcell AT, Gero JS (1996) Design and other types of fixation. Des Stud 17(4):363–383
8. Cardoso C, Badke-Schaub P (2011) the influence of different pictorial representations during idea generation. J Creative Behav 45(2):130–146
9. Kiriyama T, Yamamoto T (1998) Strategic knowledge acquisition: a case study of learning through prototyping. Knowl-Based Syst 11(7–8):399–404
10. Viswanathan V, Linsey J (2012) Physical examples in engineering idea generation: an experimental investigation. ICDC 2012, Glasgow, UK
11. Perkins DN (1981) The mind's best work. Harvard University Press, MA
12. Lawson B (1994) Design in mind. Butterworth Architecture, Oxford
13. Finke RA, Ward TB, Smith SM (1992) Creative cognition: theory, research, and applications. MIT press, Cambridge
14. De Bono E, Arzt E, Médecin I, Malta GB (1984) Tactics: the art and science of success. Little, Brown Boston
15. Shah JJ, Vargas-Hernandez NOE, Summers JD, Kulkarni S (2001) Collaborative sketching (C-sketch)—an idea generation technique for engineering design. J Creative Behav 35(3):168–198
16. Linsey J, Clauss EF, Kurtoglu T, Murphy JT, Wood KL, Markman AB (2011) An experimental study of group idea generation techniques: understanding the roles of idea representation and viewing methods. ASME Trans J Mech Des 133(3):031008-1–031008-15
17. Fu K, Cagan J, Kotovsky K, Maier JRA, Troy T, Johnston PJ, Bobba V, Summers JD (2010) Design team convergence: the influence of example solution quality case study research using senior design projects: an example application. ASME Trans J Mech Des 132(11): 111005–111011
18. Dugosh LK, Paulus PB (2005) Cognitive and social comparison processes in brainstorming. J Exp Soc Psychol 41(3):313–320
19. Perttula M, Sipilä P (2007) The idea exposure paradigm in design idea generation. J Eng Des 18(1):93–102
20. Shah JJ, Kulkarni SV, Vargas-Hernandez N (2000) Evaluation of idea generation methods for conceptual design: effectiveness metrics and design of experiments. ASME Trans J Mech Des 122(4):377–384
21. Stone RB, Wood KL (2000) Development of a functional basis for design. ASME Trans J Mech Des 122:359
22. Tabachnick BG, Fidell LS (2007) Experimental designs using anova. Thomson/Brooks/Cole, 2007
23. Kowalick J (1998) Psychological inertia. TRIZ J, August 1998, Available at http://www.triz-journal.com/archives/1998/08/d/index.htm. Access date 4 Oct 2012

Multidisciplinary Design Optimization of Transport Class Aircraft

Rahul Ramanna, Manoj Kumar, K. Sudhakar and Kota Harinarayana

Abstract A 70–80 passenger transport aircraft has been designed using third generation MDO technique. The MDO framework is used to parametrically explore the design space consisting of variables from aerodynamics, structure and control disciplines thereby studying the effects of alternative aerodynamics, structural and material concepts. It employs a panel code for aerodynamic analysis of aircraft and a finite element method for wing level structural optimization and weight estimation. The outputs from these codes are used to perform mission analysis and performance estimation. Stability constraints are implemented by calculating the amount of control power required for static stability and design concepts that do not fulfill the constraints are rejected. This integrated design environment is completely automated. The aircraft is optimized for maximum range using genetic algorithm.

Keywords MDO · Optimization · Aircraft · Preliminary design

1 Introduction

1.1 Aircraft Design

Aircraft Design is an iterative process of amalgamating various disciplinary knowledge to develop and analyze an affordable concept that can adequately satisfy customer requirements. Aircraft design can be divided into three phases [1]:

R. Ramanna (✉) · M. Kumar · K. Harinarayana
Aeronautical Development Agency (ADA), C/O Aeronautical Development Establishment (ADE) Campus, New Thippasandra, Bangalore 560075, India
e-mail: rahul.vrp@gmail.com

K. Sudhakar
Department of Aerospace Engineering, IIT Bombay, Powai, Mumbai 400076, India

A. Chakrabarti and R. V. Prakash (eds.), *ICoRD'13*, Lecture Notes in Mechanical Engineering, DOI: 10.1007/978-81-322-1050-4_10,

conceptual where a design configuration is generated and analyzed to meet given top level performance requirements; preliminary where specialists from each discipline analyze and optimize the configuration and detailed design where all components are extensively analyzed and verified to meet required criteria, redundancies are established to meet reliability levels for certification and production plans are developed.

Aircraft design apart from being an iterative effort requires extensive knowledge from disciplines such as aerodynamics, stability, controls, structures, propulsion, system layout, cabin interior planning etc. These disciplines are complex, interdependent and require a large pool of skilled engineers and scientists which makes aircraft design a highly complex task.

1.2 Multidisciplinary Design Optimization

Multidisciplinary Design Optimization (MDO) empowers a designer to simultaneously consider multiple disciplines and optimize the system as a whole. It has the potential of producing designs that are superior to those obtained by disconnected sequential optimization because it addresses and exploits the interactions between the disciplines. MDO performed at the conceptual to early preliminary stages using sufficient fidelity, can also avoid costly design changes that invariably occur at later stages. It also provides the decision makers more knowledge about the problem early in the game so that the risks and uncertainty about the program can be minimized. (refer to [2] for more information on MDO applications in preliminary design).

There are three generations of MDO technology that can be found in literature [3]. In the first generation all the disciplinary streams are integrated and analyzed. This is acceptable for simple designs but for complex designs with large number of interactions it becomes unmanageable and not viable. In the second generation medium fidelity tools with more focus on distributed analysis and analysis management was used. In the third generation MDO, higher fidelity disciplinary codes are used and more emphasis is provided to manage complexity and improving efficiency. The current paper falls under third generation MDO.

The art of designing an aircraft involves finding the best compromise between conflicting system level goals. In traditional sense, design is performed sequentially with each discipline optimizing for their disciplinary goals without considering the impact their decisions have on the overall aircraft. For example, consider a design interaction between aerodynamics and structures team. Aerodynamics team optimizes Lift to drag ratio (L/D) which has a huge influence on fuel burn and hence range and emissions of the aircraft. This is frequently accomplished by increasing AR, but there are disadvantages. Increase in AR implies an increase in the length of the wing resulting in increased bending loads. This means that the wing has to be reinforced to absorb these extra loads. Thus more fuel has to be burned to fly this extra weight along. Initially, with moderate increase in AR, the fuel burn is reduced because the weight increase is not that severe, but at some limit, the opposite occurs. The increase in AR results in such a severe weight increase that it actually increases fuel burn. This limit is difficult to establish because the increase in weight due to increased bending load is not tangible until the wing is analyzed by the structures team. An iteration loop thus ensues between aerodynamics and structures. If L/D and weight that influence system level performance metrics are not modeled with their strongly coupled nature, the resulting design will not be the optimum.

There are several other design parameters like wing area, taper ratio (TR), thickness to chord ratio (t/c) of airfoil, sweep angle, dihedral angle, tail area, tail aspect ratio, tail taper ratio, fuselage diameter, seat pitch/arrangement, placement of landing gear, fuel tank locations, placement of engines, location of wing on the fuselage which affect final system level (aircraft) objectives.

2 Design

Air traffic is increasing globally and it is estimated that India needs about 400 medium range aircraft in the next 20 years that can carry 70–100 passengers. This will increase the connectivity between tier-2 and tier-3 cities leading to all round economic growth. In the current paper, a transport aircraft that can carry 70–80 passengers is designed for maximum range. Three major disciplines: aerodynamics, structures and controls are considered and the aircraft is designed with a known propulsion system. An MDO framework is established where data flows freely between the three disciplines. More information on MDO frameworks is available in Ref. [2]. In the current paper, a two step optimization framework is used as shown in Fig. 1.

2.1 Integration Framework

An integrated environment is best suited for conceptual and early preliminary studies as the design freedom available during these stages is better exploited to obtain good designs early on in the design cycle. The tools used in this environment must have adequate fidelity to address design trades and automated to obtain results within reasonable time limits. Further, the tools adopted have to be commonly used in the organization and be endorsed by disciplinary experts to promote confidence among the engineers in the results obtained from the environment.

Commercially available integration frameworks ease the automation process by supporting an extensive list of pre-built software interfaces and scripting languages. They also provide frequently updated packages for design of experiments, optimization algorithms, surrogate modeling algorithms and post processing options which substantially reduce the time required to integrate disparate disciplinary tools and automate the environment. In the current paper ESTECO modeFrontier is used as an integration framework.

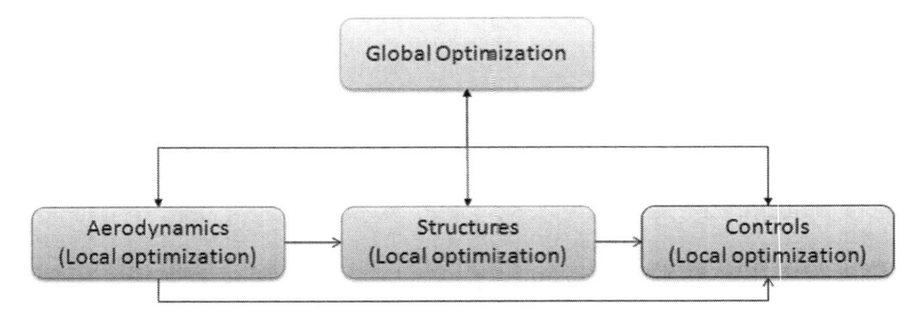

Fig. 1 Two step optimization process

2.2 Disciplines

2.2.1 Aerodynamics

Aerodynamics drives the shape of the aircraft The main goal of aerodynamics is to increase lift to drag ratio. Lift, mainly generated by wings, counters the weight of the aircraft while drag acts opposite to thrust reducing the effectiveness of the engines. Higher drag requires more energy to move the aircraft from point A to point B.

In conceptual design, aerodynamic data for a given configuration is traditionally estimated using empirical/semi-empirical relations or historical data. In current paper, higher fidelity panel method [4] is used to estimate the aerodynamics of the configuration. This methodology enables the designer to evaluate newer configurations that do not have any empirical relations or historical data. The first step in this approach is to discretise the geometry into panels (collection of surface quadrilateral patches), which, in the current paper is accomplished by the pre-processing software called Point wise [5]. The discretised geometry then forms the input to Vsaero (the panel code). The lift and drag of the aircraft are thus estimated at different angles of attack to arrive at the drag polar. The drag polar then forms one of the inputs for the performance of the aircraft. The lift is also projected onto the structural grid as loads for structural analysis.

2.2.2 Structures

The main goal of the structures discipline is to estimate the weight of the aircraft while ensuring that the airframe is designed to meet the safety and airworthiness requirements.

In the current paper, finite element methodology (FEM) [6] is used to analyze the aircraft wing structure. This involves discretising the geometry into finite elements (quadrilaterals and triangles) using Hypermesh software [7] and analyzing the structure for the aerodynamics loads projected onto it. The wing is then optimized using the software called Optistruct [8]. Advanced composite materials are used to reduce weight of the wing and the whole process is automated. The final weight, C. G and inertia information are the outputs from this module.

2.2.3 Controls

Controls ensure that the aircraft is capable of performing required maneuvers within the flight envelope. There are three main surfaces to control the aircraft along the three axes. Elevators (horizontal tail) control nose up and nose down moments, rudders control nose left and nose right moments and ailerons control rolling of the aircraft.

Trim drag is the incremental drag experienced in setting all the moments to zero by application of controls. Reducing trim drag while providing adequate controllability is one of the main goals of the stability and control discipline.

In the current design module control power required to perform 1 g Trim, basic maneuverability, steady sideslip and single engine trim are considered. This is compared to the control power available from the control surfaces of the current configuration. Only those designs that have enough control power to perform the maneuvers are considered for further stages [9].

2.3 Process

Figure 2 shows the design process. The arrows and boxes drawn using dashes represent outputs and light shaded blocks with dotted outline represent constraints.

The design process is initiated with a user chosen design of experiments (DOE) (full factorial, reduced factorial, Latin hypercube etc.). The number of cases in the DOE represents the population in the first generation of the genetic algorithm. Each case in DOE starts by invoking CATIA software to generate aero and structural geometries reflecting the current design variables. Aero geometry is processed through gridgen software and is analyzed by Vsaero. Structural geometry is processed through hypermesh and the mesh is sent to Vsaero for load mapping. The updated FEM model is analyzed in Optistruct and the thicknesses are subsequently optimized. The controls module calculates the control power required for the current DOE case using aerodynamic derivatives from aero module and weights, inertia and C. G data from structures module. There are 4 constraints in the design process. $CL(struct) < CLmax$ and $CD > 0$ are both sanity checks to ensure physics of the problem is not violated. $1 < Pcr/P < 1.01$ represents that the critical buckling load should be higher than the current load. $Cpreqd < Cpmax$ implies that the control power required must be less than the maximum control power available from this configuration.

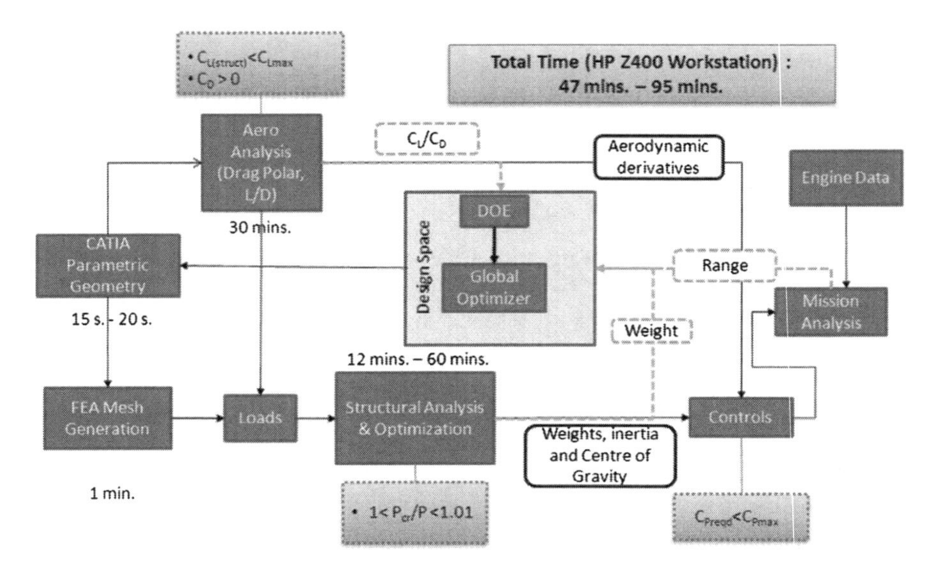

Fig. 2 Process flow diagram

2.3.1 Aerodynamic Analysis

In the current paper, aerodynamic analysis is performed on the full aircraft without landing gear and propeller (Fig. 5, note: propellers are for illustration only and not used in analysis).

Point wise software is used to mesh the geometry and the actions are recorded using the journal option. The script file can then be executed for different cases in DOE; provided the points, lines and surface numbers remain the same. This condition is satisfied by drawing a parametric geometry in CATIA. The meshed geometry is exported into Vsaero readable format and analyzed for a given flight condition. The analysis is repeated for different angles of attack to obtain the drag polar. Matlab scripts are used to extract the required values from Vsaero output file.

The structural mesh data obtained from the structures module is used to define interpolation points in Vsaero where the pressure has to be quantified. This pressure data is later applied on the mesh by the structures module for analysis and optimization.

Aerodynamic derivatives are found by unit deflection of control surfaces. These derivatives are fed into the controls module to calculate control power.

2.3.2 Structural Analysis

Structural analysis is performed on the aircraft wing. A sandwiched composite wing structure is analyzed and optimized using the design framework for a single design load case. The methodology is however amenable to be expanded to more load cases.

The quality of mesh generated using any pre-processing software is dependent on the quality of the CAD model. In the present paper, a parametric wing structural model was created using CATIA. It was ensured that all the rib-spar, spar-skin and rib-skin intersections were modeled correctly. Unnecessary fixed points were removed and a clean geometry was exported into '.igs' format. This geometry was imported into Hypermesh and the surfaces were further cleaned. The wing geometry was then meshed in hypermesh. The centroids of all elements in the FEA mesh were exported for generating pressure distribution using VSAERO.

The FEA mesh is completed by adding thickness and material data and is exported to Optistruct readable format. The aerodynamic loads are used to create a load file that is read by Optistruct using the "INCLUDE" feature. This procedure is integrated by ModeFrontier and is completely automated requiring no user intervention.

Optistruct is used for performing linear static analysis, linear buckling analysis, normal modes analysis and thickness optimization.

The composite structure was assumed to be composed of 4 types of plies (45, 0, 90 and −45 deg). The total thickness is a design variable and decides the number of plies. This approach is viable in this case because the wing configurations are compared with one another. The methodology is quite general and not dependent on the above assumption. This assumption greatly simplifies the complexity and reduces computational time. It is recommended that stacking sequence optimization be performed later in the design stage.

Another assumption made to reduce computational time without affecting the quality of results is to use parabolic thickness variation function. This reduces the number of thickness design variables in optimization.

The optimizer controls the parabola parameters which are used to create thickness zones on the wing and is analyzed by Optistruct.

2.3.3 Control Analysis

The control analysis was performed for 4 cases. 1 g trim, maneuverability, steady sideslip, and one engine out trim (as per FAR-25 [10]). The Ref. [9] gives a detailed description on the process.

The aerodynamic derivatives required to find the control power were found using a panel CFD code rather than empirical relations. This provides the environment to analyze novel and unconventional configurations while also providing required fidelity.

2.3.4 Mission Analysis

The goal of mission analysis module is to estimate the range of the aircraft as per a given flight profile and check whether the particular design meets all the constraints [1]. This was performed using an in-house code developed at ADA.

2.4 Input and Output Variables

2.4.1 Design Variables

Global: Aspect Ratio, Wing Area, Taper Ratio

Local (structural): Parabola parameters (b and c) for spars (front spar and rear spar) and wing skin, thickness of ribs, width and thickness of spar and rib caps.

2.4.2 Other Parameters

Number of thickness zones, number of ribs, sweep angle, dihedral angle, load factor for maneuver, sideslip angle

2.4.3 Output Variables

L/D, wing weight, overall weight and range of aircraft.

2.5 Optimization Algorithm

A general description on different optimization algorithms and their applications can be found in [11].

2.5.1 Non-Dominated Sorting Genetic Algorithm (NSGA-II)

The NSGA algorithm was developed by Deb et al [12] from Kanpur Genetic Algorithm Laboratory (KanGAL) at IIT Kanpur and is available in public domain. It is one of the algorithms included in modeFrontier and is widely used for complex multi-objective problems.

2.5.2 Structural Optimization

Gradient based algorithms are generally employed in structural optimization since they are quicker and provide satisfactory results for size optimization problems. Optistruct automatically chooses the best algorithm for a given problem from the following [8]:

- Optimality criteria method
- Convex approximation method
- Method of feasible directions
- Sequential quadratic programming

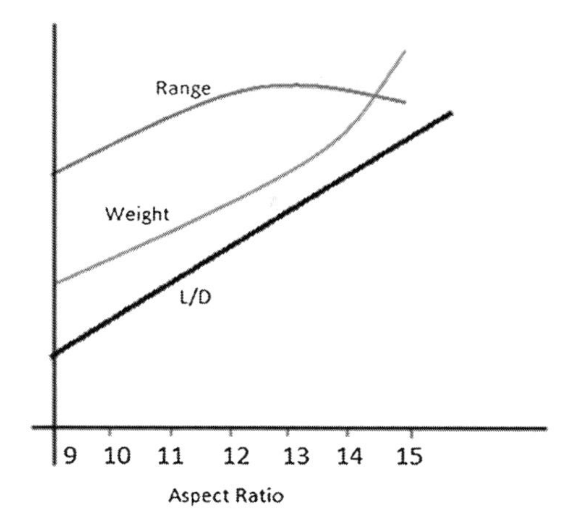

Fig. 3 Variation of metrics with Aspect Ratio

3 Results and Discussions

A notional transport aircraft was designed for maximum range using the MDO environment.

A family of designs is generated by varying AR, S and TR using a reduced factorial DOE. This is used as the initial population for the genetic algorithm. The objective function for optimization is to maximize the range of the aircraft. The genetic algorithm chooses the fittest individual (maximum range) among the initial population and propagates them to the next generation. Crossover and mutation operations are performed to provide a variety of designs for the next generation. As the optimization progresses, only those designs that give higher range are retained. The optimization is stopped when the populations of successive generations remain same or the maximum number of generations is reached.

Figure 3 shows a plot of wing weight, Range, L/D v/s AR of the aircraft. The wing weight and L/D increases with increase in AR. However Range depends on both Weight and L/D. Range is higher for lower weight and high L/D. Therefore Range first increases and then reduces with AR.

All of the designs considered in the MDO environment were of composite wing structure. This reduces the weight of the aircraft. A plot of wing weight for both composite and an Aluminum wing is shown in Fig. 4 for comparison. The aluminum wing has more weight and hence lower range.

The final aircraft configuration is shown in Fig. 5.

In the current paper a single objective function is chosen to provide more emphasis on the design process rather than on the results. However in reality there are a multitude of design objectives which are conflicting in nature. The best compromised design in such cases can be chosen using multi attribute decision making tools such as TOPSIS and AHP. The reader is referred to [13] for a good discussion on TOPSIS, AHP and other decision making tools.

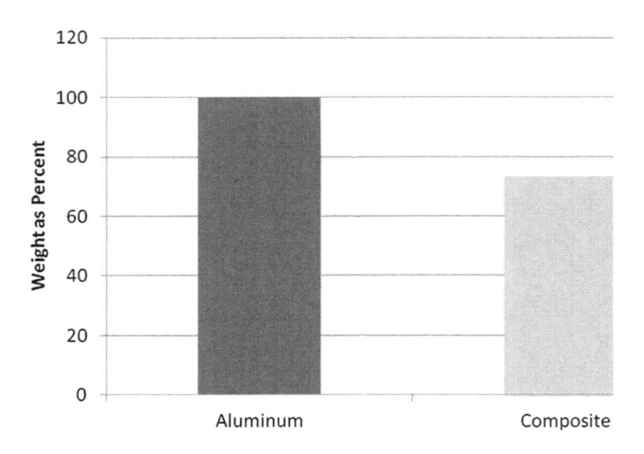

Fig. 4 Weight of wing made of aluminum and composite

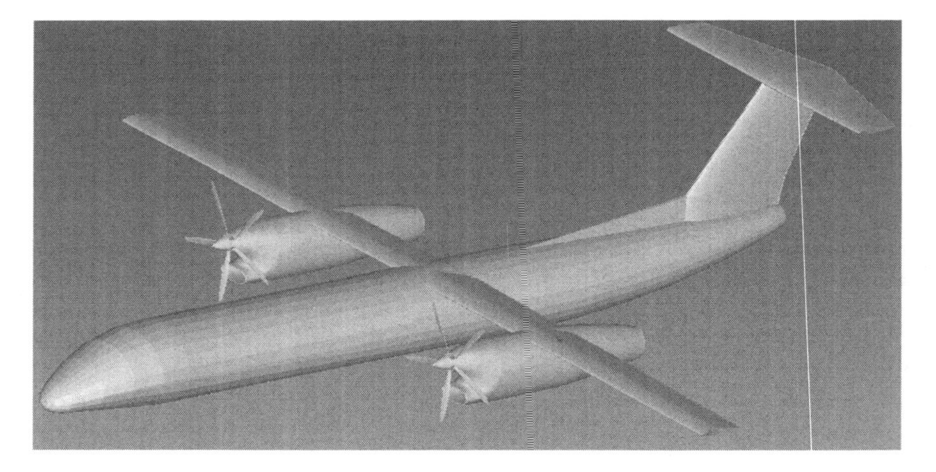

Fig. 5 Final optimized aircraft (Propellers: not used in optimization; for illustration only)

4 Summary and Future Work

A 70–80 passenger transport aircraft was designed using multidisciplinary optimization technique using a two step optimization process. Three disciplines were considered, of which, aerodynamics and structures were included into the optimization while controls was used only as a constraint. It was found that to maximize range of the aircraft, a wing with medium aspect ratio, lower area and higher taper ratio is preferred.

Our next goal would be to use multiple objectives, include controls and fuselage into the optimization loop and to prove the current design methodology for an unconventional configuration such as flying wing.

Acknowledgments Mr. Abhishek Burman, Aeronautical Development Agency (ADA) and Mr. Vikas Munjal, National Aerospace Laboratories (NAL) for their inputs on aerodynamic analysis; Dr. Vijay Patel, ADA and Mr. Mandal, ADE for their valuable inputs on controls; Dr. Pavan, NAL and Dr Shyamsundar, NAL for inputs on structures.

References

1. Raymer DP (1992) Aircraft design: a conceptual approach. AIAA Inc, Reston
2. Kroo I (1997) MDO applications in preliminary design: status and directions. AIAA Paper 97, p 1408
3. Willcox K (2008) Advanced multidisciplinary optimization techniques for aircraft design. Bangalore: Indo-US workshop on systems and technologies for regional air transportation, 2008. Indo-US workshop on systems and technologies for regional air transportation
4. AMI Corporation. VSAero User Manual
5. Pointwise Inc. Pointwise v16 user manual
6. Bathe KJ (1996) Finite element procedures. Klaus-Jurger Bathe, Cambridge
7. Altair Engineering. Hypermesh User Manual
8. Altair Engineering (2011) Optistruct user manual. Altair
9. Kay J et al. (1993) Control authority issues in aircraft conceptual design: critical conditions, estimation methodology, spreadsheet assessment, trim and bibliography. Virginia Polytechnic Institute and State University, Blacksburg
10. Federal Aviation Regulation. Part-25 airworthiness standards: transport category airplanes. www.faa.gov/regulations_policies/faa-regulations (Accessed 2012)
11. Vanderplaats (2007) Multidiscipline design optimization, 1st edn. Vanderplaats Research and Development, Inc, Monterey
12. Deb K et al. (2002) A fast and elitist multi-objective genetic algorithm—NSGA-II, vol 2. KanGAL Report Number 2000001, Kanpur, April 2002, IEEE Transactions on Evolutionary Computation, Vol. 6
13. Triantaphyllou E (2000) Multi-criteria decision making methods: a comparative study. Kluwer Academic Publishers, Dordrecht

Using Design-Relevant Effects and Principles to Enhance Information Scope in Idea Generation

Zhihua Wang and Peter R. N. Childs

Abstract This paper explores the use of a database of effects to facilitate generative activity and access to information. Some design challenges involve applications with which designers or design teams may not be familiar. This can readily be addressed by consultations with subject experts. A challenge associated with this, however, is access to expert and effective dialogue and information. Key to dialogue between a specialist expert and designer is framing of questions and understanding context. In order to improve access to expert information, a database of effects, call the Effects Database, arising from TRIZ, has been extended to enhance use across a wide range of domains by including psychological and design principles. It includes over 300 design-relevant technical effects and principles from physics, chemistry, geometry, design, and psychology. The aim of the database and associated procedures is to provide ready access to expertise at any stage within the design process.

Keywords Design · Idea generation · Knowledge accumulation · Effects

1 Introduction

It is a common view that innovation is essential in both updating original products and designing new products [1]. Based on empirical studies, innovation tends to begin with the generation of creative ideas [2, 3]. Thus, creative idea generation plays a key role in fundamentally determining the type of products which prosper modern business [4–6].

Z. Wang (✉) · P. R. N. Childs
Design Engineering, Department of Mechanical Engineering, Imperial College London, South Kensington, London SW7 2AZ, UK
e-mail: z.wang09@imperial.ac.uk

P. R. N. Childs
e-mail: p.childs@imperial.ac.uk

A. Chakrabarti and R. V. Prakash (eds.), *ICoRD'13*, Lecture Notes in Mechanical Engineering, DOI: 10.1007/978-81-322-1050-4_11,

The ability of designers in generating creative ideas is determined by many factors. Knowledge accumulation, also known as expertise, is one of the main factors [7]. Accumulation of fundamental knowledge in a field is important for the promotion of new insights in that field. This is implicit in the systems model of creativity where expertise as well as resources, motivation and communication are all essential components for creativity [2, 8]. Therefore, if a designer wants to generate creative ideas that have originality and appropriateness to a design problem, it is important for the designer to have a certain level of knowledge accumulation in the related fields.

In real design processes and opportunities, designers often face a design task that relates to fields they are relatively unfamiliar with and the knowledge accumulation they have is not specific to the design task [9]. In this situation, in order to compensate for shortages of knowledge accumulation in problem related fields, designers may learn the related knowledge by themselves or consult experts for guidance. However, with the development of science, the quantity and complexity of knowledge has led to increased diversity in the category of fields [9]. Moreover, the amount of attention in everyday life is limited [8]. Therefore, once designers face a problem that lies in unfamiliar fields, they may spend a long time on obtaining a basic accumulation of knowledge. Meanwhile, experts may be stymied by ineffective dialogue with limited understanding of subject subtleties or design context. In modern business, because time pressures for product design progress often dominate, both of these approaches are challenging and can be ineffective.

In this paper a creativity assistance tool called the Effects Database is proposed to help designers cope with the situation, and its performance is partially evaluated. A survey to investigate the reasons for time pressures in the design processes is reported in Sect. 2, along with the principle for the establishment of the database. In Sect. 3, the proposed Effects Database is illustrated. Based on a case study in Sect. 4, the effectiveness and efficiency of the Effect Database is explored in Sect. 5.

2 Theory

2.1 Survey

In order to explore approaches to design amongst novice designers, a survey was undertaken with the Innovation Design Engineering double masters students. This is a joint programme run by the Royal College of Art and Imperial College London. In this survey, there were 21 male and 10 female participants. The intake of participants is diverse with approximately one-third of the students coming from product, industrial and graphics design backgrounds, one-third coming from engineering backgrounds and the rest from diverse backgrounds. The average age on the programme is between approximately 27 and 29 years, depending on the cohort, and all of them have over 3 years design experience.

The survey results are listed in Table 1. For a 6 month project in an unfamiliar field, over 60 % of participants used over 1 month on information gathering. A significant proportion of time budget was used on problem related information gathering. The designers acquired knowledge by self-study with over 90 % of participants reporting that the time available for knowledge gathering was not enough, and over 70 % of them believed that if they had more time for knowledge gathering, their design outputs will be more creative. For situations where designers consulted experts for knowledge guidance, around 80 % of participants agreed that the consultant experts improved their work performance. Nearly 30 % of participants, however, were sometimes unsure which kind of experts they should approach when their design problems were related to unfamiliar fields. Around 88 % of participants admitted that there were communication problems between them and the experts who they consulted. According to the survey results, two principal indications are concluded as follows:

Indication 1: When designers face a design problem related to unfamiliar knowledge fields, they have to spend a significant proportion of time budget to define the correct knowledge scope for the problem, which causes time pressure for knowledge gathering.

Indication 2: When designers consult experts for help in knowledge gathering, experts may provide ineffective guidance because of limited understanding of subject subtleties or design context.

2.2 Basic Theories

One of the purposes of knowledge accumulation is for the generation of creative ideas. Creativity can be described as "the ability to invent or develop something new of value" [10]. For a design problem, a creative idea can be regarded as an idea that brings "new value" which can be of societal or financial benefit or indeed both of these. There are usually two approaches to add "new value" for design products:

Approach 1: A new design that has never appeared before is proposed.

Approach 2: Improvement updates have been undertaken on an existing design, such as integrating new functions, or improving the performance of some components.

As the majority of designs stem from new incorporations of earlier inventions [11], Approach 2 is more widely used than 1 as the main route to achieve "new value" in products and systems. The description of Approach 2 suggests the requirement that designers have to understand and be able to explore earlier inventions before making improvement updates on existing designs. Therefore, it seems valuable to have, or to build, a database that includes all earlier inventions in a field. Once designers face a task that relates to this field, they can directly receive task related knowledge assistance from the database. Thus, the first hypothesis principle for the establishment of the database could be as follows:

Table 1 Selected survey results from the IDE1 cohort

Survey question	Results (31 participants in total)				
For a 6 month project in a domain which you are NOT familiar with—normally how much time do you allow for information gathering?	Within 1 week	1–3 weeks	1–3 months		Over 3 months
	1	11	18		1
	Strongly agree	Agree	Undecided	Disagree	Strongly disagree
The time for information gathering is normally NOT enough	4	16	8	3	0
If you had more time for information gathering, would your output be better	4	20	2	5	0
Do you know which kind of expert you should approach for knowledge guidance?	5	17	6	3	0
If you go to an expert for knowledge guidance on information gathering, the work will be MORE efficient than if you work alone	5	12	7	6	1
If you go to an expert in a domain that you are unfamiliar with and the expert is unfamiliar with your design question, there are communication problems between you	1	18	8	4	0

Principle 1: The database includes all earlier inventions in a field.

However, there are many reasons to doubt the feasibility of this principle. Because the categories of fields are extremely diverse [9], it is hard to cover all earlier inventions in all fields. Moreover, because of the specification of each database, the more fields the design task is related with, the more inventions designers have to analyse, and the more time the analysis process would cost. Therefore, the principle of the database should consider the complexity of the fields and the time cost for information analysis.

It was indicated that an invention in one field can be transformed as the integration of coordinated fundamental technical effects in that field and other associated fields [12]. A technical effect is defined as any influence, transformation, phenomenon, or function that is used as a principle of a technical system for the development of the system itself. Based on this theory, Principle 1 could be upgraded into Principle 2 as follows:

Principle 2: The database includes all known effects in all fields.

Since the majority of designs are new incorporations of earlier inventions, the effects used in a new design may inherit from effects used in earlier inventions. Therefore, the total number of basic effects in a field is limited and it is possible to build such a database to include all known effects in all fields.

3 Effects Database

3.1 *Effects Conclusion*

Based on the investigation from [11], effects should be collected from the conclusions from existing domains or from design knowledge from existing designs. A successful design demonstrates the conclusion of design thinking. Principles from each subject domain are readily available in handbooks and reviews. Interdisciplinary research has also augmented knowledge in this area. In addition the research that led to the development of TRIZ identified patterns in the use of specific physical, chemical and geometric effects in successful inventions [12].

As these effects come from various fields, it may be difficult for designers to understand all of them and select problem relevant ones quickly. To provide reliable and solid definitions for effects, reference books, websites and journal papers were studied. 128 physical effects, 78 chemical effects and 28 geometric effects were selected for this initial phase of activity. For each effect, a definition, a book reference and a web reference were developed and selected. An example is shown in Table 2.

Since the effects from TRIZ are predominantly engineering-relevant, the Effects Database provides limited support outside the engineering domain. For example, in toy design for children, it is valuable to know the basic psychological effects of children to make toys more enjoyable, such as colour bias, shape bias, etc. To compensate for this limitation, the Effects Database was supplemented by further 47 psychological principles and effects and 46 design principles and effects. After it was finished, the Effects Database included over 300 design-relevant technical effects and principles from physics, chemistry, geometry, design, and psychology. Some generic effects and principles are listed in Table 3.

3.2 *System Development*

The main function of the Effects Database is to provide problem related effects and principles to assist designers rapidly define the knowledge scope of design tasks. The system is designed as an online ready access data search system. Users can access the website through the URL provided. The detailed framework has two parts: search page (a in Fig. 1) and results list page (b in Fig. 1).

Search page: The left side is a search dialogue for the input of problem related keywords and the right side is an instruction for the keywords selection.

Results list page: On the left hand side of this page, the matched effects and principles are listed. The right side shows the detailed definition of each selected result. Users can explore more information about each search result by referring to its web reference.

Table 2 An example effect

#	Physical principle	Definition	Book or Journal reference	Web reference
3	Thermal-electrical phenomena	The direct conversion of temperature differences to electric voltage and vice versa.	Serway, R., Jewett, J. W. Jr., 2002. *Principles Of Physics: A Calculus - Based Text.* 3rd ed. Press Thomson Learning.	http://www.metacafe.com/watch/1944815/you_gotta_see_this/

Table 3 Examples of selected generic effects and principles in the effects database

Category	Effects and principles
Physical effects and principles	Thermal expansion, Radiation spectrum, Curie point, Hopkinson effect, Barkhausen effect, thermal conduction, thermal convection, thermal radiation, phase transition, Joule–Thompson effect, magneto-caloric effect, electro-magnetic induction, electric heating, electric discharge, radiation absorption, shrinking, thermal insulation, reflection of light, radiation of light, photoelectric effect, X-ray, Doppler effect, nuclear reaction, magnetic field, radioactive ray
Chemical effects and principles	Thermo-chromatic reactions, chemiluminescence, endothermic reactions, exothermic reactions, explosion, electrolysis, transport reactions, synergistic effect, oxidation–reduction reaction, deoxidization reactions use of helium, dissolving bonds, photo-chemical reactions, electro-chemical reactions, biodegradable materials, phase transitions, energy transformation, self-clustering molecules, oxidation reactions, hydrophilic materials
Geometric effects and principles	Compact packing of elements, compression, construction with several floors, Mobius tape, Reuleaux triangle, Crank-cam propulsion, cone-shaped ram, paraboloids, ellipse, cycloid, transition from a linear process to a face process, eccentricity, screwing spirals
Psychological effects and principles	Childhood amnesia, Cathartic effect, baby's facial preference, baby's taste and smell, Piaget's stage theory, colour, 3D vision of human eyes, top-down lighting bias, opponent-colour theory, dark adaptation, Snellen acuity
Design effects and principles	Uniform connectedness, proximity, similarity, expectation effect, Kamin's blocking effect, evaluation apprehension, social facilitation, performance load, over justification effect, Fitts' law, open system theory, isolated system theory, Golden ratio, Fibonacci sequence, signal-to-noise ratio, Hick's law

4 Case Study

4.1 The Early Stages of the Stage-Gate™ Process

The Stage-Gate™ process has been widely used for new product development [13]. It provides effective and efficient conceptual and operational direction for the design process of new products from concepts to market [14]. After observations in more

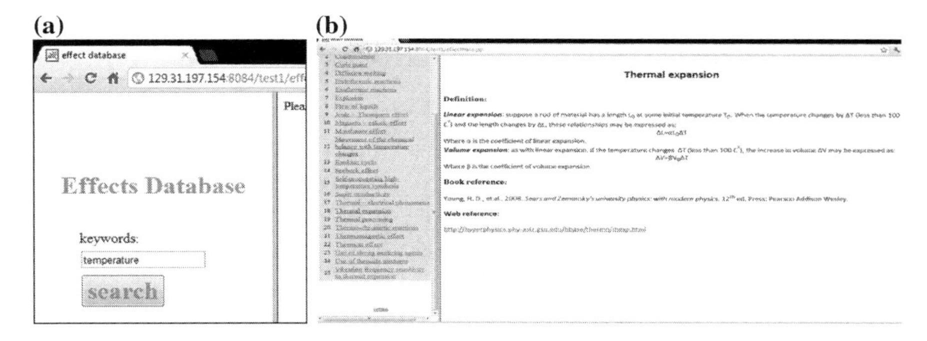

Fig. 1 The framework of the effects database. **a** Search page. **b** Result list page

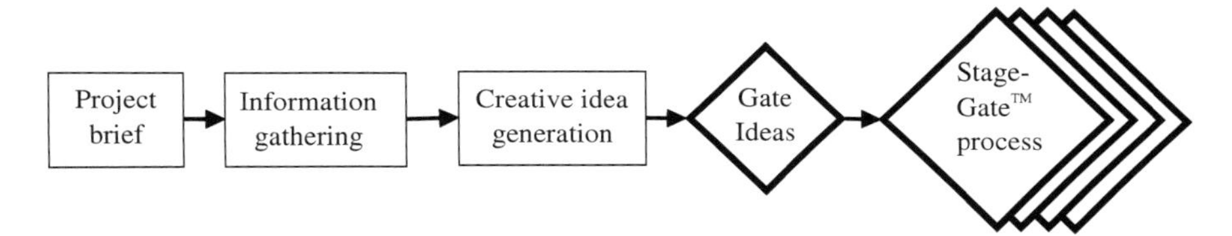

Fig. 2 The standard idea generation process in companies and industries. *Source* [7]

than 500 companies, which used Stage-Gate™ in their product developments, an Idea Discovery stage should be added at the front end of the process to supply more product ideas [13]. The investigations from [6] revealed that many companies ran idea generation sessions to directly provide product ideas. The standard idea generation process in these industries and companies is shown in Fig. 2.

Project brief: During this stage, the latest status of the project is introduced, along with the requirements and expectations for the final output.

Information gathering: In this stage, designers need to explore as many project related information sources as possible under a pre-set time budget.

Creative idea generation: This stage plays an essential role in the idea generation process. The idea pool can be populated by the creative thinking of designers which can be stimulated by using various idea generation tools.

4.2 Integration of the Effects Database into the Early Stages of the Stage-Gate™ Process

At the idea generation stages, design groups can maintain the creative performance at certain levels when they face design problems related with familiar knowledge fields [15]. However, groups fail to reach the same level of performance when they cope with design problems related with unfamiliar knowledge fields. Both the

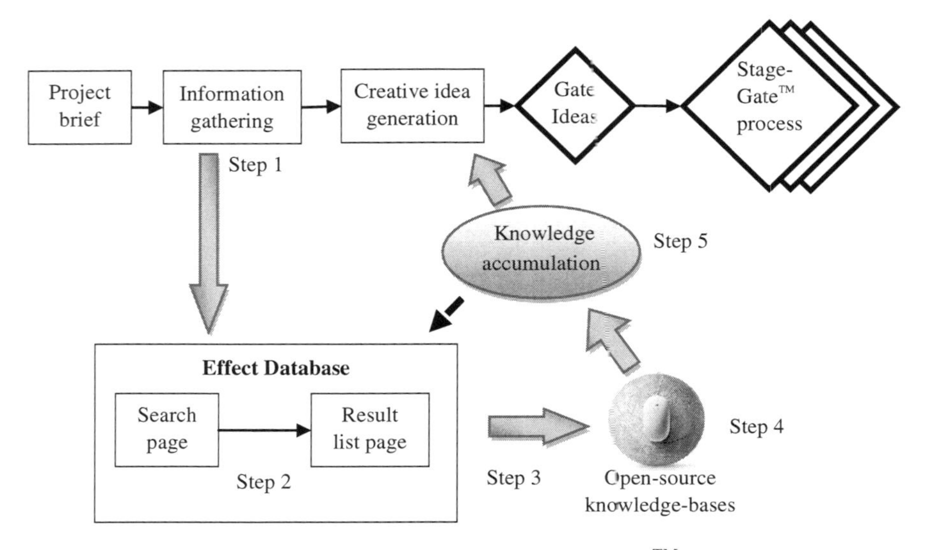

Fig. 3 Using the effect database in the first stage of the Stage-Gate^{TM} process

quality and quantity of the generated ideas can decrease dramatically. This situation is matched with the survey in Sect. 2.1. To overcome the two indications, the Effects Database is integrated with the standard idea generation process to enhance the knowledge accumulation of designers. The detail working process (shown in Fig. 3) is illustrated as follows:

Concluding problem related keywords (*Step* 1): Based on the information and facts from previous stages, some keywords are proposed. A keyword should be a simple word that is related with the design problem.

Searching the database (*Step* 2): In this step, once designers enter a keyword into the database, the background programs search keyword related effects or principles in the database and then list results obtained in the results list page.

Exploring information for each effect identified (*Step* 3): This step is for designers to briefly examine the information of each effect or principle in the results list page including its definition, book reference and web reference. After finishing examination of keyword related results, designers can revisit Step 2 to enter another keyword.

Exploring more information for the specific effects identified (*Step* 4): During step 3, designers can refer to the book reference and the web reference of an effect or principle to explore more information from open-source knowledge-bases.

Enhancing the knowledge accumulation in problem related fields (*Step* 5): After all the keywords have been entered and keyword related results examined, designers have an indication of the scope of knowledge they may need to explore in the problem related fields. Moreover designers could "inform" the experts that these effects or principles were related with their design tasks and ask for related specific knowledge guidance even though the experts may not understand the design tasks. Thus, the problem related effects and principles could be selected to enhance their knowledge accumulation.

Sometimes, the knowledge guidance from experts may stimulate designers to conclude new keywords or replace previous keywords by more appropriate ones. In this situation, designers may repeat the database searching process using the new keywords.

4.3 Case Project Details

To evaluate the effectiveness and efficiency of the Effects Database in the idea generation process, some industry-based projects from IDE alumni were explored. Four of them were selected here as examples and were shown in Table 4. The IDE course combines design with engineering, technical mastery and innovation. It has an emphasis on making students engage with technology and cultivate their innovation-focused thinking and realisation in both societal and business context. Many of their designs have been commercialised.

The Effects Database has been applied hypothetically to 4 projects using the 5 step process. It should be noted that this was undertaken for illustration processes and that the original design work did not use this approach.

At step 1, the keywords of each project were determined. For example, the 2nd task was to design a portable smoke detection system, from which the keywords "smoke" "flame" "heat" and "network" were concluded. After information on conventional smoke detectors had been explored, the keywords as "combustion" "heat" "sensor" and "portable" were also suggested.

At step 2, the keywords were used as input in the search page of the database. In the 3rd task, the "Baby's hearing" and "Baby's vision" principles were related with the "infant" keyword. The "Feature integration theory" and "Opponent-colour theory" effects were related with the "colour" keyword.

At step 3 and 4, each effect or principle was examined and its related information was also explored through the references provided.

At step 5, only problem related effects and principles were remained according to self-exploration and the guidance from experts. In the 1st project, effects such as "Bioluminescence" "Chemiluminescence" and "Electro-luminescence" were discarded because of the lack of relationship with the design question. The effects and principles "Dark adaptation", "Polarised light", "Thermo-chromatic reactions", etc., were added into knowledge accumulation for the design task.

After step 5, the knowledge scope of the design question was defined depending on the selected design-relevant effects and principles. In the 4th project, the main knowledge scope was specified on geometrical structure.

Table 4 Example use of the effects data using innovations associated with IDE alumni

#	Design question	Step 1	Step 2	Step 3	Step 4	Step 5	Product
1.	After being made, the degree of ophthalmic lens is fixed. Once eye degree changes, people have to buy a new pair of glasses. Can the degree of ophthalmic lens is also adjustable?	Light, lens, reflection, glass, focal	Bioluminescence, chemical reactions in gases that in the active area of lasers, chemiluminescence, colour, dark adaptation, electro-luminescence, highlighting, photoacoustic effect, polarized light, reactions of reversible electro sedimentation, stroboscopic motion, top-down lighting bias, thermo-chromatic reactions, reflection of light, symmetry,	Exploring information for each effect identified	Exploring more information for the specific effects identified	Dark adaptation, polarized light, thermo-chromatic reactions, reflection of light, symmetry	Adjustable eyewear
2.	The positions of smoke detectors in a home are fixed and are not always placed in ideal locations. Current portable smoke detectors are isolated from each other. Can these two systems be incorporated?	Smoke, flame, heat, combustion, sensor, portable, network	Combustion, condensation, dielectric heating, endothermic reactions, energy transformation, exothermic reactions, explosion, self-propagating high-temperature synthesis, shrinking, thermal chromes, thermal conduction, thermal conductivity, thermal insulation, thermal processing, Weber-Fechner law, periodic reaction			Combustion, Condensation, endothermic reactions, energy transformation, exothermic reactions, explosion, self-propagating high-temperature synthesis, shrinking, thermal conduction, thermal conductivity, thermal insulation, Weber-Fechner law, periodic reaction	Echo smoke detector

(continued)

(continued)

#	Design question	Step 1	Step 2	Step 3	Step 4	Step 5	Product
3.	Infants are unable to tell designers their feelings about toys. How to design infant-enjoyable toys?	Infant, toy, psychology, colour, enjoy, fun	Baby's hearing, baby's vision, Piaget's stage theory, colour, feature integration theory, highlighting, illusory conjunction, opponent-colour theory, photo chromes, photochromatic effect, thermo-chromatic reactions, trichromatic theory, use coloured materials to track the changes of some reactions			Baby's hearing, baby's vision, piaget's stage theory, colour, feature integration theory, highlighting, opponent-colour theory, thermo-chromatic reactions	Infant's toy
4.	Can wheels be foldable to save storage space?	Round, fold, curve, flexible, material	Balls, cycloid, ellipse, good continuation, spirals, use of diaphragms, Bauschinger effect, bi-metallic or bi-material construction, changes in the optical-electromagnetic properties of materials, electrostriction, input-process-output model, isolated system theory, Johnson-Rahbeck effect, Kragelski phenomenon, magnetic separation, shape-changing objects, use a magnet to influence an object or a magnet that is connected to the object			Balls, cycloid, ellipse, good continuation, spirals, use of diaphragms, Bauschinger effect, bi-metallic or bi-material construction, shape-changing objects, use a magnet to influence an object or a magnet that is connected to the object	Foldable wheel

Note That the original design work did not use the methodology developed in this paper and the examples are provided for illustrative purposes

5 Conclusions

In this paper, an Effects Database is proposed to aid designers to generate creative ideas in short periods of time compatible with modern business. Some design applications involve subjects that the designer or design team may not be familiar with. This is commonly addressed by consultations with a subject expert or subject expertise, which, however, can be time-consuming or stymied by ineffective dialogue with limited understanding of subject subtleties or design context. A survey (in Sect. 2) indicated a link between the time for knowledge gathering and time for defining appropriate knowledge scope. In addition, issues of communication between the designers and the experts have been identified with both unfamiliarity with the subject domain and understanding the context of design questions being important factors. In order to improve access to expert information, a database of effects, arising from TRIZ, has been extended to enhance use across a wide range of domains by including psychological and design principles. The Effects Database generated includes over 300 design-relevant technical effects and principles from physics, chemistry, geometry, design, and psychology. The use of the database has been illustrated by application to a series of design tasks, indicating its suitability for promoting expert relevant suggestions.

References

1. Soosay C, Hyland P (2004) Driving innovation in logistics: case studies in distribution centres. Creativity Innov Manag 13(1):41–51
2. Amabile TM (1996) Creativity in context. Westview Press, Boulder
3. Mumford M (2000) Managing creative people: strategies and tactics for innovation. Hum Resour Manag Rev 10(3):313–351
4. Bharadwaj S, Menon A (2000) Making innovation happen in organizations: individual mechanisms, organizational creativity mechanisms or both? J Prod Innov Manage 17: 424–434
5. Amabile TM (1998) How to kill creativity. Harvard Bus Rev 76(5):76–87
6. Howard TJ, Culley S, Dekoninck EA (2010) Reuse of ideas and concepts for creative stimuli in engineering design. J Eng Des 22(8):565–581
7. Mumford MD, Hester KS, Robledo IC (2011) Knowledge. In: Runco MA, Pritzker SR (eds) 2nd Encyclopaedia of creativity volume 2. Elsevier Science and Technology, Amsterdam, pp 27–33
8. Csikszentmihalyi M (1996) Creativity: flow and the psychology of discovery and invention. Harper Collins Publishers, New York
9. Lidwell W, Holden K, Butler J (2003) Universal principles of design. Rockport Publisher
10. Childs PRN (2004) Mechanical design, 2nd edn. Elsevier Butterworth-Heinemann Ltd., Oxford
11. Henderson K (1991) Flexible sketches and inflexible data bases: visual communication, conscription devices, and boundary objects in design engineering. Sci Technol Human Values 16(4):448–473
12. Orloff MA (2006) Inventive thinking through TRIZ: a practical guide, 2nd edn. Springer, Berlin

13. Cooper RG, Edgett SJ, Kleinschmidt EJ (2002) Optimizing the stage-gate process: what best-practice companies do-I. Res Technol Manage 45(5):21–27
14. Cooper RG (2008) Perspective: the Stage-Gate (R) idea-to-launch process-update, what's new, and NexGen systems. J Prod Innov Manage 25(3):213–232
15. Howard TJ, Dekoninck EA, Culley SJ (2010) The use of creative stimuli at early stages of industrial product innovation. Res Eng Design 21(4):263–274

Determining Relative Quality for the Study of Creative Design Output

Chris M. Snider, Steve J. Culley and Elies A. Dekoninck

Abstract Design creativity is often defined using the terms "*novel*" and "*appropriate*". Measuring creativity within design outputs then relies on developing metrics for these terms that can be applied to the assessment of designs. By comparing design appropriateness to design quality, this paper develops a systematic method of assessing one element of design creativity. Three perspectives from literature are used; the areas in which quality is manifest, the categories into which quality assessment criteria fall, and how well criteria are achieved. The output of the method is a relative ranking of quality for a set of designs, with detailed understanding of the particular strengths and weaknesses of each. The process of assessment is demonstrated through a case study of twelve similar designs. Through such analysis insight into the influences on quality can be gained, which in turn may allow greater control and optimisation of the qualities that design outputs display.

Keywords Creativity · Quality · Assessment · Design output

1 Introduction

There are a number of existing definitions of creativity within design literature aiming to distil the term into its necessary constituent parts. They also deal with the criteria by which creativity can indirectly be measured [1–3]. Although defined in many ways, amongst the terms most frequently used are "*novelty*", "*appropriateness*" and

C. M. Snider (✉) · S. J. Culley · E. A. Dekoninck
University of Bath, Bath, UK
e-mail: C.M.Snider@bath.ac.uk

A. Chakrabarti and R. V. Prakash (eds.), *ICoRD'13*, Lecture Notes in Mechanical Engineering, DOI: 10.1007/978-81-322-1050-4_12, © Springer India 2013

"*unexpectedness*" [4]. These are used to denote how novel an output is in terms of the field and alternative problem solutions; how suitable an output is as a solution to the specific problem that has been described; and how surprising or unusual a solution is in context of similar solutions that already exist, or the features of the output itself. Through these terms, an output that is creative is defined as a new solution that is recognised as highly and particularly suited to the problem that has been set, likely denoting it as the superior design solution available within the market.

To determine the creative properties of an output it is necessary to identify and assess these terms. To allow assessment of appropriateness of solutions, this paper presents a system of classification and comparison of design quality metrics, using existing quality assessment procedures; forming one vital part of creative assessment of design outputs.

Literature presents several methods by which creativity terms can be assessed, such as those developed by Shah et al. [3] and Sarkar and Chakrabarti [5]. However, perhaps due to the variation in terms used to describe this factor, those methods relating to the term "*appropriateness*" are often variable. As an example, Shah recommends use of the term "*quality*" and subsequent assessment according to a number of methods typically measuring adherence to product specification [3]; while Sarkar and Chakrabarti [5] recommends the term "*usefulness*", then creating a method assessing an output based on its societal importance and expected in-life use by customers. Further, if taking forward the term "*valuable*" [2, 6] as used in the sense referred to by the field of value engineering, it is necessary to consider not only the benefits of an output, but also a full breakdown of the costs [7]. This task is difficult to complete accurately and robustly at earlier stages in the design process when less information exists, or when considering more subjective criteria [8].

All of these terms refer to the same aspect of creativity [4]. While the potentially difficult and subjective tasks of determining "*value*" and "*usefulness*" do present information on how well an output solves a problem, the focus need only lie on comparison of the output with the specification and requirements that governed its design. Each output must be assessed according to how "*appropriate*" it is as a solution to the problem and its context. For example, while a solution may provide high functionality, it may also be of excessive cost to customer or company, and inappropriate for the problem that it must solve. "*Appropriateness*" can be described as a solution of appropriate quality in context of its problem, assessed through output quality as proposed by Shah et al. [3], using any existing method (such as QFD [9], or decision tables [10]).

However, the work presented in this paper shows that beyond the assessment of quality according to the above methods or otherwise, through detailed categorisation of quality manifest in a design output, significant additional understanding can be gained. Particularly in the case of creative design, additional design features may greatly improve the quality of certain elements of a design output, or indeed of generate additional parameters not originally anticipated. By detailed classification of the categories to which any parameter of any form can belong, the influence of such additional criteria can be understood and evaluated in the appropriate context. The method presented within this paper then has two functions: first to stimulate the identification of criteria which contribute to overall quality beyond those included in the original specification, and second to categorise and allow assessment with those criteria in a useful and robust manner. These are described in the next two sections; following which the paper then presents a case study demonstrating the proposed process.

It should be noted that the focus of this paper is not on the method used to assess quality, but rather on the classification of assessed criteria against one another, and the consequent understanding that can be gained. Within this paper, and due to the stage at which designs were assessed, the Pugh [11] method was most suitable, but another could be used should it provide higher detail or higher appropriateness to the assessment situation. By appropriately classifying and weighting assessed criteria, the classification method attempts to demonstrate additional understanding that can be gained, allowing more detailed understanding of final output quality between several design outputs.

2 The Assessment of Quality

2.1 The Hierarchy of Quality Criteria

As discussed by O'Donnell and Duffy [12] in their work on design performance, both the *design* itself and the *design activity* must be considered. A systematic classification of quality criteria in terms of this distinction is proposed, showing separation between criteria that are of quality in different ways. This more detailed view recognises the importance of quality both in context of the design output and of the design development process. For example, a change that greatly saves in manufacture cost without changing design output quality is valuable, but is potentially different to a change that greatly alters design output quality alone. Distinction is also made between the design output and its super-system, and the design activity and its super-system; recognising quality both specifically and induced in the design output and the design activity's environment. The proposed categories are defined in Fig. 1.

Design quality [referred to as Internal Direct (ID)]—performance characteristics of the design such as speed of operation, precision, range of operation, etc.

Design super-system quality [Internal Indirect (II)]—performance characteristics of the super-system to which the design belongs, as influenced by the design itself. For example, lower design mass may allow the super-system to operate for longer on a single charge, or to operate with higher precision. Here, lower mass does not increase quality of the design itself, but produces additional benefit in the super-system.

Design process quality [External Direct (ED)]—performance characteristics within the development and production of the design, such as in manufacturing (e.g., high use of standard parts, lower manufacture cost) or assembly (e.g., fewer assembly operations).

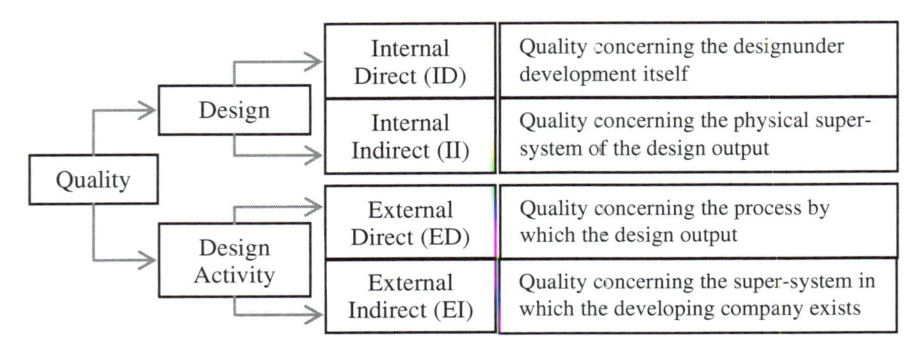

Fig. 1 Systematic hierarchy of possible categories for quality criteria

Design process super-system quality [External Indirect (EI)]—performance characteristics relevant to the company within its super-system, as induced by the design itself. For example, adherence to brand identity or conformance with environmental standards may improve the perceived quality of the company within its market.

Using such a hierarchy, it is possible to describe how a design or feature is of quality. Distinction can be made, for example, between mediocre design performance quality and exceptional manufacturing quality, an important trade-off that may need to be considered. This decomposition is particularly useful in creativity research as it enables assessment of creative quality through the product lifecycle; for example, higher quality in terms of ID is more manifest to the customer. Also of benefit is the equal credence given to high quality in terms of development, which may have little impact on the customer or product use, but is still a valid area of creative product development.

2.2 Specific Quality Metrics

The hierarchy provides understanding of the areas that assessment criteria should concern but does not state what they should be, a task requiring distinct categories of criteria.

Much literature addresses the need for categories of assessment by developing criteria from the specification [7, 8, 11, 13]. Within the work presented here, the "eight dimensions of quality" presented by Garvin [13] are used to develop specific criteria.

Performance: a design outputs' primary operating characteristics.

Features: those characteristics that supplement basic functionality.

Reliability: the probability of malfunction or failure within a specified time period.

Conformance: the degree to which an outputs' design meets established standards.

Durability: the amount of use one gets from an output before it deteriorates.
Serviceability: the speed, courtesy, competence and ease of repair.
Aesthetics: how an output looks, feels, sounds, tastes or smells.
Perceived quality: interpretation of the output through reputation.

By including criteria that relate to each of these "dimensions of quality", a highly detailed and complete assessment of the quality of a design output can be created. Together, the dimensions of quality present a thorough description of all characteristics that contribute to the overall quality of a design output. These characteristics define whether a product is of appropriate quality, and hence a vital part of the interpretation of a product as creative. The particular strength of the metrics proposed by Garvin as opposed to other metrics is in their breadth, including those relating to design output and also more subjective metrics relating to human interpretation. It is not always possible to judge criteria in a quantifiable manner, due to a lack of detail and information of the design output in question, or because the dimension of quality itself defies numeration. Considering the importance of context and human interpretation given by many to the determination of what is creative [14], such subjective criteria are vital.

It should be noted that although using the hierarchy produces a significant number of prompts for the development of assessment criteria that contribute to overall quality, not all will be relevant to each product or each company. For example, some branches of the tree require criteria that are unusual or likely irrelevant in many scenarios (such as aesthetics of manufacture process). However, as with all elements of assessment, the decision of what is relevant must be made by the assessor on a case-by-case basis.

2.3 Ensuring Quality When Meeting Criteria Output

Finally, it is also important to consider not only if a product is achieving each criteria, but also whether it is achieving them *well*. For example, while performance goals of a design output may be met, if the operation is wasteful or time consuming the output is likely not of appropriate quality. Analysis of how well a category is achieved in this work is described as performance in relation to the individual categories of assessment. Therefore, an output that achieves each category well can be said to achieve them *efficiently* and *effectively* [12, 15]. Each term is defined as such [12]:

Effectiveness: the degree to which the actual result meets the original goal.
Efficiency: the relationship between what has been gained and level of resource used.

2.4 A Combined Hierarchy of Quality

Quality is then a product of to whom it is manifest, the categories of criteria by which it is judged, and how well the criteria in those categories are achieved. Through these three levels, a hierarchy can be created that considers each aspect of design quality, and categorises in a way that gives highly detailed information of the particular strengths and weaknesses of quality and the stakeholders to whom it concerns (Fig. 2). Detailed understanding of design quality can be gained by systematically proceeding through this hierarchy when identifying criteria for assessment and when performing analysis.

2.5 The Development of a Weighting Scale

Clearly, different branches of the hierarchy will have different levels of importance in relation to the overall quality of the design output, thereby requiring weighting. In this work, weightings have been developed using the Analytic Hierarchy Process (AHP) [16], a widely used, multi-disciplinary method of ranking and assessment developed within the past two decades [17]. Through the use of standard pair-wise comparison [8], AHP produces fair and proportional values of importance for each assessed criteria in relation to every other assessed criteria. The particular strength of this system is the robust manner in which weightings are assigned in relation to the importance of each criterion, rather than the subjective weights attached by some other methods.

2.6 Using the Hierarchy to Assess Quality

To use this system for assessment of relative quality, the following steps are completed. This process is demonstrated in the case study.

1. The hierarchy must be formed depending on relevance of each category to the specific design output in question. At this point the assessor must select the branches of the hierarchy that may influence final quality in the specific case.
2. The hierarchy is populated with as many criteria as is possible by systematically working through each category, adding criteria from the specification and thinking of others that fit.
3. Each selected category is weighted according to the AHP method. This then provides as complete a set of criteria for quality assessment as possible, and places them within a hierarchy that allows easy and detailed analysis.
4. Quality assessment occurs according to the method of Pugh [11]; a datum concept is chosen (using the most complete design concept available) and all others are compared to it on a scale of better/worse/same.
5. Ratios are taken of the number of better to worse criteria in each category, multiplying by the appropriate weights during the process.

Addition of ratio values then gives a decomposable ranking of all concepts; capable of stating not only which products are of highest quality, but also to whom that quality is of importance, the specific criteria under which that quality falls, and quality of the manner in which those criteria are met.

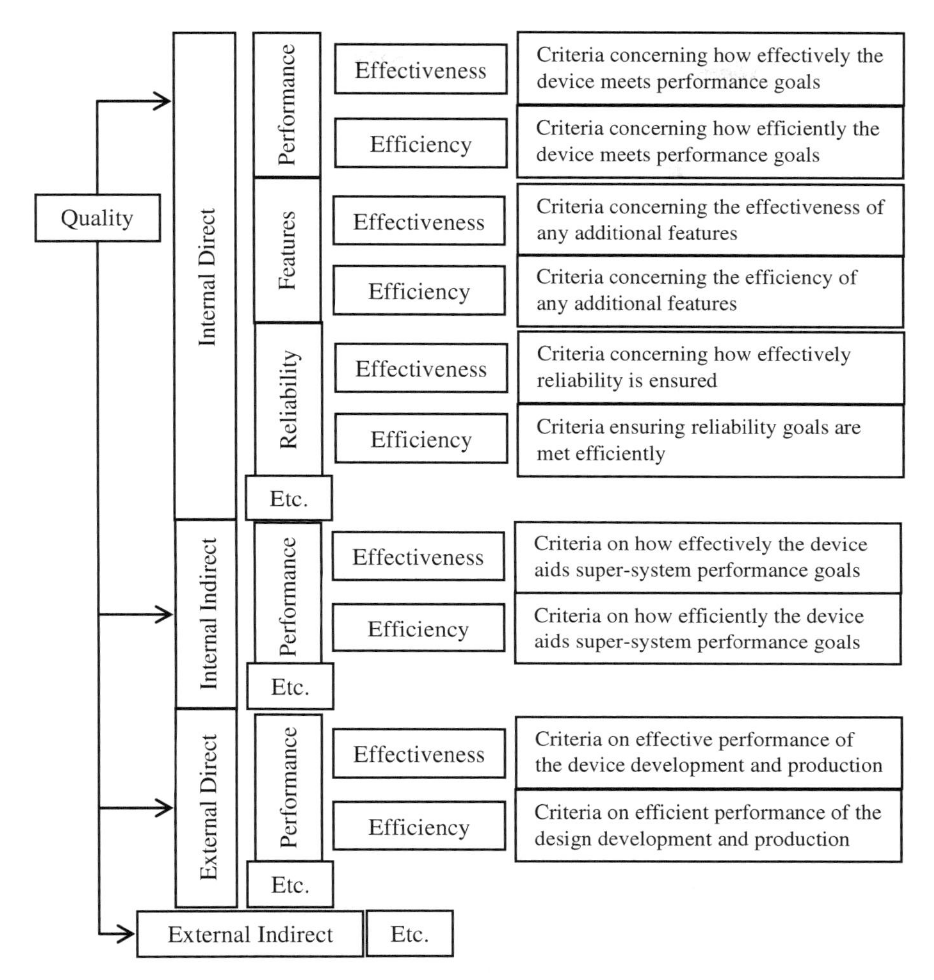

Fig. 2 Partial structure of the perspectives of quality used within this work. Complete structure contains all eight "dimensions of quality" under each category (*ID*, *II*, *ED* and *EI*)

2.7 Output of the Classification Method

The particular use of the classification method is in the breadth of additional information it provides for comparison of design outputs. Beyond assessment and comparison of single criteria (such as comparing mass of two designs), categorisation of criteria provides understanding of the broader implications of each design alternative, such as the relative influence of entirely different criteria in relation to one another.

Through the categorisation and weighting method it is possible to understand the summation of different quality criteria and their influence on the final design output. For example, although one design may perform better, it may also be more difficult to manufacture. Through the proposed hierarchy it is possible to quantify and assess this trade-off accurately, and choose the final solution that has the highest final output quality.

3 Case Study—Assessment of Quality

An example of the method in use is provided through assessment of 12 designs produced during a previously reported experiment [18], each designed to solve the same problem.

The problem was to design a hanging camera mount to be placed beneath a balloon, with controllable hemi-spherical motion pointing downwards. The design was to accept any amateur camera and be controlled remotely. Designers received identical briefs, and were given 90 min to individually produce their design to a high level of detail (Fig. 3).

3.1 Assessment

Assessment occurred according to the process within Sect. 2.6. Each level of the hierarchy in Fig. 2 was considered with respect to the design problem in order to prompt the generation of as many criteria (that could be applied to all designs) as possible, and to categorise each in a manner that enhanced understanding. Due to relevance to the brief and the development stage at which the designs were compared, only categories shown in Table 1 could be assessed. In all, 24 criteria were assessed.

Weight was assigned to each category using the AHP method. To judge importance, four engineering assessors (between 7 and 45 years' experience, average 25) were presented with the relevant categories and asked to perform pair-wise comparison between. Comparison achieved a sufficient value of Krippendorff's alpha of 0.74 (a measure of inter-coder reliability), and the resulting weights were averaged (Table 1).

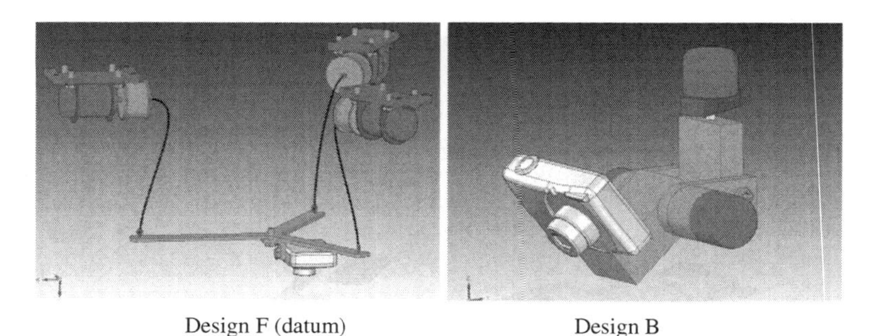

Design F (datum) Design B

Fig. 3 Example designs produced within the experiment

Table 1 Hierarchy used for assessment of designs (with selected examples from total of 24)

Hierarchy of quality	Quality metrics	Quality within criteria	Weight	Example criteria
Internal direct	Performance	Effectiveness	0.179	Range of motion
		Efficiency	0.179	Length of single charge
	Features	Effectiveness	0.0798	Protect cargo
	Reliability	Effectiveness	0.143	Security of connection etc.
Internal indirect	Performance	Effectiveness	0.306	Operational mass of balloon etc.
External direct	Performance	Effectiveness	0.292	Manufacture using existing tooling
External indirect	Insufficient information for assessment in this case			

As the experts asked within this study have only a general experience of the output and its context, this work demonstrates only the process and reliability of the weighting process. In reality, weighting would occur through experts with extensive understanding of the design, market, and other important influences. Weighting in reality would likely vary depending on the background of the experts; e.g., should they be heavily involved in manufacturing, then that would likely receive a higher value. Hence, although the weightings presented here represent the opinions of highly experienced engineers and so are valid; those used by a company may vary. It is for the company to decide the importance of each category depending on priority.

3.2 Results

In all, the designs were compared on 24 criteria fitting into these categories. One of the designs (Design F) was selected as a datum, against which all others were compared for each criteria on a better/same/worse scale [as shown for the Internal Direct (performance) criteria within Table 2]. Ratios of better scores to worse scores (in relation to the datum) are then taken for each individual category and multiplied by the weight, to produce a final weighted ratio for each design (Table 3). The value for this weighted ratio determines the ranking of quality of designs in relation to the original datum.

3.3 Discussion of the Results

By this method, each design can be ranked according to the criteria that contribute to its interpretation as of appropriate quality. Extra insight can also be gained; Design F performed poorly in terms of ID criteria, therefore showing a poorer

Table 2 Comparison with datum for all criteria within internal direct (performance) category

Criteria	A	B	C	D	E	F	G	H	I	J	K	L
Range of motion	+	+	0	+	+	Datum	+	−	+	+	−	+
Speed of operation	−	−	−	−	−		−	−	−	−	0	−
Stability in position	+	+	+	+	+		+	+	+	+	+	+
Ease of connection	−	0	−	0	0		−	0	−	−	0	0
Operational range	0	0	0	0	0		0	0	0	0	0	0
Type of control	0	0	0	0	0		0	0	+	0	0	0
Power requirement	0	0	+	0	0		+	0	+	+	+	+
Mass	−	+	−	+	+		+	−	0	+	0	−
Volumetric size	0	+	0	+	+		+	+	0	+	+	0
Length of single charge	+	+	+	+	+		+	+	+	+	−	+

Table 3 Ratios, weighting and final ranking

Criteria	A	B	C	D	E	F	G	H	I	J	K	L
ID (performance)	3:3	5:1	3:3	5:1	5:1	Datum	6:2	3:3	5:2	6:2	3:2	4:2
ID (features)	1:1	1:0	1:1	1:1	0:0		1:1	0:0	1:1	1:1	0:0	0:0
ID (reliability)	0:0	0:0	0:2	1:1	0:0		0:0	0:0	0:0	1:0	1:0	0:1
II (performance)	0:1	3:1	0:1	3:1	1:0		3:1	2:2	0:0	3:1	0:0	2:2
ED (performance)	0:4	1:3	0:5	1:3	0:4		0:4	0:5	0:4	1:4	1:3	0:5
Overall ratio (decimal)	0.44	2.0	0.31	1.6	1.2	1.0	1.3	0.50	0.86	1.5	1.0	0.55
Weighted ratio (decimal)	0.30	1.6	0.21	1.5	0.89	1.0	1.1	0.44	0.61	1.3	0.79	0.46
Ranking	11	1	12	2	6	5	4	10	8	3	7	9

design in itself, but excelled in terms of ED, therefore being more appropriate to company capabilities such as manufacturability and assembly. Conversely, Design L has strong ID performance capabilities, but is particularly poor in terms of manufacturing and assembly.

4 Discussion of the Process

While information from ranking aids selection of designs for development, it is for additional information that this method was designed. Many research opportunities result from breaking down the manner in which quality is displayed. Deeper understanding can be gained by comparing how quality appears with designer behaviour, the prescribed design process and brief, and the ways in which designers are creative, for example. Information about these potential relationships will provide a valuable insight into preferable behaviours leading to different forms of output with different forms of quality. This will perhaps allow designers to focus their work to the specific priorities of the company and the brief that they are presented, or will allow greater understanding and optimization of the way in which outputs are developed to maximise their quality in an appropriate manner. Such analysis has begun, and will continue in further work.

5 Conclusions

The aim of this paper has been to present a systematic, hierarchical system to generate criteria and assess quality, in the context of what is appropriate to the design problem.

This is with the goal of allowing assessment of *appropriateness* in design outputs (a fundamental part of the interpretation of creativity) described here as the appearance and recognition in a design output of quality appropriate to the problem that it must solve. Through the categories used, the method also creates deeper understanding about the quality of the design output, which can in turn be used to better inform design decisions, or as a research tool to better understand quality development through the design process.

By assessing through separate perspectives used within multiple fields of literature, this method stimulates the assessment of quality in terms of to whom the design output is of quality, the specific "dimensions" in which its quality is manifest, and how well those dimensions are achieved. Thus, when used to as full an extent as is feasible, this method ensures that all criteria affecting quality are considered.

In assessment, this method uses the well-established process of AHP to determine weights for each category, then assigning these weights to the ratios used for ranking. Weighting is flexible, occurring based on the companies discretion and priorities.

Through the presented example, this assessment method has shown capability in ranking of relative quality of multiple designs, as well as the ability to produce additional information of how that quality appears. Following further validation and comparison of quality with traits of designers and the design process (as will occur in further work), understanding can be gained of the relationships and dependencies of designer, process and quality; information that can then be used to enhance methods of designer support.

Acknowledgments The work reported in this paper has been undertaken with support from the Engineering and Physical Sciences Research Council (EPSRC) at the Innovative Design and Manufacturing Research Centre (IdMRC) University of Bath.

References

1. Chakrabarti A (2006) Defining and supporting design creativity, in design 2006. In: The 9th international design conference 2006. Dubrovnik, Croatia
2. Gero JS (1996) Creativity, emergence and evolution in design. Knowl-Based Syst 9(7):435–448
3. Shah JJ, Smith SM, Vargas-Hernandez N (2003) Metrics for measuring ideation effectiveness. Des Stud 24(2):111–134
4. Howard TJ, Culley SJ, Dekoninck EA (2008) Describing the creative design process by the integration of engineering design and cognitive psychology literature. Des Stud 29(2):160–180

5. Sarkar P, Chakrabarti A (2011) Assessing design creativity. Des Stud 32(4):348–383
6. Simon HA (1979) Models of thought. Yale University Press, London
7. Ullman DG (1997) The mechanical design process, 2nd edn. McGraw-Hill, London
8. Cross N (2000) Engineering design methods—Strategies for product design, 3rd edn. Wiley, Chichester
9. Kolarik WJ (1995) Creating quality: concepts, systems, strategies, and tools. McGraw-Hill, London
10. Pahl G, Beitz W (1984) Engineering design: a systematic approach. Springer, London
11. Pugh S (1990) Total design: integrated methods for successful product engineering. Prentice Hall, Harlow
12. O'Donnell FJ, Duffy AHB (2002) Modelling design development performance. Int J Oper Prod Manage 22(11):1198–1221
13. Garvin DA (1987) Competing on the eight dimensions of quality. Harvard Bus Rev 65(6):101–109
14. Boden M (2004) The creative mind: myths and mechanisms. Routledge, London
15. Neely AD et al (2002) Strategy and performance: getting the measure of your business. Findlay Publications, Horton Kirby
16. Saaty TL (1990) How to make a decision: the analytic hierarchy process. Eur J Oper Res 48(1):9–26
17. Vaidya OS, Kumar S (2006) Analytic hierarchy process: an overview of applications. Eur J Oper Res 169(1):1–29
18. Snider CM et al. (2012) Variation in creative behaviour during the later stages of the design process. In: The 2nd international conference on design creativity (ICDC2012), Glasgow, Scotland

Development of Cognitive Products via Interpretation of System Boundaries

Torsten Metzler, Iestyn Jowers, Andreas Kain and Udo Lindemann

Abstract Cognitive products use cognitive functions to work autonomously and reduce the amount of interaction necessary from the user. However, to date no method exists to support the integration of cognitive functions in common products. This paper presents a method that supports designers when exploring ideas for new cognitive products. The method is based on functions/actions that humans perform while using a product, as well as functions/actions performed by the product itself, all of which can be consistently modelled in an activity diagram. Initially, the system boundary of the product is drawn around the functions/actions performed by the product. Cognitive functions are then identified that are currently performed by the user, and can possibly be integrated into a new cognitive concept. The resulting concept is specified systematically by interpreting the system boundary of the product to include cognitive functions. This method has been verified via design projects performed by interdisciplinary student design teams, and an example of this work is presented.

Keywords Cognitive products · Functional modelling · Activity diagrams

T. Metzler (✉) · I. Jowers · A. Kain · U. Lindemann
Institute of Product Development, Technische Universität München, Munich, Germany
e-mail: metzler@pe.mw.tum.de

I. Jowers
e-mail: jowers@pe.mw.tum.de

A. Kain
e-mail: kain@pe.mw.tum.de

U. Lindemann
e-mail: lindemann@pe.mw.tum.de

A. Chakrabarti and R. V. Prakash (eds.), *ICoRD'13*, Lecture Notes in Mechanical Engineering, DOI: 10.1007/978-81-322-1050-4_13,

1 Introduction

Cognitive products are tangible and durable things with cognitive capabilities that improve robustness, reliability, flexibility and autonomy [1]. They meet and exceed customer expectations by using cognitive functions e.g. *to perceive*, *to learn*, *to plan*, etc., to reduce the need for human input, for example when such input is difficult or repetitive. However, no method exists to support the integration of cognitive functions into common products. This paper presents such a method, with the intention of supporting designers as they explore ideas for new cognitive products.

The research presented here is concerned with identifying how cognitive functions can be included in the functional modelling process. Functional modelling is core to many product development activities, and numerous methods have been introduced that result in a holistic representation of a product according to its functional structure [2]. The resulting functional models are often represented as flow diagrams with functions described according to some taxonomy, e.g. [3], and linked according to the material flows between them e.g. [4]. In the method presented here, the functions of a product are represented as actions in an activity diagram. The diagram is then extended to include the actions of the user during product use, with the system boundary of the product surrounding the actions performed by the product. The user actions are compared to a taxonomy of cognitive function and flows [5] to identify those that could be integrated into the functionality of the product. Finally, the system boundary of the product is interpreted to include those cognitive functions that are identified and have the potential to improve the product as a cognitive product. This systematic variation of the system boundary results in gaining a holistic perspective which can support the design of cognitive products, and the method has been validated via design projects conducted by interdisciplinary student design teams, as described in [6].

The next section provides an overview of the role of functional modelling in product development, and an overview of modelling with cognitive functions. In Sect. 3, the problem of using cognitive functions in product development is presented, and in Sect. 4 a method is introduced which seeks to overcome this problem. The method is illustrated with reference to a cognitive washing machine which was developed by a student design team. The paper concludes with a discussion exploring the potential of the method to support cognitive product development, and an outlook towards future research.

2 Background

2.1 Functional Modelling in Product Development

Functional modelling is central to many product development activities, particularly conceptual design [4]. It supports a systematic, top-down approach to product definition, starting from a description of the required core functionalities of the product. These can then be sequentially decomposed into lower-level sub-functions, resulting in an abstract specification of the product that describes how the required functionalities can be realised by sub-functions and the relations between them, e.g. [7]. There are various approaches to formally representing the resulting functional models, a review of which are provided by Erden et al. [2]. A common approach, and the one that is employed in this paper, is to represent functions according to a flow-oriented model [4]. In particular, functions can be defined as a general input/output relation that is used to perform a task, and can be described by verb-noun pairs, e.g. *mix water and detergent*. The relations between these functions can then be defined as flows characterised according to types e.g. *material*, *energy* or *signal*, and the resulting functional models are represented as flow diagrams, as illustrated in Fig. 1. Here, the functionality of a washing machine is presented as a system of functions and sub-functions, and the flows of material, energy and signal between them.

Functional models are well suited for supporting modern design processes, in which multi-disciplinary teams collaborate to develop complex products [2]. They provide a common representational framework for defining a product as a system of functions and sub-functions, which is accessible to all members of the team, regardless of engineering discipline. A functional model provides an abstract but holistic view of the system which allows designers to better understand the complex products with which they are working, individually and in collaboration with team members [8]. If a functional model is constructed based on an accepted language of functional descriptions then this reduces potential ambiguity in the model, increases uniformity and increases the potential to reuse the model either manually or in an AI-based system. For example, the NIST Reconciled Functional Basis is a taxonomy in which functions are input/output relations connected via flows and represented in flow-oriented models [3].

In systems engineering, flow-oriented functional models can be represented as activity diagrams. These are flow diagrams representing activities, which are defined according to constituent actions and their inputs/outputs [9]. Activities/actions are an abstract formalism for describing behaviour in the same way that functions/sub-functions are an abstract formalism for describing behaviour [7].

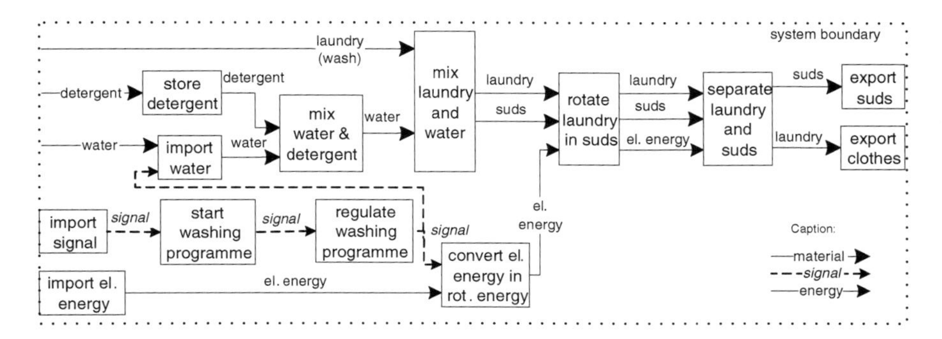

Fig. 1 Flow-oriented functional model for a washing machine

So activity diagrams can be used to model the functions carried out by a system as actions, but with additional capabilities beyond those provided by a flow-oriented model. For example, control nodes such as fork/join nodes or decision nodes can be included in a diagram to represent control logic and provide additional constraints on the timing and order in which actions execute [9]. The combination of object flows with control flows is a powerful formalism for modelling products as systems of functions. Also, the additional capabilities provided by activity diagrams mean that when they are defined formally, using a language such as SysML, they can be mapped to executable constructs which in turn support evaluation, for example through simulation. For these reasons activity diagrams will be used in the remainder of this paper, to represent the functional structure of products and support development of cognitive products via introduction of cognitive functions.

2.2 Functional Modelling for Cognitive Products

Cognitive products are tangible and durable things with cognitive capabilities that consist of a physical carrier system with embodied mechanics, electronics, microprocessors and software. The surplus value is created through cognitive capabilities enabled by flexible control loops and cognitive algorithms. Customer needs are satisfied through the intelligent, flexible and robust behaviour of cognitive products that meet and exceed customer expectations. Cognitive products have all or a subset of capabilities of Cognitive Technical Systems (CTSs) and the solid grounding of an everyday product that meets user needs and desires [1].

What makes products cognitive are their special properties stemming from the integration of CTSs. The implementation of cognitive capabilities results in high-level performance. In particular, in contrast to products which have deterministic control methods, cognitive products do not only act autonomously, they do so in an increasingly intelligent and human-like manner. This means that they are more robust than non-cognitive products since they are able to adapt to a dynamic environment, such as human living environments. Cognitive products should be able to maintain multiple goals, conduct context sensitive reasoning, and make appropriate decisions. This results in higher reliability, flexibility, adaptivity and improved performance.

Incorporation of cognitive capabilities in a product concept requires specific descriptions of cognitive functions, as discussed by Metzler and Shea [1]. The term *cognitive function* is used to refer to basic functions that enable cognition, e.g. *learn*, *perceive*, *understand* or *decide*. Such functions perform operations on flows of information, or they process data flows to create information. However, there is no commonly agreed list of cognitive functions that are required for a cognitive system, neither human nor artificial. Typically, researchers in particular areas compile their own list of cognitive functions. Metzler and Shea [5] present a comprehensive set of cognitive functions and flows structured in a taxonomy that incorporates views from engineering and cognitive sciences. The taxonomy of cognitive functions and flows is tailored for mechanical engineers and supports consistent functional modelling through a standardised representation. It can be used to model a wide range of cognitive products and is used throughout this paper. As discussed, functional modelling results in an abstract representation of a product that is useful for multi-disciplinary concept development phase. An example of using the taxonomy of cognitive functions and flows to define flow-oriented functional models is presented in Metzler and Shea [5].

3 Problem Identification

There are many factors that are driving the introduction of cognitive functionality in today's consumer products, i.e. functionality that introduces cognitive capabilities so that products can operate with robustness, reliability, flexibility and autonomy. The need for companies to differentiate themselves from the competition means that they are constantly looking for opportunities to develop their products in innovative ways and they often want to be seen to be on the cutting-edge of technological development. Also, consumers expect more functionality from their products and want the user experience to be as enjoyable as possible. AI algorithms and methods have reached a high-level of maturity which means that they can be reliably incorporated into cognitive products. And, the steady reduction in the cost of components necessary to utilise these algorithms, such as CPUs, digital cameras, actuators, etc. means that they are cost effective. Cognitive products use cognitive functions to enable products to work more autonomously so that they can reduce the amount of interaction necessary from the user, while exceeding their expectations. For example, iRobot's Roomba is an autonomous robot vacuum cleaner that uses cognitive functions to map, navigate and plan routes and significantly reduces the need for human interaction [10].

Although there is a drive to incorporate cognitive functions in consumer products it is not always obvious how to include such functions in a design. As discussed in the previous section, functional modelling is central to many product development activities, but cognitive functions are rarely considered. The taxonomy of cognitive functions and flows (as described in Sect. 2.2) defines a language for describing the required cognitive functionality of a product in the same way that the NIST Reconciled Functional Basis defines a language for describing non-cognitive functions. Despite this, it is not obvious how to incorporate the taxonomy when introducing cognitive functions to an existing product concept.

The difficulty that arises with respect to developing cognitive products was observed during a series of student projects, as described by Metzler and Shea [6]. Since 2007, 6 projects have been set in which 16 teams of students were tasked with inventing or adapting household products that address user needs by using cognitive functions. The students who participated were from varied backgrounds including mechanical engineering, electrical engineering or computer science, and worked in multi-disciplinary teams of 3–5. They were tasked with developing cognitive products that address a general problem, e.g. *saving energy* or *recycling*, by incorporating cognitive functions into existing products. For example, the washing machine design that is used throughout this paper was developed by one student team in response to a project where they were asked to introduce cognition into a household product so that it can be more easily used.

Before starting the projects, the students were introduced to a user-centred process that supports early phases of development of new products that are useful and usable [11]. The process was adapted to the context of cognitive product development to assist in the identification of user needs; to aid in the development of a product concept; and to support the building of a functional prototype. The students were also encouraged to use functional modelling to support the development of a product concept and were presented with the taxonomy of cognitive functions and flows to aid in the specification of cognitive functions.

During the projects the performance of the students was observed, and it was found that the teams were able to identify product market opportunities and needs, and they constructed functional prototypes of adequate quality. However, it was also observed that many of the teams had difficulties incorporating cognitive functions into a product concept. The major difficulty for the students was on the one hand translating the user needs into cognitive functions and on the other hand identifying and incorporating functions related to or required by the cognitive functions in the new product concept. Most teams did not use a systematic approach that could assist in the integration of cognitive functions into product concepts. The result was product concepts that had to be adapted in the following development phases, resulting in additional iterations of the design process and delayed development of the cognitive product.

4 A Method for Incorporating Cognitive Functions

The method described in this section addresses the incorporation of meaningful cognitive functions into existing product concepts. The goal is to turn existing product concepts into cognitive product concepts by identifying and incorporating cognitive functions that are currently carried out by the user. To achieve this goal four steps have to be carried out, as described in Fig. 2. These steps are explained in the following sub-sections.

4.1 Model a Product Concept as an Activity Diagram

Step 1 of the method is concerned with creating an activity diagram as a model of the product concept into which cognitive functions will be incorporated. The model can be derived from a product already existing in the physical world or from a product concept under development. This makes the method applicable to new

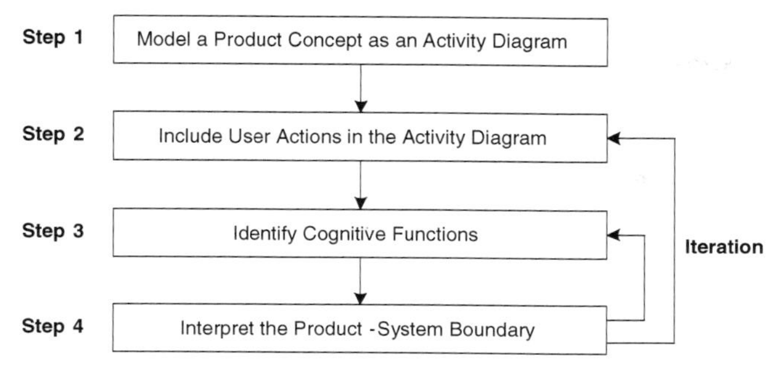

Fig. 2 Procedural model of the method

product development as well as incremental product development. As discussed in Sect. 2, activity diagrams are an abstract formalism to describe behaviour. They represent activities as flow diagrams, defined according to constituent actions and their inputs/outputs. When applied to product modelling they are used to represent the product according to object flows which represent the input/output of functions represented as actions, and control flows which represent the chronology of the actions. If a functional model of a product concept already exists, such as the flow-oriented functional model of the washing machine illustrated in Fig. 1, then it is a trivial task to translate this into an activity diagram. In Fig. 1, the functionality of the internal operations of the washing machine is described with active verbs and objects from the NIST Reconciled Functional Basis. In the translated model, these functions are represented by actions, and the functionality of the product is represented as a system of actions and the flows between them.

4.2 Include the User Actions in the Activity Diagram

In Step 2, the user's interaction with the product (during intended use) is considered, and the system of actions represented in the activity diagram is extended to include the actions of the user. To achieve this, the user experience is considered chronologically. First, start and end nodes are included; these indicate the beginning and the end of product use. Next, all possible (within the limits of the design) user actions are added to the activity diagram. Finally, these are connected by flows, which are either object flows, such as those commonly used in flow-oriented functional modelling, i.e. material, energy, signal, etc., or control flows that model the chronology of actions by specifying when and in which order actions are executed [9]. The authors recommend using the taxonomy of cognitive functions and flows and the NIST Reconciled Functional Basis to model the user actions. For example, Fig. 3 shows an extended activity diagram of the washing machine that includes, in addition to the functionality of the washing machine, illustrated in Fig. 1, the user's pre-wash actions. The post-wash actions of the user

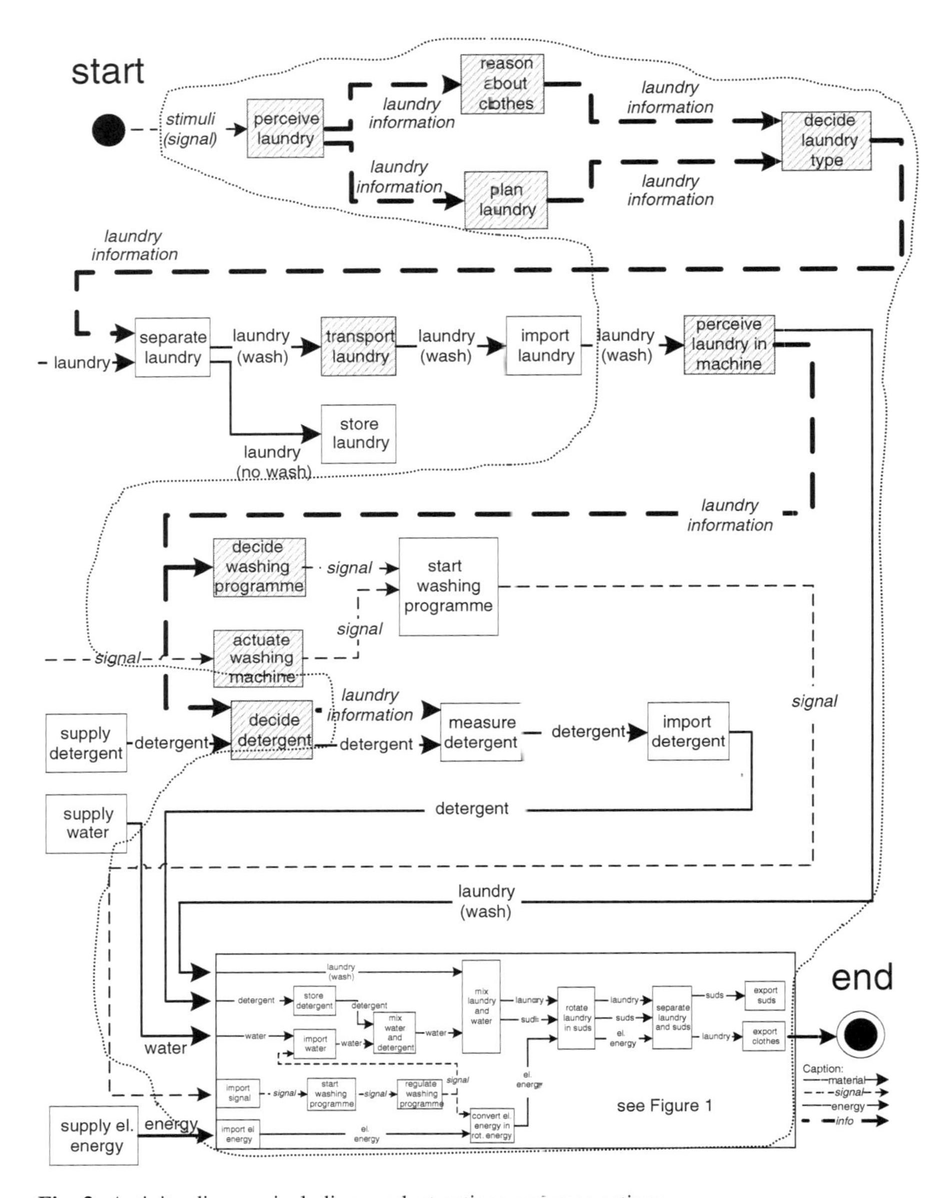

Fig. 3 Activity diagram including product actions and user actions

are not included due to space limitations, and the activity diagram ends when the wash cycle terminates. The functionality of the washing machine was presented in Fig. 1, and in Fig. 3 is enclosed in one block as a subsystem. The actions included describe the user's process of planning and executing a washing programming, for example deciding the type of laundry to wash, separating the laundry, putting the laundry in the machine, choosing appropriate detergent, etc. The flow between these actions describe the materials that are needed for input/output, and also structure the actions chronologically.

4.3 Identify Cognitive Functions

After incorporating the user actions into the activity diagram, Step 3 is concerned with identifying the cognitive functions currently implemented by the user. As discussed in Sect. 2, these are basic functions that enable cognition by performing operations on flows of information, or by processing data flows to create information. Such functions offer a great potential for innovation by shifting complicated or often repeated tasks from the user to the product. The taxonomy of cognitive functions, defined by Metzler and Shea [5] is an aid to identifying cognitive functions and provides a comprehensive set for comparison. For example, comparison of the user actions specified in the use of the washing machine in Fig. 3 with the taxonomy of cognitive functions gives rise to a list of cognitive functions performed by the user during interaction with the product. In Fig. 3, these cognitive functions are highlighted as striped blocks, and include *perceive laundry*, *decide laundry type*, *decide washing programme*, etc.

4.4 Interpret the Product-System Boundaries

In Step 4, the product's system boundary is interpreted to include some of the cognitive functions that were identified in Step 3. Currently these functions are actions of the user, but can potentially be an action of a new cognitive product concept. Here, the critical decision is to decide *which* of the identified cognitive functions to include in the new product. The following questions could assist in this decision:

1. How close is the cognitive function relative to the product?
2. How many flows link the cognitive function to other functions?
3. How difficult is it to technically realise the cognitive function?
4. How annoying is the cognitive function for the user when carried out?

Question 1 can be addressed by considering the activity diagram defined in Step 2. Here, the distance between functions is measured according to any intermediate functions. In the activity diagram, the cognitive functions closest to the original product concept are likely to be suitable for integration in the product. This is because the closer a function is to the original product, the more likely it is that this function is associated with the product, and would be acceptable as part of the product. Conversely, the more functions that have to be carried out by the user between a cognitive function and the functions carried out by the product, the less it is associated with the product.

Question 2 can also be addressed by considering the activity diagram. In the activity diagram, the more a cognitive function is linked to other functions, the more complex the functional model becomes. If there are many flows linking a cognitive function to the original product concept it is expected that many components in the concept will have to be adapted to accommodate the new functionality. Similarly, if there are many flows linking a cognitive function to other cognitive functions outside the original system boundary it is expected that the user has to strongly interact with this cognitive function.

Question 3 relates to the technical feasibility of implementing a cognitive function. Its answer relies on the expertise available to realise such a function in a physical device, and also on the current state of the art, since some cognitive functions may not be realisable using currently available technology.

Question 4 focuses on the user and asks which actions would make the experience of using a product more enjoyable, if they were implemented by the product. A study investigating product market opportunities and user needs is an appropriate method to explore this question.

After answering the questions a pair-wise comparison can help to identify which cognitive function(s) should be integrated into the new cognitive product concept. This results in an interpretation of the product's system boundary to include the identified cognitive functions, as well as other required non-cognitive functions, as part of the product's functional structure. For example, in Fig. 3 cognitive functions have been incorporated into the system boundary for a new cognitive washing machine concept, as identified by the new system boundary represented by the dotted line. The new washing machine concept perceives the laundry and, based on how much is available of each type, e.g. colours, whites, delicates, etc., decides which laundry should be washed. The machine suggests to launder the most homogeneous laundry group with the highest capacity utilisation first. However, the final choice of which type of laundry to wash is made by the user; this avoids dissatisfaction due to the paternalism of the washing machine. The user then separates this laundry and places it in the machine. The machine perceives which laundry is inserted and adapts its behaviour to the user's decision. For example, the washing machine determines the ideal washing programme or which detergent suits best, how much detergent is needed for the current laundry group and if softener is needed. In case the user mixes two different laundry groups accidentally the washing machine can output a warning to inform the user and avoid staining. This new concept improves the experience of clothes washing by carrying out some of the tedious and repetitive actions usually carried out by the user. It was implemented as a physical prototype by the multi-disciplinary team of students who designed it.

5 Discussion

The method described in the previous section was applied by student teams during the development of cognitive products, including the cognitive washing machine concept represented in Fig. 3. These initial applications provide evidence for the usefulness of the method as a way of identifying cognitive functions that are involved in the use of an existing product concept. The method is visual in nature, allowing the system boundary between the user and the product to be interpreted according to identified cognitive functions. This visual nature means that the approach is intuitive for the designer, and easy to communicate to other members of a multi-disciplinary design team. A more thorough evaluation of the method, including comparisons with other approaches and a control group, remains to be conducted.

When applying the method, cognitive functions could easily be identified by following Steps 1–3. As discussed, in these steps cognitive functions are identified as user actions modelled in an activity diagram. However, Step 4 involving the decision of *which* cognitive functions to incorporate is more difficult, and required input from experts with sufficient experience in CTSs to make informed and realistic decisions. The fact that the realisation of cognitive functions in hardware and software is difficult is known and further research is being carried out to improve the decision making process and support implementation of Step 4. This includes the definition of design catalogues that provide patterns of how to realise cognitive functions. Also, analysis of aspects of a generated activity diagram, e.g. according to number, type and direction of flows, may be sufficient to estimate the feasibility of incorporating a cognitive function.

In addition to the difficultly of carrying out Step 4, there are other open issues that remain to be investigated in further research. For example, it is not known if the initial product concept has to be modelled and how detailed this model should ideally be. It may be beneficial to use a black box to represent the initial product concept, with input and output flows indicated. User actions could then be modelled around the black box as illustrated in Fig. 3. This may be sufficient to inform and motivate the designers of cognitive products, without having to take the time and effort to model the existing product in detail. This could also avoid issues with design fixation, which may arise through consideration of the original functional structure of the product. However, this approach may be detrimental, since there is no available information about how the flows link to structure and how the structure has to be changed when incorporating a certain cognitive function.

6 Conclusion

The method described in this paper provides a systematic approach for extending product concepts to include cognitive functions that would otherwise be implemented by the user. This results in products which implement cognitive functions previously carried out by the user and therefore require less interaction and provide a more enjoyable user experience. The approach was illustrated with reference to a cognitive washing machine concept which was developed and built by a multi-disciplinary team of students. This initial use of the method is promising and suggestive of its potential as an aid in cognitive product development.

Acknowledgments This research is part of the Cluster of Excellence, Cognition for Technical Systems -CoTeSys (www.cotesys.org), funded by the Deutsche Forschungsgemeinschaft (DFG).

References

1. Metzler T, Shea K (2010) Cognitive products: definition and framework. In: 11th international design conference—DESIGN 2010, pp 865–874
2. Erden MS, Komoto H, van Beek TJ, D'Amelio V, Echavarria E, Tomiyama T (2008) A review of functional modeling: approaches and applications. Artif Intell Eng Des Anal Manuf 22(2):147–169
3. Stone RB, Wood KL (2000) Development of a functional basis for design. J Mech Des 122(4):359–370
4. Pahl G, Beitz W, Feldhusen J, Grote K-H (2007) Engineering design: a systematic approach, 3rd edn. Springer, Berlin
5. Metzler T, Shea K (2011) Taxonomy of cognitive functions. In: International conference on engineering design (ICED11), pp 330–341
6. Metzler T, Shea K (2011) Lessons learned from a project-based learning approach for teaching new cognitive product development to multi-disciplinary student teams. In: Proceedings of the ASME IDETC/CIE, DETC2011-48168
7. Umeda Y, Tomiyama T (1995) FBS modeling: modeling scheme of function for conceptual design. In: Proceedings of 9th international workshop on qualitative reasoning, pp 271–278
8. Ponn J, Lindemann U (2008) Konzeptentwicklung und Gestaltung technischer Produkte, Springer, Berlin
9. Friedenthal S, Moore A, Steiner R (2008) A practical guide to SysML: the systems modeling language. Morgan Kaufmann, Burlington
10. Jones JL (2006) Robots at the tipping point, the road to the iRobot Roomba. Robot Autom Mag IEEE 13(1):76–78
11. Cagan J, Vogel CM (2002) Creating breakthrough products: innovation form product planning to program approval, 1st edn. FT Press, Upper Saddle River, NJ 07458

A Design Inquiry into the Role of Analogy in Form Exploration: An Exploratory Study

Sharmila Sinha and B. K. Chakravarthy

Abstract Cross transfer of ideas from one target domain to another with analogies as triggers to generate new ideas is being studied in various fields. But the role of analogy in the creative process for both generating new ideas and exploring high degree of novelty in product form generation needs to be investigated. Using an analogy can help transcend the obvious to the unexpected by forming a confluence of two mediated thoughts, guided by creative interpretations. The important aspect of generating creative ideas is the ability to use the source analogy for a resultant design solution by interpreting it in terms of the design task at hand. The study analyses the creative ideas that emerges from the design solution, specifically focusing on the use of analogy for form exploration. By using the analogical inference, transition from one domain idea to another opens the exploration of novel viewpoints and numerous alternatives. Such a malleable approach of creative generation can yield novel ideas and forms. The paper explores how the use of analogy can facilitate idea generation and allow form exploration. This research examines idea generation issues of creativity by the use of multiple source analogies to help develop multiple domain concepts threw case study method. The study collects empirical data from observations and design outputs during a design session. The aim is to articulate the influencing effect of the use of analogy on factors of form exploration. The goal is to create a value proposition that designers can use to develop novel ideas as well as expressive form generation through radical exploration.

Keywords Analogy · Idea generation · Form exploration

S. Sinha (✉) · B. K. Chakravarthy
Industrial Design Centre, IIT Bombay, Mumbai, India
e-mail: sharmila.sinha@iitb.ac.in

B. K. Chakravarthy
e-mail: chakku@iitb.ac.in

A. Chakrabarti and R. V. Prakash (eds.), *ICoRD'13*, Lecture Notes in Mechanical Engineering, DOI: 10.1007/978-81-322-1050-4_14, © Springer India 2013

1 Introduction

Cross transfer of ideas from one target domain to another with analogies as triggers to generate new ideas is being studied in various fields. But the role of analogy in the creative process for both generating new ideas and exploring high degree of novelty in product form generation has scope for further investigation. Using an analogy can help transcend the obvious to the unexpected by forming a confluence of two mediated thoughts, guided by creative interpretations.

The important aspect of generating creative ideas is the ability to use the source analogy for a resultant design solution by interpreting it in terms of the design task at hand. The study analyses the creative ideas that emerges from the use of analogy towards the design solution in form exploration. The use of analogical inference, opens the exploration of different viewpoints and numerous alternatives. Thus allowing a malleable approach of creative generation that can yield novel ideas and forms.

The proposed study investigates how novice designers use analogy, to explore innovative product forms during the creative process, and how it can be a means to generate unique designs and help effective product representation. As the use of analogies is an associative process, it can lead to unique solutions by providing either direct or abstract triggers [1]. Koestler [2] explains the act of creation as combining process in 'the bisociation of unrelated matrices'.

The study has been built on the hypothesis, that use of analogy in idea generation can allow radical form exploration by novice designers. This research examines idea generation issues of creativity and the mental sourcing of ideas threw case study method. The study collected empirical data from observations and design outputs during the design session. The aim is to articulate the influencing effect of the use of analogy on factors of form exploration, and the representation of the expression on the product form. The effect of this approach has, scope of empirical investigation to establish certain key issues regarding its application as procedures in design practice. The study is undertaken during a design module, 'Studies in Form' for first year design students. It discusses the strengths and limitations of its outcome in generating ideas for innovative form exploration.

1.1 Aim of the Study

This study proposes to explore the possibilities of the varied, novel and expressive form exploration with the use of analogy. Csikszentmihalyi [3] has elaborated that one connects to objects through the feeling it emotes. The paper explores the effect on the representation of form of the product with the use of analogy for idea generation. It is hoped that the results can benefit student designers to train for novel idea exploration and expressive form design.

The process of exploration while designing is fundamental to finding new and creative ways to design a given object. The notion of idea generation has been explored in the field of psychology and in the field of management, computer sciences, engineering, linguistics etc. Significant work in the AI sector is exploring this phenomena to translate the functioning of the mind and simulate it as discussed by Holyoak and Thagard [4]. In neurosciences it is said that the structuring of perceptions gathered from various experiences help in problem solving by bringing them to the foreground as and when needed [5]. The role of analogy in this process of representation is visible, but very little structured approach is available. This paper is a part of a larger study that aims to build knowledge on the role of analogy in the process of idea generation for product design.

The intent of the study is to establish the role of analogy as an integral part of creative/disruptive idea generation and its utility in innovative product design. Thus exploring the relation of analogy to idea exploration and its influence on the form expression. This can be useful towards methodology development for pedagogical interventions, to enhance idea exploration and its manifestation in the product form and break fixations.

1.2 Analogy in Design

Analogies are used within the design context, it maps the causal structure between the source object in one domain to the target design task/problem by building a relation. A few formal methods have been developed to support design-by analogy such as Synectics [6], Word tree [7] and Random input [8]. Methods basing analogies on the natural world are seen in use in Biomimetic concept generation, using the powerful examples nature provides for design. Analogies are used to project a structural relation from one domain to another, revealing new information and insight, further catalysing new analogy, to move further out of the problem and help map radical new solutions. According to Gero [9], unexpected design solutions are a product of the confluence of two schemas mediated through an analogy.

As creativity is a stimulated aspect of human thinking, it is also associated with the capacity to look critically at reality, explore unconventional alternatives, and perceive situations from innovative perspective [10]. Design being an amalgamation of both creativity and rational functionality, analogy seems to fit into it quite easily. Analogies can support understanding by abstracting the important ideas from the mass of new information by using the power of analogical relationships that is said to be based in their potential to comprise an entire set of associative relationship between features of the concepts that are compared [11].

1.2.1 Analogy in Idea Generation

Analogies are used when we want to say that something is like something else (in some respects but not in others) making it a key feature of many approaches to creativity and proposed as an underlying mechanism [12]. It is extensively used in Bionics, which follows the systematic use of biological and botanical analogies to solve novel engineering problems.

Literature reveals that stimulation, fixation, time, parameters have subsequent effect on idea generation as a whole in positive as well as negative way. It is stated that to set the flow of ideas and start the process of ideation, external stimulation can be helpful and particularly more heterogeneous stimuli form multiple categories are found to facilitate the production of more diverse ideas [13]. But on the downside the stimulation can lead to a fixation of reproducing parts of the given example to the created design, which will be discussed in the later part of the paper.

2 Method

The study is conducted during a course module "Studies in Form" as a case for data collection. The method of document analysis is used. From the choice of analogy to breaking up of its attributes for idea exploration by mapping on to the target product and how it is relating to the form expression is analysed through documents of the process. The study uses an explorative approach to show how the exploration of ideas with the use of analogy, generates novel forms in a classroom experiment with 15 design students. This paper has tried to look at the relationship between analogies, its workings and influence on the idea generation process during form creation. The study is done as a precursor to the main research of measuring the various influences of use of analogy in idea generation.

2.1 Description of Research

The sample consists of 15 first semester M Des. students of product design and interaction design from various engineering and architecture background. There were 7 female and 7 male students. The mean age is 25 years. Their experience ranged from fresh graduates to 3–4 years industry experience. The Data was generated from documenting class activities, sketches and thought outline during the design exploration process (unstructured interviews to get personal info was also done). The investigation was in the natural environment of the studio/classroom during the module "Studies in form". Protocol was specifically ruled out as it was felt that it would hamper the flow of thinking as the samples were novices-thus emphasis was given on thought links and sketches.

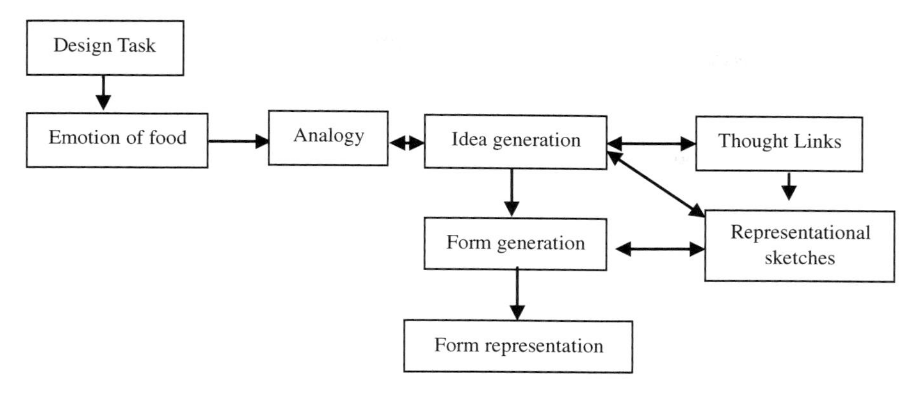

Fig. 1 The form exploration process using analogy

The task given was to design a 2 compartment plate connoting an expression with the use of analogy from nature. The study of the retrieval and representation of source data on the target form was done.

Values of plate: Experience plate. It had to be a 2 compartment plate with a minimum depth of 1 inch and maximum size 14 inches diameter, to be produced through a specific manufacturing process of vacuum forming.

A workshop on idea generation techniques with the use of analogy was conducted within the module. Students were taught to identify the source analogy and use the process of attribute mapping and transfer to the target, then translate it into the designing of the form (Fig. 1).

2.1.1 Procedure and Task

First of all, the teacher (i.e. researcher) provided the brief and the descriptions of the design task within the module "Studies in form: designing with analogy". The steps to be followed:

1. *Expression depiction*
2. *Take an analogy from nature and source visuals*
3. *Identify attributes of the source analogy*
4. *Representation of analogy to target plate design and its form*

3 Data Collection

This study draws from Cross's designerly way of thinking and idea generation by novice and professional designers [14], which investigated how designers can ideate through analogy. It is stated that the mind always looks for associative links to build ideas, through continually challenging abstract representations against visual representations during designing. This paper draws upon a single design case study, which is very limited in scope, in order to generalise the role of analogy in supporting ideating and development of novel product forms. Though the insights gained can be useful for further research.

The design problem given was to design a 2 compartment plate with the theme being any expression related to food with the use of analogy from nature, with size and process of manufacturing being specified. The study only focuses upon a subset of design activity the idea generation for exploring form.

The data in this paper is drawn from a 2 week long module with sessions on abstract form generation and sessions in idea generation with analogy, which the participants used to design the 2 compartment plate. Their thought flow and line representations were taken as document data for analysis. Students were asked to develop ideas by using analogy and develop line representation. They used river clay and prototyping materials for their 3D exploration of the ideas generated for form exploration and refinement.

3.1 Analysis

The technique of document analysis was used to identify typical ways in which analogy was used to support exploration (Fig. 2).

This exploratory approach of the study was used to help formulate an understanding of the thinking with analogy activity during the design process. This paper does not attempt to statistically proof any claims, because of the experiments small sample size. The design examples presented are the design outcomes of the experiment on which the data was collected as depicted in Table 1.

The levels of analogy used in the exploration process and the representations manifested in the product form was rated and plotted to observe the relationship that emerged (Fig. 3).

3.1.1 Findings and Discussions

The findings of the inquiry showed that idea exploration, with the use of analogy has a compelling effect in product designing. It gives physical representation to conceptual thought, and translates abstract emotions into tangible forms, by breaking predetermined perceptions and transcend to the next level of abstraction. It can be manipulated to have different association in differing contexts, as the same theme can be represented radically differently in the form with the use of different analogies.

Analogy from nature that was easily understood was adopted thus aiding and supporting ideation and novel form exploration. The use of direct analogy helped as an idea starter by offering a point of association. It helped to steer from direct to

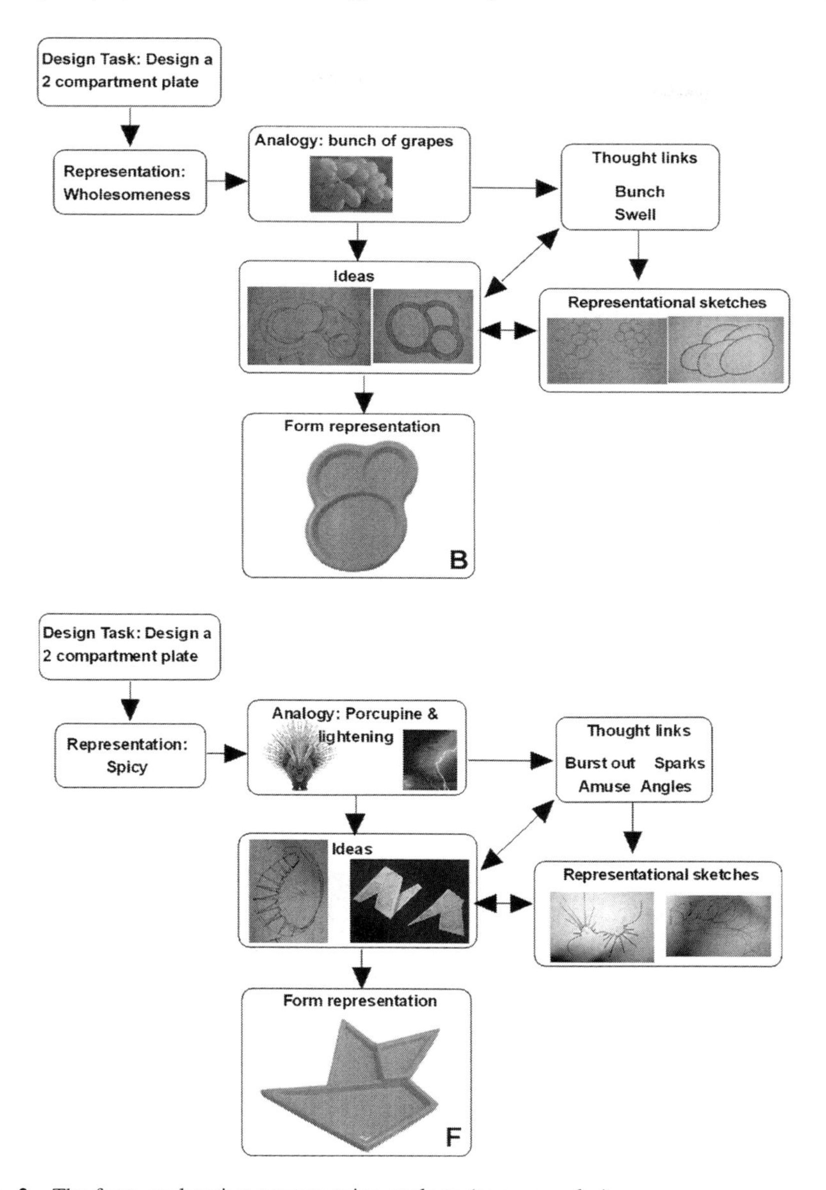

Fig. 2 The form exploration process using analogy (two examples)

abstract thoughts, leading to radical representation through attribute mapping and help as a prop to idea exploration. Once the students were able to break out of their conditioned thinking by choosing an analogy from nature for the expression they wanted to represent in the plate design they were able to freely explore radical forms, where the analogy worked as a catalyst to trigger multiple thoughts and help in developing novel designs.

Table 1 Form exploration with use of analogy to represent an expression

Expression	Analogy	Thought links	Visual representation in form exploration	Final Form
Amusement Plate	Honey drip Surf 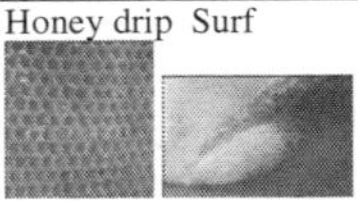	Pleasure Exuberance Lush Curling	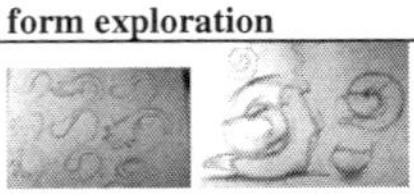	A 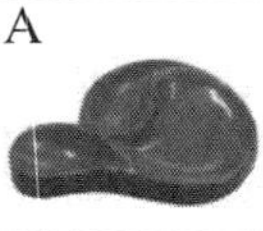
Appetite Plate	Grapes bunch 	Wholesome Bunch Swell	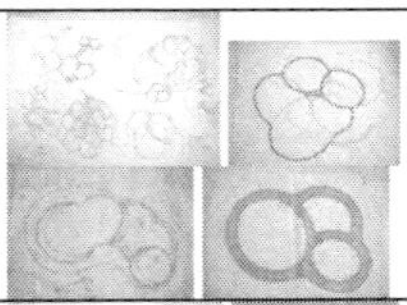	 B
Healthy Plate	Pumpkin Elephant 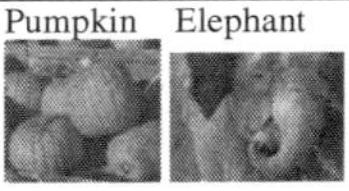	Bulge Volume Full whole	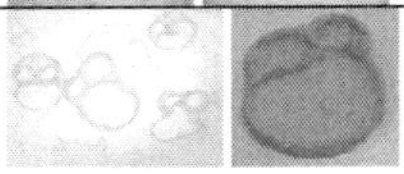	 C
Elegant plate	Pearl 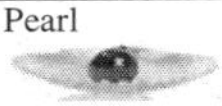 Snake	Exotic Style Attractive Smooth Flow Balanced	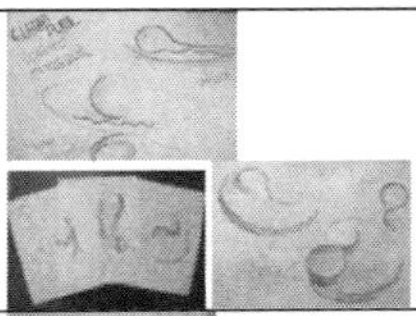	D 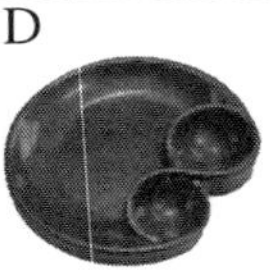
Happy plate	Flower & Bud 	Bloomed Bright Fresh	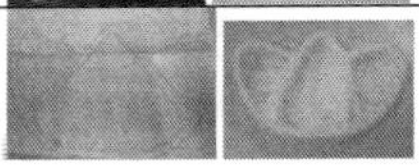	E 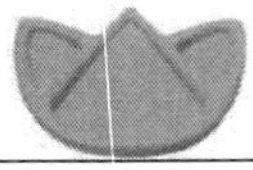
Spicy plate	porcupine Lightening 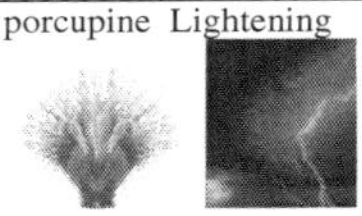	Amuse Excitement Burst Angles Open	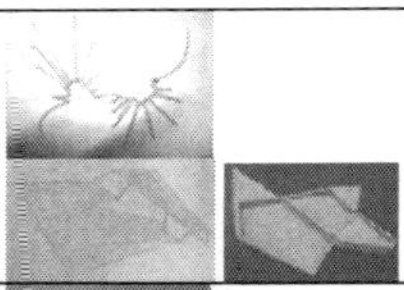	F 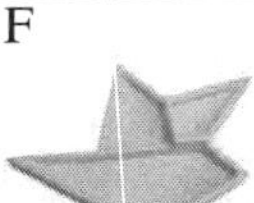
Aroma plate	Sprouting	Freshness Movement Upwards	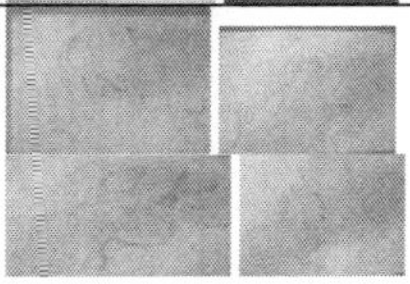	G 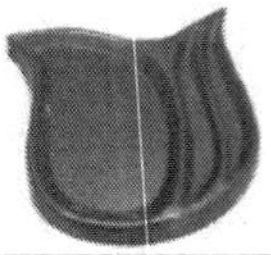

(continued)

Table 1 (continued)

Healthy plate	Apple Green leaves 	Bulgy Fullness Freshness Happy	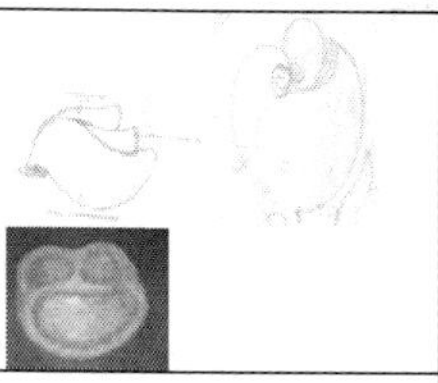	H 
Fresh plate	Lemon Water 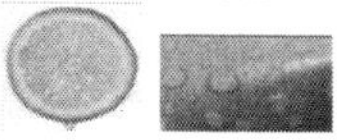	Radiating Lively Splash		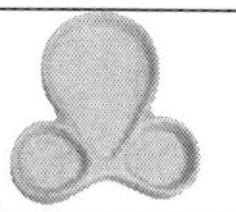 I
Secure plate	Bud Peanut 	Enveloping Overlapping Compact Freshness Blooming	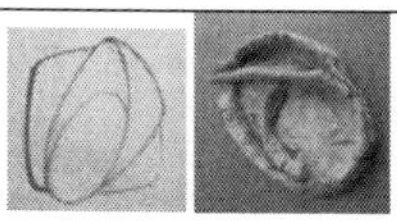	J 
Soft plate	Cloud Cotton 	Curvilinear Smooth light	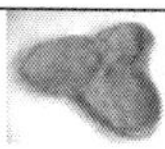	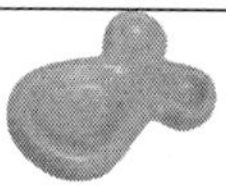 K
Poise plate	Swan  Agley 	Balance Slender Order Grace Style Elegant	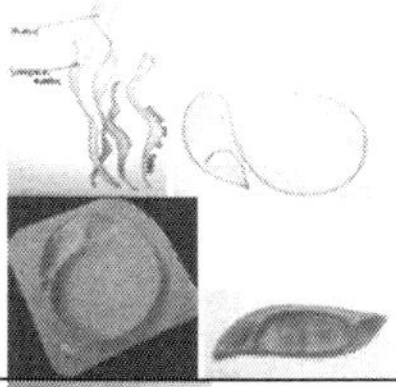	L 
Strong plate	Alligator Tortoise 	Power Heavy Hard	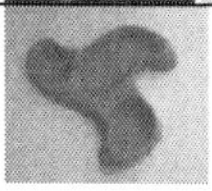	 M
Musical plate	Ripples 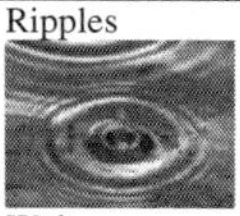 Web	Spread Resonance Repetition Harmony	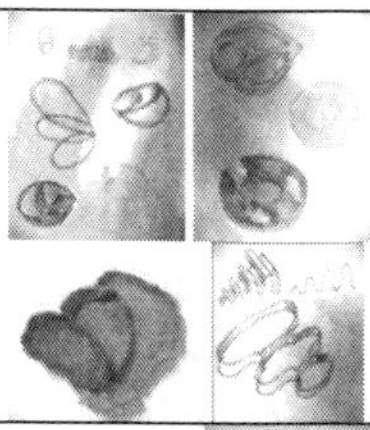	N 
Crisp plate	Lotus 	Organized Symmetry Precise	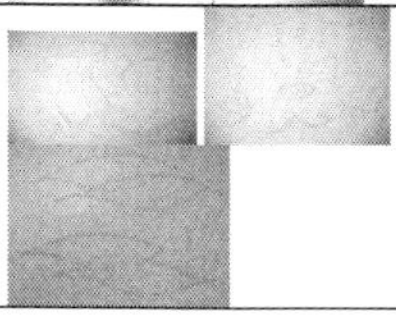	O 

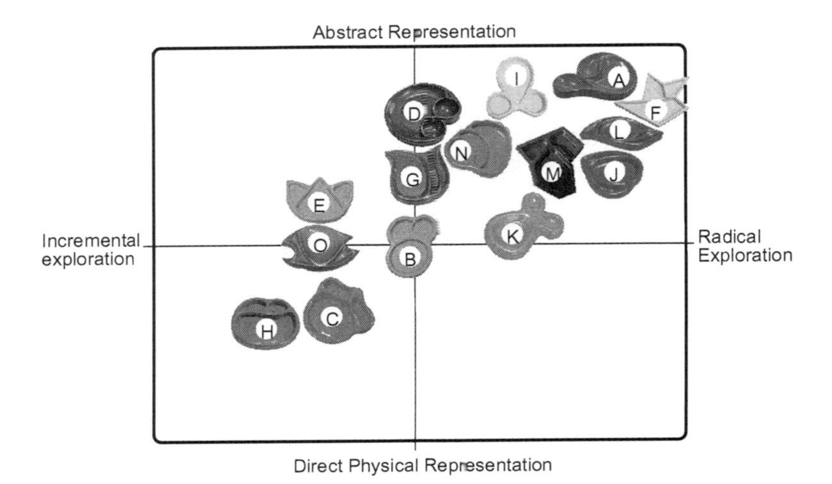

Fig. 3 Form exploration and levels of representation

Observing and mapping attributes from analogy allowed simple representation of abstract concepts as well as abstract representation of simple attributes and helped evolve unanticipated design solution and form generation. This can be seen specifically in forms A, F, L, J, G, M, I, N, D.

In a few cases the use of analogy were rather bounded to a physical representation only. In the case of forms H, O, E exploration remained in the realm singular physical attributes usage thus restricting next level of exploration. The contextual aspect of the use of analogy revealed various associative links to similar themes, allowing infinite possibilities of exploration. As seen in forms D and L where the theme is elegance but due to varied analogies the form exploration is diverse.

Use of analogy helped in integrating emotions to the physical qualities of the analogy to the form and bringing about a expressive representation. This helps to expand and extend the designers exploratory field. From the perspective of using analogy in form exploration it can be argued that the exploratory nature of analogies and its associative functions, significantly facilitates idea exploration and novel form generation. The use of multiple representations extracted from the analogies and its abstraction is seen to effect radical form generation in an encouraging way. Thus the assumption that radical form generation by novice designers can be enhanced with the use of analogy and can bring benefit to assignments in hand has been observed. By understanding how using analogy in the thinking process enhances form exploration, this study can help to identify the incremental to radical exploration of form and its relation to the abstraction of analogical attributes mapped.

The key role that analogies can play during the design exploration stage is in building sense through the process of representational transfer and facilitate the mind to think differently. Despite the apparent benefits that analogy provides for idea exploration, the possibility of fixation is an area that needs to be given considerable attention. Thus a structured guided pedagogical intervention can be useful. Further work in the area has to be done to formulate it.

4 Conclusion

This paper has articulated with examples, the role of analogy in idea exploration and in facilitating expressive form generation.

The fundamental findings of our inquiry are that:

1. The use of analogical representation in form generation is able to move the ideas to the next level. Design ideation can be stimulated with analogical inspirations to multiple directions, allowing a more versatile field of exploration of ideas and deliver novelty.
2. Using analogies to explore ideas and form generation helps students to break out of their trained perception of form generation and discover new expressions in form exploration.
3. The use of analogy as a training for associative thinking is useful for exploratory designing. As designers unconsciously source ideas analogously, but by understanding its deeper implication in the design representation can help them to enhance their practice of designing. This gives scope to further this study to way all the implications and develop a framework for creativity, and validate new methodology through more robust experiments. In the training process it is important to convey that the analogue is not meant to be a copy of all the features of the target [15], rather its value lies in the abstraction of the representation.

In today's competitive market the ability of analogies to produce novel forms by aiding idea exploration during the design process, can be considered as an important tool. This study focuses on use of analogical stimuli to evoke radical form generation, and shows that more abstract themes can be translated to novel form representation, through the use of multiple analogies. Thus reflecting a scope for further study to develop a structured pedagogical intervention for training idea exploration and form realisation from abstract themes and arrive at novel products.

Further, this practice can facilitate design at multiple levels of formulation, representation and implementation of ideas, leading to the form expression. Thus further examination of the specific types of analogy usage [6] and its specific implementation may help progress the use of analogy to show radical effect on original design outputs of novice designers.

Acknowledgments I'm grateful to my guide Prof. B K Chakravarthy for his guidance and for allowing me to conduct my experiment within the module, Studies in Form at IDC. The cases cited here are assignments of first semester MDes. students of the 2010 batch. I'm grateful to the students for sharing their work and cooperating with my data collection.

References

1. Boden MA (1994) Dimensions of creativity. MIT Press, Cambridge
2. Koestler A (1964) The art of creation. Hutchinson, London
3. Csikszentmihalyi M (1992) Flow: the psychology of happiness. Rider, London
4. Holyoak KJ, Thagard P (1995) Mental leaps: analogy in creative thought. MIT Press, Cambridge
5. Christoff K, Keramation K (2006) Abstraction of mental representations: theoretical considerations and neuroscientific evidence. Oxford University Press, Oxford
6. Gordon WJJ (1961) Synectics: the development of creative capacity, Harper & Row, Publishers, Incorporated, New York
7. Hey J, Linsey J, Agogino A, Wood K (2007) Analogies and metaphors in creative design. Proceedings of the mudd design workshop VI, Claremont
8. De Bono E (1970) Lateral thinking. Penguin Books, Harmondsworth
9. Gero JS (1992) Creativity, emergence and evolution in design In: Gero JS, Sudweeks F (eds) Preprints computational models of creative design, department of architectural and design science, University of Sydney, pp. 1–28
10. Csikszentmihalyi M (1997) Creativity: flow and the psychology of discovery and invention. Happer Perennial, New York
11. Glynn S (1991) Exploring science concepts: a teaching with—analogies model. The psychology of learning science. Erlbaum, New Jersy
12. Finke RA, Ward TB, Smith SM (1992) Creative cognition: theory, research and application. MIT Press, Cambridge
13. Howard-Jones PA, Murray S (2003) Ideation productivity, focus of attention, and context. Creativity J 15(2–3):153–166
14. Cross N (2006) Designerly ways of knowing. Springer, London
15. James M, Scharmann L (2007) Using analogies to improve the teaching performance of preservice teachers. J Res Sci Teaching, 44 (4):565–585

Supporting the Decision Process of Engineering Changes Through the Computational Process Synthesis

Florian Behncke, Stefan Mauler, Udo Lindemann, Sama Mbang, Manuel Holstein and Hansjörg Kalmbach

Abstract Engineering changes (ECs) are considered as cost and time-consuming. Based on the understanding of EC as a specific representative of cycles within development processes, the implementation of ECs is initiated by a target deviation, which leads to a decision over different alternatives of the implementation process of the particular EC. This decision is based according to literature and industrial practices within the field of process design methods on an expert discussion, without a formalized and explicit consideration of different change options. This paper presents computational process synthesis (CPS) as a support for the decision making of change options. The CPS is embedded within a procedural model for the decision process of ECs. Besides, this paper presents a case-study, which describes the application of the presented procedural model as well as the CPS.

Keywords Cycles · Changes · Process design · Design synthesis

F. Behncke (✉) · S. Mauler · U. Lindemann
Institute of Product Development München, Technische Universität München, Munich, Germany
e-mail: behncke@pe.mw.tum.de

S. Mauler
e-mail: stefan.mauler@mytum.de

U. Lindemann
e-mail: lindemann@pe.mw.tum.de

S. Mbang · M. Holstein · H. Kalmbach
Daimler AG, Sindelfingen, Germany
e-mail: sama.mbang@daimler.com

M. Holstein
e-mail: manuel.holstein@daimler.com

H. Kalmbach
e-mail: hansjoerg.kalmbach@daimler.com

A. Chakrabarti and R. V. Prakash (eds.), *ICoRD'13*, Lecture Notes in Mechanical Engineering, DOI: 10.1007/978-81-322-1050-4_15, © Springer India 2013

1 Introduction

1.1 Motivation

The development towards a worldwide buyer's market driven by globalization, leads to an intensified individualization of industrial goods [1, 2]. The results are a growing number of products and especially a growing variety of these products [3]. As a consequence development efforts combined with higher expectations on products increase, while market-induced development periods are reduced at the same time [2]. So cycles of development and innovation need to abbreviate at increasing frequency to keep the company competitive [4]. Therefore, it is inevitable to adapt a company's processes to these new requirements. Engineering changes and their implementation are a vital sign of this adaption. Reorganizing processes is very expensive and time-consuming. The outcome is a static process-map, which is unable to react on unplanned changes [3]. The effects of engineering changes on the process flow are difficult to evaluate. In particular external cycles, induced by changing customer preferences, technological or economic developments, are barely predictable [5]. Kleedörfer [6] suggests a better crosslinking between the sub-processes, which can be transferred to the implementation of engineering changes. Based on this, Heinzl et al. [7] developed a concept of defined process building blocks with standardized interfaces, in order to design a flexible development process.

At Daimler it was the objective to create a process which reacts flexibly on unpredicted engineering changes and can be adapted to new requirements. With "Computational Process Synthesis" (CPS) an approach was developed to answer the central research question of this paper: *How can computational process synthesis support the decision process of engineering changes*?

This paper describes the fundamentals of decision making (see Sect. 3.1) and of process design (see Sect. 3.2). Section 4 shows the development of the concept of CPS and Sect. 5 its implementation at the Daimler AG.

1.2 Background of Research

Complex products are characterized and influenced by a number of company-internal and external cycles. These cycles are objects of investigation of the Collaborative Research Centre (SFB 768)—'Managing cycles in innovation processes—Integrated development of product-service systems based on technical products'. Cycles are reoccurring patterns (temporal and structural), which are classified by phases. As a result, a cycle is always connected with repetition, phases, duration, triggers and effects. Moreover, cycles could include retroactive effects, interlockings, interdependencies (within cycles and between cycles), hierarchies and further influencing aspects [5, 8]. Engineering changes are, due to their reoccurring process, a specific occurrence of cycles in development, which are focused within the SFB 768 by a subproject on the cycle oriented planning and coordination of development processes. Triggers cause the repetition of a pattern and result of a deviation of the current and the intended status of an object. In order to resolve this deviation, changes are required. However, the characteristics of change options vary in terms of their effects, which emphasize the relevance of a systematic creation and decision on change options.

2 Research Methodology

The development of a procedure for the computational process synthesis is based on established literature of process design and decision making in complex systems considering the specific requirements of engineering changes. Through applied science the procedural model is refined iteratively and further developed by using real process examples of the tool-making department in the Mercedes-Benz factory Sindelfingen. The procedural model and its results in terms of evaluated change options (alternatives) were evaluated by expert interviews at Daimler AG and are in place for a validation within succeeding student project.

3 State of Technology

3.1 Decision Making

Roy [9] describes three basic kinds of decisions: decision by selection, decision by ranking and decision by classification and sortation. Decision by selection means to choose a group or a single alternative from all possible alternatives of action, while decision by ranking means to arrange the alternatives in sequence, in order to determine the ability for certain demands. Decision by classification and sortation corresponds to a common preselection. Decisions get more challenging when more objectives need to be matched with that decision. Still more effort is necessary if not all alternatives are known to the decision maker. In that case we talk about a multi-criteria decision problem, with an additional identification problem. These kinds of complex decision processes need to be supported methodically and conceptually [10].

3.2 Process Design

Process Design is a complex problem. Established procedures suggest a discussion-based process-mapping performed by project groups of experienced employees from the relevant process areas [11].

Together they document the current process of a change option using modeling methods like sign-posting. The group develops an ideal process as long-term objective, which the group's members expect to be perfect in the work-flow as well as in its results. Based on these two processes the target-process is mapped. It is a compromise between the perfect solutions of the ideal process and those solutions that are realizable under real conditions. They discuss the workflow step-by-step and determine the target-process with a sequence of decisions by selection. The number of alternatives is based on the solutions of the current process, the ideal process and the know-how of all involved experts, as well as their creativity. The criteria used for decisions aim mostly at local optimization, because the impact on the complete process cannot be evaluated with the given methods.

Methods like 'Program Evaluation Research Task' (PERT) or Critical Path Method (CPM) provide a procedure to derive an optimal process sequence out of given tasks, mainly focusing on the lead time. The process sequence is event-driven and considers uncertainties of single tasks [12]. However, multiple criteria are not used for the derivation of the process sequence. They are mandatory to consider optional tasks, which for example are performed to increase the quality of the output of specific tasks.

4 Computational Process Synthesis

CPS offers methodical support for the identification of change options as well as for the multi-criteria decision on the process-mapping of those change options. Conventional approaches use a sequence of local decisions for the selection, while CPS uses integrated decisions by ranking complete process-paths from the solution space. CPS is composed of three steps (Fig. 1). First, required input-data for the simulation is collected. All sub-processes with their inputs and outputs are determined and formalized. Afterwards, an excel-based tool spans the solution space, generating all possible process-paths through combinatorics. The third step is a two-stage assessment to choose the optimal solution from the solution space. Using defined K.O.-criteria, the number of solutions is reduced to a manageable amount. With a customized set of criteria the optimal process-path for the specific situation is determined.

4.1 Process-Mapping of Change Options

Objective is a process-map, which includes all process-paths for the implementation of the different change options. Like in the conventional approach a project group is established consisting of experts from all relevant areas [11]. The discussion is focused on the inputs of the sub-processes, as their performance and the quality of their results depend significantly on the quality of the process-inputs [13].

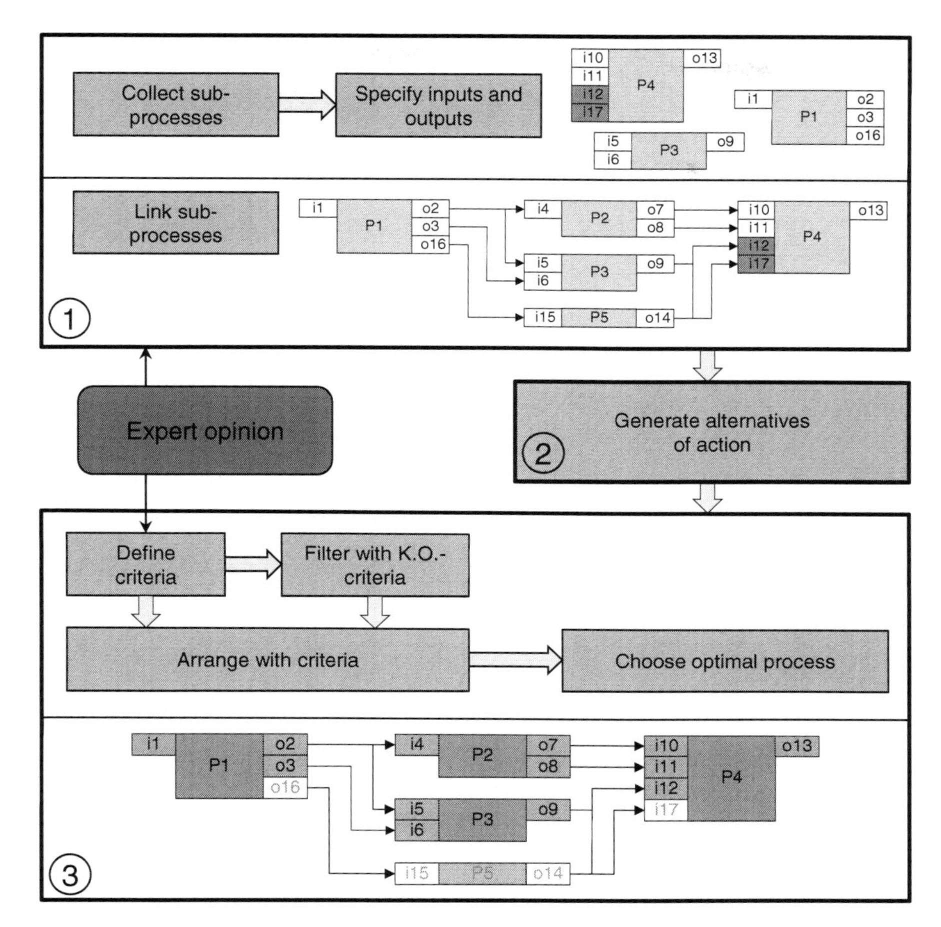

Fig. 1 Procedural model of CPS

The specification of the outputs results from linking inputs to their corresponding providers. This procedure avoids a deduction of unnecessary work results. Solely the processing time, needed from process-start to completion, is allocated to the outputs.

First all sub-processes are analyzed, according to their inputs, used in the current process. Here three kinds of inputs are distinguished: fixed, optional and alternative. Fixed inputs are those that are necessary to perform the sub-process at all, optional inputs complement the fixed ones, while alternative inputs replace other alternative parameters. Each input can be provided by several different suppliers.

In addition, new options and alternatives for the target-process are developed. This step holds high potential for innovation, however, it is very elaborate. Customer orientation is of high importance to prevent a misguided development [14]. When customers suggest additional input to raise the quality of their sub-process or its results, providers need to evaluate, whether or under which conditions they can deliver the demanded input. It is crucial, that the demand-pull on the customer side coincide with the technology push on the provider side [15]. The provider has to deliver what the customer needs, and vice versa. The previously documented sub-processes get expanded with this additional data. Using the defined links between inputs and outputs a process-map is drafted.

4.2 Spanning the Solution Space

This step determines the possible process-paths of change options by an automated excel-based tool. The characteristics (see Sect. 4.1) are available for the program in a formalized state. For sub-process holding *optional* or *alternative* inputs, the tool determines all variations of the sub-processes. Thereby a group of *optional* inputs can be interpreted as a binary number, where every input represents a digit that is either used '1' or it is not '0'. The number of variations V with n *optional* inputs in one sub-process is calculated after the rules of combinatorics, as shown in Eq. 1:

$$V = 2^n \tag{1}$$

A group of m alternative inputs expands the number of variations with a multiplication by the factor m as shown in Eq. 2:

$$V = 2^n \times m \tag{2}$$

Afterwards the path generation is performed, following the pull-principle just like in Sect. 4.1. Because of the focus on the inputs and the generation of variations using them, this approach is necessary to achieve an explicit set of relevant paths. The tool proceeds backwards, it starts with the last sub-process. Each sub-process requests those parameters from its predecessors, which are needed in its current variation. If a sub-process holds more than one variation, new paths are generated. The tool ends, as soon as all paths are determined. If a new path is applied, the attribute *open* is assigned. The path is attributed as *closed*, when it reaches the first (initial) sub-process and all its sub-processes are *closed*. The sub-processes on the other hand count as *closed,* as soon as all their inputs are linked to a predecessor. The final result is the whole solution space. It holds the entity of all paths through the process-map, which are able to achieve the requested outcome.

4.3 Two-Stage Assessment

The assessment occurs in two steps. It is based on substantial expert knowledge about the workflow and the individual operation steps. First the possible solutions are narrowed down using K.O.-criteria. In the second step the remaining paths are evaluated with a custom-designed set of criteria in a quantitative way. The set of criteria consists of quantified parameters, which allow making a statement about the capabilities of the process and the quality of its products. All criteria are derived from general requirements for the process. They indicate the degree of fulfillment of these requirements, which originate from the process owners themselves, from the management or from literature.

K.O.-criteria reduce the solution space to a manageable amount of relevant process-paths, either using a critical value or with Boolean statements about certain process properties. The second stage of assessment works with various criteria, set in weighted relation to each other. All criteria have to be quantified, to allow a computer-aided assessment. One possibility to use qualitative criteria as well, is score evaluation, or weighted score evaluation for higher significance.

For dynamic and flexible process design, critical values as well as the weighting of the criteria is to be defined specifically for each situation. As a result, the optimized process-paths for each situation can be found and implemented.

5 Case-Study

The tool-making department in the Mercedes-Benz factory Sindelfingen develops, constructs and produces dies and tools for deep-drawing operations of sheet metal parts for Mercedes-Benz cars. The increasing number of models and their variations, in addition to the market-induced shortening of development cycles, leads to severe deadline pressure and a shortage of resources. Out of this motivation an in-house-consulting project (IHC) was launched, performing a classical process optimization via sign-posting within a project group of experts. The following approach was developed in parallel to this project, to support the established approach. It was implemented in a pilot study, with which the specific actions are described in the following.

5.1 Process-Mapping of Change Options

The process-mapping took place within the project meetings of the IHC-project. The different variations of input compositions and alternative providers for the inputs resulted from discussions about the current, ideal and possible target-process. Figure 2 shows an example of an anonymized sub-process.

Inputs i1 and i2 are fixed parameters, which are required to perform the sub-process. i3 was provided sporadically in the past, to improve the quality. However, it is an *optional* input. i4 is requested by the process-owner, intending to improve his results. Another sub-process is identified as possible provider, whose owner agreed to provide the requested parameter under specific conditions. So i4 is charted and marked as *optional*. i5 and i6 serve the same purpose. Nevertheless, they show different characteristics. One of the two parameters is needed to perform

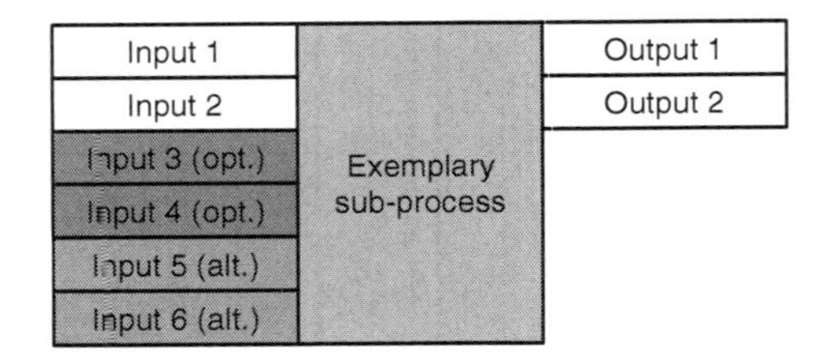

Fig. 2 Exemplary sub-process with inputs and outputs

the sub-process. That's why both are marked as *alternative*. Outputs o1 and o2 are requested by other sub-processes. The process-owner agreed to this request and defined processing times for both parameters. All sub-processes are connected, by linking their inputs and outputs. In the case-study 25 sub-processes are defined, holding 113 inputs and outputs. To keep clarity in the following a simplified exemplary process-map is used to describe further actions. Figure 3 shows this process-map. It consists of five sub-processes, with a total of 17 inputs and outputs. Two of them, i12 and i17 are *optional*.

The linking is conducted from input to output. Customers choose their providers. Every input needs at least one predecessor—except i1, which is an initial input—otherwise the process would stop at this point. Several predecessors for every input in the map are possible, if the customer considers more providers to be suitable. That way i12 for example can be provided over o9 by sub-process P3 or over o14 by P5. If one output has more than one successor, e.g. o2, which delivers to i4 and i5, it must be ensured, that the required goods are available in a sufficient amount. Such constellations in the case-study turned out to be unproblematic as most parameters were digital data like CAD-construction data, which are available.

5.2 Generate Alternatives of Action

The inputs and outputs' properties are formalized, documented in an excel-list and provided to the tool. Based on this, the tool derives all possible process-paths. In the exemplary process-map (Fig. 3) only P4 holds more than one variation (Fig. 4).

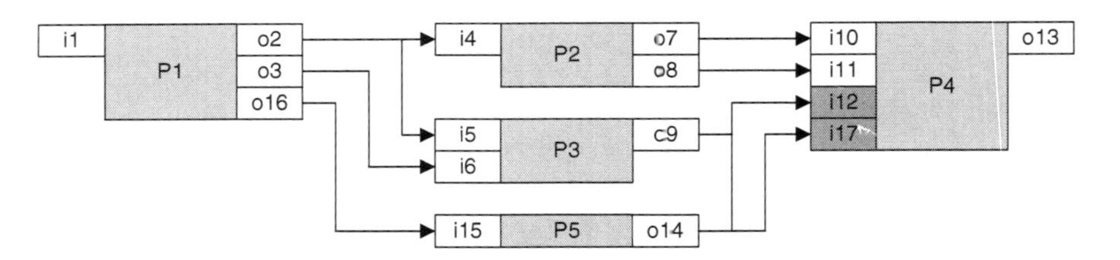

Fig. 3 Exemplary process-map

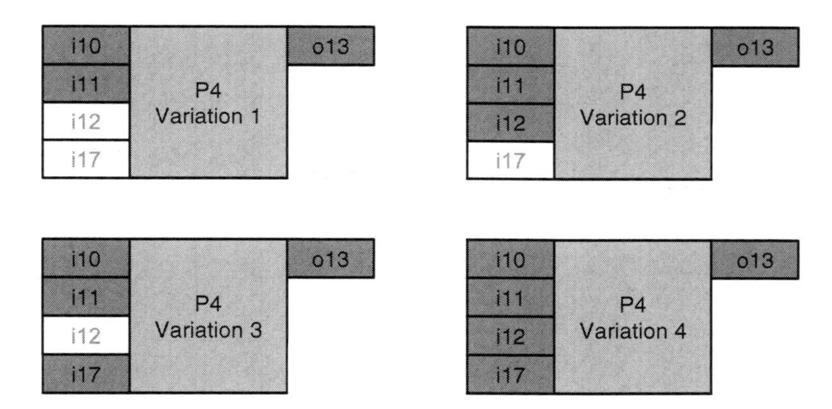

Fig. 4 Variations of sub-process P4

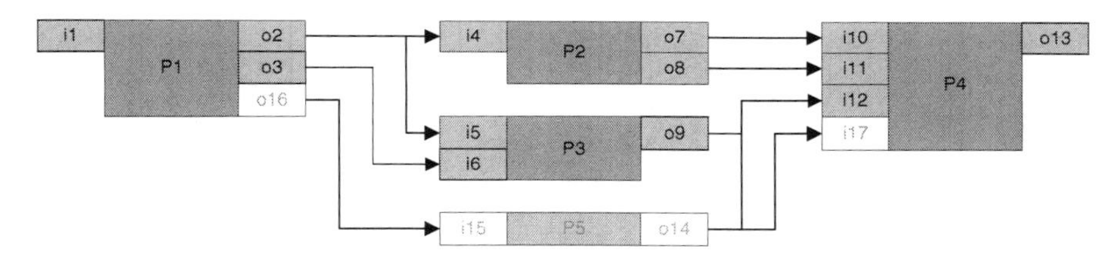

Fig. 5 Path no. 2 of the exemplary process-map

Following the earlier defined rules, two optional inputs cause an amount of $2^2 = 4$ different variations. The exemplary sub-process in Fig. 2 would double the amount because of the two additional alternative inputs to a total of $2^2 \times 2 = 8$. The path generation starts with o13, meaning at P4. Four paths originate here, caused by the four described variations of P4. As i12 holds two predecessors, every path containing i12 generates another path to display the additional alternative. This way for variation 2 and 4 there is one path providing i12 through o9 and one using o14.

Altogether there are six possible paths how to get through the exemplary process-map. Figure 5 shows path no. 2 of the exemplary process-map as a sample of the solution space. It contains P4's variation 2 and uses P3 to supply i12.

5.3 Two-Stage Assessment

The case-study's real process-map caused a total amount of 19,584 possible paths through the map. To handle this amount, the solution space was narrowed down using the two-stage assessment, described earlier.

Table 1 shows a selection of requirements for the process and the derived criteria for the process-assessment. As mentioned earlier the product-push in the

Table 1 Requirements and criteria for the assessment

Requirement	Criteria
Reduce processing time	Processing time (K.O.) T_p
Reduce processing cost	Operational hours (a.o.) T_o
Parallelize operations	Parallelism: T_o/T_p
…	…

automobile industry creates increasing deadline pressure. Therefore, one requirement was to reduce the entire processing time, to enable the workers to meet their deadlines. If deadlines cannot be kept, the process-path is not suitable. This shows, that processing time is a perfect K.O.-criterion. Process-paths that need less time than the critical value for this criterion are kept as possible solutions. The critical value in the case-study was set to 200[1] days, which lead to a reduction of the solution space to 2176[2] process-paths. However, the selection of the criterion and its critical value depends on the specific project.

All further criteria are assessment-criteria serving a decision by ranking, which have to be interpreted and weighted individually for every project by a group of experts. A basic requirement for all business processes is cost reduction. Among others, this is assessed with the operational hours for every cycle, which are equivalent to the labor costs of the process. So less operational hours are evidence for less labor cost or at least less tied-up working capacity. Simultaneous engineering claims parallelized operations. This can be assessed with the quotient of operational hours and processing time. This ratio, called parallelism, makes a statement about how much work is achieved during the processing time of the process. So a high ratio indicates a process, which performs several sub-processes in a short period of time. The connection between the first two criteria in the third one causes a conflict of objectives, which allows no process to achieve optimal values for every criterion. Weighting the criteria preforms a prioritization of the requirements according to their importance.

6 Results

Figure 5 demonstrates that all process-paths containing P5 are sorted out due to the K.O.-criterion. The processing time exceeded the critical value, because of the high processing time of P5. As o9 causes more operational hours than o7 and o8, P3 extends the complete processing time as well. However, its involvement improves the quality of the outcomes and the parallelism of the processes. As best

[1] Numbers were changed for this publication.

[2] Numbers were changed for this publication.

compromise between processing time and practicable work content, the shown second path is chosen from the solution space.

In the case-study the conventionally elaborated process was identified within the solution space and compared to other reasonable alternatives. As recommendation one path was determined, that has a shorter processing time and more operational hours at once. The parallelism rises from 1.36 to 1.62. This way the principles of simultaneous engineering are implemented consequently. Work content along the time-critical path is reduced as far as necessary, while parallel to this path validation content is performed, in order to use the available time effectively and to reduce critical changes during the stages of implemented hardware. Aspects like the time-critical path could not be considered, using the conventional approach, until the process planning was complete. Iterations for optimization are very elaborate. Using the approach of CPS, the results of the process organization can be improved, while the effort is reduced at the same time.

7 Summary and Outlook

This paper presents the CPS as procedure to support the decision making of engineering changes. Based on the fundamentals of decision making and process design, the procedure depicts the relevant sub-processes and their inputs and outputs for a specific change through structured interviews. The combinatorics of the sub-processes, using an excel-based tool, derives the different change options, which are fed to a two-step assessment method. First the change options are preselected by K.O.-criteria, before these options are assessed through a weighted evaluation method. Finally, the research results were evaluated within a case-study at the tool-making department in the Mercedes-Benz factory Sindelfingen and interviews with corresponding experts.

A first step of future work is the extension of the excel-based tool for the creation of change options, in order to improve its capabilities. Thereby, the processing speed and the graphic representation of the change options are focused. Moreover, checklists are to be prepared to support the user, while identifying K.O.-criteria as well as evaluation criteria. Based on the evaluation of the procedural model in a case-study at the tool-making department in the Mercedes-Benz factory Sindelfingen, we aim at a validation of the approach within a student project.

Acknowledgments We thank the German Research Foundation (Deutsche Forschungsgemeinschaft—DFG) for funding this project as part of the collaborative research centre "Sonderforschungsbereich 768—Managing cycles in innovation processes—Integrated development of product-service-systems based on technical products". Furthermore our thanks are given to Daimler AG, especially to the tool-making department BM of the Mercedes Benz factory Sindelfingen and all employees who supported this study.

References

1. Porter ME (1998) Competitive strategy: techniques for analyzing industries and competitors. The Free Press, New York
2. Cooper RG, Edgett SJ (2005) Lean, rapid, and profitable new product development. Basic Books, New York
3. Ehrlenspiel K (2009) Integrierte Produktentwicklung. Hanser, München
4. Griffin A (1993) Metrics for measuring product development cycle time. J Prod Innov Manag 10:112–125
5. Langer S, Lindemann U (2009) Managing cycles in development processes—analysis and classification of external factors. In: International conference on engineering design (ICED '09), vol 17. pp 539–550
6. Kleedörfer R (1998) Prozeß- und Änderungsmanagement der Integrierten Produktentwicklung. Dissertation, TU München
7. Heinzl J, Reichender J, Schiegg H (2001) Prozeß- und Änderungsmanagement der Integrierten Produktentwicklung. In: PMaktuell 2/2001
8. Langer SF, Knoblinger C, Lindemann U (2010) Analysis of dynamic changes and iterations in the development process of an electrically powered go-kart. In: International design conference (design'10), vol 11
9. Roy B (1996) Multicriteria methodology for decision aiding. Kluwer Academic Publishers, Dordrecht
10. Krishnan V, Ulrich KT (2001) product development decisions: a review of the literature. Manag Sci 47(1):1–21
11. Becker J (2005) Prozessmanagement. Springer, Berlin
12. Malcolm DG, Roseboom JH, Clark CE, Fazar W (1959) application of a technique for research and development program evaluation. Oper Res 7(5):646–669
13. Schöneberg U (2010) Prozessexcellence im HR-management. Springer, Heidelberg
14. Lukas BA, Ferrell OC (2000) The effect of market orientation on product innovation. J Acad Mark Sci 28(2):239–247
15. Hauschild J (2011) Innovationsmanagement. Vahle, München

Concept Generation Through Morphological and Options Matrices

Dani George, Rahul Renu and Gregory Mocko

Abstract The use of morphological analysis as a tool to aid concept generation is examined. Two principal limitations of the method are highlighted; (1) the lack of details generated for system concepts and (2) the explosion of combinatorial possibilities in the use of morphological matrices. The authors propose a method to support the generation of detailed conceptual ideas through functional combinations and use of options matrices, facilitating an intelligent exploration of the design space. In the options matrices, functions that are highly coupled are grouped together and idea generation is performed on the functional combinations based on identified innovation challenges. A subset of highly coupled functions are extracted from the morphological matrices and systematically integrated to form system level concepts. The resulting system concepts have greater design details compared to those generated through traditional morphological analysis techniques, allowing a designer to make informed decisions regarding their feasibility for the design purpose. An example of the proposed method is provided in the design of a seating chassis for automotive applications.

Keywords Concept generation · Morphological matrix · Options matrix · Innovation challenges · Functional coupling

D. George · R. Renu · G. Mocko (✉)
Department of Mechanical Engineering, Clemson University, Clemson, South Carolina, U.S.A
e-mail: gmocko@clemson.edu

D. George
e-mail: dgeorge@g.clemson.edu

R. Renu
e-mail: rrenu@g.clemson.edu

A. Chakrabarti and R. V. Prakash (eds.), *ICoRD'13*, Lecture Notes in Mechanical Engineering, DOI: 10.1007/978-81-322-1050-4_16, © Springer India 2013

1 Introduction

Research in engineering design over the past few decades has resulted in the formulation of various theories, perspectives, models and methodologies for performing design activities [1–5]. Common to the various models and theories of design is the importance of creativity and the need for generation of good product concepts in order to satisfy customer needs, reflect their preferences and generate revenue. It is well understood that a poor design concept resulting from the product conceptualization process cannot be compensated for by 'bandaging' or quality of manufacturing [6]. Hence, it is paramount that designers focus time and effort in generating ideas, evaluating alternatives and performing concept selection appropriately.

Idea generation during product conceptualization can be roughly grouped into: methods to support the generation of ideas (means) to perform the individual functional requirements of the system and methods to support the combination of means to generate system level concepts [1, 2]. Several tools have been developed to support these idea generation activities [7]. The focus of our research is on the combinatorial aspects of concept generation.

Different combinatorial tools used to support concept generation can be classified as intuitive tools (such as storyboarding and affinity method) and systematic combination tools (such as check-listing and action-verbs) [7]. Morphological matrices may be used as a systematic combination or an intuitive combination tool. The concept of morphological thinking is essentially the method of systematic combination used to explore the complete set of possible relationships within any multi-dimensioned problem that can be decomposed into its' constituent sub-problems [8]. However, it also supports an intuitive approach to combination of means. A detailed review of the development and use of the morphological matrices is presented in [9].

Since the use of the morphological matrices is in the conceptual design stage, it typically contains means that are not detailed—the means may be high level working principles, non-dimensioned sketches, or vague ideas. Consequently, during combination of means into system concepts, it becomes difficult to identify which means are compatible with others to support their physical combinations. Therefore, simply choosing one means for each functional requirement may not yield a system concept if the means cannot physically be integrated into a working mechanism. Hence, the first limitation of the use of morphological matrices is the challenge of identifying compatible means to perform a system level combination.

Three techniques can be used to combine means into system concepts in a morphological matrix—a computational/quantitative technique, a qualitative technique and a fusion of the two. The computational technique is essentially based on calculating the estimated performance of the system that is generated as a result of combination of the means [6, 10–14]. The main challenge with this is that it requires the definition of performance parameters and a significant amount of design detail for each of the means, requiring a lot of time and effort from the designers, although the computations themselves may be automated in some cases.

The qualitative techniques [12, 15] and the mix of qualitative and computational technique [16] can be employed to generate concepts using three approaches—a systematic combination of all possible means (a full factorial approach), random combinations of means, and intelligent combinations of means.

The systematic combination approach systematically identifies all possible combinations, thus allowing the designer to choose the optimum system concept from the entire set. However, the major limitation of this approach is the number of combinatorial possibilities that must be explored. A small design task that is decomposed into 5 functions with 10 means identified to fulfill each function has 10^5 (100,000) combinatorial possibilities. In practise, may not be possible to explore all these combinations to identify the optimum system concept. The combinatorial possibilities become even larger for large complex design problems. The demonstrative example described in Sect. 3 had over 600,000 combinatorial possibilities despite having only four functions.

The approach of random combinations identifies one means from each row randomly to generate system concepts. This approach can result in the combination of unexpected means that force the designers to think deep into how the combinations can be achieved, thus leading to innovative system concepts. However, the challenge is that the randomness can also result in the exclusion of potentially good system concepts. Although the randomness can be biased toward good quality means to yield improved results, the adaptation may yet fail to identify the most complimentary sets of means to generate good system concepts.

The intelligent combination approach conceptually lies in between the previous two approaches. Some of methods using this approach are described in [12, 15, 16]. However, these methods still propose high level combinations of the means within the morphological matrices, without the generation of significant details to allow a designer to make informed decisions to explore the design space effectively.

In light of these limitations, this paper proposes a method of qualitative exploration of the design space with morphological matrices using a focused ideation technique based on identification of functional coupling and innovation challenges. The method proposes a strategic exploration of the design space and combinatorial possibilities are explored based on informed decisions that are influenced by the designers' domain knowledge, experience and technological challenges. The next sections explain the proposed method, demonstrate an application to design a seating chassis for an automotive application, outline our conclusions, discuss the limitations of the method and identify future work.

2 Proposed Method for Concept Generation

The proposed method for concept generation aims to reduce the number of combinations that need to be explored to identify good system concepts, generate detailed information on promising combinations of means, encourage the exploration of design alternatives, and facilitate innovation in design. As a prerequisite, the functional decomposition of the design task and the generation of individual means that achieve each function should be performed and listed in a morphological matrix. A summary of the proposed method is illustrated in Fig. 1.

Figure 1 illustrates the high level steps required to perform a focused detailed exploration of the design space. The means in the initial morphological matrix are grouped and filtered (Step 1) before identifying functional combinations to perform focused ideation using options matrices (Step 2). The crosses in the initial morphological matrix represent the filtered means. The specific combinations explored in the options matrices (Step 3) are carried forward to subsequent higher level morphological matrices that list the functional modules against the sub-system concepts. The sub-system concepts are subsequently combined again using pairwise combinations using options matrices and higher level morphological matrices to generate system concepts (Step 4). The process is repeated to generate alternate system concepts (Step 5—not shown in Fig. 1).

Options matrices are proposed for several reasons including: (1) not all functions are identified in the initial morphological matrix, (2) some sub-functions can only be identified when combining specific means and thus are not relevant to all the means in the morphological matrix, (3) there may be several different geometric and physical combinations of means within the morphological matrix, and (4) the combination of means in the morphological matrix are explored at a high level of detail.

Step 1: Grouping and preliminary filtering of the individual means

The individual means from the morphological matrix are organized and grouped according to their similarities based on similar working principles, similarity of components, or other explicit design specific criteria such as strength, reliability or complexity. Clustering the means into specific groups help to identify strategies for combining the means across functions and stimulate generation of additional means. Affinity diagramming is a useful technique that can be used to perform grouping of the means. The grouped means are then analyzed individually

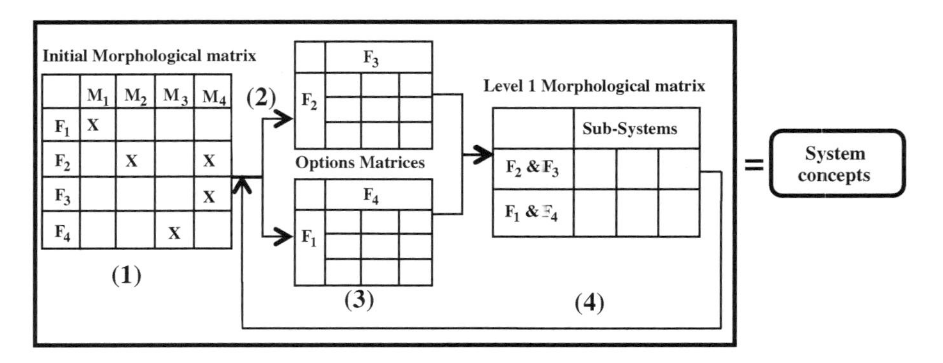

Fig. 1 System concept generation using morphological matrices and options matrices

and as a group with respect to feasibility to flag obviously infeasible ideas. The preliminary filtering helps to reduce the number of potential combinations to consider with minimal danger of losing feasible ideas. The result of this step is the generation of a morphological matrix with a reduced set of means that are grouped to identify similar working principles and design strategies.

Step 2: *Identification of functional combinations and innovative means for each function*

Subsets of the functions from the morphological matrix are extracted along with the means pertaining to those functions from the morphological matrix. Each subset consists of a pair of coupled/compatible/anti-functions that can be combined to form desirable or innovative sub-systems. These functions may be related to each other temporally, geometrically, or logically. Theory of Inventive Problem Solving (TRIZ) principles may also be used to identify function combinations representative of high level conflicts to identify the subsets [17]. The selection of functional combinations to form subsets may be subjective, i.e., designer specific, or design task or design focus specific.

Different functional combinations may be possible and desirable. However, an initial set of function subsets is chosen to specifically explore the possible ways in which the respective means can be combined, so that the functional coupling desired within each subset is physically realized. Different combinations of functions may be explored later during generation of alternative system concepts.

Innovative or promising means pertaining to each function are also chosen to carry forward for detailed exploration. The functional integration within each subset is explored through these means. The remaining means are not considered for further detailing at this point, although they are explored to generate alternative system concepts. The result of this step is the generation of functional combinations pertaining to possible design strategies and the identification of innovative/interesting ideas.

Step 3: *Generation of sub-system concepts*

The functional subsets and the promising means are exported to an options matrix, where focused idea generation based on explicit innovation challenges is performed on each subset, to understand how the various combinations of means can physically realize the functional integration. The innovation challenges are stimuli designed to provoke ideas for combining the means using different perspectives. They may be the result of provocative questions asked by the designer of the design task statement, requirement list, functional decomposition or the identified design space. TRIZ principles can serve as a basis to generate innovation challenges through forcing designers to consider different perspectives such as segmentation, merging, or mechanical inversion [17]. An options matrix is a two-dimensional matrix where means for one function are listed against the means for a second function. Every options matrix is constructed to explore the combinations of the means of two functions using a distinct, explicit innovation challenge.

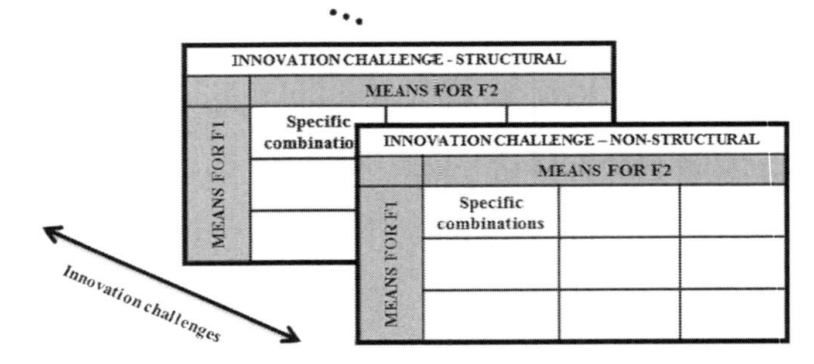

Fig. 2 Example representations of options matrices

A specific functional combination may be explored using different innovation challenges using independent options matrices as illustrated in Fig. 2.

Figure 2 illustrates the generation of distinct options matrices to explore the combinations of functions (F1 and F2) using two innovation challenges—structural and non-structural. For example, the two options matrices can be used to explore how the different means can be combined to create a mechanical sub-system where the components are (1) structural members or (2) non-structural members.

Each identified functional subset is explored in detail using the options matrices to generate sub-system concepts. The specific characteristics, positives, negatives and discussions regarding the sub-system concepts are explicitly captured in design documents.

Although the addition of the innovation challenges to the options matrices will result in additional combinatorial possibilities to explore, they are essential to push designers to explore non-obvious combinations, identify implicit assumptions, question the design space boundaries and facilitate innovation. The innovation challenges are generated from the information that is known about the system—design statement, list of requirements, function model and the design space. Innovation challenges may result in additional exploration of the solution space, necessitate a functional redefinition where an additional sub-function may be generated to support the combination of two functions, challenge the requirements thus questioning the implicit assumptions, question perceived design boundaries, identify functional combinations or stimulate exploration of additional configurations of combinations of means. The result of this step is the generation of small functional sub-systems that offer the designers a first glimpse of how the sub-system modules could be integrated into a complete system.

Step 4: *Generation of system concepts through sub-system combinations*

The functional modules and the sub-system concepts obtained through the options matrices are combined using a modified morphological matrix (Level 1 morphological matrix) as illustrated in Fig. 3.

Functional module pairs are extracted from the level 1 morphological matrix using a similar process as explained in Step 2. The sub-system concepts corresponding to the functional modules are then populated in level 2 options matrices to generate higher level sub-system concepts. These are fed back into the next level of morphological matrix and the process is continued until all the functions have been combined to form complete system concepts (Fig. 3).

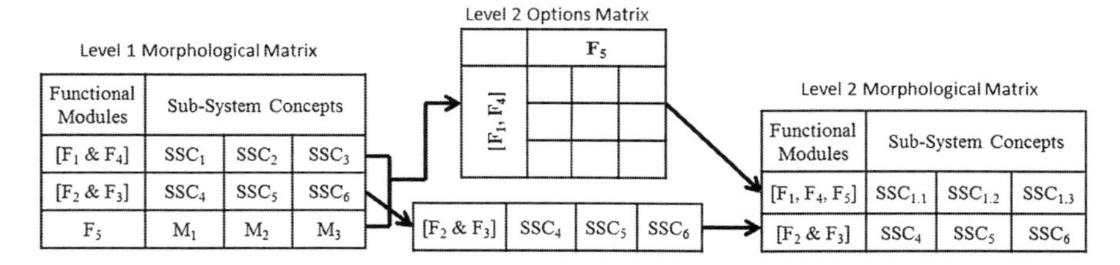

Fig. 3 Illustration of modified morphological matrices with functional modules

Additional levels of morphological matrices and options matrices are used depending on the size of the design task. A formal hierarchy of system decomposition needs to be established in order to apply the method to large design tasks so that the combinations formed at the various levels of the system may be appropriately termed. The decomposition described in [18] is particularly useful. The result of this step is the generation of complete system concepts and an understanding of the effects of specific functional combinations on the system characteristics.

Step 5: *Iteration to identify alternatives*

Once an initial set of system concepts are identified, alternative means and approaches to combinations are used to generate additional system concepts. Using the same method to generate system concepts, alternative designs can be created by altering functional combinations, using means that were previously unconsidered, and using different strategies for combinations. The result of this step is the generation of alternative system concepts based on solution characteristics resulting from functional combinations that affect the feasibility of the generated concepts.

3 Example Application of Proposed Method for Design of Automotive Seat

The proposed method was applied to the design of a seating chassis for the front driver/passenger of an automobile. The design task was decomposed into four functions and ideas were generated to perform each function. These were then populated in a morphological matrix as illustrated in Fig. 4.

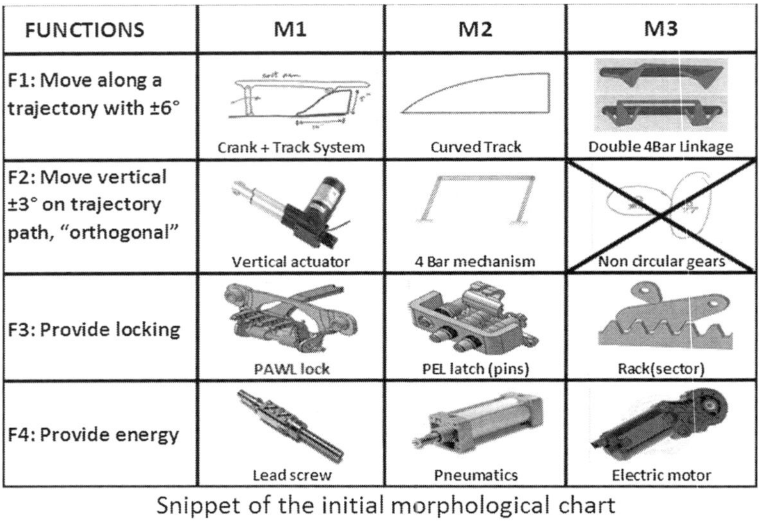

FUNCTIONS	M1	M2	M3
F1: Move along a trajectory with ±6°	Crank + Track System	Curved Track	Double 4Bar Linkage
F2: Move vertical ±3° on trajectory path, "orthogonal"	Vertical actuator	4 Bar mechanism	Non circular gears
F3: Provide locking	PAWL lock	PEL latch (pins)	Rack(sector)
F4: Provide energy	Lead screw	Pneumatics	Electric motor

Snippet of the initial morphological chart

Fig. 4 Snippet of the filtered morphological matrix

Step 1: *Grouping and preliminary filtering of individual means*

Figure 4 illustrates a snippet of the initial morphological matrix and an example of one of the options matrices that was explored. The initial ideas in the morphological matrix were grouped according to the types of means that were generated for each function. For example, the means that were generated for providing the locking mechanisms were grouped according to the locking principles—positive interaction mechanisms, friction based mechanisms, geared mechanisms, and miscellaneous mechanisms. Once the groupings were complete, a preliminary filtering was performed on the generated ideas to discuss the individual means and ascertain their feasibility. Some clearly infeasible ideas such as use of tank tracks and complex gear trains (crossed out in the figure) were suspended from further detailing due to cost, manufacturability, and complexity concerns.

Step 2: *Identification of functional combinations and innovative means for each function*

The combination of functions 1 and 4 was one of the interesting functional combinations that was explored. The designers wanted to generate a unique and innovative power-assisted movement mechanism through integration of the two functions. Functions 2 and 3 were not combined initially because the designers wanted to focus on generating innovative power assisted movement mechanisms. The most promising ideas from the set of means for F1 and F4 were selected to carry forward to perform focused ideation, some of which are illustrated in the options matrix from Fig. 5.

Step 3: Generation of sub-system concepts

Functions 1 and 4 were paired in an options matrix with the selected means for each function as illustrated in Fig. 5.

Focused ideation was performed on the options matrix using the perspective of the innovation challenges—structural and non-structural, so that both structural and non-structural mechanisms could result from the combinations. After some geometric configurations of the paired mechanisms were identified and sketched, the designers were encouraged to generate more configurations using the perspectives of additional innovation challenges obtained using TRIZ principles (such as segmentation—instead of using one mechanism to achieve the movement along trajectory, a union of two mechanisms was explored, and mechanical inversion—designers were forced to consider how the physical combinations would be affected if the mechanisms switched positions in the sketch). This resulted in the generation of additional configurations, identification and discussion of the limitations of the means that had been identified for the locking function (F3), and a redefinition of the functional decomposition of the system.

Step 4: Generation of system concepts through sub-system combinations

The functional module consisting of functions 1 and 4 was then fed into a level 1 morphological matrix along with the separate functions 2 and 3. Subsequent

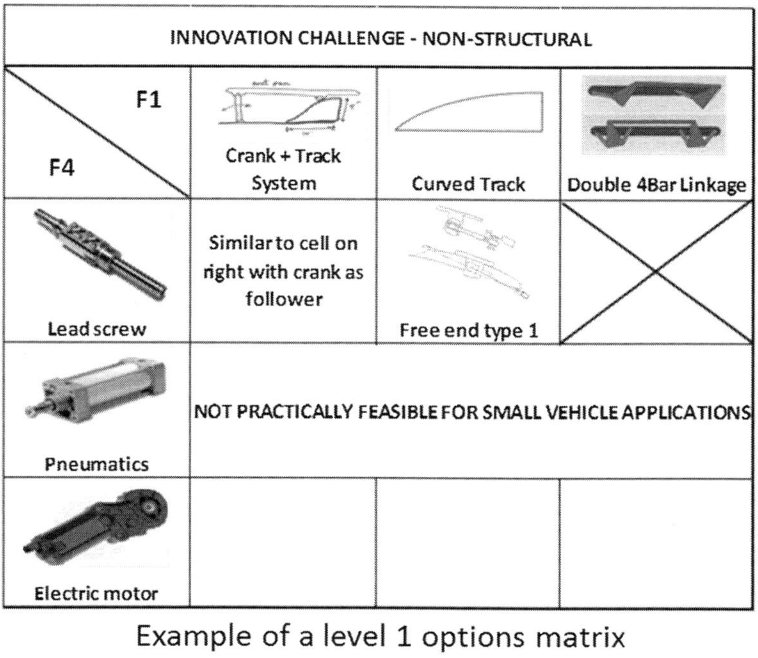

INNOVATION CHALLENGE - NON-STRUCTURAL			
F1 / F4	Crank + Track System	Curved Track	Double 4Bar Linkage
Lead screw	Similar to cell on right with crank as follower	Free end type 1	X
Pneumatics	NOT PRACTICALLY FEASIBLE FOR SMALL VEHICLE APPLICATIONS		
Electric motor			

Example of a level 1 options matrix

Fig. 5 Snippet of a level 1 options matrix for F1 and F4

options matrices saw the coupling of the locking mechanisms with the functional module of the power-assisted movement mechanism for the main trajectory and finally the overlay of the orthogonal movement mechanisms into the structure to generate complete system concepts.

Step 5: Iteration to identify alternatives

Additional system concepts were generated using different means, different innovation challenges and different order of functional combinations. For example, the movement mechanisms were coupled with the locking function, generating sub-system concepts that had a greater emphasis on managing the load paths through the structure to cater to crash load requirements. Other functional combinations and different orders of functional combinations were explored to generate additional system concepts.

4 Conclusions and Future Work

A method is proposed that provides detailed guidelines on how to efficiently generate system level concepts from individual functional solutions through focused detailed exploration of functional combinations using morphological and options matrices, and intelligently limiting the number of combinatorial possibilities explored. The method allows a designer to explore the specific characteristics of combinations of means or sub-systems, while generating detailed understanding of the design space, thereby enabling the generation of system concepts that better adhere to the design purpose and improving overall design feasibility. An example application of the method is provided in the design of a seating chassis for automotive application.

Further, research is currently underway to address the limitations of the method. First, the method implies some selection and filtering based on a high level understanding of the means. This may sometimes result in the inconsideration of potentially good means. Second, the individual means are filtered before consideration of their combinatorial effects. It is possible that two lower performing means are good in particular combinations. Also, the method proposes exploration of the design space through pairwise combinations. In other words, the order of identifying pairs and the pairs that are considered have an effect on the generated solutions. Finally, the method relies on the identification of explicit innovation challenges to perform innovative combinations of means. However, forcing the designers to identify the combinatorial perspectives (innovation challenges), questioning the design task and design boundaries, and looking to identify implicit assumptions will result in thorough understanding of the design task and help the detailed exploration of the design space.

Work is being done to refine the proposed method using feedback from practicing designers and to validate the steps of the method through user studies. The authors believe that the method is also complimentary to and capable of being coupled with the more recent utilization of design repositories [19–21], and ontological frameworks [22, 23] to support engineering design. A software tool is also being developed using ontological frameworks that supports the proposed approach while providing added functionality.

Acknowledgments This material is based upon work supported by Johnson Controls Incorporated (JCI). Any opinions, findings, and conclusions or recommendations expressed in this material are those of the author(s) and do not necessarily reflect the views of JCI.

References

1. Pahl G, Beitz W, Feldhusen J, Grote KH (2007) Engineering design a systematic approach. Springer, London
2. Ulrich KT, Eppinger SD (2008) Product design and development. McGraw Hill, New York
3. Ullman DG (1997) The mechanical design process. McGraw-Hill, New York
4. Buede DM (2000) The engineering design of systems: models and methods. Wiley, New York
5. Otto KN, Wood KL (2001) Product design—techniques in reverse engineering and new product development. Prentice-Hall Inc., New Jersey
6. Yan W, Chen C-H, Shieh M-D (2006) Product concept generation and selection using sorting technique and fuzzy c-means algorithm. Comput Ind Eng 50(3):273–285
7. Shah JJ (1998) Experimental investigation of progressive idea generation techniques in engineering design. In: Proceedings of DETC98 1998 ASME design engineering technical conference, Atlanta, GA
8. Zwicky F (1948) Morphological astronomy. Observatory 68:121–143
9. Ritchey T (2006) Problem structuring using computer-aided morphological analysis. J Operational Res Soc 57(7):792–801
10. Chen Y, Feng P, He B, Liz Z, Xie Y (2006) Automated conceptual design of mechanisms using improved morphological matrix. J Mech Des 128(3):516–526
11. Ölvander J, Lundén B, Gavel H (2009) A computerized optimization framework for the morphological matrix applied to aircraft conceptual design. Comput Aided Des 41(3):187–196
12. Huang GQ, Mak KL (1999) Web-based morphological charts for concept design in collaborative product development. J Intell Manuf 10(3–4):267–278
13. Tiwari S, Teegavarapu S, Summers J, Fadel G (2007) Automating morphological chart exploration: a multi-objective genetic algorithm to address compatibility and uncertainty. Int J Prod Dev 9:111–139
14. Matthews PC (2008) A Bayesian support tool for morphological design. Adv Eng Inform 22(2):236–253
15. Lin C-C, Luh D-B (2009) A vision oriented approach for innovative product design. Adv Eng Inform 23(2):191–200
16. Lo C-H, Tseng KC, Chu C-H (2010) One-step QFD based 3D morphological charts for concept generation of product variant design. Exper Syst Appl 37(11):7351–7363
17. Altshuller G (2005) TRIZ keys to technical innovation. Technical innovation center
18. Shishko R, Aster R (1995) NASA systems engineering handbook. In: Cassingham R (ed) NASA, Washington
19. Bohm MR, Stone RB, Simpson TW, Steva ED (2008) Introduction of a data schema to support a design repository. Comput Aided Des 40(7):801–811
20. Fenves SJ (2002) A core product model for representing design information. National Institute of Standards and Technology, NISTIR 6736

21. Schotborgh WO (2009) Knowledge engineering for design automation. Ph. d thesis, faculty of engineering technology, University of Twente
22. Darlington MJ, Culley SJ (2008) Investigating ontology development for engineering design support. Adv Eng Inform 22(1):112–134
23. Kitamura Y, Kashiwase M, Fuse M, Mizoguchi R (2004) Deployment of an ontological framework of functional design knowledge. Adv Eng Inform 18(2):115–127

Understanding Internal Analogies in Engineering Design: Observations from a Protocol Study

V. Srinivasan, Amaresh Chakrabarti and Udo Lindemann

Abstract The objective of this research is to understand the use of internal analogies in the early phases of engineering design. Empirical studies are used to identify the following: type and role of analogies in designing; levels of abstraction of search and transfer of analogies; role of experience of designers on using analogies; and, effect of analogies on quantity and quality of solution space. The following are the important results: analogies from natural and artificial domains are used to develop requirements and solutions in the early phases of engineering design; experience of designers and nature of design problem influence the usage of analogies; analogies are explored and unexplored at different levels of abstraction of the SAPPhIRE model, and; the quantity and quality of solution space depend on the number of analogies used.

Keywords Analogy · Novelty · Variety · SAPPhIRE model · Experience

V. Srinivasan (✉) · U. Lindemann
Institute of Product Development, Technische Universitat of Munich, Munich, Germany
e-mail: srinivasan.venkataraman@pe.mw.tum.de

U. Lindemann
e-mail: lindemann@pe.mw.tum.de

A. Chakrabarti
Centre for Product Design and Manufacturing, Indian Institute of Science, Bangalore, India
e-mail: ac123@cpdm.iisc.ernet.in

A. Chakrabarti and R. V. Prakash (eds.), *ICoRD'13*, Lecture Notes in Mechanical Engineering, DOI: 10.1007/978-81-322-1050-4_17, © Springer India 2013

1 Introduction

Design-by-analogy is used to produce creative solutions [1–3], in particular to enhance novelty and number of solutions [4, 5]. In designing, analogies aid [6] and inhibit fixation [7]. Design-by-analogy involves the transfer of analogous knowledge from a source domain to a target domain, to solve problems in the target domain. The following types of analogies are identified based on different criteria: domain of analogies: natural or biological and artificial analogies; apparent distance between the target and source domains: close domain and distant domain analogies; representation of analogy: verbal-, image- and video-based [3, 7]. In this paper, another category is identified based on development of analogies: internal and external analogies. An internal analogy is created using only the cognitive abilities of designers, mostly based on past experiences. An external analogy is created using an external source like a book, database, computer-based tool, etc. It could be argued that the external analogies also involve cognitive abilities of the designers; however, these kinds of analogies are created primarily due to the use of the external source. Current research on analogies in designing focuses only on understanding and supporting external analogies. As a precursor, it is important to understand the use of internal analogies. Further, this understanding should be the basis for understanding and supporting external analogies. Therefore, this research focuses on understanding internal analogies in designing.

2 Objective and Research Questions

The objective of this research is to understand the use of internal analogies in the early phases of engineering design. Specifically, the following research questions are posed:

1. What are the purposes and types of analogies used in the early phases of engineering design?
2. At what levels of abstraction are analogies searched for in the target domain, created in the source domain and, implemented in the target domain?
3. What is the role of experience of designers on the use of analogies?
4. What is the effect of analogies on the quantity and quality of solutions developed?

3 Research Methodology

Protocol studies of eight design sessions from earlier research [8, 9] are used to answer the research questions. Each design session consists of a designer, experienced (E1-E4) or novice (N1-N4), solving a design problem (P1 or P2),

Table 1 Design sessions (values available from [9])

	E1, P1	E2, P2	E3, P2	E4, P1	N1, P1	N2, P1	N3, P2	N4, P2
VCS	4.44	3.88	3.75	3	2.42	3.14	4.54	3.69
NCS	3.89	3.13	2.92	2.57	1.58	2.14	4	3.54
VIS	255	85	92	72	89	46	132	109
N_{ideas}	103	38	37	32	43	21	40	39

individually, by following a think-aloud protocol, under laboratory conditions (see Table 1). The objective of P1 is to develop solutions for a machine for making holes in any direction in three dimensions, subject to the following machine constraints: (a) change direction while making a hole; (b) make holes of different sizes; (c) make holes in metal, plastic, or wood; and (d) simple, small and portable. The objective of P2 is to develop solutions for a device to clean utensils subject to the following device constraints: (a) meant for urban middle-class family of maximum 10 members; (b) clean all kinds of utensils like tumbler, dining plate, pressure cooker, mixer-grinder, etc.; (c) clean utensils made of all general kinds of materials like stainless steel, porcelain, glass, plastic and aluminum. Before the commencement of designing, all the designers are instructed to develop requirements and as many solutions as possible. They are also instructed to explain on how solutions are developed, but are not told anything about analogies. The time for designing is unconstrained. The transcriptions and measures of solution space—variety of concept space (VCS), novelty of concept space (NCS), variety of idea space (VIS) and number of ideas (N_{ideas})—of these design sessions, available from [9] (see Table 1), are used for the following: (a) identify the analogies and, determine their domains and levels of abstraction, (b) identify the levels of abstraction in the target domain from which the search for analogies commenced and, (c) identify the levels of abstraction in the target domain at which the analogies are implemented. Since the SAPPhIRE model (see Sect. 4.2) can describe outcomes at several levels of abstraction in the early stages of engineering design, the levels of abstraction in the source and target domains are assessed using this model. To assess the effect of analogies on the solution space, Pearson's correlation, from Microsoft Excel^{TM}, is used to correlate novelty of concept space, variety of concept space, variety of idea space and number of ideas, with the number of analogies used to develop that solution space.

4 Literature Survey

In this section the relevant literature is organized according to the following topics.

4.1 Analogies

Several researchers studied the role of analogies in designing. The effect of experience of designers on the use of analogies, at the conceptual and detail design stages, is explored through empirical studies in an aerospace industry in [10]. The impact of the different kinds of representations of triggers on the representation and creative quality of design solutions inspired by those triggers in engineering design is studied in [3]. The effect of timing and similarity of analogies during idea generation is studied in [11]. The effect of the apparent distance between the source and target domains on the solutions developed is assessed in [7]. The effects of using no, biological- and engineering-based analogies in idea-generation are studied using an empirical study in [12].

4.2 SAPPhIRE Model

The *S*tate change, *A*ction, *P*art, *P*henomenon, *I*nput, *OR*gan, *E*ffect (SAPPhIRE) model is developed as a model of causality to explain the working of engineered and biological systems [13]. The model is observed to describe outcomes at several levels of abstraction in the early phases of engineering design [14]. Phenomenon is defined as an interaction between a system and its environment (e.g., displacement of an object over a surface). State change is defined as the change in property of the system due to the interaction (e.g., change in position of the object). Action is defined as the interpretation or high level abstraction of the interaction (e.g., movement of the object). Effect is the principle underlying the interaction (e.g., second equation of motion, $x = u \times t + 0.5 \times a \times t^2$). Input is a physical quantity, which comes from outside the system boundary, required for the interaction (e.g., acceleration on the object). Organ is a set of properties and conditions of the system and its environment, also required for the interaction (e.g., degree of freedom of the object in direction of acceleration, acceleration applied for a finite time, Newtonian properties of the object, etc.). Part is a set of components and interfaces that constitute the system and its environment (e.g., object lying on a surface).

4.3 Novelty and Variety

All the definitions and findings in this section are taken from [9]. A concept is defined as an overall solution which is intended to satisfy most of the identified requirements. An idea at a level of abstraction is defined as a constituent of a concept and is intended to satisfy only some requirements. Variety of a concept in a concept space is defined as a measure of a difference of that concept from the concepts developed earlier in that concept space. Variety of a concept space is defined as the average of the variety of all the concepts in that concept space. Novelty of a concept in a concept space is defined as a measure of the difference of that concept from: (a) concepts developed earlier in the same concept space, and

(b) concepts in the other existing concept spaces that satisfy the same overall function. Novelty of a concept space is the average of the novelty of all the concepts in that concept space. Both, variety and novelty of concept space are found to depend on the number of ideas explored at the different levels of abstraction; higher variety and novelty are observed when ideas at higher levels of abstraction are explored more. Variety of idea space is a measure of the difference of the ideas from each other in that idea space. It is also found that variety of idea space correlates well with the variety of concept space, which in turn correlates well with the novelty of concept space.

5 Results

The following observations are made from the analysis of the transcriptions and are supported with utterances from them. Even though the designers are not told anything about analogies, they are found to be used by both, novice and experienced designers. Since no external support is used during the designing, it could be reasoned that the analogies are based on the past experiences of the designers. This is supported by the utterances of E1: *It is a sort of electro-chemical erosion. I have read it somewhere in some manufacturing technology handbook. If you have a book…I do not remember erosion exactly. It is electrochemical erosion. There are other methods in this but that (electrochemical erosion) can be used. It has many limitations.* Here, the designer uses an analogy of electrochemical erosion for removing material and making a hole, and remembers reading about it. Table 2 shows the list of analogies used in all the design sessions. The fact that designers use analogies without being instructed to use them signifies that in studies involving the use of external analogies from a support, designers may be developing both, internal and external analogies. These analogies need to be distinguished, especially while studying the role of the support and experience of designers, on the use of analogies. It is also seen that solutions and requirements previously developed during the same session are also a source of analogies. This is epitomized by the utterances of E1, *That idea (laser) triggered the second idea of water jet because I felt laser might be little expensive. Water jet is, I think, a little cheaper and, I know that there is a process, so it was not very difficult correlating these two processes.* E1 develops a solution of water-jet machining by using the analogy of laser-jet, which is developed as a solution earlier.

Designers use analogies from both, natural and artificial domains (see Table 2). This observation is illustrated through the following utterances of E1 and N3 for natural and artificial, respectively. *It (tool head) can stick to any surface it wants to stick to locally, then why don't it have a have a, you know, lizards can stick upside down because of vacuum sort of, so why don't I use that principle here. I want to stick inside it* (work-piece). *It has some sort of vacuum pads it sticks wherever it wants.* Here, E1 uses the analogy of *vacuum principle of lizards* (natural domain) in the tool head, to enable it to fix itself. *So if we have to do that, why not dip the fresh utensil in a solution? This idea I got from Fevicol. When you apply Fevicol on the hand and after it gets dried, it comes out as a layer and we can peel off, comes with dirt or say everything.* N3 uses the analogy of Fevicol, a brand of adhesive, which after coming in contact with the hands, forms a layer after drying and can be easily peeled. This analogy can be used for cleaning utensils—utensils are dipped in a special solution, a layer of the material of the solution is formed on the utensils and the layer with the leftovers can be separated after the use of the utensils. A total of 12 and 36 analogies are used from the natural and artificial domains, respectively. All the designers, except N2, use more analogies from the artificial domain than natural domain. This is because all the designers have engineering or architecture backgrounds, and so their previous experiences are based more on the artificial rather than the natural domain. This signifies that, to exploit the rich and diverse knowledge of nature, designers with non-natural backgrounds need assistance.

It is observed that design problems also affect the use of analogies. All the designers solving problem, P1, use more analogies than those solving problem, P2. A total of 36 (11 natural and 25 artificial) analogies are used while solving P1, while an aggregate of 12 (1 natural and 11 artificial) analogies are used while solving P2 (see Table 2). Design problem, P1, is less conventional than problem, P2, therefore, P1 should be tougher to solve than P2. So, solving P1 should require more analogies than solving P2.

It is seen that analogies are used for developing both, requirements and solutions. Analogies for developing requirements is supported through the utterances of E1, *For example, you consider trees, trees grow in all directions in three dimensions, there is no fixed pattern as such and that's the kind of hole I am trying to achieve. So it is basically, there is an open space and, the branches and leaves are growing in different directions, so material is added into space. So, if I think of this problem in a different direction—it would be similar to adding material in open space and achieving my goal of creating cavity. In space I create the material from zero.* Here E1 uses the analogy of three-dimensional growth in tress to develop the requirement of adding material in space, instead of removing material from a given material. Out of a total of 48 analogies, only 4 are used for developing requirements while the rest are used for developing solutions (see Table 2). It has to be noted that the designers are instructed to only explain how the solutions are developed, not requirements. It is reported in literature that analogies can assist in the following: design problem search, identification, interpretation, elaboration, decomposition, and reformulation; solution refinement, evaluation; and evaluation criteria interpretation [2, 3, 15].

Experienced designers use more analogies than novice designers; on average, experienced and novice designers use 7.5 and 4.5 analogies, respectively (see Table 2). No experienced designer, except E1, uses any analogies from the natural domain, but prefer to use more analogies from the artificial domain. While the experienced designers use more analogies from the artificial domain, the novice designers distribute the analogies between the natural and artificial domains. These findings show that experienced designers need assistance for using analogies from

Table 2 Analogies created by designers and their domains, abstraction levels, purposes and implementation

Designer, problem	Analogy number	Analogy used	Domain	Abst. level	Req / sol	Imp. / unimp.
E1, P1	E1-1	Electrochemical erosion	Artificial	Ph	Sol	Imp
	E1-2	Arms of table lamp	Artificial	P	Sol	Imp
	E1-3	Decorative food items in exhibitions	Artificial	P	Sol	Imp
	E1-4	Beetles and insects	Natural	P	Sol	Imp
	E1-5	Rapid prototyping	Artificial	Ph	Req	Imp
	E1-6	Powder metallurgy	Artificial	Ph	Sol	Imp
	E1-7	Random 3-d growth in trees	Natural	A	Req	Imp
	E1-8	Casting	Artificial	Ph	Sol	Imp
	E1-9	Small creature	Natural	P	Sol	Imp
	E1-10	Light	Artificial	P	Sol	Imp
	E1-11	Cost, feasibility, laser jet	Artificial	P	Sol	Imp
	E1-12	Melting	Artificial	Ph	Sol	Imp
	E1-13	Injecting gadgets	Artificial	P	Sol	Imp
	E1-14	Rain water	Natural	P	Sol	Unimp
	E1-15	Cavities in cake	Artificial	P	Sol	Unimp
	E1-16	Chemical etching in PCBs	Artificial	Ph	Sol	Unimp
	E1-17	Complex shapes of human intestines	Natural	P	Sol	Unimp
	E1-18	Bull dozer	Artificial	P	Sol	Imp
	E1-19	Vacuum principle in legs of lizards	Natural	E	Sol	Imp
	E1-20	Globular creatures	Artificial	P	Sol	Imp
E2, P2	E2-1	Thread of bottle cap	Artificial	P	Sol	Imp
E3, P2	E3-1	Shoe polish	Artificial	P	Sol	Imp
	E3-2	Car wash	Artificial	Ph	Sol	Unimp
	E3-3	Washing toilet	Artificial	Ph	Sol	Imp
E4, P1	E4-1	Laser	Artificial	P	Sol	Imp
	E4-2	Flexible arm of robot	Artificial	P	Sol	Imp
	E4-3	Endoscopy	Artificial	Ph	Sol	Imp
	E4-4	Etching	Artificial	Ph	Sol	Imp
	E4-5	Tunnel digging	Artificial	Ph	Sol	Imp
	E4-6	CD burning	Artificial	Ph	Sol	Imp
N1, P1	N1-1	Drilling a tunnel	Artificial	Ph	Req	Imp
	N1-2	Earthworms	Natural	P	Sol	Imp
	N1-3	Insects	Natural	P	Sol	Unimp
	N1-4	Operating inside a human body using small robots	Artificial	A	Req	Imp
	N1-5	Giant-wheel with buckets attached on its periphery	Artificial	P	Sol	Imp
N2, P1	N2-1	Pneumatic guns and pneumatic actuators	Artificial	P	Sol	Imp

(continued)

Table 2 (continued)

Designer, problem	Analogy number	Analogy used	Domain	Abst. level	Req / sol	Imp. / unimp.
	N2-2	Earthworms and rats	Natural	P	Sol	Imp
	N2-3	Tunnel boring	Artificial	Ph	Sol	Imp
	N2-4	Penetration of roots of plants	Natural	Ph	Sol	Unimp
	N2-5	Growth in plants	Natural	Ph	Sol	Unimp
N3, P2	N3-1	Whirlpool	Artificial	Ph	Sol	Imp
	N3-2	Cats and dogs	Natural	P	Sol	Unimp
	N3-3	Fevicol (adhesive glue)	Artificial	P	Sol	Imp
	N3-4	Processes in beauty parlor	Artificial	Ph	Sol	Unimp
N4, P2	N4-1	Sweeping	Artificial	Ph	Sol	Unimp
	N4-2	Hairbrush	Artificial	P	Sol	Imp
	N4-3	Vacuum cleaners	Artificial	P	Sol	Unimp
	N4-4	Evaporators	Artificial	P	Sol	Unimp

the natural domain, while novice designers need assistance for using analogies from both, natural and artificial domains, to be on par with the experienced designers.

From Table 2 the following are observed. All the designers use analogies at the level of abstraction of *part* and, with the exception of E2, *phenomenon*. Analogies at no other levels of abstraction—*action, state change, input, effect* and *organ*—of the SAPPhIRE model are found to be used with the same intensity. This lack of exploration could be because designers do not understand these levels of abstraction as well as the other levels, to explore them with the same intensity. Another set of empirical studies shows that designers do not explore all the levels of abstraction of the SAPPhIRE model with the same intensity, but explore *part* of the SAPPhIRE model with greater intensity [14], although all the levels of abstraction contribute to variety and novelty [9]. This underlies the need for a support to assist creating analogies at the unexplored levels of abstraction.

Table 3 shows the contents in the target domain and their levels of abstraction with which the analogies in the source domain are searched. All the designers search for analogies at the level of abstraction of *action*. Less searching is done at the levels of abstraction of *phenomenon, organ* and *part*. No search is found at the other levels of abstraction of *state change*, *input* and *effect*. All the designers use analogies which are at the same or lower levels of abstraction than the contents in the target domain with which searching is done (see Tables 2 and 3). In other words, the transfer from the target domain to the source domain is always from a higher to the same or lower level of abstraction. The definition of analogous designs in [1] suggests that the search for analogies can happen at function-, behavior- and structure-levels. In the biomimetic design process in [16], search is performed using functions. In the biomimetic design process in [17], a problem is framed in biological terms, and this is used for searching analogies. Four classes of transfer in biomimetics based on the SAPPhIRE model are reported in [18]: copy parts, transfer organs, transfer attributes, and transfer state change.

Table 3 Search, abstraction level of search and implementation of analogies

Analogy number	What was searched	Abstraction level of search
E1-1	Make hole/remove material	A
E1-2	Flexibility	R
E1-3	Make hole	A
E1-4	Material removal	A
E1-5	Make cavity	A
E1-6	Lay material	A
E1-7	Lay material	A
E1-8	Lay material for metals	A
E1-9	Remove material	A
E1-10	Digging material	Ph
E1-11	Remove material	A
E1-12	Make material soft and remove material	A, A
E1-13	Remove material	A
E1-14	Make cavity	A
E1-15	Make cavity	A
E1-16	Remove material	A
E1-17	Make cavity	A
E1-18	Remove material	A
E1-19	Stick at a desired position	Ph
E1-20	Motion of insects	Ph
E2-1	To grip	A
E3-1	To sprinkle	Ph
E3-2	To clean	A
E3-3	To clean	A
E4-1	Material removal	A
E4-2	Flexibility	R
E4-3	Material cutting, removal, etc	A
E4-4	Material removal	A
E4-5	Material removal	A
E4-6	Material removal	A
N1-1	Drill hole	Ph
N1-2	Drill hole and change direction	Ph, A
N1-3	Remove material; size of hole and tool to make hole	A
N1-4	Drill hole and change direction	Ph, A
N1-5	Expanding and contracting tool diameter	Ph
N2-1	Flexibility and stiffness	R; R
N2-2	Make hole in desired direction	A
N2-3	Make hole in desired direction	A
N2-4	Make hole in desired direction	A
N2-5	Make hole in desired direction	A
N3-1	Relative motion between utensil and fluid in contact	A
N3-2	To clean	A
N3-3	To clean	A

(continued)

Table 3 (continued)

Analogy number	What was searched	Abstraction level of search
N3-4	To clean	A
N4-1	Cleaning	A
N4-2	Scrubber	P
N4-3	Cleaning	A
N4-4	Cleaning	A

Among all the analogies developed by the designers, some of them are not implemented into requirements or solutions in the target domain (see Table 2). For instance, designer, E1, uses an analogy of "cavities in cake" for "make cavity", but the designer finds that this analogy cannot be implemented because the direction of making cavity cannot be controlled for the given materials and so, this analogy is not implemented as a solution. Utterances to support are, *I thought cake, cake is porous and has cavities in random direction, but it is made out of baking process, so baking process creates the random cavities, I won't be able to bake metal.* In another instance, designer, N2 develops an analogy of "penetration of roots of plants underground" and "growth of plants in direction of sunlight" to make a hole in the desired direction—*I am thinking of the roots of the plants, it (root) penetrates and goes inside. But they do not have pre-defined path. They go in search of water in the ground.* and *Even the plant grows such that it gets maximum sunlight.* For experienced and novice designers, a total of 5 out of 30 analogies (17 %) and 8 out of 18 analogies (44 %) remain unimplemented. This shows the difficulties that the novice designers face while translating analogies from the source domain to the target domain. This shows that designers, in particular novice designers, need assistance in transferring analogies from the source domain to target domain.

To assess the effect of analogies on the variety and novelty of solutions, the variety and novelty of concept space, variety of idea space and number of ideas, all known from earlier research in [9] (see Table 1), are correlated individually, with the number of analogies, as shown in Table 4. The high correlation values between: (a) variety of idea space and number of analogies and, (b) number of ideas and number of analogies, indicate that the use of the internal analogies has positive effects on the quality and quantity of ideas. However, this positive effect is

Table 4 Correlation values

Correlating variables		Correlation value
Variety of concept space	Number of analogies	0.3201
Novelty of concept space	Number of analogies	0.3234
Variety of idea space	Number of analogies	0.8960
Number of ideas	Number of analogies	0.9067

not translated into concepts, as seen by the correlation values between: (a) variety of concept space and number of analogies and, (b) novelty of concept space and number of analogies. Nonetheless, all the correlation values are positive, which show the positive effect of the use of analogies on the quality and quantity of solutions. Analogies help build associations between the target and source domains, which are different from each other. These associations, not possible without the use of analogies, help develop solutions (ideas and concepts), which are different from the existing solutions including those developed earlier, thus enhancing the chances of variety and novelty.

6 Summary and Conclusions

This research helps understand the use of internal analogies in the early stages of engineering design through existing empirical studies. It is found that analogies from natural and artificial domains are used to develop both requirements and solutions in the early phases of engineering design. The experience of designers and nature of design problems influence the usage of analogies. Analogies are observed to be searched, developed and implemented at a few levels of abstraction of the SAPPhIRE model, while the other levels of abstraction are unexplored. The use of analogies has a positive effect on the variety of concept space, novelty of concept space, variety of idea space and number of developed ideas. This research gives directions for developing an assistance to support the use of analogies in the early phases of engineering design.

References

1. Qian L, Gero J (1996) Function-behaviour-structure paths and their role in analogy-based design. AIEDAM 10(4):289–312
2. Goel A (1997) Design, Analogy and Creativity. IEEE Expert Intell Syst Appl 12(3):62–70
3. Sarkar P, Chakrabarti A (2008) The effect of representation of triggers on design outcomes. AIEDAM 22(2):101–116
4. Young L (1987) The metaphor machine: a database method for creativity support. Decis Making Support Syst 3(4):309–317
5. Kletke M, Mackay J, Barr S, Jones B (2001) Creativity in the organization: the role of individual creative problem solving and computer support. Int J Human Comput Stud 55(3):217–237
6. Jansson D, Smith S (1991) Design fixation. Des Stud 12(1):3–11
7. Lopez R, Linsey J (2011) Characterizing the effect of domain distance in design-by-analogy. ASME IDETC/CIE design theory and methodology conference
8. Sarkar P, Chakrabarti A (2007) Understanding search in design. International conference on engineering design (ICED07)
9. Srinivasan V, Chakrabarti A (2010) Investigating novelty-outcome relationships in engineering design. AIEDAM 24(2):161–178

10. Ahmed-Kristensen S, Christensen B (2008) Use of analogies by novice and experienced design engineers. ASME IDETC/CIE, New York
11. Tseng I, Moss J, Cagan J, Kotovsky K (2008) The role of timing and analogical similarity in the stimulation of idea generation in design. Des Stud 29(3):203–221
12. Wilson J, Rosen D, Nelson B, Yen J (2010) The effects of biological examples in idea generation. Des Stud 31(2):169–186
13. Chakrabarti A, Sarkar P, Leelavathamma B, Nataraju BS (2005) A functional representation for aiding biomimetic and artificial inspiration of new ideas. AIEDAM 19(2):113–132
14. Srinivasan V, Chakrabarti A (2010) An integrated model of designing. JCISE 10(3)
15. Christensen B, Schunn C (2007) The relationship of analogical distance to analogical function and pre-inventive structure: The case of engineering design. Memory Cognit 35(1):29–38
16. Gramann J (2007) Problemmodelle und Bionik als Methode. Ph.d Thesis, Technical University Munich
17. Helms M, Vattam S, Goel A (2009) Biologically inspired design: process and products. Des Stud 30(5):606–622
18. Sartori J, Pal U, Chakrabarti A (2010) A methodology for supporting "transfer" in biomimetic design. AIEDAM 24(4):483–506

Craftsmen Versus Designers: The Difference of In-Depth Cognitive Levels at the Early Stage of Idea Generation

Deny W. Junaidy, Yukari Nagai and Muhammad Ihsan

Abstract This paper investigates the in-depth cognitive levels at the early stage of idea generation for craftsmen and designers. Examining this early stage may explain the fundamental thoughts in observing and defining design problems. We conducted an experiment using think-aloud protocol, where verbalized thoughts were analyzed using a concept network method based on associative concept analysis. Furthermore, we identified semantic relationships based on Factor Analysis. The findings showed that craftsmen tended to activate low-weighted associative concepts at in-depth cognitive level with a smaller number of polysemous features, thus explaining their concerns about tangible-related issues, such as proportion and shape. Designers, however, activated highly weighted associative concepts with more polysemous features, and they were typically concerned with intangible issues, such as surroundings context (i.e., eating culture) and users' affective preferences (i.e., companion, appeal).

Keywords In-depth cognitive level · Early stage of idea generation · Designers · Craftsmen

D. W. Junaidy (✉) · Y. Nagai
Japan Advanced Institute of Science and Technology, 1-1 Asahidai, Nomi, Ishikawa 923-1292, Japan
e-mail: denywilly@jaist.ac.jp

Y. Nagai
e-mail: ynagai@jaist.ac.jp

M. Ihsan
Institute of Technology Bandung, Jl. Ganesa 10, Bandung 40132, Indonesia
e-mail: ihsan@fsrd.itb.ac.id

A. Chakrabarti and R. V. Prakash (eds.), *ICoRD'13*, Lecture Notes in Mechanical Engineering, DOI: 10.1007/978-81-322-1050-4_18, © Springer India 2013

1 Introduction

This study focuses on the early stage of idea generation to capture the associative concepts at the in-depth cognitive levels of craftsmen and designers. Examining the early stage of idea generation may explain the fundamental thoughts in observing and reframing design problems. Many attempts have been made to capture users' affective preferences based on users as subjects. However, we examined the in-depth cognitive levels of the creators (craftsmen and designers), who attempt to grasp users' feelings when producing successful impressions of products [1, 2]. We conducted an experiment using think-aloud protocol, where verbalized thoughts were analyzed using a conceptual network based on associative concepts and semantic relation analysis.

1.1 Early Stage of Idea Generation

Idea generation, which consists of observation and ideation, is the essential step in the design thinking process. It is the interplay of cognitive and affective skills that lead to the resolution of a recognized difficulty [3]. Following are general steps of design thinking; the early stage of idea generation is the step mainly discussed:

1. Imagination (early stage of idea generation): the stage to observe and reframe the design problem.
2. Ideation (later stage of idea generation): the stage employing sketches, graphs, or paper models to generate ideas visually.
3. Prototyping: the stage of making a rough model to convey an idea concretely.
4. Evaluation: the stage to acquire user's feedback by evaluating affective preferences.

The next step after the design thinking process is realization or production for commercial purposes.

The early stage of idea generation is one of observation by craftsmen and designers through first-hand experiences. This stage is associated with greater diversity of ideas [4]; therefore, it is reasonable to assume that one's fundamental thoughts are captured fairly at this point.

1.2 In-Depth Cognitive Level

It is generally known that designers cannot express their thoughts explicitly; their latent sensitivity is widely researched in cognitive psychology. It is known as implicit cognition, which is understood to be that which is not explicitly

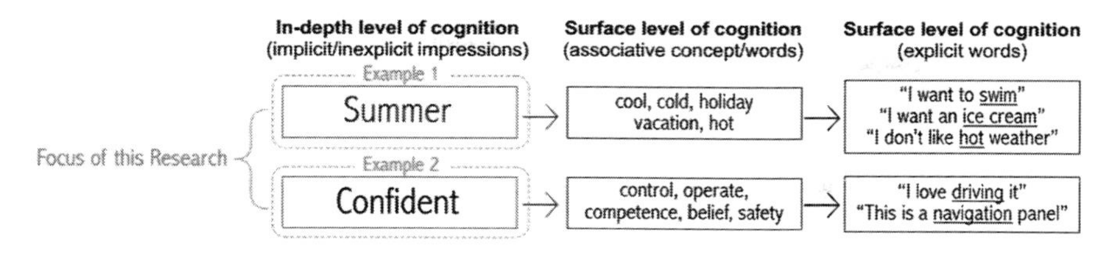

Fig. 1 Focus of this research: capturing the in-depth cognitive level

recognized or verbalized [5]. Explicit expression, which is presumably a shallow analysis, is referred to as surface-level cognition, and underlying cognition that is difficult to express is referred to as in-depth cognitive level (e.g., feeling, taste, impression) [2, 6, 7] (Fig. 1).

Taura et al. [7] explain that implicit impressions could exist in the feelings and are implied underneath explicit impressions that are related to deep impressions. Humans establish extremely rich metaphorical concepts (within in-depth impression) as key features of cognition in creative design; thus, a designer is able to capture a profound understanding of an object [2]. Previous studies have focused on capturing in-depth cognitive levels (impressions) of users based on created artifacts, but our study focuses on the creators of the artifacts (craftsmen and designers).

1.3 Verbalized Thoughts and Associative Concept Analysis

To examine the structure of thoughts from subjective experiences, a think-aloud, as a part of protocol analysis was employed to produce verbal reports of the thinking process [8]. Subjects were instructed to describe their thoughts and observations and reframe design problems through verbal expression. Verbalized thoughts reflect some aspects of the regular cognitive process [8]; for this study, they were reconstructed using a computational model to reproduce observable aspects of the in-depth cognitive level.

Associative concept analysis captures concepts of an expression associated with the individual's mental state. The associative concept is comprised of six subtypes: connotative, collocative, social, affective, reflected, and thematic [9]. It is latent within implicit cognition. Therefore, a conceptual network is suitable as an associative analysis tool for exploring the latent links among concepts. In the field of psychology, the conceptual networks depict human memory as an associative system, where a single idea can contain multiple meanings (polysemous). The concept dictionary utilized in conceptual network is from the University of South Florida Free Association Norms database (USF-FAN) [10, 11].

2 Aim

The aim of this research was to capture the differences in associative concepts at in-depth cognitive levels of craftsmen and designers at the early stage of idea generation in design thinking. Thus, we conducted an experimental study using think-aloud protocol, where designers and craftsmen freely expressed their ideas verbally.

3 Methodology

In this study, we used a concept network method based on the associative concept dictionary to extract verbalized thoughts. The framework of this research was comprised of the following steps (Fig. 2):

1. Two craftsmen and two product designers conducted a think-aloud protocol. They were instructed to imagine designing a fruit basket/container and freely express their ideas verbally without necessarily drawing or observing the object. Verbal data were recorded, and the sorted verbal expressions were then transcribed into English.
2. The verbal data, which consisted of explicit words, were transferred onto vector graphs (conceptual network on the basis of the USF norms database) to obtain extraction of highly weighted associative words indicated by the out-degree centrality score (ODC).
3. Differences in the concept network structures were identified by analyzing the following:
    a. Density of connection, which exhibits the property of idea within the associative concept network.
    b. Semantic relation, which finds the characteristics of the associative concepts at the in-depth cognitive levels using an orthogonal semantic map based on factor analysis.

4 Experiment

4.1 Subjects

Four subjects (two Indonesian craftsmen and two Indonesian designers in the age range of 27–51 years) participated in this experiment. The two designers were university graduates with experience in craft design and concern for natural material utilization. Each of the two craftsmen, known as master craftsmen, who

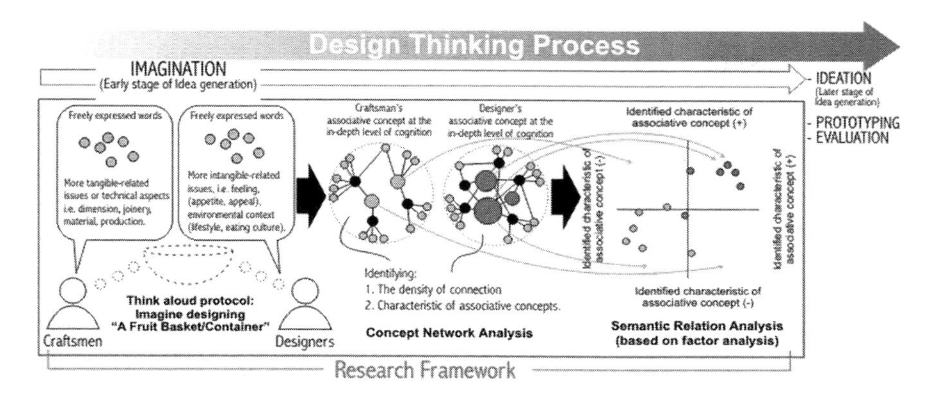

Fig. 2 Research framework: capturing craftsmen's and designers' associative concept at in-depth cognitive levels at the early stage of idea generation

has acquired special skills in artistry and apprehends design as an artistic or decorative creation. They gained their special expertise, passed down from one generation to another in the local village's traditional bamboo crafts.

4.2 Procedure

The experiment was set up simply; the subjects (craftsmen and designers) were not required to engage in specific activities, such as drawing or observing stimuli. They were deliberately conditioned with minimum instruction to be able to capture fundamental associative concepts. Rigid instructions about determining design theme, market segmentation, or design function were avoided since they might provide excessive information that would be unfair and misleading. Minimum instruction maintains a fair stage for noting craftsmen's and designers' first-hand experiences in observing and reframing design problems. There were no constraints on the subjects for expressing their ideas verbally and engaging in spontaneous thinking.

All procedures were recorded as verbal data and transcribed word by word. Grammatical rules were followed for connecting words, such as prepositions, a few general verbs, articles, and pronouns; also, other less relevant explanations were omitted [6]. Finally, the sorted verbal data consisting only of nouns, adjectives, adverbs, and verbs were translated into English and further analyzed according to the concept network method on the basis of the USF norms database (also visualized as graphs using Pajek 2.05 with the algorithm of Fruchterman Reingold) [12].

4.3 Concept Network Analysis

At the first stage of analysis, 107 sorted verbal expressions (nouns, adjectives, adverbs, and verbs) were obtained from craftsmen; 102 were sorted from

Table 1 Sorted verbal expressions

Category	List of sorted verbal expressions (partly shown)
Craftsmen	**Capacity, dimension, measure, standard, super, big, count, size, leg, height, thin, shape, square, position, part, head, stack, body, solid, base, width, top, long, oval, three-dimensional, thick, centimeter, box,** design, container, fruit, duck, salt, egg, adjust, buyer, function, capable, form, set, color, supply, bamboo, scar, spot, glue, mark, sandpaper, etc.
Designers	**Place, kitchen, pluck, tree, shop, sensation, reap, pick, preservation, tropical, rotten, fresh, delicious, interaction, inform, remind, children, invite, accommodate, people, way, salad, commercial, habit, crowd, appeal, appreciate, attractive, dignity, snack**, put, table, hang, fruit, wood, appear, stand, durian, banana, apple, orange, watermelon, grape, etc.

designers. Expressions of craftsmen tended to focus on tangible aspects, such as technique, material, and production (bold text). Designers, however, paid more attention to intangible-related issues, such as users' affective preferences and the environment (bold text) (Table 1).

The sorted verbal data were further visualized as graphs of conceptual network analysis (Fig. 3). Craftsmen's conceptual networks generated 1,941 vertices (nodes), and designers' networks generated 1,662 vertices (nodes). The networks were too dense and complex for analysis; therefore, it was necessary to simplify the created networks by a reduction method. Systematic reduction was based on considerations that not all the words from verbalized protocols contribute to an in-depth cognitive level, and surface-level cognition is overemphasized. The following indicate low scores associated with explicit words/surface-level cognition (bold).

- Craftsmen (total: 1,941 words): **0.000–0.010 = 1,462 words**; 0.020 = 352 words; 0.030 = 94 words; 0.040 = 27 words; 0.051 = 6 words.
- Designers (1,662 words): **0.000–0.010 = 1,259 words**; 0.021 = 293 words; 0.032 = 77 words; 0.043 = 23 words; 0.054 = 8 words; 0.065 = 1 word; 0.076 = 1 word.

5 Analysis and Results

5.1 Conceptual Network Analysis (After Reduction)

Application of the simplified concept reduced the words that were less important to the networks so that the extraction of the associative concept within the in-depth cognitive level was apprehensible (Fig. 4a, b). The reduction omitted <50 % words with lower ODC scores to get an observable network diameter [13] (i.e., craftsmen:

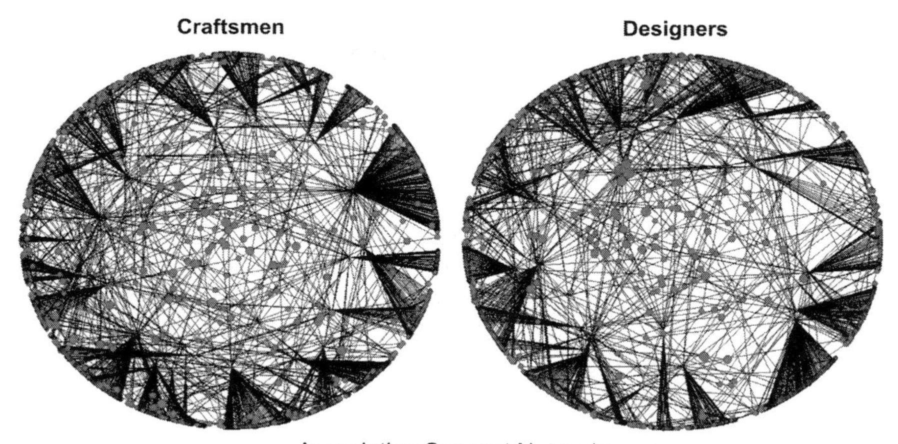

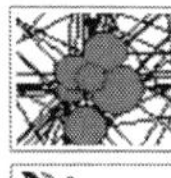

A big node/vertex represents highly-weighted associative concept/word/idea
(In-depth level of cognition) containing rich meanings.
(indicated by Out-Degree Centrality/ODC score).

A small node/vertex represents associative concept/word/idea
(In-depth level of cognition) containing several meanings.
(indicated by Out-Degree Centrality/ODC score).

A point of arc is explicitly-expressed word that leads to node/vertex
(Surface level of cognition) containing shallow meanings.
(indicated by Out-Degree Centrality/ODC score)

Fig. 3 Associative concept networks of craftsmen's and designers' before reduction (words and scores not shown due to complexity)

50 % × 0.051 ODC = > 0.025 ODC score; designers: 50 % × 0.076 ODC = > 0.038 ODC score).

The reduction was applied independently to each group where the highly weighted associative words were identified at the in-depth cognitive level with ODC scores as follows (bold text) (Table 2a, b):

- Craftsmen (total: 202 words): 0.000 = 75 words; 0.040 = 94 words; **0.053 = 27 words**; **0.067 = 6 words.**
- Designers (total: 81 words): 0.000 = 48 words; **0.083 = 23 words**; **0.104 = 8 words**; **0.125 = 1 word**; **0.146 = 1 word.**

Hereafter, we selected the top 10 highly weighted associative words from each group for further analysis.

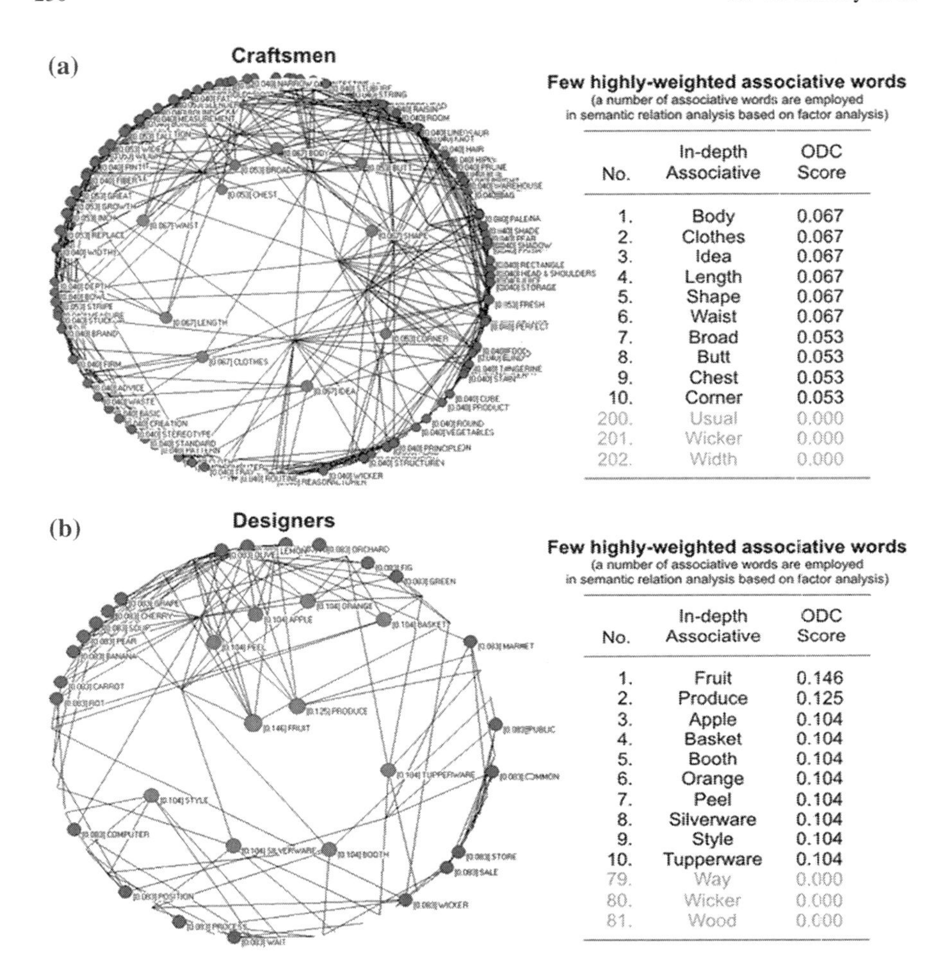

No.	In-depth Associative	ODC Score
1.	Body	0.067
2.	Clothes	0.067
3.	Idea	0.067
4.	Length	0.067
5.	Shape	0.067
6.	Waist	0.067
7.	Broad	0.053
8.	Butt	0.053
9.	Chest	0.053
10.	Corner	0.053
200.	Usual	0.000
201.	Wicker	0.000
202.	Width	0.000

No.	In-depth Associative	ODC Score
1.	Fruit	0.146
2.	Produce	0.125
3.	Apple	0.104
4.	Basket	0.104
5.	Booth	0.104
6.	Orange	0.104
7.	Peel	0.104
8.	Silverware	0.104
9.	Style	0.104
10.	Tupperware	0.104
79.	Way	0.000
80.	Wicker	0.000
81.	Wood	0.000

Fig. 4 **a** Simplified concept networks of craftsmen's in-depth cognitive level. **b** Simplified concept networks of designers' in-depth cognitive level

Up to this stage, data extraction according to the associative model suggested that craftsmen tended to activate low-weighted associative concepts, as demonstrated by the surface-level cognitive score of 169/202 (83.6 %). Designers, however, activated more highly weighted associative concepts concerning issues linked to the presence of the fruit basket/container, as significantly demonstrated by a high ODC score and lower surface-level cognitive score of 48/81 (59.2 %). Following are identified characteristics of craftsmen's and designers' associative concepts after reduction.

5.2 Analysis of Semantic Relation Based on Factor Analysis

We distributed 120 associative words corresponding to identified characteristics of associative conceptual structures; ODC scores ranged from highest to lowest (Table 3). Identified characteristics were *proportion, shape, operation, companion,*

Table 2 Extracted verbal expressions of in-depth cognitive level (after reduction)

Category (a)	List of 202 extracted verbal expressions (ordered by the highest ODC score)
Craftsmen	**Body, clothes, idea, length, shape, waist, broad, butt, chest, corner**, creativity, exercise, fresh, great, grow, grown, growth, ideal, impression, inch, oval, plaid, portion, replace, sample, size, slender, stripe, suggestion, tall, tight, weigh, wide, advice, bag, etc.
Category (b)	**List of 81 extracted verbal expressions (ordered by the highest ODC score)**
Designers	**Fruit, produce, apple, basket, booth, orange, peel, silverware, style, Tupperware,** banana, carrot, cherry, common, computer, fig, grape, green, juice, lemon, market, olive, orchard, pear, position, process, public, rot, sale, soup, store, wait, etc.

Table 3 Identified characteristics of craftsmen's and designers' associative concepts

Category	List of identified characteristics
Craftsmen	(*Proportion*) length, inch, oval, portion, size, tall, tight, wide, centimeter, width, thin, thick, form, rectangle, measurement, narrow, weight, etc.
	(*Shape*) body, shape, waist, butt, chest, corner, round, leg, hip, giant, cube, prism, etc.
	(*Operation*) exercise, grow, replace, advice, bold, blend, bond, decision, firm, fit, perfect, stain, form, combine, cover, tie, trace, use, etc.
Designers	(*Companion*) fruit, apple, orange, peel, banana, carrot, cherry, fig, grape, green, lemon, olive, orchard, pear, etc.
	(*Appeal*) salad, peel, juice, soup, process, produce, display, method, rotten, put, save, buy, shop, stand, fresh, etc.
	(*Scene*) booth, silverware, tupperware, market, public, store, crowd, leaf, tree, wood, etc.

Table 4 Rotated factor matrix

Adjectives (+)	Adjectives (−)	F1	F2
Scene	Less scene	0.942	0.009
Appeal	Less appeal	0.932	0.193
Companion	Less companion	0.891	−0.199
Proportion	Less proportion	−0.757	0.613
Shape	Less shape	−0.722	0.636
Operation	Less operation	0.140	0.912
Eigenvalue (after rot):		3.66	1.6
KMO:		0.571	

appeal, and *scene*—six variables used in factor analysis. Furthermore, the correlation among variables was extracted into two factors; the KMO score of 0.571 was significant. The factor matrix and corresponding names are as follows: (Tables 4 and 5).

For Factor 1, *Scene*, *Appeal*, and *Companion*, hereafter referred to as *Surroundings*, were associated with the presence of the fruit basket/container. For Factor 2, *Less Proportion*, *Less Shape*, and *Operation*, hereafter referred to as *Object-Oriented*, concerned technical aspects of the fruit basket/container. Furthermore, factors were displayed on an orthogonal map to investigate the semantic relationships of the identified characteristics of craftsmen's and designers' associative concepts (Fig. 5).

Table 5 Corresponding name

Factor	Adjectives	Eigenvalue	Factor name
F1	Scene, appeal, companion	3.66	SURROUNDINGS
F2	Less proportion, less shape, operation	1.6	OBJECT-ORIENTED

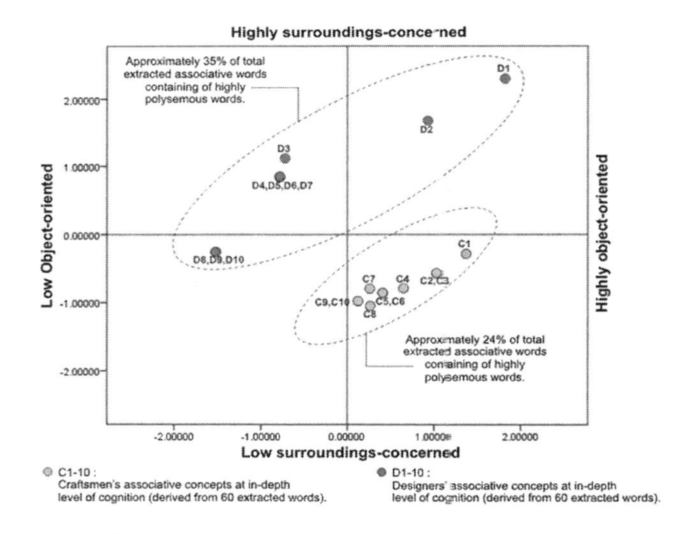

Fig. 5 Semantic relation map

6 Discussion

From the initial stage, the sorted verbal expressions showed that craftsmen paid attention to appearance and technical aspects of the fruit basket/container. They described such features as height, length, stack, capacity, standard, coating, and form (e.g., duck, heron). The extracted words obtained from concept network analysis were identified, as well as the characteristics of *proportion, shape*, and *operation.* In contrast, designers' sorted verbal expressions concerned issues pertaining to the presence of the fruit basket/container. Descriptive words included *place*, *kitchen*, *hang*, *pluck*, *tree*, *wood*, *inform*, *remind*, *children*, *salad*, and *dignity*. Some interesting comments were, "I don't want to put it on the table; I want to hang it," "I want it to be inviting so the children will reap its fruits," "How attractive to serve a salad in a fruit container," and "It's like a traditional banana-leaf container with prestige". The extracted words were identified along with the characteristics of *Companion, Appeal,* and *Scene.*

We identified that craftsmen tended to activate low-weighted associative concepts, as demonstrated by the high surface-level cognitive score of 169/202 (83.6 %). Designers activated more highly weighted associative concepts, as demonstrated by the high ODC score and lower surface-level cognitive score of 48/81 (59.2 %). We referred to the Associative Gradient Theory, which proposes that the more closely associated or "stereotypical" representations may lead to less creativity. The greater the number of associations, the greater the probability of reaching a creative solution, because remote associations (highly weighted associative concept) are best suited to such solutions [14–17]. We also found that approximately 24 % of 202 extracted words derived from craftsmen and 35 % derived from designers were highly polysemous. As Yamamoto et al. [18] argue, the polysemy of a design idea has significant correlation with its originality. It indicates that designers' in-depth cognitive levels have greater probability of reaching creative solutions.

The findings of this research suggest that the roles of closely and remotely associated concepts at the in-depth cognitive level during the early stage of idea generation are different for craftsmen and designers as they observe and define design problems. Craftsmen's in-depth cognitive levels, with fewer polysemous features, explain their concerns about tangible-related issues, such as proportion and shape. Designers' in-depth cognitive levels, with more polysemous features, concern intangible issues, such as surroundings context (i.e., eating culture) and users' affective preferences (i.e., companion, appeal). The semantic relation map confirms that craftsmen focus on the physical properties of an artifact instead of the surroundings and the user's affective preferences. Designers, on the contrary, are much more concerned about issues pertaining to the presence of the artifact and less attentive to physical properties.

7 Conclusion

In general, we can easily differentiate between artifacts created by craftsmen and designers by describing their appearance. However, it is difficult to describe the nature of creative cognition that influences the respective design thinking processes. This study has revealed the differences between in-depth cognitive levels of craftsmen and designers at the early stage of idea generation. Further, these findings can be developed as a reference for a co-created educational program (design training) that suits craftsmen's creative cognition.

References

1. Cross N (2006) Designerly ways of knowing. Birkhauser, Switzerland
2. Nagai Y, Georgiev GV, Zhou F (2011) A methodology to analyze in-depth impressions of design on the basis of concept networks. J Des Res 9(1):44–64
3. Houtz JC, Patricola C (1999) Imagery. In: Runco MA, Pritzker SR (eds) Encyclopedia of creativity, vol 2. Academic Press, Sand Diego, CA
4. Leijnen S, Gabora L (2010) An agent-based simulation of the effectiveness of creative leadership. In: Proceedings of the annual meeting of the cognitive science society, Portland, 5 Aug 11–14 2010
5. Reingold E, Colleen R (2003) Implicit cognition, in encyclopedia of cognitive science. Nature publishing group, London, pp 481–485
6. Georgiev GV, Nagai Y (2011) A conceptual network analysis of user impressions and meanings of product materials in design. Mater Des 32(8–9):4230–4242
7. Taura T, Yamamoto E, Fasiha MYN, Nagai Y (2010) Virtual impression networks for capturing deep impressions, design computing and cognition DCC'10. In: Gero JS (ed) pp xx–yy © Springer
8. Ericsson KA, Simon HA (1993) Protocol analysis: verbal reports as data. MIT Press, Cambridge
9. Mwihaki A (2004) Meaning as use: a functional view of semantics and pragmatics. Swahili Forum 11:127–139
10. Nelson DL, McEvoy CL, Schreiber TA (2004) The university of South Florida free association, rhyme, and word fragment norms. Behav Res Methods 36:402–407
11. Maki WS, Buchanan E (2008) Latent structure in measures of associative, semantic and thematic knowledge. Psychon Bull Rev 15(3):598–603
12. Batagelj V, Mrvar A (2003) Pajek—analysis and visualization of large networks. In: Jünger M, Mutzel P (eds) Graph drawing software. Springer, Berlin, pp 77–103
13. Leskovec J (2008) Dynamics of large networks. PhD thesis, Carnegie Mellon University, Pittsburgh
14. Mednick MT, Mednick SA, Jung CC (1964) Continual association as a function of level of creativity and type of verbal stimulus. J Abnorm Soci Psychol 69(5):511–515
15. Baer J (1993) Creativity and divergent thinking: a task specific approach. Lawrence Erlbaum Associates, Hillsdale, NJ
16. Eysenck HJ (1997) Creativity and personality. In Runco MA (ed) Creativity research handbook vol 1. Hampton Press, Cresskill, NJ pp 41–66
17. Martindale C (1995) Creativity and connectionism. In: Steven SM, Ward TB, Finke RF (eds) The creative cognition approach. The MIT Press, Cambridge, MA, pp 249–268
18. Yamamoto E, Mukai F, Fasiha MYN, Taura T, Nagai, Y (2009) A method to generate and evaluate creative design idea by focusing on associative process. In: Proceedings of the ASME2009. California

Part III
Design Aesthetics, Semiotics, Semantics

A Comparative Study of Traditional Indian Jewellery Style of *Kundan* with European Master Jewellers, a Treatise on Form and Structure

Parag K. Vyas and V. P. Bapat

Abstract Jewellery has universal appeals that transcend borders of countries and cultures. There are examples of Indian jewellery being influenced by European tools, technology and in turn European jewellery drawing inspirations from Indian jewellery motifs and culture. Indian and European jewellery styles due to this mutually wholesome relationship have common grounds for a comparative study, observing similarities and understanding differences. A traditional Indian goldsmith works in anonymity, rarely seeking personal name or recognition. Therefore, in India a style begets a name that is not associated with a particular design house, yet has a distinctly different identity. *Kundan* is one such example that uses an intricate frame of gold for setting minimally polished diamonds. A European master jeweller on the other hand has a style synonymous with their design houses. Names such as Cartier, Van Cleef and Arpels and Tiffany are few such examples. They are identifiable from other comparable styles by use of distinctive motifs, treatment of form, usage of specific cuts of gemstones and types of setting. These characteristics give a logical basis for comparison and understanding features of form. Based upon these, Indian style of *Kundan* is compared with European contemporary styles. A treatise on the subject elucidates how a typical Indian jewellery style is analogous to European Master Jewellery styles. It draws parallels between these styles and provides a structure for studies pertaining to form in domain of jewellery. Such studies in domains of design are new, this subject gains importance by sharing deep insights from pioneering research in subject matter.

Keywords Traditional Indian jewellery · European master jewellers · Design · Studies in form

P. K. Vyas (✉) · V. P. Bapat
Industrial Design Centre, Indian Institute of Design Bombay, Powai, Mumbai, India
e-mail: paragvyas01@gmail.com

A. Chakrabarti and R. V. Prakash (eds.), *ICoRD'13*, Lecture Notes in Mechanical Engineering, DOI: 10.1007/978-81-322-1050-4_19, © Springer India 2013

1 Introduction

Kundan is traditional Indian jewellery style that is elaborately embellished with gemstones and colourful enamel. It has a character of its own that has remained unchanged over a period of time. Motifs [1, 2], often identifiable by their individual names are used in this particular style. Motifs are frequently mimicking nature, liberally adopting from flora and fauna [3, 4]. These smallest semantic units are the fundamental building blocks and used in combination with each other make a form cluster. In turn, these form clusters make patterns which are repetitive and appear through the body of jewellery as a coherent theme. There are manners of construction of these form clusters that are passed down from generation to generation, from master to an apprentice. An article of kundan jewellery is distinctly identifiable by its characteristic appearance [2] as shown in Fig. 1. This technique of setting, from which it derives its name does not belong to a particular design house but encompass different jewellers and their style of working in one coherent style. The features of form are use of specific shape of smallest semantic units [5] and the type of setting [6]. Practitioners spread across working in different locations independent of each other are joined by a common invisible link, making kundan distinctively identifiable.

A European master jeweller on the other hand has a style synonymous with their design houses. In comparison to kundan, the working style is limited to the house in terms of articles they specialise in and design themes. Names such as Cartier, Van Cleef and Arpels and Tiffany are few such examples. They are identifiable from other comparable styles by use of distinctive motifs, treatment of form, usage of specific cuts of gemstones and types of setting. For example a six prong tiffany setting, signature of this style is distinctively different from the Van cleef and Arpels mystery setting. Though both styles use formal brilliant cuts, their treatments have marked differences as reflects in following sections.

Fig. 1 An article of kundan jewellery, typical example

2 Observations and Articulation of Select Styles

For the purpose of understanding the selected styles are articulated briefly, so that at a glance the visual appreciation and comparison is plausible.

Cartier—though founded in America they in their manner of working is similar to European master jeweller and has a particular style that drew inspiration from animal forms, typically cheetah, birds and elephant [7, 8]. By virtue of their Geneva office, they do follow a European style and traditions. It is known for its exquisite craftsmanship and high quality of finish. This style is exemplified by pictures as in Fig. 2.

Van Cleef and Arpels—this house is specially known for 'invisible' setting where no metal is visible between two adjoining gemstone, this mysterious way of holding begets the name mystery setting [9, 10]. An additional cut under the girdle of the gemstones to be set makes this plausible. This style is their signature and is patented by them. This style is exemplified in Fig. 3.

Tiffany—is known for its six prong setting where the gemstone is set for maximum visibility and an effect is created by raising and supporting the stone in a manner where it appears levitating [11, 10]. This style is exemplified in Fig. 4.

Kundan—though it does not belong to a particular house, it is a distinct style that uses flush stone close setting [12] to hold minimally polished flat stones [13, 14]. The stones are set in an intricate framework of gold by compressing gold foils all around the stone to make a firm fit. This style is exemplified in Fig. 5.

They are identifiable from other comparable styles by use of distinctive motifs, treatment of form, usage of specific cuts of gemstones and types of setting, which constitute the characteristics or a visual identity. Characteristic of each style that are synonymous with a particular style are tabulated as under for Comparative

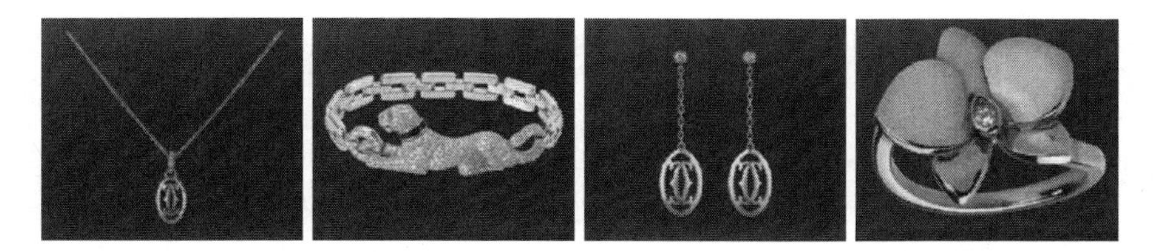

Fig. 2 Cartier as a style exemplified

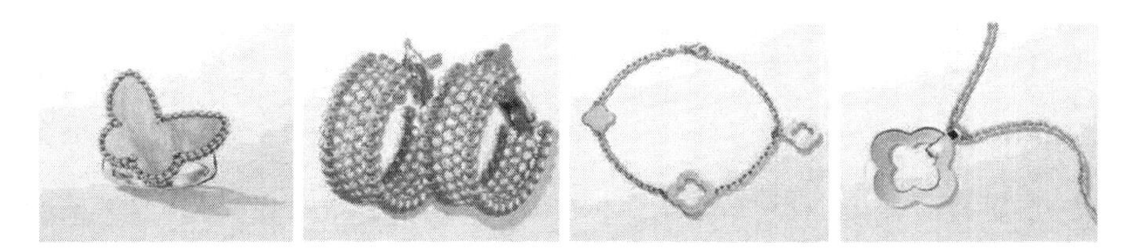

Fig. 3 Van Cleef and Arples as a style exemplified

Fig. 4 Tiffany as a style exemplified

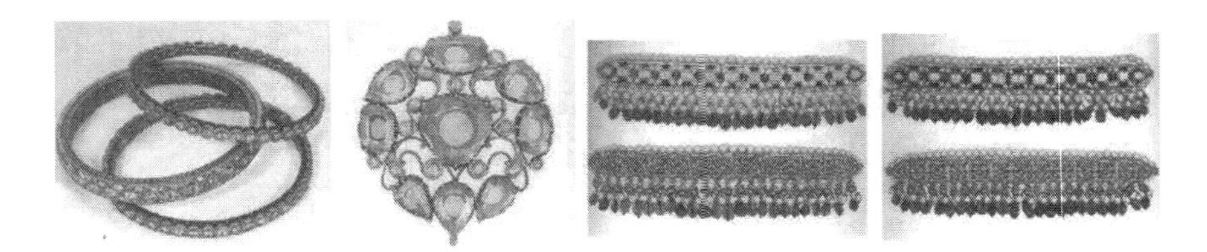

Fig. 5 Kundan as a style exemplified

study. This particular style of comparison is used by design houses for trend tracking as well as development of a storyboard for new trend style. It is readily adopted as a research tool and applied for this application.

The limitation of this style is that it does not follow a ruthless mathematical objectivity, however, design rarely demands that level of mathematical interpretation and often a gentle guidance in the direction is adequate [15]. This model can be adapted for such research.

Following is a tabulation of formal aspects that are taken into consideration while appreciating the products visually and articulating discussions and conclusions, as in Table 1.

3 Methodologies

In search for a suitable methodology for visual research, first inspiration was taken from visual research in domain of product form, [16] proposed a structure containing presentation plates, well suited for such purposes. For comparison of features of form products are arranged for visual comparison on plates providing a common frame of reference. This is in concurrence of views with [17, 18], where factor approach is propounded for dimensional classification and intercorrelation of items as shown in Fig. 6. Using this frame for reference articles from these master jewellers and jewellery tradition are visually appreciated and analysed. This approach is in tandem with evaluation approach by experts. The subsequent plates on following pages are schematically represented in the dimensionality model. The plates arranged in this manner create a frame of reference with respect to a datum. It is in light of this frame of references that form can be studied and interpretations can be made more objectively.

Table 1 Formal aspects taken into consideration for visual appreciation

		Comparison of master jewellers/tradition			
Sr. No.	Feature/formal aspect	Cartier	Van Cleef and Arpels	Tiffany	Kundan
1.	Motifs used	Leopard and elephant	Floral, post modern, art deco	Geometric	Flora and fauna/nature
2.	Treatment of form	High mechanical precision in metal.	Very high mechanical precision in metal and stones.	Geometric and precise forms.	Precision achieved visually using very irregular stones.
3.	Usage of specific cuts of gemstones	Round brilliants/other brilliant cuts	Square/nicked under girdle	Round brilliant/tear drops	As found/minimally faceted
4.	Type of setting	Pave/prong	Invisible or mystery setting	Six prong	Flush stone close setting
5.	Surface appearance	High lustre metal	More use of colour stones	Equitable amounts of metal/stones	Hand worked metal
6.	Geometric precision	High emphasis	Very high emphasis	Emphasis on type of prong settings	–
7.	Embellishment	–	–	–	Backside richly enamelled
8.	Representative picture	As in table	As in table	As in table	As in table

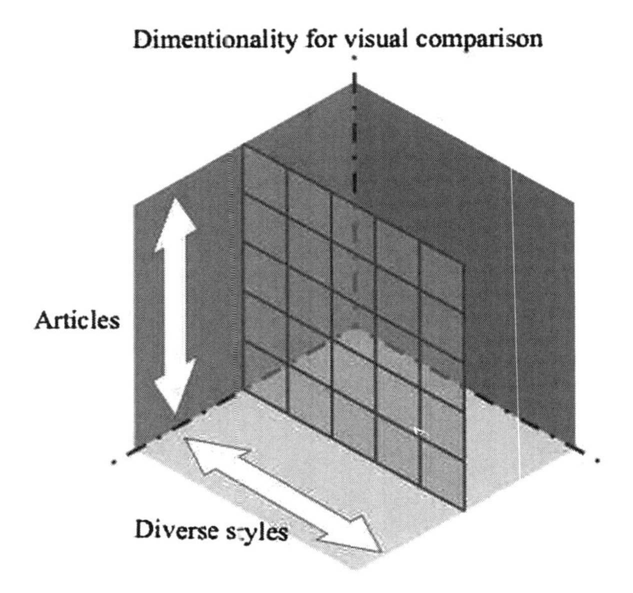

Fig. 6 Illustrated dimensionality of the comparative study

A mental model is created by observer in everyday life in order to understand everyday world. Perceptions and hence the observations on which a theory is based is shaped by interpretive structure of human brain. A model therefore is a help to understand a complex thing with relative ease. However, this simplification is by way of compromise, so proposed models have their applications and limitations.

The articles from four broad categories were selected namely, rings, earpieces, neck pieces and miscellaneous articles. These are tabulated and pertinent observations and comments were duly noted under each category for discussions and interpretation, as shown in Table 2. As methodical research are in their initial phases, they need validation by experts for checking argumentative fallacies and misinterpretation of terms. Study was presented to experts and practitioners. Their valid comments and point of views were incorporated in study.

4 Discussions

The European jewellery by master jewellers uses predominantly floral and butterfly as motifs in a very strong way. Both derive their form from nature and this is common factor with Indian jewellery too, drawing inspiration from nature. However, the significant difference is avoidance of dragonfly and butterfly as they are insects from the family of lepidoptera family, or insects that transform. Owing to a general aversion to the use of insect forms in indian jewellery, such occurrences rare in indian jewellery, unless they are bought abroad or made as an imitation of a particular article.

Table 2 Articles tabulated for visual appreciation and comparison

Sr. No.	Article	Tiffany	Van Cleef & Arpels	Cartier	Kundan
1	Rings				
2	Ear Rings				

(continued)

Table 2 (continued)

Sr. No.	Article	Tiffany	Van Cleef & Arpels	Cartier	Kundan
3	Pendents				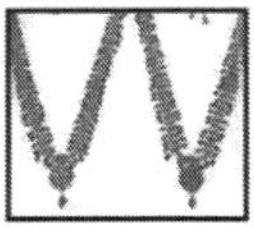
4	Misc. Accesories				

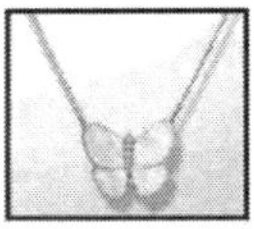

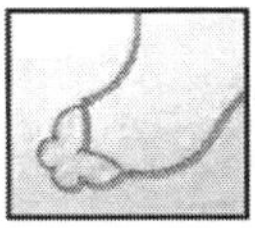

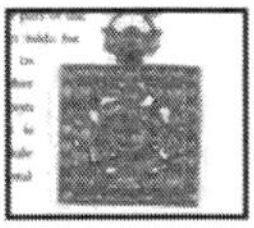

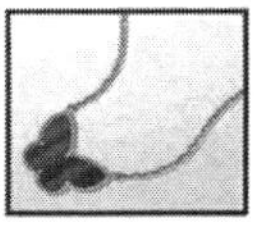

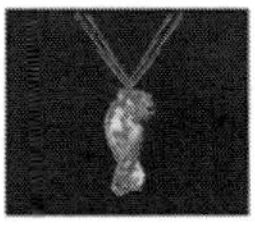

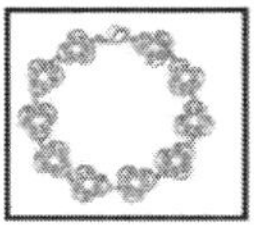

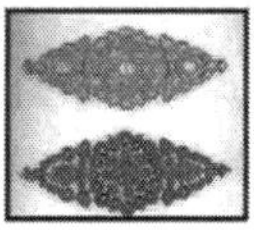

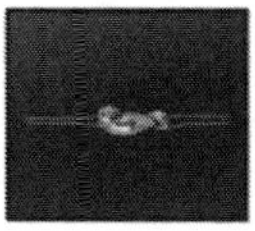

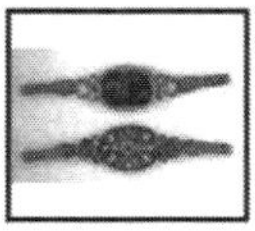

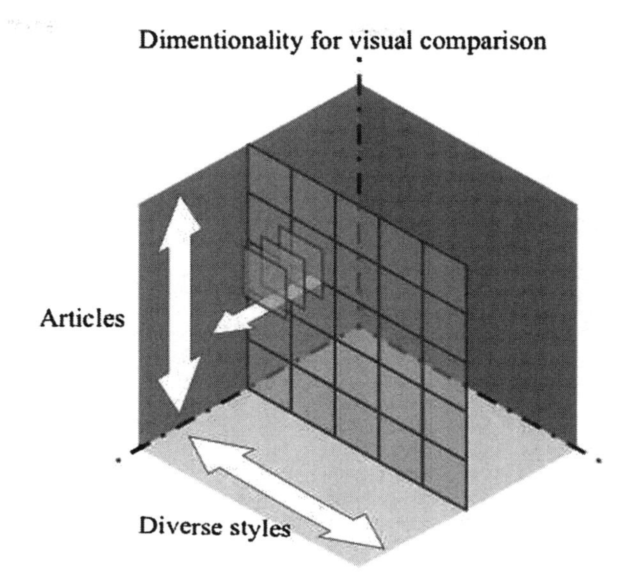

Fig. 7 Addition of new dimensionalities and parameters in the present study

The common element between western and Indian jewellery is use of lion head in some jewellery, with a minor change from leopard to lion in Indian jewellery. Sometimes in India the nails or the canine teeth of the animal used as an amulet with a belief to propitiate animal spirits or invoke divine protection. It is these belief system that govern the use of motifs in a particular article of jewellery [19]. For example, fish is a fertility motif and flowers depict blossoming, hence, find an adaptation in both eastern as well as western jewellery forms.

Geometric forms are common to both trends and freely adapted for use in Indian as well as western jewellery. The common frequently used forms are, tear drops, circle, oval and a variety of polygons both regular and irregular [20] and they appear consistently across diverse time and space in jewellery forms. This can be attributed to the simplicity and universal acceptance of polygons as readily adaptable forms for a variety of product forms. Familiarity with object, ease of definition and manufacturing are the factors that contribute to their universal usage and acceptability.

5 Conclusions

The conclusions are drawn under two heads drawing out similarities and dissimilarities. In general European inclination is towards a simple form that has a high level of precision, while Indian forms have a high level of detail per unit area with less emphasis on precision. There is a high level of symmetry and aesthetically pleasing appearance in Indian jewellery with minimal polishing, which clearly reflects in forms.

5.1 Similarities

1. Both European as well as Indian jewellery techniques draw inspiration from nature [3] often mimicking nature.
2. Polygons and familiar forms are common features, however, the overall treatment of form may be different on grounds of precision.
3. Diamond and its various cuts are preferred for its hard wearing quality in both Indian and European styles.
4. Indian styles prefer one of five major gemstones along with diamonds for setting and therefore are more colorful. On the other hand western styles are more liberal in trying out semi precious stones.
5. Emphasis on simple things done with perfection in European style, large details packed in a small unit area is the defining feature of Indian style.

5.2 Dissimilarities

1. Though adapted from nature, Insects are generally avoided in Indian styles while readily accepted in European.
2. Overall symmetry is preferred in Indian jewellery, while asymmetry and deliberate use of visual provocation is characteristic of European styles.
3. Use of stones in Indian style is minimally faceted stones retaining original form and weight while European styles go for formal and brilliant cuts compromising weight.
4. Emphasis on hand work in indian style demands acceptance of imperfections or individuality of a craftsman. While the mechanical precision of European style is independent of a person.
5. European craftsmanship heavily relies on technology, which to an extent dictates form. While *kundan* relies more on skill and creativity in use of material.
6. Conformance to original floral or animal motif is high in European jewellery while a high level of abstraction in form is characteristic of *kundan*.

6 Future Directions

1. New dimensions and dimensionalities can be added as shown in Fig. 7 to the proposed construct of the model and different parameters can be added either locally or following a grid approach [19].
2. A frame of reference can be extended in a parametric way and a ordinal scale can be incorporated for managerial as well as mathematical interpretation of things, that are visual in nature.

3. A tabulation is scalable in its construct and various columns containing styles and rows containing articles can be added to fortify the study.
4. An approach this way is better than an ad hoc approach, in the sense it gives a more rational approach comparatively.

7 Limitations

Study is a proposed model concept and there can be other ways to observe and compare styles, however, design is undemanding of mathematical precision and often only a gentle hint in the direction is good enough and serves the purpose. Hence, this model gains importance in quick market research and understanding things and bring in a level of objectively.

References

1. Balakrishnan UR, Kumar MS (2004) Dance of the peacock: jewellery traditions of India. India Book House Pvt. Ltd, India, pp 123–212
2. Untracht O (1997) Traditional jewellery of India. Thames & Hudson Ltd, London, pp 103–269
3. Adorno TW (2011) Aesthetic theory. University of Minnesota Press, USA, p 118
4. Sharma RD, and Varadarajan M (2008) Handcrafted Indian enamel jewellery. Lustre Press, Roli Books, India, p. 21–33
5. Vyas P, Bapat VP (2010) Identification and classification of semantic units used in formation of patterns in *kundan* jewellery, a methodical approach, design thoughts. IDC, IIT, Bombay
6. Untracht O (1985) Jewellery concepts and technology. Doubleday & Company, Inc., Garden City, p 580
7. Cartier "Jewellery" in http://www.cartier.co.uk/#/show-me/jewellery
8. Coleno N (2009) Amazing Cartier jewellery design since 1937. Published by Flammarion, Spain, p. 62
9. Van Cleef, Arpels "Jewellery" in http://www.vancleefarpels.com/ww/en/
10. Snowman AK (1990) The master jewellers. Thames & Hudson Ltd., London, p 215
11. Tiffany "Jewellery" in http://www.tiffany.com/Shopping/Default.aspx?mcat=148204
12. Balakrishnan UR, Kumar MS (2006) Jewels of the Nizams. India Book House Pvt. Ltd, India, pp 60–224
13. Tait T (2006) 7000 years of jewellery. The British Museum Press, London, p 174
14. Elgood R (2004) Hindu arms and ritual. Mapin Publishing, Ahmedabad, p 177
15. Kirsch I (2009) The Emperor's new drugs, exploding the antidepressant myth. The Bodley Head, Random House, London, p 51
16. Mokashi RP (2009) Study of user preferences for the visual domain of product form, IIT Guwahati
17. Shajahan S (2004) Research methods for management. Jaico Publishing House, India
18. Madhusudanan N, Chakrabarti A (2011) A model for visualizing mechanical assembly situations research into design. Research Publishing, India

19. Vyas P, Bapat VP (2011) Design approach to *kundan* jewellery-development of a tool to study preferential likeness of articles using mother grid and form clusters in a methodical manner, research into design. Research Publishing, India
20. Vyas P, Bapat VP (2011) investigation of form clusters made of smallest semantic units and patterns they create as building blocks of *kundan* jewellery, research into design. Research Publishing, India

A Structure for Classification and Comparative Study of Jewellery Forms

Parag K. Vyas and V. P. Bapat

Abstract Indian jewellery has a character of its own. It is rich in form and crafted in intricate details. Motifs often identifiable by their individual names are the fundamental building blocks. They are used in combination with each other to make a form cluster. These form clusters are repetitive and appear throughout the body of jewellery as a coherent theme. Complex Structures in gold, forms created using these motifs, are numerous and are made in a variety of ways. They track construction details to conform to body contours, for best presentation view they are often worn on a junction such as neck or wrist. An article therefore, is best understood in form with reference to anthropometric dimensions used as an underlying framework. The articles specifically chosen for study are neckpieces, as they are the largest in size and central part of a particular set. They are directly in line of sight for visual appreciation and therefore gain further importance as lead pieces in jewellery design. This study expounds on diverse types of forms and their characteristic features. A structure is presented by a comparative study that is expected to provide orientation, define key aspects of form and parameters. Outcome of this study is for benefit of jewellers as well as clients by better articulation of jewellery forms and thereby clear understanding of clients' expectations.

Keywords Form clusters · Complex structures · Jewellery design · Visual appreciation

P. K. Vyas (✉) · V. P. Bapat
Industrial Design Centre, Indian Institute of Design Bombay, Powai, Mumbai, India
e-mail: paragvyas01@gmail.com

A. Chakrabarti and R. V. Prakash (eds.), *ICoRD'13*, Lecture Notes in Mechanical Engineering, DOI: 10.1007/978-81-322-1050-4_20,

1 Introduction

Kundan is traditional Indian jewellery style that is elaborately embellished with gemstones and colourful enamel [1]. It has a character of its own that has remained unchanged over a period of time. Motifs, often identifiable by their individual names are used in this particular style. Motifs are frequently mimicking nature, liberally adopting from flora and fauna. [2]. These smallest semantic units are the fundamental building blocks and used in combination with each other make a form cluster. In turn, these form clusters make patterns which are repetitive and appear through the body of jewellery as a coherent theme. There are manners of construction of these form clusters that are passed down from generation to generation, from master to an apprentice. The knowledge is typically passed down verbally and little written material is available, typical of Indian traditions.

Our concepts of reality as Model dependent realism [3] makes it simple and comprehensible by having a structure for understanding things around us [4]. The beauty of a model is in its simplicity and how closely they agree with observations. Both of these are met by a way of partial fulfilment of objectives, or by finding a golden balance. One can use whichever model is more convenient in the situation under consideration. Model dependent realism applies to conscious and subconscious mental models we create in order to interpret and understand the everyday world.

In absence of well documented and dependable material it is difficult to have a structure for scientific studies in any domain, in particular area of form based studies in jewellery it becomes more severe as the field is very little explored [5, 6].

This primary objective of study is on providing a structure for comparative study of jewellery forms. The secondary objective is to articulate terms used in description of various aspects of jewellery. This is to provide a common standard vocabulary set for efficient and effective communication between client and designer.

2 Structure to Study Jewellery Form

Form of an article of jewellery which is the first visual interaction with a prospective client has a high influence on the client's decision in selection process [7]. It is one of the most important aspects that influence the preferential choice of the client. When it comes to study forms in a constructivist approach, literature is silent on the subject. It is of importance therefore to illuminate this potentially rich area of research.

A necklace is chosen for study as it is the largest article of jewellery amongst a collection, typically called 'a set' comprising of bangles or bracelets, necklace and ear pieces that follow a coherent theme. It is worn on the neck, in line of sight for

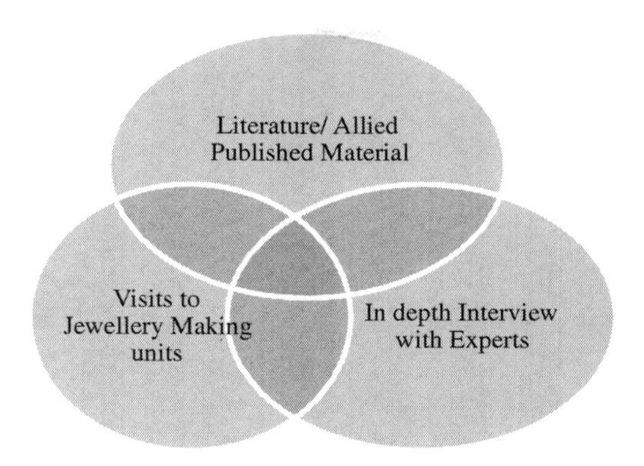

Fig. 1 Approach for study and validation by experts

visual appreciation and presents a good view, as compared to bangles which is worn on wrist and relatively small in size. The motifs are comparatively large and therefore it is dominant and most expensive article, ideally suited for such study. It was observed during visits to shops that the clients selects a neckpiece (or a design for the same first) and the jeweller makes the remaining articles of aforesaid set in line with this as a reference.

In search of a structure for a comparative study a three pronged (Fig. 1) approach was adopted firstly literature search, secondly visits to jewellery making units and thereafter informal in depth interview with experts for validation of findings. The proposed structure for classification evolved from observations on shop floor where a generating curve, often a circle of certain diameter is used as a reference to arrange form clusters.

3 Structure to Study Jewellery

There are varieties of forms existing in jewellery; the choice available to a potential buyer is enormous. With variants and creative embellishments *kundan* neckpieces present virtually limitless options to the observer [8]. A large variety of objects demand some classification for methodical study, for example the modern form of periodic table and its predecessors gave some sort of a structure for studying the elements. Despite their limitations and shortcomings they serve a purpose by providing a model for classification and cataloguing. This in turn makes it easy for us to comprehend large number of objects by compartmentalisation into similar groups and categories. A typical neckpiece is illustrated for its components in the following illustration.

In search for a rational for classification of jewellery articles, insights came from anthropometric studies [9] where the most basic differentiation can be made based on the manner in which they are worn. These can be broadly classified in two categories as follows:

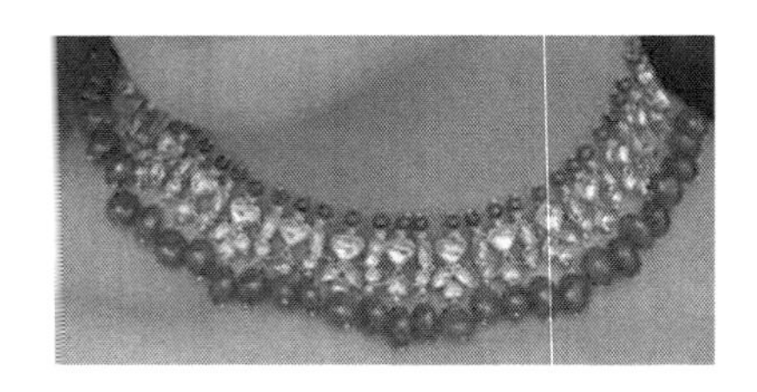

Fig. 2 Neckpieces worn close to the body as chokers

1. Neckpieces Worn close to body as chokers that are conforming to the dimensions and contours of the neck and its proximity. As they are close fitting and worn tightly, gravity has little effect on their overall appearance. It has a slider and tassel that slides back and forth to facilitate the fit. As shown in Fig. 2.
2. Neckpieces Worn as chains or strings of a particular length. These are free hanging and the appearance is formed by weight and gravity. It has a hook and eye arrangement for opening that facilitates wearing. As shown in Fig. 3.

Both the types are made up of repetitive form clusters that appear throughout the body of the neckpiece. The free hangings do not necessarily conform to body form and dangle freely under gravity. The current trend is inclined towards the first type that is worn close to neck. This is due to two reasons; firstly they are worn against skin with a firm fit giving a neat appearance. They do not move as freely

Fig. 3 Neckpieces worn as chains or strings

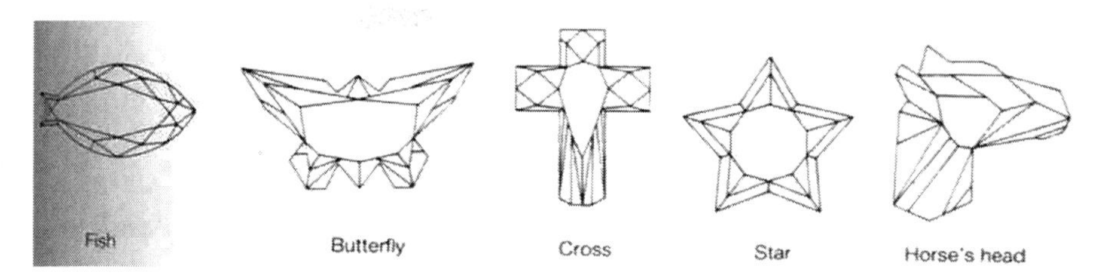

Fig. 4 Resemblance to a known object is used for Classification of fancy cut diamonds illustration adopted from Vyas and Bapat [6]

with movements and therefore can be worn both formally and casually. Secondly, their linear length is relatively small as compared to the free hanging ones, thereby making them less expensive and light. The design of a form cluster, however, for both the types may be same. Owing to this prevailing factor the first type is taken for further study and elaboration.

In search for further examples of classification structures that can be adopted for this purpose, reference was found in cataloguing of brilliant cut diamonds and other gemstones. Resemblance to a known object as a reference is used for Classification of fancy cut diamonds that come in a variety of shapes ranging from heart, butterfly, horse [10]. A similar approach is adapted for classification of necklaces that provides a frame of reference. Words used on shop floor are freely accepted and incorporated as they communicate that particular aspect most efficiently (Fig. 4).

These neckpieces can be classified in five categories based on composition of form clusters with respect to each other. Their characteristic features are summarised as follows:

1. Simple form clusters: these are the type of necklaces made up of form clusters that are same in size and appear repetitively through the body of the article at equal intervals or cluster pitch. These are easy to make as they have same sized smallest semantic units constituting them. They can be batch produced and subsequently assembled to form a necklace. As shown in Fig. 5.
2. Chhed Uttar: (or uniformly tapering form clusters) are the type of necklaces made up of form clusters that have a large form cluster at the centre and as the necklace progresses to either side the clusters decrease in size. This uniform gradient is called chhed uttar or a gradually decreasing cluster size. A variable cluster pitch is the typical characteristic of this style. This decrease in size, though appealing and interesting to eyes, requires each form cluster to be made using smallest semantic units that themselves are reducing in size from the previous adjoining cluster. This type of work therefore is far more intense in terms of time and money. Slowly the uniform sized clusters for its ease of manufacturing are replacing this type of work. As shown in Fig. 6.
3. Simple form clusters with pendent: these are the necklaces that have a simple form cluster, fixed cluster pitch as mentioned in first type. Besides this it has a

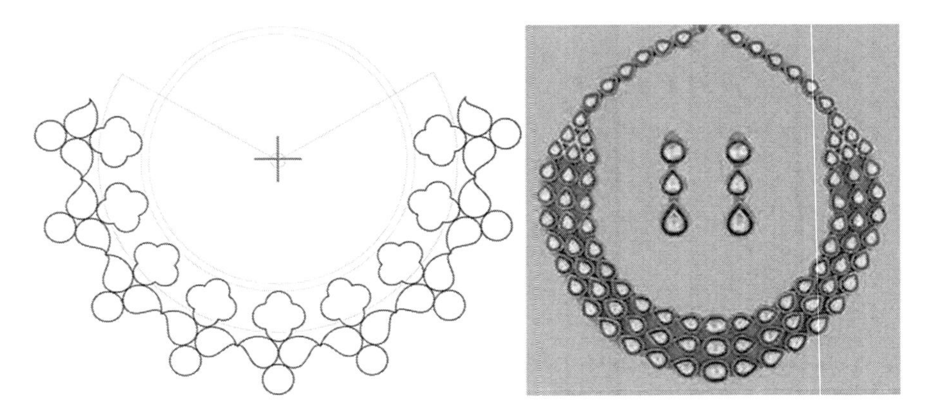

Fig. 5 Examples of simple form clusters, illustration and picture

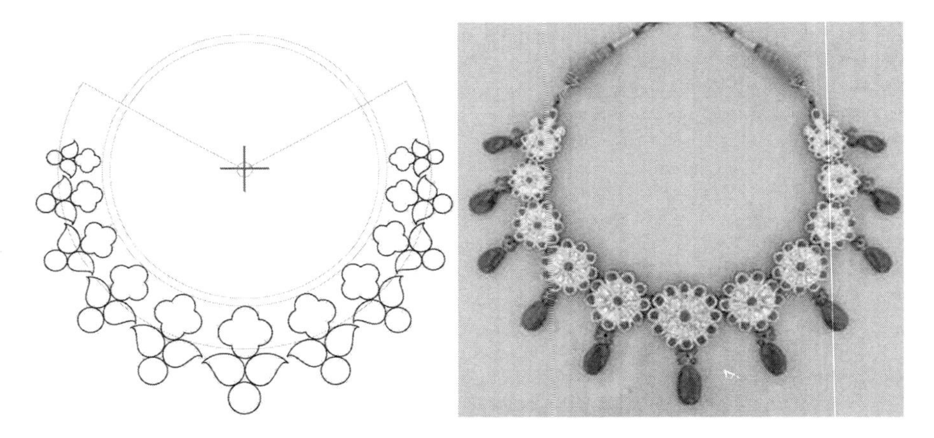

Fig. 6 Examples of *Chhed Uttar*, illustration and picture

pendent or 'padak' a typical form cluster hanging from the central unit. This cluster may have same combination of semantic units alternatively a different design that gels well with the rest of the structure may form this medallion. As shown in Fig. 7.

4. Chhed uttar with pendent: it follows same construction details as chhed utaar, but has a medallion similar to previous description. One typical variant of this type is where the pendent is integrated within body and rest of the clusters follow the gradient. As shown in Fig. 8.
5. Random paved: are the necklaces that appear as if paved with different shapes of smallest semantic units. Even though the sizes are random, yet an overall pleasing composition is achieved by carefully organising the adjoining units. As shown in Fig. 9.

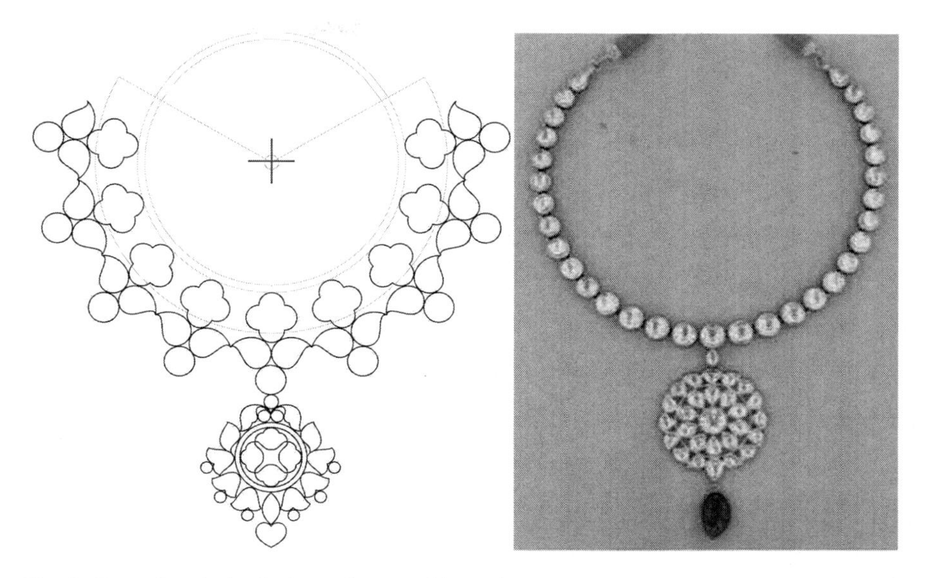

Fig. 7 Examples of simple form clusters with pendent

Fig. 8 Example of Chhed uttar with pendent

A system diagram can therefore be formed for understanding at a glance (Fig. 10). With this structure it is possible to compare the forms of neckpieces inter group as well as intra group. An objective study therefore can be based on the articulated factors.

Proposed further is a conceptual structure for an inter group comparison of simple form clusters that takes into consideration three factors, number of semantic units constituting a form cluster, cluster pitch and number of form clusters constituting a necklace as in Fig. 11.

Fig. 9 Example of clusters randomly paved with shapes

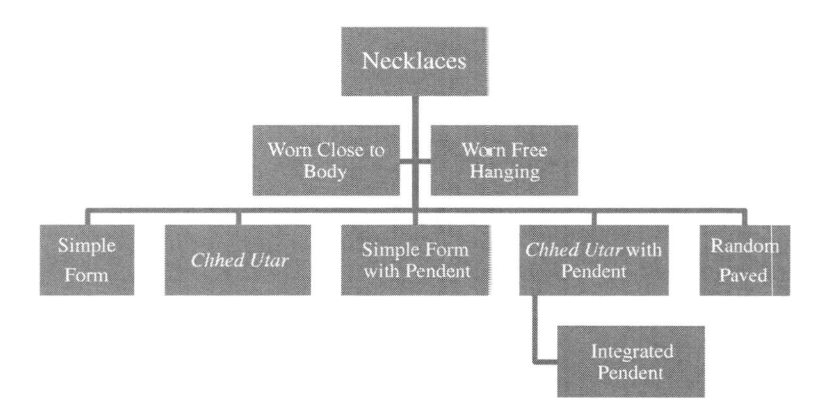

Fig. 10 Structure diagram of form cluster based classification of neckpieces

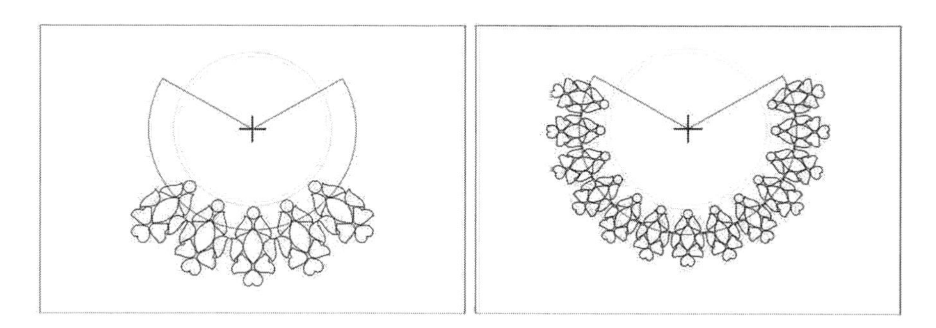

Fig. 11 Formation of form clusters for display and visual comparison

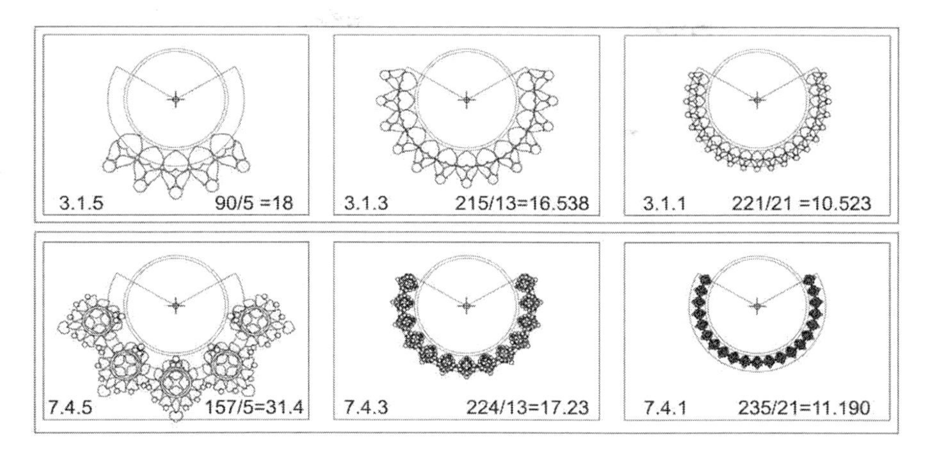

Fig. 12 Mathematical interpretation of cluster pitch

Arranged in form of a grid these can be used for mathematical interpretation of various parameters as shown in Fig. 12 as well as a three dimensional virtual model to assess preferential likeness of particular articles by using it in an adaptable appropriate form such as a display book or catalogue to have an insight into a clients choices and the inclinations.

This structure can also be used by designers for intra group comparison by two intra group grids being presented to a subject. Say for example, a simple form cluster can be compared with a chhed uttar type of form cluster with same number of clusters in each necklace.

4 Advantages

The beauty of a conceptual model is in its simplicity and adoptability. The proposed model structure can also be used for a variety of purposes like inter group and intra group comparisons. It can be used for assessing preferences of diverse and specific client profiles. As shown in the scalable construct in Fig. 13.

Assays sensitivity of the experiment being designed can be increased or decreased depending on the requirement [11]. Say for example for a quick survey a three by three grid can be used while for a fine survey a five by five grid can be used. The grid can only be increased in one particular dimension in case a fine measurement is needed for one particular aspect.

This model is minimalistic in its construct and has scalability in all desired dimensions. The model can also be used for trend tracking by making surveys time to time to observe changes that are taking place in preferential likeness of clients over time as shown in Fig. 14. This trend tracking, however, may include temporal

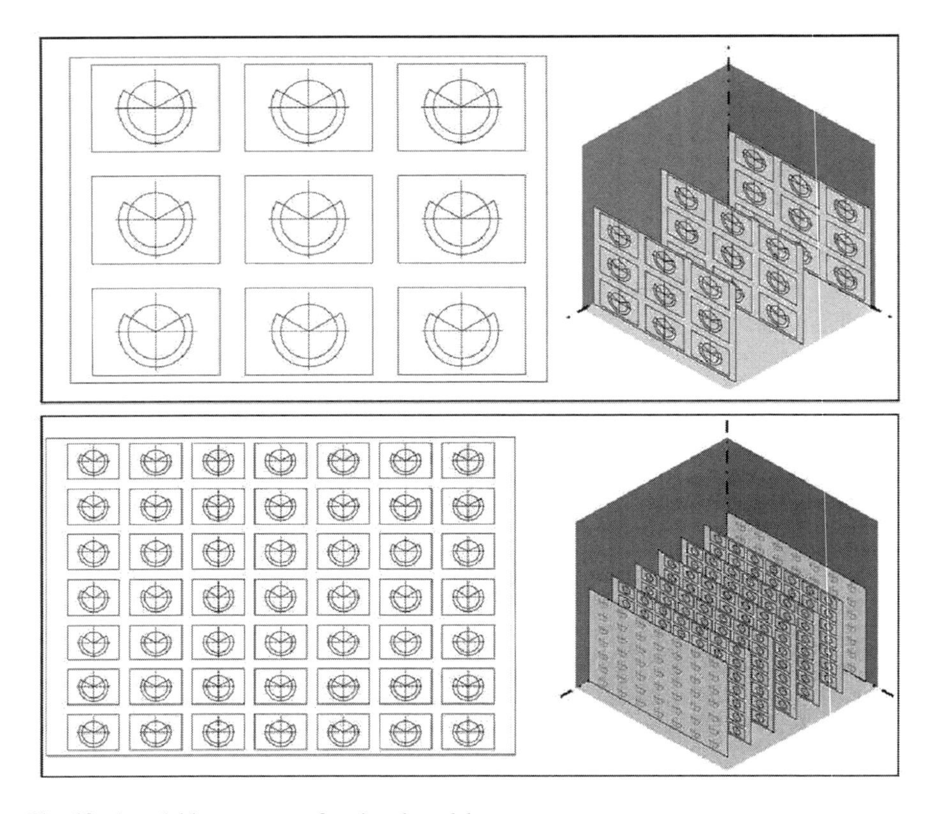

Fig. 13 A scalable construct of a virtual model

studies spread over a long time frame. The proposed model can accommodate other parameters to be tested in specific areas.

5 Limitations

The limitations can be best understood by quoting Hawking and Mlodinow [12], who propound that all models that we create will be off beam in some aspects. Nonetheless, as ground-breaking attempt this is a practicable structure. The success of a model is in how closely it matches to reality.

Also the field has vast potential for research and there may be more ways to classify and catalogue jewellery by presentation of different models which may be more effective in illuminating some other aspects.

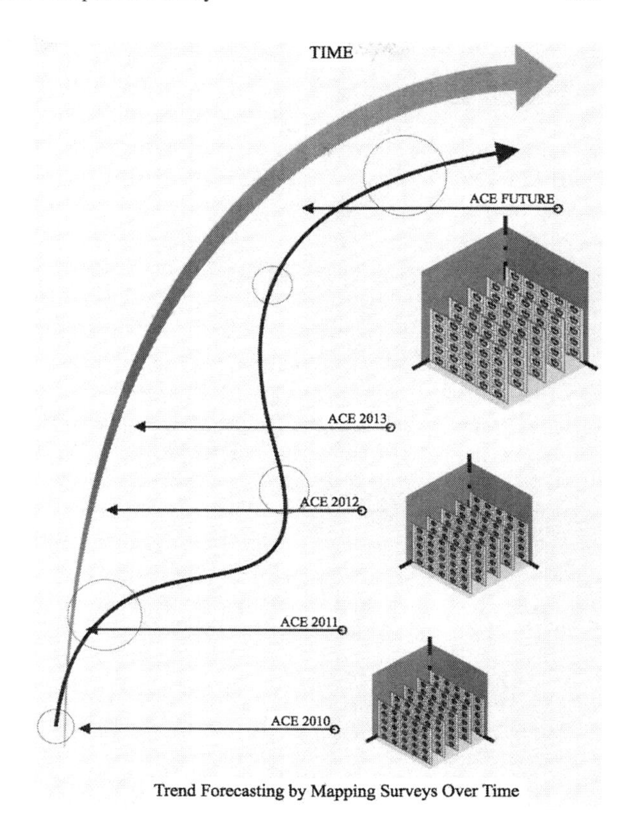

Fig. 14 Trend tracking across time

6 Future Directions

This study can be further extended to take into considerations other relevant factors such as cost, colour of gemstones being set, a combination of coloured gemstones for a particular article. Moreover, this study can be periodically administered to study trends and see if there are any emerging patterns.

Over a long period of time this data can be used for understanding changes that kundan designs are going through or if any transformation is happening in the basic design style itself.

Trend forecasting is a very critical aspect of jewellery design, as the survival of many jewellers depends on accurate assessments of potential needs of a future market. The tool with a good data base generated over years can be used to a limited extent to predict future behaviour; this can be based on mathematical parameters and not opinions as it happens in this domain presently.

The proposed model can be readily adapted for other types of jewellery and miniature products for paired comparisons and studies that require features of form to be compared.

Acknowledgments The authors greatfully acknowledge Vinay S. and Manu Panchal for illustration and photographs, Prof. G. V. Sreekumar and Nitin S. for valuable inputs on construct of trend tracking and forecasting model.

References

1. Untracht O (1985) Jewelry concepts and technology. Doubleday and Company, Inc., Garden City, New York, p 614
2. Sharma RD, Varadarajan M (2008) Handcrafted Indian enamel jewellery. Lustre Press, Roli Books, India, pp 36–41
3. Hawking S, Mlodinow L (2010) The grand design. Bantam Dell, London, p 46–51
4. Hawking S, Mlodinow L (2010) The grand design. Bantam Dell, London, p 117
5. Vyas P, Bapat VP (2011) Design approach to kundan jewellery—development of a tool to study preferential likeness of articles using mother grid and form clusters in a methodical manner. Research into design, Research Publishing, India, p 775–783
6. Vyas P, Bapat VP (2010) Identification and classification of semantic units used in formation of patterns in kundan jewellery, a methodical approach. Design Thoughts, IDC, IIT Bombay, p 59–72
7. Vyas P (2007) Understanding features of form in fine jewellery, by overt eyes. Dev Agric Ind Ergonomics 1:101–108
8. Balakrishnan UR, Kumar MS (2006) Jewels of the Nizams. India Book House Pvt. Ltd, India, p 81–203
9. Chakrabarti D (1997) Indian anthropometric dimensions, for ergonomic design practice. National Institute of Design, Ahmadabad
10. Pagel Theisen V (1993) Diamond grading ABC handbook for diamond grading. Rubin & Son Bvba, Antwerpen
11. Shajahan S (2004) Research methods for management. Jaico Publishing House, India
12. Hawking S, Mlodinow L (2005) A briefer history of time. Bantam Dell, New York, p 23

Product Design and the Indian Consumer: Role of Visual Aesthetics in the Decision Making Process

Naren Sridhar and Mark O'Brien

Abstract The research paper aims to examine the influence of 'Product design' on buying decision making in the Indian urban market sector, focussing on visual product aesthetics; the characteristics that create a product's appearance and have the capacity to affect observers and consumers [1]. Product design, specifically the 'Visual Aesthetic' [Visual Aesthetics (VA), for the purpose of this study, is defined to entail the colour, size and proportions, materials and design expression of the designed product] has been recognised as a key strategic variable in securing or defending a marketplace advantage. This question will be examined in the context of Indian social, cultural and economic systems and with regards to the relative position of visual aesthetics in the decision making process of young adults in the Indian urban consumer market.

Keywords Product design · Indian consumer · Social psychology

1 Introduction

'Aesthetics proper is a recent discipline, born of a real revolution in our perception of the phenomenon of beauty'—Ferry.

Ferry [2] comments in his article—'The Origins of aesthetics', specifically the visual aesthetics in the field of art, and this same understanding prevails in the

N. Sridhar (✉) · M. O'Brien
University of Huddersfield, Queensgate, Huddersfield HD1 3HD, UK
e-mail: naren_vasista@yahoo.com

M. O'Brien
e-mail: m.a.obrien@hud.ac.uk

A. Chakrabarti and R. V. Prakash (eds.), *ICoRD'13*, Lecture Notes in Mechanical Engineering, DOI: 10.1007/978-81-322-1050-4_21,

relatively younger creative areas such as Graphic design and Industrial design. This comment can be construed such that the understanding of beauty or the subjective perception of beauty and in a broader and more relevant context, general visual aesthetics is congenital in all humans and is then subject to recall.

The demands of today's consumer marketplace continuously influence how retail, wholesale, distribution and consumer product manufacturers operate. As a result, these organizations need to understand and anticipate customer needs in order to remain competitive and provide innovative, differentiated products.

Marketers realise now, that the reactions of the target consumer is foremost at the visual level, especially in the tangible consumer goods market and the measure of this as a factor in the success of the product is mostly in the marketer's domain of control. "It is important to understand and measure these individual differences relating to design for several reasons: First, individual differences in responsiveness to visual aesthetics may underlie a number of other well-established consumer behaviour variables such as product involvement, brand loyalty, materialism, innovativeness, self image congruence, choice, and usage behaviour" [3].

In such markets, developing a product image in a consumer's mind becomes an essential marketing tool; an image once in the consumer's long-term memory is more likely to be purchased when a need for that product arises [4]. However, this need is primarily addressed by the functional aspect of the product. The visual value of the product has been, in the recent past, given its due as the premier and most immediate reaction drawing factor in the cognitive and attention processes. The obvious shift in the paradigm of marketers is evident in the evolving role of product image, which includes the visual aesthetics, the packaging, branding and the communication methods in the ambit of marketing strategy.

A good product image may also increase the level of product equity (the value that consumers assign to a product above and beyond the functional characteristics of the product) [5]. This is where the design of the product comes into the fray. A well designed product brings with it an edge over the competing products that can be the defining criteria in the decision making process of the consumer.

With the market clustered with competing products that are very similar to each other in terms of technicalities and functionalities, the deciding factors for the consumer in choosing one product over another could dictate success or failure of a brand. This decision making has moved away from the basic variables such as pricing, performance to more intangible yet influential factors such as brand and product image, communication methods and product design. These factors independently or in groups become the stimulus inputs or cues for the consumers. In a lot of cases, these inputs impact sub consciously on the consumer. Wilkie [4] indicated that consumers translate stimulus inputs into mental identification. This process is called perceptual categorisation. For consumers, this process works extremely rapidly and is usually not perceived at a conscious level.

In this cluttered market, the key word for marketers and businesses becomes 'Differentiation', the process of creating an individual image to the product, distinct and attention grabbing, while confirming to the overall general perception of the product and hence not creating a dissonance with its unfamiliarity. "While manufacturing costs always decline with the use of commonality, the firm's overall profits may decline because of reduced differentiation" [6].

One such consumer market, India, has been steadily seeing this kind of consumerism and its associated effects and is predicted to grow at an ever increasing pace, with total domestic consumption growing from $370 to $1,500 billion by 2025 and with the Gross Domestic Product (GDP) growth of India on the rise and set to reach a high of 7.3 % increase by 2025, and with 'Thos macroeconomic forecast' indicating that the Indian economy which has till date been a developing economy is set to continue into a new stage of development in comparison with other Asian and world economies, it becomes imperative to understand this market. The socio-economic and cultural factors of the region present a unique opportunity and a necessity to examine it as an independent case study.

2 Economics of Product Design

The Indian consumer market has never had it better. Higher disposable incomes, the development of modern urban lifestyles and an increase in consumer awareness have affected buyer behaviour—in cities, towns and even rural areas. According to a 2007 report by McKinsey & Co., India is set to grow into the fifth largest consumer market in the world by 2025 [7].

In an excited market atmosphere such as this, stand out products were hard to come by for a long time, though it was a necessity. Products tended to resort to the comforts of tested and usually conservative market strategies with the intention of market survival overpowering market dominance. This was especially the case during the period of the initial years post globalisation or the opening of the economy to foreign businesses and investments in 1991, till when India had pursued an economic strategy based on import substitution [8]. The first decade or more after the economic liberalisation saw a number of market entrants, especially in the consumer market forum, but the product differentiation was not the primary focus of the marketers, the attention was more on the registration of the brand in the minds of a relatively infantile consumer market. There has, however, been a sustained shift towards underlining the individuality of the product since recent times and this required the understanding of Indian consumer psychology.

Products, which have been newcomers to the consumer market not long ago (e.g., mobile phones, laptop etc.) are now becoming mature products. Though the market has opened, the sales are not seeing the proportional increase. Quick model changes, technical updates or price reductions in order to improve the turnover have not been sufficient in terms of net sales per potential market increase. This new environment has created a situation where the consumers have more say in what they want and for what they pay. An increasing number of people want to express their individuality and even mass produced products have to be adaptable to individual demands regarding form, design and function [9].

However, in the present charged and turbulent international economic state, India has relatively resisted the meltdown but has still been measurably affected. The recommended and urged strategies to counter this crisis include departure from the dominant economic philosophies of neo-liberalism, dominant free market capitalism and shift from manufacturing to agriculture [10]. With such drastic expected potential changes, the value of 'Design' or more specifically 'Visual Aesthetics' (VA which incorporates under its brackets facets such as shape, colour, visual image, form and features) amongst the chosen market segment and the resulting shifts in paradigms needs to be monitored and interpreted.

3 Socio-Economic Factors

The process of demystification of the behaviour process of a consumer has to acknowledge the social and economic atmosphere surrounding the studied segment. Social psychology, vitally concerned with the behaviour of individuals and groups and the relationship between them, has been used in order to discover and recognise relationships between customers' attitudes, personality and motives, and the purchase decisions.

The effects of family, culture, peer groups, reference groups, subcultures, external cultural influences are all unique and respective to individual societies and countries.

The sociologist George Simmel felt that the deep changes occurring in man's psychic structure were a consequence of the socio-economic development of capitalism [11].

The developing countries are now under the system of consumer capitalism, which affects more and more people; the process of consumption and its related services has become the dominant social activity, even its philosophy. This has progressively led to the creation of its own: own characteristic and philosophical framework, social ranks and exclusivities; to the extent that membership and identification is gauged by the level of consumer participation. Richards [12] terms this 'Commodification' and defines it—'treating people and things only in terms of their market value—the degree of integration into this ritual of consumption being indicative of a person's social worth and his sense of social and individual responsibility'.

Marx's 'Commodity fetishism' describes this shift in social ideology as 'the fetishism of commodity is a definite social relation between men themselves' [13] or more vividly, 'to the producers... the social relations between the private labours appear... as material [*dinglich*] relations between persons and social relations between things'. According to Marx, this is the fetish associated with the products of labour as soon as they are produced as commodities, and is therefore inseparable from the production of commodities [14]. This sociological change is evident in the Indian consumer market.

India is the second most populous country in the world with an estimated population of more than 1.2 billion [15] and it may at the rate of growth, be the most populated country within a decade. As a developing country which saw economic liberalisation in the early 1990s and a recent economic growth surge, with rapid socio-economic changes taking place in India, the country is witnessing the creation of many new markets and a further expansion of the existing ones. With above 300 million people moving up from the category of rural poor to rural lower middle class between 2005 and 2025, rural consumption levels are expected to rise to current urban levels by 2017 [16]. India's market potential is greater than that of many countries in Western Europe with more than 150 million middleclass consumers earning more than $4,000 annually in local purchasing power [17].

Beginning in the late 1980s, the Indian government has taken a series of steps to liberalise the economy and ease restrictions on imported goods. As a result, the Indian market today is flooded with imported products from many countries. This has increased collaborations, indigenous brands and international brands all competing to grab the growing market opportunities. This process of globalisation has been an integral part of the recent economic progress made by India. Globalisation has played a major role in export-led growth, leading to the enlargement of the job market in India, especially in the urban centres such as Bangalore, Pune and Mumbai. However, with globalisation, corresponding social changes have developed in the country. This is primarily due to the lowering of the employed age group. This has a domino effect in the increase in disposable incomes, younger earners, nuclear families, younger consumer market, shift in consumption choices, and shift in spending patterns.

This assortment of at times congruous and at others conflicting agents in a dynamic socio-cultural zeitgeist makes for keen and necessary junctures of observation and recording. That Product Design is affected by and affecting many of the observed changes is crucial to be registered.

4 Psychology of Product Design

"Products are often instinctively perceived, and product choice is rarely just an exercise in logic"—Macdonald [18].

According to Macdonald [18], the buying behaviour of the consumer primarily and ultimately depends on how the product has been perceived by the individual and the reasoning behind this perception. Chapman [19] proposed that *"we are consumers of meaning, not matter and products provide a chassis that signify the meanings to be consumed"*. Between these two statements, the importance of 'meanings' and 'perceptions' in the buying behaviour process can be evaluated.

The visual appearance of products is a critical determinant of consumer response and product success. Judgements are often made on the elegance, functionality and social significance of products based largely on visual information [20]. These judgements relate to the *perceived* attributes of products and frequently centre on the satisfaction of consumer wants and desires, rather than their needs [21].

Previous research studies indicate that the relationship between cognitive response and product appearance can be classified into three elements: [22].

- Aesthetic impression—This is the emotional response resulting the aesthetic attractiveness of the product. This could be positive or negative responses.
- Semantic interpretation—This can be defined as what the product intends to describe about its function, performance and qualities. This can also be a response to function which the form is trying to underline.
- Symbolic association—This is the perception of social status and significance of the product, the design represents what the product says about its owner or user. The personal significance of the design might overcome the social significance.

However, these aspects of response are not exclusive to each other in most cases, they do not operate independently, but are highly inter-related; each one influences the others. For example, evaluation of the product and its features can be influential and modify judgements on the aesthetic response of the product and also the social value or symbols it may connote. The relative importance of the elements and the combinations may vary depending on the situation [22].

The decision making process with regards to the product design or visual value can be affected and influenced by visual references such as stereotypes, metaphors, clichés etc. [22]. The viewer makes sense of the visual information which the product presents based on these visual references [21]. Visual referencing may also influence the semantic interpretation allowing the viewer to categorise the product easily and associate it with familiar concepts and similarly can influence the symbolic value by allowing the viewer to connect them with products that already have associated social meanings [23].

Product design can be viewed from the perspective of psychoanalysis, especially the form of the product, which provides an ample area for subjective perceptions. We concentrate on the behavioural approach to marketing: Social psychology, vitally concerned with the behaviour of individuals and groups, has been used in order to discover relationships between customers' attitudes, personality and motives, and his or her purchase decisions.

"Nothing we can learn about an individual thing is of use unless we find generality in the particular. The endless spectacle of ever new particulars might stimulate but would not instruct us" [24].

The premise to begin with is that consumers are rational subjects, capable of governing their thoughts and actions by the principles of reason when faced with purchase decisions. According to the view that became dominant with the enlightenment, human nature is divided into a rational part, the faculty of reason, and a non-rational part comprised emotions, appetites and desires [25]. These two parts are distinct and opposed. Reason is disinterested, universal, objective and autonomous in its operation. The emotions and appetites, by contrast, are partial, particular and subjective. They are a force hostile to reason in human life [25].

All emotions are experienced and learnt in the interactions between self and society. Emotions exist only through reciprocal exchanges in social encounters. Emotions cannot be studied without paying attention to the "local moral order" and the existence of "those concepts (emotions) in the cognitive repertoire of the community" [26]. This suggests that the individual emotion is often affected by the collective expectations of society or as referred to previously, by the Subjective Norms. This is explained by Sabini and Silver [27] as the relationship of 'selves' to the 'moral objective universe'.

The essence of these arguments is the observation that emotions are necessarily crucial to the understanding of behaviour of most kinds, including the behaviour arising from the process of decision making. However, the occupying argument is the viewpoint that 'emotions' could stem from a collective or social sub-conscious and the standoff between the 'rational' and 'irrational' schools of thought in the understanding of 'emotions'.

Another paradigm which is pertinent to the overview of product design and consumer psychology is the dominance of the 'collective' sub-conscious in the social psychology. Jung divided Freud's 'Unconscious' into two very unequal levels: the more superficial 'Personal', and the deeper 'Collective'. Those mental contents that the ego or the conscious part of the psyche does not recognise fall into the 'Personal Unconscious', which is composed of many contents that vary from person to person and from time to time. But the significance of the contents lies in its partial personal and partial 'collective' nature. The unconscious contents are much more extensive than the conscious contents [28]. Everyone has their own 'Personal Unconscious'. The 'Collective Unconscious' in contrast is universal. It cannot be developed by the conscious observations and learning like one's personal unconscious is; rather, it predates and in most cases, is beyond the control of the individual. It is the repository of all the religious, spiritual, and mythological symbols and experiences. Its primary structures are the deep structures of the psyche, which were referred to by the term 'Archetypes' by Jung and Hull [29].

Jung says "we do not think of distrusting our motives or of asking ourselves how the inner man feels about the things we do on the outside" [29].

This can be utilised in the understanding of the visual communication of the products and its relationship with the human psyche; the visual archetypes are not questioned by the conscious human mind and thus the reasons for the associations and affects of the visuals is not consciously analysed.

Product design provides the consumer with this sense of metaphorical opposites, the sense of being a part of the collective infinite and also the individual identity. Thus, according intimacy and a connection to others which is the primary motivation in human beings is confirmed to and succeeding which, pleasure that is rather a secondary motivation derived from this more primary motivation [30].

Jung theorised that we are tied to a much greater archaic collective unconscious mind that emits universal symbols and processes we all share [31].

Therefore, the hypothesis is that this is the symbolic meaning associated with all products and their forms. These symbols can be deliberately controlled so that the association of the product to its intended buyer works in the collective sub-conscious strata of the mind, orchestrating the desired market behaviour. The process of the symbols communicating with its audience and the behavioural response towards these symbols happens at the unconscious level and this process is known as 'Symbolic interactionism'

The concept of symbolic interactionism is based on the premise that individuals interact with society at large and with reference groups, and this usually occurs at the sub-conscious levels of the human psyche, in determining the structure of the individual behaviour to determine. Individuals are, therefore, assumed to relate to objects or events based on their symbolic meaning given by society [32].

5 Summary

This study employs these theories of Social Psychology in its attempt to test the hypothesis that the VA in Product design is a 'Universal', 'Collective' and sub-conscious process between the designers and the consumers and is reciprocal and cyclic in nature. This is achieved through a systematic theoretical understanding of the various categories of influencing factors and their inter-relationships leading to a final developed theory that is grounded in the behaviour, words and actions of those under study [33].

This is aided by the employment of the Grounded Theory; method is usually used to generate theory in areas where little is already known, or to provide a fresh slant on existing knowledge about a particular social phenomenon.

This method allows the researcher to examine the various facets concerned such as Social, cultural and economic factors, Situational factors, Methods and channels of communication of design, Intention and reception of design in India. These were explored in the primary stage of the study, based on the philosophy of the chosen research paradigm 'Grounded Theory' along with the relationship between the different theoretical 'categories' such as Consumer Behaviour, Product Design and Social Psychology towards the direction of a 'core' theory.

The study is examining Product Design and its associated affecting components in the Indian context with a focus on its socio-psychological relationship with the consumer. The theories of psychology and sociology explored are the paradigms juxtaposed with India and Product Design to provide us the ability to intrinsically and critically analyse them in a fresh perspective.

The research study is currently in the second phase where working alongside the theory building and empirically complimenting it, will be an in-depth case study of 'Titan Industries' and specifically Titan wrist watches. It is envisaged that this study will contribute to the general understanding of the Indian consumer providing an insight into the consumer psyche and cognitive reactions to Product Design.

References

1. Lawson B (2006) How designers think: the design process demystified. Elsevier, Westefield
2. Ferry L (1990) The Origins of Aesthetics. Available: http://findarticles.com/p/articles/mi_m1310/is_1990_Dec/ai_9339050. Last accessed 12 Sep 2011
3. Bloch H, Brunel F, Arnold T (2003) Individual differences in the centrality of visual product aesthetics. J Consum Res 29:551–565 (Trice Award Winning Article, College of Business)
4. Wilkie WL (1994) Consumer behavior, 3rd edn. Wiley, New York
5. Hawkins DI, Best RJ, Coney KA (1998) Consumer behavior: building marketing strategy. McGraw-Hill, New York
6. Desai P, Kekre S, Radhakrishnan S, Srinivasan K (2001) Product differentiation and commonality in design: balancing revenue and cost drivers. Manage Sci 47(1):37
7. Ablett J, Baijal A, Beinhocker E, Bose A, Farrell D, Gersch U, Greenberg E, Gupta S (2007) The bird of gold: the rise of India's consumer market. The McKinsey quarterly: the online journal of McKinsey & Co, May 2007. Available: http://www.mckinsey.com/insights/mgi/research/asia/the_bird_of_gold/. Last accessed 04 Feb 2012
8. Kesavan KV (2003) Economic liberalization in India and Japan's wavering response. Ritsumeikan Annu Rev Int Stud 2003(2):129–142
9. Shimizu Y, Sadoyama T, Kamijo M, Hosoya S, Hashimoto M, Otani T, Yokoi K, Horiba Y, Takatera M, Honywood M, Inui S (2004) On-demand production system of apparel on basis of Kansei engineering. Int J Clothing Sci Technol 16:32–42
10. Bhatt RK (2011) Recent global recession and Indian economy: an analysis. Int J Trade Econ Finance 2(3):2011
11. Slattery M (2003) Key ideas in sociology. Nelson Thornes, Cheltenham
12. Richards B (1989) Crises of the Self. Free Association, London
13. Marx K, Mandel E, Fernbach D, Fowkes B (1991) Capital: a critique of political economy. Penguin Classics, London
14. Balibar E, Turner C (1995) Race, nation, class: ambiguous identities. Verso, London
15. Sala-i-Martin X (2006) The world distribution of income: falling poverty and convergence, period. The quarterly journal of economics, May 2006 121(2). Available: http://qje.oxfordjournals.org/. Last accessed 03 Dec 2011
16. Falk S (2009) Shift of power or multipolarity?. Zeitschrift für Politikberatung 2(4)
17. Kulkarni VG (1993) Marketing: the middle-class bulge. Far Eastern Econ Rev 156(2):44–46
18. Macdonald AS (1998) Developing a qualitative sense human factors in consumer products. Taylor & Francis Ltd, London
19. Chapman J (2005) Emotionally durable design. Earthscan, London
20. Coates D (2003) Watches tell more than time: product design, information and the quest for elegance. McGraw-Hill, New York
21. Lewalski ZM (1988) Product aesthetics: an interpretation for designers, design and development. Engineering Press, Carson
22. Crilly N, Moultrie J, Clarkson PJ (2004) Seeing things: consumer response to the visual domain in product design. Des Stud 25(6):547–577
23. Postrel VI (2003) The substance of style: how the rise of aesthetic value is remaking commerce, culture, and consciousness. Harper Collins, New York
24. Arnheim R (2004) Visual thinking. University of California Press, Berkeley
25. Damasio R (1994) Descartes' Error: Emotion, Reason and the Human Brain. Vintage Digital, London
26. Harré R (1986) The Social construction of emotions. Blackwell, Oxford
27. Sabini J, Silver M (1986) The social construction of emotions. Blackwell, Oxford
28. Mattoon MA (2005) Jung and the human psyche: an understandable introduction. Routledge, New York
29. Jung CG, Hull RFC (1959) The Archetypes and the collective unconscious. Routledge & Kegan Paul, London

30. Robbins BD (1999) A brief history of Psychanalytic thought—and related theories of human existence. Available: http://mythosandlogos.com/psychandbe.html. Last accessed 17 Nov 2011
31. Small J (1994) Embodying spirit: coming alive with meaning and purpose. HarperCollins, New York
32. Mead GH, Morris CW (1967) Mind, self, and society: from the standpoint of a social behaviorist. University of Chicago Press, Chicago
33. Goulding C (1999) Grounded theory: some reflections on paradigm, procedures and misconceptions. Working Paper Series WP006/99, University of Wolverhampton, Telford

Effective Logo Design

Sonam Oswal, Roohshad Mistry and Bhagyesh Deshmukh

Abstract Today's world is becoming increasingly visually oriented and logos have become a prime asset of companies. Logos are absolutely essential in branding and brand building and aims to facilitate cross-language marketing. The aim of this paper is to translate and expand the concept of logo effectiveness in the field of visual logo designs by identifying and understanding various factors that influence consumer perception of logos. The current research investigates numerous factors which influence the effectiveness of logo design including human perception factor. The study reveals that effective logo design adheres to seven elemental standards that all designers should be aware of. These principles may serve as key elements for judging whether the design and style systematically delivers the message to its potential audience. Degree of logo effectiveness is function of simplicity, versatility, memorability, relevance, timelessness, quality and appropriateness where as the psychological factor acts as multiplier of any one or more of the above factors. Logo design is an explicit function of shape factor (S) Color factor (C) and Font factor (F) and Human Perception Factor (P). The research includes case studies from different industrial backgrounds including automobile, sports, food products and beverages. The research classifies and studies different types of logos such as ones based on shapes, words, letters and Graphics. The work also addressed the issue of Good and bad logos. The study developed a set of guidelines for the effective logo design.

S. Oswal (✉) · R. Mistry · B. Deshmukh
Mechanical Engineering Department, Walchand Institute of Technology, Solapur, Maharashtra, India
e-mail: sonamoswal2907@gmail.com

R. Mistry
e-mail: roohshadmistry@gmail.com

B. Deshmukh
e-mail: dbhagyesh@rediffmail.com

A. Chakrabarti and R. V. Prakash (eds.), *ICoRD'13*, Lecture Notes in Mechanical Engineering, DOI: 10.1007/978-81-322-1050-4_22, © Springer India 2013

Keywords Logo effectiveness · Logo design factors · Good logos · Bad logos · Guidelines for effective logo design

1 Introduction: Need for Effective Logo Design

A logo conveys the entire essence of the owner or of the product represented and a good graphical realization. Effective logos successfully communicate a brand's style and personality. A logo is a distinct and recognizable identity of an organization.

A logo identifies a company or product via the use of a mark, flag, symbol or signature. Logo is basically a statement of business to all over the world. Logos give the first impression of a product or a company and hence play a very big role in the presentation of a company or an organization. A good logo can not only influence a customer's decision to buy a particular product but also attract the right kind of customers. Good logo expresses the organizations vision, its values and outlook.

Logos help in making a company more distinguishable and memorable because target audience easily gets connected with the organization through them. It creates an impact in consumers mind and represents the company's personality in most favorable manner. Logos are very effective in communicating a brand's message to consumers.

2 Objective of Study

1. Discuss factors affecting logo effectiveness.
2. To understand the relation of shape, color, and fonts on logo effectiveness.
3. To provide designer with a guidelines to select the optimum features including color, font and shape for logo design.

3 Literature Review

Chakrabarti and Kumari [1] studied the effect of structure sharing on logo design and to enhance the effectiveness of it. The degree of structure sharing and resource effectiveness on logo design was done. Kohli et al. [2] provided their thoughts on the two facets of logo design: content and style where content referred to the elements contained in the logo, including text and graphic representation and style referred to how these elements are presented in a graphical sense. Hem et al. [3] found that logo representativeness and design were important determinants of logo success. Bottomley et al. [4] investigated color appropriateness and showed the effects of colors and products on perceptions of brand logo appropriateness. Doyle et al. [5] studied the effect of font appropriateness on brand choice. It was found that brand presented in appropriate fonts was chosen more often than brands presented in inappropriate fonts. Saleh [6] has investigated the relationship between consumer's perceptions of a logo the organization it represents and the organization's performance. The relationship between the perceived image of the logo and organization perceptions was found to be positively significant. Fang et al. [7] studied the effect of a logo design on attitude toward the firm and the perception of the firms' modernness. They found that the respondents had a better attitude toward the firm for a round logo versus an angular logo. In terms of a logo change, Kohli et al. [2] provided some guidelines and stated that if a logo is changed, the change should be made in content. However, the changes to the logo should be kept to a minimum. Walsh et al. [8] has studied consumer's attitudes toward a logo change or redesign and investigated the effect of different degrees of redesigned logos on consumers brand attitude toward them, with the moderating effect of brand commitment.

It has been observed form the above literature survey that research has been done on factors such as logo content and style, logo appropriateness, logo redesign, color appropriateness, font appropriateness but not a so much on effectiveness.

4 Scope of the Study

The study covered logos from different business domains and comparative study was done on the basis of shape, font and color and was ranked on the basis of their effectiveness. Analysis of good and bad logos is carried out considering its simplicity, relevance, appropriateness and quality.

5 Methodology

5.1 Study of Various Logos for Understanding the Effectiveness

Various logos belonging to companies from various business domains were selected for evaluating effectiveness. The logos which scored high on effectiveness were considered for further analysis. After analyzing logos those parameters affecting the effectiveness are established. The methodology is to expand the concept of effective logo design. The study reveals that effective logo design adheres to seven elemental standards that all designers should be aware of.

Most logo designs can be broadly classified into three types: text, emblem, emblem and text, graphical or mascot type. In most cased the emblem or icon is accompanied with the company name. As a general rule icon or emblem based logos the following rules apply;

- Best visual appeal.
- Likely to appeal to a large and diverse audience. More memorable.
- Good for merchandising.
- Can serve as a trademark effectively.
- Lacks uniqueness and versatility especially if the business group is diverse (Fig. 1).

Fig. 1 Icon based logos (Google Chrome, Apple Computers and Nike)

Pure text logos are also the simplest but at the risk of becoming monotonous. The following features for text based logos have been noted:

- Most appropriate for companies who don't have a physical product viz. Banks, service sector companies, legal firms etc.
- Do not stand out.
- Esoteric appeal (Fig. 2).

Fig. 2 Text based logos (Walt Disney Entertainment Company and NASA)

Some logos contain a mascot or some graphical illustration. These are quite complex and require a skilled artist and designer. The points noted about these logos are:

- Must choose the character or mascot carefully.
- Suited for entertainment industry, food and beverages and toy manufacturers and service oriented companies.
- Can be memorable if designed properly.
- Not suited for manufacturing industry, legal firms and banks (Fig. 3).

Fig. 3 Graphic logo (Starbucks Coffee)

As a general rule icon or emblem based logos the following rules apply:

- Best visual appeal.
- Likely to appeal to a large and diverse audience. More memorable.
- Good for merchandising.
- Can serve as a trademark effectively.
- Lacks uniqueness and versatility especially if the business group is diverse.

In majority of cases icon and company name is used. The rules which apply for icon based and text based logos apply here as well (Fig. 4).

Fig. 4 Icon and text logos (Puma and Burger King)

Above logos were studied in detail and following 7 elemental standards [9, 10] are explained in detail related to logo design:

(a) *Simplicity* A logo must be simple and this is true especially when on an insignia or an emblem is used as a logo as in the case of Nike and Apple. Very intricate and geometrically complicated logos must be avoided. For example the flag of Japan depicting the land of the rising sun appropriately represents the nation. A logo must be simple and this is true especially when on an insignia is used as a logo as in the case of Nike and Apple. Often the simplest things are also timeless and memorable. This is the reason that most of the logos which are memorable and timeless are also simple.
(b) *Relevance* Logos must reflect clearly the company's message, values, beliefs and business domain. For example using the emblem of a Jaguar in Jaguar cars does qualify on the basis of relevance. The same is not be true for the logo Peugeot especially given the fact that Peugeots product lineup includes mainly low to medium budget cars where as the logo is that of a Lion.
(c) *Memorability* The memorability of logos can vary from person to person country to country and depends upon the individual's cultural political geographic background and his own experiences. Hence a rule of thumb is difficult to establish but can be generalized for a certain populous. For example the Apple has immense philosophical implications for western people but is just a fruit for most Asians. For a logo to be memorable it must be unique and also simple. Nike and Apple logos are the best examples. Coco Cola and McDonalds logos also fall in this category.
(d) *Quality* Quality can be very effectively expressed by use of certain shapes fonts and colors. Use of circles is preferred followed by ellipses and squares must be avoided. For example Honda, BMW and Mercedes Benz all make excellent quality cars. The Honda logo however does not reflect this as compared to the BMW and Mercedes logo. Quality is often best expressed in simple sharp fonts. Cursive fonts must be avoided.

(e) *Appropriateness* Logos must reflect the product business environment or company policy appropriately. For example the Burger King logo appropriately represents the product 'Burger'. Radio Mirchi logo represents the company name appropriately and so does the apple logo. But they do not represent the company product appropriately.
(f) *Versatility* Many large business houses operate different business and manufacturing different products. In such case it is better to have logo which reflects the group rather than the individual product. In such cases the logo must be versatile. Mitsubishi logo is effective representation for automobiles and manufacturing but is not so effective in representing the group which includes banking insurance etc. A good example of a versatile logo is TATA and GENERAL ELECTRIC.
(g) *Timelessness* This is a combined effect of the above factors and it is difficult to establish a rule of thumb for the same. Generally speaking logos exhibiting simplicity and memorability exhibit timelessness; the logos of Coco Cola, Nike and Apple for to speak.

5.2 Testing the '7 Elements of Logo Effectiveness'

These 7 elements are further tested for Color, Shape, and Font for the logo of Talk more. In 'Structure sharing in logo design' [1] the study was done on structure sharing and little work on effectiveness was done and structure sharing exist when there is more than one function fulfilled by the same structure at the same time. It also need FM tree to show structure sharing. The focus was more on structure sharing and least on effectiveness.

Logo of talk more as shown in Fig. 5; a telephone service from Kingston Communication is taken for analysis. Talkmore uses symbology in the form of quotation marks to replace the letters A and E, creating a clever image that gives a graphical representation of the words and meaning of the brand. The touch of color enhances this effect making the logo stand out. Simple and clean use of typography is observed to communicate professionally with the potential clients.

Fig. 5 The Talkmore logo

A 10 point scale was used to rate the Color, Shape, Cultural Impact and Human perception Factors as shown in Table 1.

A Tree structure of various factors of logo effectiveness is drawn as shown in Fig. 6.

An equation for logo design effectiveness is written as,

$$\text{Logo Design Effectiveness} = f(\text{Colour (C)} + \text{Font (F)} + \text{Shape (S)}). \quad (1)$$

Table 1 Analysis of various factors affecting logo effectiveness

Font (F)	Color (C)	Shape (S)	Cultural impact factor rating (point)	Human perception factor rating (point)
Sanserif	Grey	Constant (C)	10	9
Serif	Blue	Constant (C)	8	8
Handwritten	Black	Constant (C)	7	6

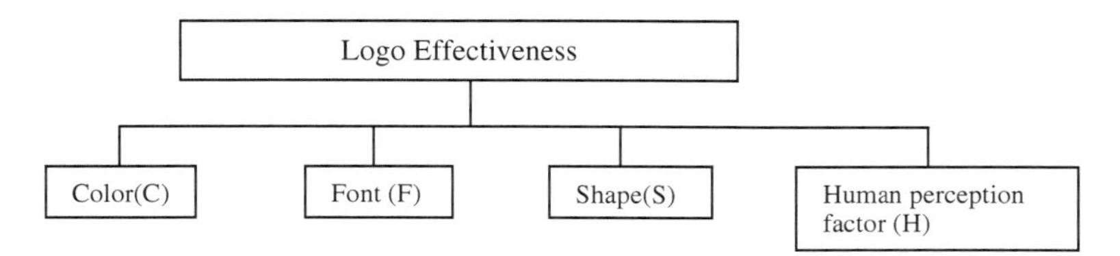

Fig. 6 Logo effectiveness factors

5.3 *Applying Shape, Font and Color Factors to the Logo of Various Sectors to Study Effectiveness*

On the basis of shape factor, color factor and font factor logos have been analyzed in for automobile, beverage and sports good industry. The ratings for shape, color and font will depend upon the Human perception factor. Keeping the variable shape factor (S) as constant. Effect of color on logo design is due to cultural Impact factor.

Case I *Automobile Industry logo*

Various automobile logos were studied as shown in Table 2, and given a rating from 1 to 10 on a ten point scale.

It is seen from the above ratings that the Circular shape, Simple font and blue or black and white color is the effective combination for the logo design.

Case II *Beverage industries logo*

Various beverage industries logos were studied and some of them are as shown in Table 3, and given a rating from 1 to 10 on a ten point scale.

It is seen from the above ratings that the Circular shape, Simple-Cursive font and red, blue and white color is the effective combination for this type of logo design.

Case III *Sports and Shoe Wear Industry logo*

Various Sports and Shoe Wear Industry logos were studied and some of them are as shown in Table 4 and given a rating from 1 to 10 on a ten point scale.

It is seen from the above ratings that the Distinguishing shape, Simple-Cursive font and black color is the effective combination for this type of logo design.

Case IV *Food Industry logo*

Various Food Industry logos were studied and some of them are as shown in Table 5 and given a rating from 1 to 10 on a ten point scale.

Table 2 Order of ranking of effectiveness for automobile Industry

Automobile industry	Shape	Font	Color	Rating (point)
1. BMW	Circle	Non-serif	Blue, black and white	10
2. Mercedes Benz	3 pointed star	Uncomplicated typeface	Metallic grey	09
3. Ford	Oval	Cursive style	Blue	08
4. Renault	Rhombus	Simple bold	Yellow	07
5. Mitsubishi Motors	3 equally shaped diamonds	Simple	Red	06
6. Fiat	Letter mark	Improper typography	Red	05

Table 3 Logo Effectives of beverage industries

Beverage industry	Shape	Font	Color	Rating (point)
1. Pepsi	3-Dimension globe	Simple, roman, italic	Red and blue	10
2. Coca Cola	Rectangular shape/ circular	Cursive style	Red and white	8

Table 4 Logo effectiveness in sports and shoe wear industry

Sports and shoe wear industry	Shape	Font	Color	Rating (point)
1. Nike	Swoosh icon	Simple	Strong black	10
2. Adidas	3 parallel lines	Simple bold	Black	8

Table 5 Logo effectiveness in food industry

Food Industry	Shape	Font	Color	Rating (point)
1. Mc Donald	2 golden arches	Simple font	Golden and red	10
2. KFC	Mascot	Simple bold	Red	8

It is seen from the above ratings that the Distinguishing shape, Simple font and Golden-Red color is the effective combination for this type of logo design.

5.4 Analysis of Good and Bad Logos

Rating among good and bad (which also includes complicated) logos on the basis of elemental standards and given rating from 1 to 20 point scale where simplicity, relevance Quality and Appropriateness contributes a maximum of 5 points each.

Good Logos

Automobile industry	Simplicity	Relevance	Quality	Appropriateness	Rating (point)
1. BMW	4	5	5	5	19
2. Mercedes	5	3	5	3	16
3. Jaguar	3	5	5	5	18

Bad Logos

Automobile industry	Simplicity	Relevance	Quality	Appropriateness	Rating (point)
1. Honda	4	2	2	1	09
2. Hyundai	4	1	2	1	08
3. Toyota	2	2	3	1	09
4. peugot	3	5	2	1	11
5. Porshe	1	2	4	1	07
6. Cadillac	1	2	3	2	07
7. Scania	2	1	3	1	07

(a) (IRS, London Olympics) **(b)** (Honda, Toyota, Hyundai, Subaru)

Fig. 7 Bad logos. **a** IRS, London Olympics. **b** Honda, Toyota, Hyundai, Subaru

Examples of bad logos (Fig. 7).
Bad logos are characterized in various shapes and font.

- Unable to recognize the logo as to which domain it belongs.
- The use of too many shapes used together makes logo very difficult to comprehend.

6 Set of Guidelines

(a) **Guidelines for selective most effective shapes for logo**

- Prefer circular or oval shape that can quickly and easily be seen.
- Tall and skinny or overly wide format may take longer for the human eye to see and recognize the information.
- Avoid controversial icons and shapes. For example Swastika is widely accepted in India but abhorred in the West. Use rectangles over squares.

(b) **Guidelines for selecting most effective fonts**

- The theme of your logo is important in deciding what font to be used.
- If it is a more futuristic logo, stick with sci-fi fonts.
- For legal firms and manufacturing industry choose simple neutral fonts like Arial, Sans, etc.
- Avoid using italics for manufacturing sector.
- Choose the size of the text such that it does not overwhelm the icon.
- Complicated fonts (like grunge or fancy fonts) either don't look good or are illegible when they are too small so consider the size of the text while choosing font.

(c) **Guidelines for choosing most appropriate color**

- Universal use of green, yellow/amber, and red to label safety status.
- Avoid using green for automobiles and manufacturing.
- Blue is the best color for automobiles followed by red.
- Do not use more than two colors.
- The contrast between text and its background shall be sufficiently high to ensure readability of the text.
- In all cases the luminance contrast and/or color differences between all symbols, characters, lines, or all backgrounds shall be sufficient to preclude confusion or ambiguity as to information content of any displayed information.

(d) **Guidelines for effective logo design**

- Identify what kind of product the logo is for and its area.
- Benchmark the competitors, 'project budget' for the logo.
- Identify the most important thing the potential customers should think of while looking at the logo.
- Find out all potential implementations of the logo design letterhead design, business card design.
- Choose the right shape, font and color based on the respective guide lines.

7 Conclusion

The study reveals that effective logo design adheres to seven elemental standards that all designers should be aware of. These principles may serve as key elements for judging whether the design and style systematically delivers the message to its potential audience. Degree of logo effectiveness is based on its simplicity, versatility, memorability, relevance, timelessness and quality. For a Logo design to be effective shape factor (S) Color factor (C) and Font factor (F) and Human Perception Factor (P) must be considered while designing. Also a set of guidelines for effective logo design have been developed.

References

1. Chakrabarti A, Kumari MC (2011) Structure sharing on logo design ICoRD 2011 international conference, Indian Institute of Science, Bangalore, pp 838 ISBN: 978-981-08-7721-7
2. Kohli C, Suri R, Thakor M (2002) Creating effective logos: insights from theory and practice. Bus Horiz 45(3):58–64
3. Hem LE, Iversen NM (2004) How to develop a destination brand logo: a qualitative and quantitative approach. Scand J Hospitality Tourism 4(2):83–106
4. Bottomley PA, Doyle JR (2006) The Interactive Effects of Colors and Products on Perceptions of Brand Logo Appropriateness. Mark Theory 6(1):63–83
5. Doyle JR, Bottomley PA (2004) Font appropriateness and brand choice. J Bus Res 57(8):873–880
6. Abdulaziz S Consumers perception of rebranding: case of logo changes. Ph.D Thesis
7. Fang X, Mowen JC (2005) Exploring factors influencing logo effectiveness: an experimental inquiry. Adv Consum Res 32:161
8. Walsh MF, Page KL, Mittal Vikas (2006) Logo Redesign and Brand Attitude: The Effect of Brand Commitment ACR 2006 Conference. September Orlando, Florida
9. http://www.logobird.com/the-principles-of-good-logo-design/
10. http://www.johnnyflash.net/creative-and-effective-logo-designs/

Effect of Historical Narrative Based Approach in Designing Secondary School Science Content on Students' Memory Recall Performance in a School in Mumbai

Sachin Datt and Ravi Poovaiah

Abstract Use of Narratives for teaching science and technology at secondary school level has gained strength in recent years. Many approaches and explorations have been done by researchers like Arthur Stinner, Aaron Isabelle, Stephan Klassen and Yannis Hadzigeorgiou. With the spirit of extending existing work in domain of teaching science using narratives, a General Narrative Schema, (also known as Epistemological Narrative Schema or ENS) for describing a scientific inquiry event in context of cultural tradition of science was developed to assist secondary school science content writers in designing narratives to be used for explaining science concepts. Control group experiments were conducted with secondary school students in Mumbai to test the effect of ENS on certain aspect of science concept learning. The control group was taught using lesson from their existing text book while the Experimental group was taught the same content but modified and delivered in the Form of a Narrative designed using the Epistemological Narrative Schema. The posttest for evaluating difference in short term memory recall of students of the delivered lesson showed significant difference in the mean score of Experimental group versus the Control group. The experimental group scored significantly better than the Control group in Posttest results. This experiment strengthens the case for introducing narrative based approaches for designing science lessons at secondary school level.

Keywords Narrative · Epistemology · Science curriculum · Content design · Independent sample t-test

S. Datt (✉) · R. Poovaiah
Industrial Design Center, Indian Institute of Technology, 400076 Powai, Mumbai, India
e-mail: sachindatt9@gmail.com

R. Poovaiah
e-mail: ravi@iitb.ac.in

A. Chakrabarti and R. V. Prakash (eds.), *ICoRD'13*, Lecture Notes in Mechanical Engineering, DOI: 10.1007/978-81-322-1050-4_23, © Springer India 2013

1 Epistemological Narrative Schema and a Brief Justification for Its Development

In order to find the relation between Narrative structure and scientific inquiry event, a search into literature on Theory of Knowledge, Theory of Learning and Theory of Narratives was commenced. The final outcome of the search was the development of the Epistemological Narrative Schema. This framework is meant to assist content developers in designing science lessons in a Narrative format by having a better understanding of the scientific process of knowledge creation. The final outcome of this research was development of the Epistemological Narrative Schema.

A search into theory of knowledge and specifically the work of John Dewey revealed that Science is a narrative process with certain discrete events distributed in the five step general narrative structure defined by Existing Situation, Doubt, Reasoning, Suggestion, and New situation. It was realized through further search into Theory of Knowledge, that the components given above are linked with some other components associated with a scientific Inquiry.

The basic five steps in Narrative schema of science have further divisions and the overall components of a scientific inquiry process are believed to have the following constituents (Fig. 1):

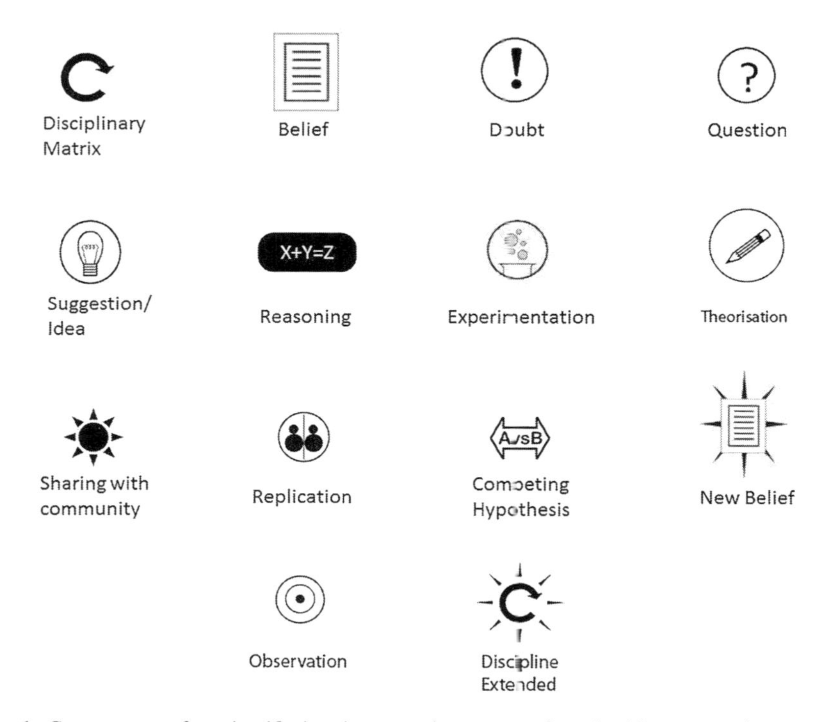

Fig. 1 Components of a scientific inquiry event in context of a scientific community

The first point of justification is supported from John Dewey's explanation of the Narrative nature of a scientific inquiry event. The narrative schema presented in Fig. 2 is an extension of the Narrative process of a scientific inquiry presented by John Dewey. Dewey's schema of a scientific inquiry is given in Fig. 2.

> John Dewey went as far as to believe that the 'reality' itself which philosophers and scientist have been attempting to understand (since birth of cognitive abilities in the human mind in prehistoric ages), is essentially Narrative in nature. Dewey believed that something appearing as permanent and rigid as a mountain, in a geologist's reality, is a scene of a drama of birth, decay and ultimate death. A flash of lightening to a layman may appear as a single event but for a scientist has a prolonged narrative history, with the growth of science, the tale of why lightening happens, becomes longer [1].

The second justification for believing in the truth of Epistemological Narrative Schema presented in Fig. 1 comes from the picture of a scientific community painted by Thomas Kuhn in his papers on *The Structure of Scientific Revolutions.* Kuhn explained that the compilation of a theory is not one person's work. Development of scientific ideas is directly related to the organization of the scientific community as a whole with its tradition of common language for knowledge sharing among members of a community within the context of a particular paradigm or disciplinary matrix [2].

The two ideas of the role of individual research and the role of a social community in advancement of science can be seen from another point of view which is presented in literature on Theory of Learning. Piaget's Constructivist theory of learning can be compared with an individual's process of acquiring knowledge in the four step process of Existing Schema (Assimilation), Expectation breaking event, Adjustment of new information into existing schema through Accommodation and the establishment of New Schema [3]. However, constructivist theory is limited to an individual's learning process. This limitation is rectified in Vygotsky's theory of social constructivism where learning is also a function of the socio-cultural environment in which the student learns which not only includes the teacher but, friends and family background [4]. For these reasons, the cultural aspect of science and its routine ways of sharing knowledge through community presentations and publications are part of the Epistemological Narrative Schema.

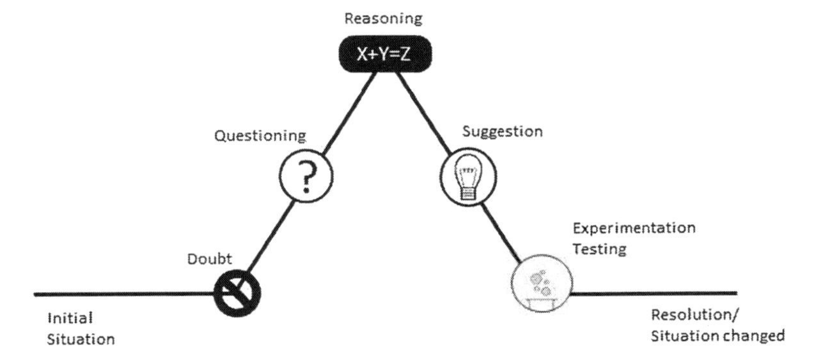

Fig. 2 John Dewey's narrative process of a scientific inquiry (Dewey, 1955, p. 105)

2 Existing Literature on Experiments Conducted for Testing Story Based Teaching Methods in Science Learning

Kokkotas et al. [5] conducted experiments in teaching the concept of Electricity and Electromagnetism with sixth grade primary school students to test the impact of a narrative based teaching intervention. The authors believe that telling a coherent story may be best way of learning, remembering and re-telling of science. They conducted a study in which the objective was to access impact of storytelling on student learning of Electricity and Magnetism. Within this, they wanted to study the impact on three factors (a) Importance of evidence and creative thought in the development of scientific theories (b) Consider how scientific knowledge and understanding need to be supported by empirical evidence and (c) Relate social and historical contexts to scientific ideas by studying how a scientific idea has changed over time. Their experiment result showed significant effect on student's learning on these factors after the narrative based teaching intervention.

In another study Casey, et al. studied the impact of story telling technique in teaching geometry skills to kindergarten students. Their experiment result showed a significant improvement in the performance of student. They also found that the impact of story intervention was greater on girls compared to boys [6]. A comparative study was done between the experimental group which was given story telling context and geometry component intervention and the control group which was only given the geometry component intervention. Their results showed significant improvement in the learning of students of the experimental group.

Cunningham set up an experimental design to test the hypothesis that narrative text structure would be more interesting than expository text structure and would motivate in learning. It was conducted on five secondary school classes with a lesson from history textbook. The results did not show any significant difference in the students' level of performance among the control and experimental group however they showed positive attitude towards the narrative text over the expository text [7].

3 Design of Experiment for Testing Effectiveness of ENS on Short Term Memory Recall: Experiment No. 1

The experiment was mainly centered on testing recall of chapter content after lesson delivery. We would like to differentiate between recall and rote memorization. In rote memorization, repeated exposure to material to be memorized is given. While short term memory is a cognitive process, which does not include repeated exposure. The recall in these experiments confirms a cognitive process rather than rote memorization as the students have to answer questions within a few seconds after only one recitation of the lesson. According to Bloom, 'remembering' is at level one of bloom's taxonomy of cognitive domains [8]. Hence, one can say that whatever the students recall after the lesson, it is due to some degree of cognition of content rather than rote memorization.

3.1 Experiment: Aims and Objectives

Objective of experiment was to find the difference in short-term memory recall of students who were taught with chapter designed using the ENS approach against those taught with current NCERT textbook approach. A study was conducted on class VII students from Powai English Medium School at Powai, Mumbai. Two sections of the same class very selected. One section was treated as the Control group, the other as the Intervention or experiment group. The control group had 52 students and Experiment group had 50. The experiment was performed in the natural classroom setup during the regular school hours. The experiment was conducted by a third person, the researcher collected data in the form of response to achievement test.

3.2 Design

The study was a between group design comparing two types of lesson delivery approach namely Current NCERT Textbook versus Chapter designed using Epistemological Narrative Schema. The science content to be taught was chapter on electricity taken from class VII NCERT book. A single variable design was adopted for this experiment where effect of one independent variable was tested on the dependent variable Fig. 3.

Following is the description of the variables.

Teaching Method (Textbook Based TB or Story Based) = Independent Variable

Achievement Score = Dependent Variable

Operational definition of each variable is as follows:

Teaching method is the way a lesson is delivered to a group of students. Textbook based (TB) teaching method means that a chapter from current NCERT textbook is taken as it is and is read out to the students. The Textbook based independent variable constitutes the control group. Story based (SB) teaching method means that the lesson was delivered in the form of a narrative. The significance of difference in the mean score of both achievement tests groups was calculated using independent sample t-test, which is the index used to find the same [9]. The test is called independent because there is no relationship between the control and experimental groups.

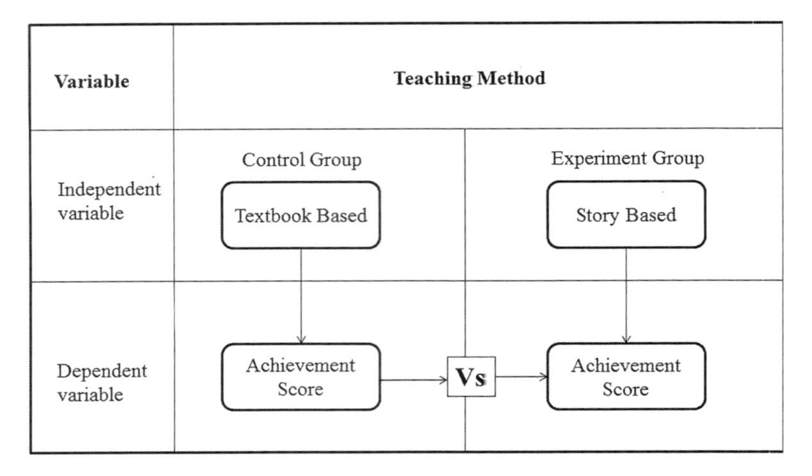

Fig. 3 Pretest-posttest nonequivalent control group design

3.3 Material and Procedure

Each group was delivered a lesson from a class VII NCERT textbook on electricity and how the Voltaic cell came into being. The layout and font size of both the chapters were kept the same. The lesson was taught to each group from their respective booklets. The same instructor read out the chapter to the groups on two consecutive days. The group which was chosen for experimentation had no prior exposure to the contents of the chapter. This was ensured through a screening test. The achievement test was evaluated by giving scores to students on each correct answer.

3.4 Instrument: Achievement Test

An achievement test in the form of a set of long and short essay type questions was given to both the groups before and after the lesson delivery. The achievement test was meant to measure the difference in mean performance of Control versus Experiment group. To find if the difference was significant, an Independent sample t-test was conducted with the help of SPSS software. The long essay type question simply asked the students to write the summary of chapter on Electricity in their own words in the case of Control group while the Experiment group was asked to write the story of Volta and the Voltaic cell in their own words. The short essay type question were a total three in number the first question asked the structure of Voltaic cell and What a Voltaic cell is made of? The second question was, What is the use of a Voltaic cell. The third question was to cross check with the second question whether the response was result of understanding. So it was the reversed form of second question which is how electricity could be generated. Same questions were given before and after the test to know the difference in response. The pre-test was meant to check their prior knowledge of subject which was supposed to be none as the chapter on electricity is taught to class 8th student and the test was given to class 7th. So the pre-test confirmed that they had almost no prior knowledge of the subject matter. So we can assume that the difference in post test would be because of the lesson delivery alone. The Subjective essay test questions prepared were from the list of 16 types of questions in textbooks of science outlined by Francis D. Curtis [10].

3.5 Results

The independent sample t-test was performed two times to know about two different aspects of the experiment outcomes. These two type are:

1. Score difference in mean of all Short essay type questions taken together.
2. Score difference in Long essay type question
1. The mean score for Short summary type for NCERT was 3.2 (out of 15) while for Experimental (Story group) was 6.3 and the difference is significant at $p < 0.001$ with confidence level of 95 %. The data is shown in Table 1.
2. The mean score for Long summary type for NCERT was 0.54 (out of 8) while for Experimental (Story group) was 3.88 and the difference is significant at $p < 0.001$ with confidence level of 95 %. The data is shown in Table 2.

3.6 Interpretation and Discussion

Since the achievement score on recall is significantly higher in the Experimental group, it can be concluded that the effect on student's recall because of story intervention is significantly higher than the control group. This implies that students were better able to remember the concepts in STORY group as against the NCERT group. This is not only true for Short answer essay type, but also for long answer essay type question. Within short essay type question, the question about the structure of voltaic cell has received almost equal response from both the groups that is because there is no significant difference in this case between the groups. The score of short answer type questions is significantly higher for STORY group because of significant mean difference of question two and three.

Table 1 Mean difference is significant at $p < 0.0005$ for short essay type test with Powai school

Group	N	Mean	Std. deviation	t-test Sig. (two tailed)
NCERT (control group)	52	3.201	2.751	0.000
STORY (experimental group)	50	6.300	3.157	

Table 2 Mean difference is significant at $p < 0.000$. Long essay type test for Powai English medium school

Group	N	Mean	Std. deviation	t-test Sig. (two tailed)
NCERT (control group)	52	0.538	0.821	0.000
STORY (experimental group)	50	3.880	2.791	

The mean of Story group is significantly higher than the NCERT or control group. However since the std. deviation is also positive, the curve seems to slide towards the higher side of mean. This means that a higher mean may be caused by few students scoring exceptionally well. The reason of this could be that students with good writing ability may be in a better position to write the answers while those with poor writing abilities may be at a disadvantage in answering the questions. Even in the story group, students with poor writing abilities may have scored low. An additional oral test might give a better picture of overall student performance.

4 Conclusion

A significantly increased performance in short term memory recall does not absolutely validate that narratives designed using Epistemological Narrative Schema enhance overall cognitive development of students, however it does indicate that Narrative approach to lesson design may be enhancing some level of cognitive development. Separate set of experiments need to be designed to test whether student's other cognitive skills like reasoning and understanding are also effected in the same way as short term memory recall. Testing of such skills can be done by allowing students to design Narrative to explain evolution of a concept using the constituents of scientific inquiry event as presented in the Epistemological Narrative Schema. The content of narratives designed by students can reveal their level of reasoning and understanding of a science concept. The experiment strengthens the case for organizing secondary school science lessons in the form of narratives of evolution of scientific concepts.

References

1. Dewey J (1955) Logic: The theory of Inquiry. George Allen & Unwin Ltd., London
2. Kuhn TS (1962) The Structure of Scientific Revolutions. In: Kuhn TS (ed) International Encyclopedia of Unified Science Vol 2. The University of Chicago Press,Chicago
3. Leonard DC (2002) Learning theories, A to Z. Greenwood publishing group, Westport
4. Ivic I, Vygotsky LS (1994) UNESCO: International Bureau of Education, vol 24(3/4), pp 471–485

5. Kokkotas P, Rizaki A, Malamitsa K (2010) Storytelling as a strategy for understanding concepts of electricity and electromagnetism. Interchange 41(4):379–405
6. Casey B, Erkut S, Ceder I, Young JM (2008) Use of a storytelling context to improve girl's and boys' geometry skills in kindergarten. J Appl Dev Psychol 29(1):29–48
7. Cunningham LJ, Gall M (1990) The effects of expository and narrative prose on student achievement and attitudes toward textbooks. J Exp Educ 58(3):165–175
8. Bloom B (1956) Taxonomy of educational objectives, Handbook I: the cognitive domain. McKay Co Inc., New York
9. Ary D, Jacobs LC, Razavieh A (1972) Introduction to research in education. Holt, Rinehart and Winston, Inc., New York
10. Ebel RL (1966) Measuring educational achievement. Prentice-Hall of India Private Ltd., New Delhi

The Home as an Experience: Studies in the Design of a Developer-Built Apartment Residence

P. K. Neelakantan

Abstract The developer-built apartments in Mumbai are advertized as ultimate symbols of iconic lifestyles. Referred to by evocative names, these are self-sufficient, gated microcosms. These advertisements claim uniqueness in terms of amenities; each attempting to outdo the other by projecting a more exclusive lifestyle. They declare: Purchase houses, acquire lifestyles. This paper intends to examine the relationship between lifestyle conception and house composition. A case-study of the design process of an apartment-housing project in Mumbai offers insights into this lifestyle-architecture combination at the moment of conception. The assumption here is that the space and moment of design allow unselfconscious forms to become apparent and be subjected to interrogation. Also, design itself might offer critical positions to view the relationship between *where we live* and *how we live*. The results of the case-study analysis point to how *new* lifestyles require the incorporation of *other* diverse spaces usually never associated with urban residences. The shape of this lifestyle emerges in the form of the super-built-up space, beyond the individual dwelling unit; the *value-addition* space that belongs to all and none in particular. A particular lifestyle-architecture configuration also seems to posit specific relations between the individual dwelling unit, the collective apartment and the urban neighborhood. Also brought forth is the underlying assumption of aesthetic congruence between lifestyle and house design.

Keywords Housing · Architecture · Experience · Lifestyle

P. K. Neelakantan (✉)
Industrial Design Centre, Indian Institute of Technology Bombay Powai, Mumbai, India
e-mail: neel.blu@gmail.com

A. Chakrabarti and R. V. Prakash (eds.), *ICoRD'13*, Lecture Notes in Mechanical Engineering, DOI: 10.1007/978-81-322-1050-4_24, © Springer India 2013

1 The Apartment-Block

As a type, the apartment-block is a single 'house' belonging to a community of dwellers. This house is further divided into smaller dwelling compartments called flats. It can also be thought of in terms of a vertical stack of a series of peculiarly interior 'houses' called flats having different owners. The type is an interior structure of a form/principle. It is capable of shared meaning and infinite variation. Therefore the tension between invention and form would be evident in the manifestation of any type.

What role does the house play in the apartment-type? This question also pertains to the relationship between the house-space and the home-space. The house-space is part of the material-spatial realm, in comparison to which the home-space is the more amorphous socio-politico-economic realm. Various relations have been posited between the house and the home: the house as a register of socio-cultural changes in the home-notions/practices [1]; as a place-bound aspect, the house acts like a geographic centre for the inhabitants [2]. The household in the form of the fusion of the house and the home, has been considered a socio-political unit [3]. The house has been also seen as a container of deep structures of the home [4]. If territory is the critical distance between two beings of the same species [5], the house also maintains territory through a number of elements like floors, walls, gaps, stairs, passages, etc. The act of housing is thus a complex act, and it depends upon how it decides/intends to receive the home-space. This would also mean imagining the home-space in advance.

The background for the question is this: In urban India, for instance in Mumbai, the phenomenon called apartment-block, predominantly involves someone else (the developer) in building the house. This house is purchased and then made into/called home. In Urban India, especially metropolises like Mumbai, the notion of the home cannot be thought of without bringing to mind images of both the vertical stack of compartments called the apartment-block and the horizontally dense slums. Mumbai in the recent times has witnessed a spate of developer-built housing. Referred to by evocative names, these are usually either a single or group of tall residential buildings. While the slums have been discussed extensively with respect to their development, homelessness, etc., there is dearth of literature on the home in relation to the apartment-block itself.

2 Designing Housing

Architecture brings into being types of dwelling [6] —an ordering and diagramming of the social. The architect's role is to transform this ideal diagram into a physical model. Design is an intentional, strategic and calculative practice. It is a moment of conceiving and assembling. It would thus offer an insight into the vectors that would actualize the house. The designing of housing would allow unselfconscious forms to become apparent and be subjected to interrogation. Design practice requires that the forms of sociality and home be mobilized, thought about, reinforced, altered, articulated, translated, and reflected upon. This paper wishes to study the 'designing' of housing which would allow the house to be seen as an aggregate of multiple parts. It would bring forth the concerns associated with each part. The biggest challenge is the home as an affair of prospective inhabitants. It's personal, intimate and unique nature poses a challenge to the mass-housing nature of developer-built apartment residences.

3 Apartment Layout: A Case

The negotiate tactics of 'housing' are examined, through the case-study of the design process of an apartment-housing project in the city of Mumbai. A reputed architectural firm in the city, offering design solutions in areas of architecture, urban design and interior design, had been invited to compete, by one of the well-known construction companies and developers, for the design of a residential apartment-complex in a central suburb of Mumbai. This of course limits the case to the conceptual, master-plan level, at the cost of detailing the individual unit.

The developer's brief to the architects, other than providing details of the location, land and amenities to be provided, also highlights the development philosophy and theme. It discusses the historic aspects of the site, highlighting its intention to restore and integrate some of the existing architectural elements. Also attempted is a profile outline of prospective inhabitants. The brief describes the immediate surroundings as being 'not up market and having small towns look', devoid of 'high' structures and planned development. What is expected is a spatial remedy for this place. To compensate for the mundane surroundings, therefore, they demand an 'iconic' architecture and landscape to produce a 'feel of being in a different world' to attract prospective inhabitants.

The project site of approximately 25 acres is a redundant textile mill compound, located in a suburb of Mumbai. The amenities comprise of common facilities for Health, Sports and Leisure, which include Olympic size swimming pools, gymnasiums, skating, amphitheatre, etc. The project components include residential, substantial built-to-suit offices, convenience retail, school and auditorium. The residential mix consists of 2, 2.5, 3 and 3.5 bhk, (the unit areas are extremely well-defined—the external areas are not so well-defined). The time period allotted to them was about three weeks. The firm put four senior architects on the job. They produced one alternative each for the project (see Fig. 1). All the designs went through a process of iterative refinement and detailing. This happened through the team's discussions with and guidance from the chief architect. All the important design decisions were taken through a discursive process where every one of the team members was present. The researcher had access to all the formal discussions. He had a contact person in the architectural firm, a senior architect engaged in the project. It was at his behest that the researcher was

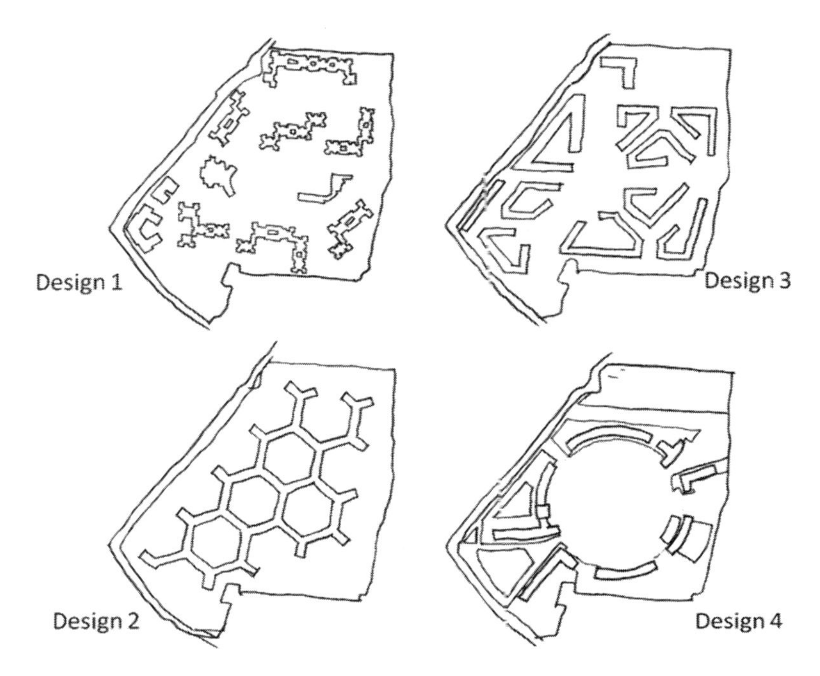

Fig. 1 Diagrams showing the four alternative schematic plans for the apartment-complex. *Source* Author

permitted into the meetings, to make notes of the discussions and access the work being prepared for the presentation.

These discussions were studied for how the 'role of the house' was being thought about. In the design-process, since the house is being thought about in the multiple, the discussions on the role of each of these parts were also examined. The analysis of the material provides us with the following three themes: Spacing's, which deal with the distributions of architectural space, especially between the individual dwelling-unit and the collective housing. Placings are about the entexturing of the spatial distribution through place-making. Ambiences engage with the quality of immersion in the space of architectural containment.

4 Spacings

The design team's conceptual framework involves categorizing space into micro-meso-macro scales. It ranges from the individual dwelling unit (micro) to the apartment-complex (macro) respectively. This scaling shows the interior-flat's intimate relationship with the flat-collective and of the project to the city. For instance, through the 'views', the individual unit possesses the vast community space laid out ahead. The view has to be 'housed' in the apartment-block. The view is a linkage between the individual dwelling unit and the community space constituted by the super-built-up. The tension between the two brings out the dialectical nature of privacy and the necessity of articulating it differently for different contexts.

Conventional practices are not only employed in evaluating the design alternatives, but are themselves questioned and redefined. For instance, 'is the balcony private or semi-private?'—this was the subject of a significantly intense debate. The argument for privacy was based on the point that the dwelling unit was private space. The argument for 'semi-privacy' claimed that the balcony looked onto the public space because of which it wasn't really private. The glass-house was the highlight of the above discussions. One member of the design team, as part of the public versus private argument, wondered aloud: Is a glass-house private or public? This was posed as a hypothetical, designerly manner of questioning in order to grasp architecturally, the terms of the private–public dialectic and territorial interface. The glass house question is posed as a special, exaggerated scenario emerging from the private autonomy of the individual flat-house. It pits the 'one' against the collective. The tension between the one and the collective, is brought into play through this question. Emerging from the practices of spacing as against that of locality/place, the absurd glass-house question interrogates place-based conventions. The hypothetical glass-house question, by challenging the inside-outside, private–public dialectic, challenges the apartment-block typology itself.

The concern for 'active' streets or 'streets for interaction' emerges from the desire to house a community. The notion of an average occupant is at work when the community is discussed. The attempt is to discuss specificities in terms of an average, or a group of people who for instance, "celebrate festivals like Ganesh Chaturthi" or "spend time in coffee shops". The design discussions reveal that the architects not only work with a certain notion of the average inhabitant ('people in Mumbai, 'people in India', 'people here', etc.), but also with that of the stereotype ('Mr. Tendulkar', 'Mr. Shah', etc.) employed in terms of its 'difference' with the average. This housing can be thought of at two levels: one at the more form-based, affordance-level, generic aspect, and the other which is more culturally nuanced. Much of the discussions refer more to the anthropomorphic figure than the cultural figure. This use of the anthropomorphic figure, allows certain architectural 'aesthetic' of form relatively autonomous of the milieu. It is an abstraction not colored by culture. This concern regarding the apartment's belongingness is also evident in their intention to move away from depicting inhabitants in the manner in which they are depicted in the developer's brochure—as slim, fit, young, foreign-looking couples. They wanted to put 'real people'; people from 'here' in the images. They referred to a free online repository of Indian, non-model-like, ordinary-looking people to characterize the residents.

5 Placings

Since place is something that happens through inhabiting, here it is employed as a semantic device. It fleshes out the spatial structure. It creates 'effects' of place. These places are meaning-chunks. They contribute in creating atmospheres. The effort by the group was to create the effect of place in/on the apartment-block. The intention was to bring 'places' into and contain it within the home-world. Actual non-residential places (like Kala Ghoda, Azad Maidan, etc.) are referred to depict the quality of spaces to be included in the housing design. According to Tuan, architects discuss space and place interchangeably [7]. The attempt seems to be to incorporate the attributes of these places, like a patchwork of places. The apartment-typological aspect of design is more space-based. The domestic aspects dealing with the home-notion are dealt with at the level of place. The spatial organization could question, innovate and also threaten place-based practices. Boundaries also were the subject of heated debates. The arguments around the decision to allow a public shopping street right through the private residential area are an example. In this sense, the technical-spatial aspect of the place and the private home-nature, especially in apartments might be at odds with each other [8]. For instance, lamenting on the market-based neutrality and loss of identity that discourages place-bound architecture in Mumbai, one of the participant architects critiqued the 'placeless' nature of the apartment-complex.

By describing/naming space in terms of the 'look and feel of a mohalla' or a 'mohalla-like space', or the typologies of 'here', associations with certain patterns, places, activities, practices, people are evoked, providing meaning to the abstract geometrical aspect of organizing space. The historical architectural remnants of the factory and warehouses on mill land, were made part of the common public space and lifestyle and 'shared' amongst hundreds of dwelling units. As a fragment, it is not only a spectacle but also points to another place—like collectively owning a piece of history.

This focus on the outside is extended in the design team's discussion with an environmental designer. When shown the site-plan on Google Earth, he asked for a zoomed out view so as to gauge the surroundings of the site. His proposal was to develop the space around the site. The fact that the environmental designer considers the surrounding area beyond the site for development makes it seem like the super-built-up would go one-notch to become a *super* super-built-up area!

6 Atmospheres

Different ways of organizing/distributing the dwelling-units lead to different shapes of super-built-up area. The intention of the team was to present a 'menu card' of spaces corresponding to distinctive 'spatial qualities' and multiple life-style options. As a mark of spatial distinctiveness, they refer to American Architect

Steven Holl's work in Netherlands and the residential project 'Amanora' of the Dutch firm MVRDV in Pune. The chief architect hints at the fact that the 'extra' value-addition in the form of amenities in the super-built-up space would help in differentiating from other such projects in the market. Given most interior-flat layouts across various residential projects are the same (2, 3 bhk), he suggests that the real space of distinction would be the super-built-up, and the corresponding lifestyle offered.

The current trend of separation and hiding of the service space (plumbing, electrical, etc.) and its disadvantages were discussed. The service-space as being squeezed out of the visible sphere and relegated to the infra-structure (the structure which is not seen). They also observed a shift in the trend from 'Hafeez-like frills' (Hafeez Contractor is a very popular architect in India known once for his very ornate style of architecture) to amenities—from a visual to an experiential aesthetic; from an ornamental to a more diffuse, subtle effect. This directly points to the shifts in what has to be 'housed' for a particular lifestyle. The image ability of the project rests on capturing this aesthetic. According to a team-member, a 'walk-through', for instance, simulates the quality of space—the 'inter-building' and the 'intra-building' spaces—revealed through strategic positioning of 'camera angles'. Three-dimensional walk-throughs with 'some music' would act as an experiential substitute for the spatial surround.

The architectural solution tucks in the infrastructural services like plumbing and electricity. But it also attempts to incorporate services which manage the household. This service space is integral to generating a truly leisurely, conflict free, smooth space. This structural sorting, frees up interior space from the technicalities of territory. It opens it up to various ambient expressions. Towards the more expressive aspects of atmospheric manipulation, opens it up to all kinds of ambient expressions. Sightlines and views strategically constitute the relationship between the landscape and the interior as that of foreground-inside thus interiorizing the scenic outside. Owned interior-out, since the foreground is an extended interior, the apartment-complex induces in each individual unit, a sense of possessing more than its share of super-built-up space. Township developments are ways of interiorizing the city/urbanity.

The advertisements attempt to depict this ambient aspect. The image has to convey or rather perform this ambience. Lifestyles are a certain way of being immersed in that space. Therefore, the lifestyles, part of the placings ultimately are a performance of this ambience. The discussion on the 'complexion of space', the 'quality of space', and the 'richness of space' point to the experiential dimension of space. The richness of space means that the multiple employment of the same space is possible. The complexion of space also points to these multiple elements. The larger complexes, the more the collective space and the more the thought that can be given to the atmosphere.

7 Spacings, Placings and Atmosphere

An underlying assumption seems to be at work in the above discussion—that of the 'everyday' being too close and too normal to be considered spectacular. The home's association with the everyday would also make it seem non-spectacular. Special experiences are not usually identified with the everyday. An attempt is made to bring this special feeling into the everyday. So, one doesn't have to 'seek out' an experience. The experience could be simply had at home. The intention of the housing seems to be to offer a 'new' place, and thus a 'new' lifestyle; the exciting possibility of new habit-formations, along with the maintenance of some old ones. The aim is to provide an 'out-of-the-world' experience—the conception of a self-sufficient and complete world, everything within convenient reach. The housing thus attempts to bring the new experience indoors. The home as much as it is a place for possessions, is also a space that is possessed. In this case, the house-composition allows for the possession of new experiences. So the apartment-complex has to project the containment of a world of experiences.

The case-study discussions also engage with the shifting formal stylistics—the fact that the experiential aspect is more contemporaneous than the surface decoration on the façade of the built-form. The packaging of cities as commodities produces the city as a set of scenographic sites, a 'non-place urban realm'—a sovereign space of atmospheric effects [9]. The apartment is a similar scenographic site promising intense experiences. The space of being together is articulated in terms of 'new' experiences of hyper-leisure. The housing thus assembles new experiences pieced together from various fragments and place-effects. The dweller is encouraged to incorporate the spectacular 'other' spaces and programs into how he lives and discover 'out-of-the-world' experiences in his own home. This ingesting of the extra-ordinary within the ordinary space of the everyday life is intensified in the space-choked urbanity of Mumbai. Through the common space and all the amenities ('no need to go out during the holidays' says a developer's hoarding), the city is also interiorized, domesticated, made 'homely'.

Bhartiya Group, DLF, and a host of other developers are planning townships which are 'integrated', 'futuristic' and 'well-managed' across the country. These developments claim to have dedicated green spaces. There are many manicured urban villages sprouting across the country in the outskirts of the city. These are fashionably gentrified enclaves. They are projected as idyllic. They are marketed furiously. They assure a variety of lifestyles. The individual gated community property developments (single apartment-blocks), unlike the township, lack the infrastructure to provide the 'experience' of lifestyle. The larger the common collective spaces, the greater the possibility of ambient, immersive modes of housing.

Acknowledgments I would like to acknowledge Uday Athavankar, under whose able supervision this study was conducted. I would also like to thank Nina Sabnani, Ramesh Bairy, Sharmila Sreekumar and Kushal Deb for having contributed significantly to the development of this study.

References

1. Srinivas T (2002) Flush with success: Bathing, defecation, worship and social change in south India. Space and Culture 5(4): 368–386 http://sac.sagepub.com/content/5/4/368.full.pdf+html
2. Dovey K (1985) Home and homelessness: introduction. In: Altman I, Werner CM (eds) Home environments, human behaviour and environment: advances in theory and research, vol 8. Plenum Press, New York
3. Saunders P, Williams P (1988) The constitution of the home: towards a research agenda. Hous Stud 3(2):81–93
4. Hillier B, Hanson J (1989) The social logic of space, Cambridge University Press, Cambridge
5. De Landa M (1995) Homes: meshwork or heirarchy? Mediamatic http://www.mediamatic.net/page/5914
6. Nesbitt K (1996) Theorizing a new agenda for architecture: an anthology of architectural theory 1965–1995, Princeton Architectural Press, New York
7. Tuan Y (1977) Space and place: the perspective of experience, University of Minnesota, Minneapolis, pp 1–10
8. Kaika M (2004) Interrogating the geographies of the familiar: domesticating nature and constructing the autonomy of the modern home. Int J Urban Reg Res 28(2):265–286
9. Ruthesier C (1996) Imageneering atlanta: making place in the non-place urban realm, Verso, London

Meta-Design Catalogs for Cognitive Products

Torsten Metzler, Michael Mosch and Udo Lindemann

Abstract This paper presents a concept of meta-design catalogs for cognitive products. Mechatronic and cognitive products usually demand multidisciplinary hardware and software solutions and while catalogs exist to support domain specific-design, to date there is no support for finding non-obvious and alternative solutions for cognitive products. The meta-design catalogs for cognitive functions proposed in this paper provide a link between abstract functions and hardware/software making it possible to find non-obvious and alternative solutions. They also make re-use of existing solutions possible by abstracting from the specific to an abstract pattern.

Keywords Design catalog · Design pattern · Cognitive product · Cognitive product development

1 Introduction

Design catalogs and catalogs of design patterns, such as [1–3] were created to support engineers in finding non-obvious solutions, alternative solutions and to avoid inapplicable solutions or repeated solutions. They exist for different domains but address mainly domain-specific problems.

T. Metzler (✉) · M. Mosch · U. Lindemann
Institute of Product Development, Technische Universität München, Munich, Germany
e-mail: metzler@pe.mw.tum.de

U. Lindemann
e-mail: lindemann@pe.mw.tum.de

A. Chakrabarti and R. V. Prakash (eds.), *ICoRD'13*, Lecture Notes in Mechanical Engineering, DOI: 10.1007/978-81-322-1050-4_25, © Springer India 2013

This paper presents a concept of meta-design catalogs for cognitive products that is appropriate to classify all types of solution patterns and thereby helps to transfer cognitive functions, e.g. *perceive, learn, plan,* etc., into cognitive products enabling them to act in an increasingly intelligent and human like manner. Meta-design catalogs for cognitive products are necessary because cognitive functions currently are realized only through interdisciplinary solutions and can not be realized by domain-specific solutions alone. A complete decomposition of cognitive functions into elementary functions is not possible yet due to a lack of understanding of cognitive processes. Instead, solution patterns systematically stored in design catalogs provide the possibility to include knowledge of different domains and support the conceptual design of cognitive products because they can be re-used. By directly addressing cognitive functions a further decomposition into elementary functions is usually unnecessary.

After a brief definition of terminology, domain-specific patterns are compared with the aim to identify how design catalogs from different domains are structured, what they have in common and how they differ. Based on this, a holistic framework of meta-design catalogs for mechatronic and cognitive products and systems is derived. This structure includes different meta-design catalogs differentiated by the type of design catalogs, complexity and granularity. They are appropriate to classify all types of solution patterns and thereby help to transfer cognitive functions into cognitive products. To demonstrate how solution patterns can be identified in existing cognitive products and generalized for re-use in conceptual design of future cognitive products an example is presented. One solution pattern is explained at different levels of abstraction and allocated to the framework of meta-design catalogs for cognitive products.

2 Background

Design catalogs and catalogs of design patterns are information sources supporting the conceptual design process by re-using solutions. They help to find patterns that realize certain functions used in function structures and models describing a system- or product-concept. They exist in several domains and vary mainly by the functions they provide patterns for and their level of abstraction. Alexander et al. [1] say that "each pattern describes a problem which occurs over and over again in our environment, and then describes the core of the solution to that problem, in such a way that you can use this solution a million times over, without ever doing it the same way twice".

In mechanical engineering, design catalogs are tailored to systematically support the conceptual product development process and have to fulfill the following requirements: quick accessibility of information, ease of use, customizability, integrity within given boundaries, validity, upgradeability and consistency [4]. They are subdivided in three categories according to [2–4]: object catalogs, process catalogs and solution catalogs. Object catalogs contain available objects, e.g. bearings and screws, and are independent from specific design problems. They do not contain all principle solutions for a design problem but describe objects and their characteristics generically. Process catalogs contain processes, rules and process steps and are related to objects. Each solution catalog contains a variety of patterns for specific design problems and constitutes a source for alternative solutions [2]. Obviously, the three catalog types are related to each other, e.g. objects of an object catalog can help to generate new patterns for a solution catalog according to a process described in a process catalog. However, one object may be included in several patterns in different solution catalogs and adaption may be possible using different processes.

Design catalogs in mechanical engineering have mainly been developed for elementary functions, e.g. to convert, to increase/decrease, to mix/separate, etc. because it is assumed that all functions can be decomposed into them [2, 3]. However, they also include frequently re-appearing functions [3]. Homogeneous internal catalog structures are important with respect to convenience and clarity [1]. Therefore, most catalogs are structured in a similar way. They consist of an index structuring the content, a main part describing the solutions and an access part explaining the properties of the solutions. Recent research extends the access part by adding disturbances and robustness ratios to physical effect catalogs in order to consider them in the conceptual design phase of a product while avoiding additional effort and cost [5].

The development of electric and electronic systems and products is, when compared to mechanical engineering, more object-oriented. In electronic systems, objects, e.g. resistors, transistors and integrated circuits, are stored in object catalogs. Solution catalogs in this domain contain for example adaptable circuit diagrams for operational amplifiers that amplify a differential input voltage to a much higher output voltage. General circuit diagrams in solution catalogs are adapted according to the particular problem by defining properties of basic hardware components, e.g. resistors and capacitors. According to design rules and processes that are stored in process catalogs (or even computer tools), for example the width of conductor lines or the spacing between lines can be calculated. These rules, processes and especially the tools allow inexperienced electrical engineers to design electronic components, for example "Very-Large Scale Integration Systems", because the design methodology is based on the electrical behavior of circuit elements [6] and universal design rules. It is assumed that design catalogs for electrical engineering can be structured like those for mechanical engineering due to the above mentioned characteristics even though only object catalogs were found.

In computer science, design patterns are considered re-usable elements of software supporting the conceptual software design. Patterns in object-oriented software provide generic solutions in terms of objects and interfaces. They vary in granularity and level of abstraction. Similar to design catalogs design patterns are characterized by the "pattern name", the "problem" they address, the "solution" they provide and the "consequences" of their application. Further, they are classified by their purpose and their scope and grouped in families of related patterns [7].

Experts re-use successful solutions and base new designs on prior experience without having to re-discover the whole problem [7]. Design patterns help inexperienced software developers to do the same based on the experts sharing their knowledge.

Looking at the status-quo of design catalogs and catalogs of design patterns several limitations exist. First, design catalogs in [3] are still mainly paper-based even though Derhake [8] proposed a system to computationally create and represent design catalogs. Computational search for patterns is limited and synthesis is not adequately supported. For this reason, the ease of use is low and the speed of search is slow. The authors assume a negative impact for existing paper-based design catalogs regarding the acceptability and usability due to these issues. In addition, design catalogs in the mechanical engineering domain are predominantly available in German language hindering broad usage. Koller et al. [2], Roth [3] and VDI 2222 [4] demand a consistent and conceptually accurate catalog structure. This demand must be extended to design catalogs and catalogs of design patterns of all domains in mechatronics to enable a search for multidisciplinary patterns. The use of solution patterns was already proposed for mechatronic systems in [9].

3 Terminology

In this section the terms used throughout the paper are explained. This is important since every domain uses their own terminology with respect to re-usable solutions and the paper intends to be understandable for interdisciplinary product development teams.

A *design catalog* contains knowledge about engineering design. It is systematically developed and makes knowledge available for everyone, independent of the experience and knowledge of individuals [3]. A difference between a collection of solutions and a design catalog is drawn regarding the level of completeness, the systematic structure and the accessibility within the design process, all of which are more restrictive for design catalogs [2, 3]. The term design catalog is predominantly used in the electro-mechanical domain with a strong focus on mechanical design.

Design patterns describe general design problems and their solutions [1, 7]. According to Gamma et al. [7] a design pattern "names, abstracts, and identifies the key aspects of a common design structure". They can vary in granularity and the level of abstraction that is influencing the reuse of every design pattern. Design patterns with a high level of abstraction are likelier to be reused. The term design pattern is mainly used in the domains of computer science and architecture with increasing interest in mechatronic design.

Catalogs of design patterns organize design patterns to make systematic and efficient retrieval possible. They correspond to design catalogs in the electro-mechanical domain.

The term *solution patterns* is used in this paper to describe a generic solution of a problem that is solved in a domain-spanning way and is appropriate for re-use. Typically, the implementation of cognitive functions requires a domain-spanning solution. The new term "solution pattern" is introduced to avoid confusion with terms that have different meanings in different domains and to point out when solution patterns for cognitive functions are discussed. Solution patterns can vary in complexity, granularity and type of solution.

Solution patterns are stored in *meta-design catalogs for cognitive products*. The meta-design catalogs provide a framework for all solution patterns, similar to design catalogs and catalogs for design patterns. In addition they are linked with domain-specific design catalogs and catalogs for design patterns allowing a breakdown into domain-specific subfunctions. In contrast to all types of above mentioned patterns a *solution* is one problem-specific occurrence of a pattern. The terms *cognitive product* and *cognitive function* are defined in Sect. 4.2.

4 Meta-Design Catalogs for Cognitive Products

This section first compares design catalogs available in engineering disciplines and catalogs of design patterns from architecture and software design. Based on this comparison, functions in meta-design catalogs are described. Finally the framework of meta-design catalogs for cognitive products is presented.

4.1 Comparison of Design Catalogs and Catalogs of Design Patterns

A comparison of design catalogs and catalogs of design patterns from different domains leads to the following assumptions. They all serve one main purpose: to find solutions for general design problems and avoid repeating the same work. In general, they use a semi-formal description of the initial situation, e.g. elementary functions in [2], electrical behavior in [6] or pattern name in combination with problem description in [7]. The solution pattern usually is broken down into a description including elements like name, problem, solution and consequences.

The common goal of design catalogs and catalogs of design patterns is to make the re-use of successful designs and architectures easier and so help designers to find design alternatives quickly [2, 7].

Graphical notations solely are not sufficient to represent solution patterns; neither in engineering design nor in object-oriented software design [2, 7]. Nevertheless, they are important and useful to foster understanding of an abstract textual description in all domains and can provide concrete examples.

4.2 Functions in Design Catalogs

To date, design catalogs and catalogs of design patterns address mainly design problems related to single domains. Nevertheless, element design and system design are inseparable [6] and systems engineers manage the development process of complex engineering projects. This paper is about meta-design catalogs for cognitive products. *Cognitive products* are tangible and durable things consisting of a physical carrier system with embedded mechanics, electronics, microprocessors and software [10]. The surplus value is created through cognitive functions, e.g. to perceive, to learn and to act [11], enabled by flexible control loops and cognitive algorithms. *Cognitive functions* are the elementary functions enabling cognition as a whole and heavily rely on a software component but nevertheless are regularly realized through the combination of solution-elements from different domains. This already indicates that common, domain-specific design catalogs and catalogs of design patterns are not appropriate to search for high-level solution patterns of cognitive functions. First order cognitive functions, e.g. perceive, learn and think, are very abstract and neither support a straightforward search for solution patterns nor allow an easy decomposition into subfunctions with known solution patterns. Nevertheless, solution patterns are needed for each cognitive function that can be adapted to specific design problems in the conceptual design of cognitive products.

Koller et al. [2] and Gausemeier et al. [9] show that product functions can be decomposed into elementary functions, e.g. according to Pahl et al. [12]. Is it possible to similarly decompose cognitive functions into elementary functions? If so do these correlate with common elementary functions that can be found in existing design catalogs of different domains? If such a decomposition is possible, it is not intuitive and even cognitive scientists or neuroscientists can not precisely tell what the elementary functions are that are involved in cognitive processes in human beings. For example, Rees [13] says that "seeing is not perceiving" but can not name the extra "bit" required for perception. The authors assume that a full decomposition of human cognition into elementary functions is, with the current knowledge of human cognitive processes, not possible and for this reason not realized in a cognitive technical system to the same extent yet. However, cognitive functions are imitated and realized in CTSs and cognitive products.

Functional decomposition of imitated cognitive functions always points to solution patterns in catalogs of different domains without considering interrelations among them. By developing meta-design catalogs for cognitive functions interrelations can be considered among corresponding sections of the catalogs. It is also expected that, in future, objects will be available off the shelf that conduct cognitive functions instead of elementary functions. Therefore new object catalogs are required, capable of linking abstract solution patterns to real solutions.

4.3 Framework of Meta-Design Catalogs for Cognitive Functions

In this paper a pragmatic way to support the design of cognitive products is proposed. The approach is to develop a framework of meta-design catalogs for cognitive products that links generic solutions to related subsets of solutions in domain-specific design catalogs and catalogs for design patterns. Relating the meta-design catalog to domain-specific design catalogs is important because it allows product developers to use design catalogs they are familiar with, they do not have to be updated independent from the domain-specific catalogs and it is less demanding computationally.

The catalog types used in mechanical engineering already provide a meaningful catalog classification for design catalogs of cognitive products as well as other domains. The above mentioned catalog types constitute the first dimension of Fig. 1: type of catalog. Process catalogs for cognitive products contain processes, rules and process steps describing for example how to develop a cognitive product, how to connect cognitive functions or how to decompose cognitive functions into cognitive subfunctions. Object catalogs for cognitive products are empty at the moment because integral objects accomplishing cognitive functions independent from specific design problems are not available. For this reason process catalogs are not yet linked to object catalogs targeted at cognitive products. In the future, integral objects accomplishing cognitive functions may exist and will be integrated in the object catalogs. Solution catalogs for cognitive products contain solution patterns for specific design problems related to artificial cognition and constitute a source for alternative solution patterns. The authors currently work on a modeling approach for cognitive products using the systems modeling language (SysML) and expect the identification of patterns realizing different cognitive functions. Instead of linking process catalogs and solution catalogs for cognitive products with empty object catalogs directly, they are linked to domain-specific design catalogs including object catalogs.

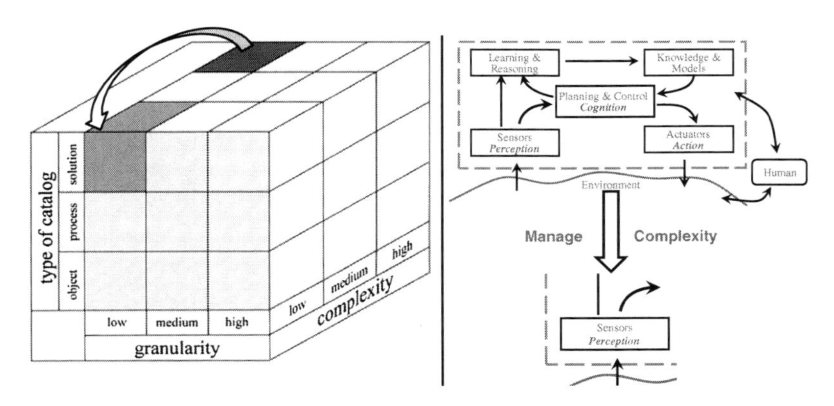

Fig. 1 Framework of all design catalogs for cognitive products (*left*) and the management of complexity (*right*)

Further, [3] and [4] distinguish design catalogs regarding their complexity whereas [7] use the term granularity. The authors consider both differentiations important for the following reasons. Roth [3] and [4] determine complexity by the number of relations among elements in a system or product. This means that a cognitive product with several interlinked cognitive functions is usually more complex than a system with an isolated cognitive function. Granularity describes how detailed a system is broken down or decomposed into subfunctions and therefore a higher granularity is characterized by an increase of interlinked functions in the model. In conclusion it is assumed that a higher granularity increases the model complexity but not the inherent system complexity because only the level of abstraction is changed. In order to create holistic design catalogs for cognitive functions different levels of complexity as well as different levels of granularity have to be covered. This helps to manage system and product complexity and to increase model granularity by breaking down functions into elementary functions that can then be linked to single-domain design catalogs.

Figure 1 (left) shows all possible design catalogs for cognitive products structured according to catalog types, complexity and granularity. Solution patterns with high complexity and low granularity, e.g. in solution catalogs, are linked to design catalogs with a lower complexity and low granularity. This is visualized in Fig. 1 (left) with the black arrow pointing from the left side of the building block in the back to the left side of the building block in the front. An example for a solution pattern with high complexity and low granularity is for example a generic system architecture for CTS as proposed in [14], see Fig. 1 (right).

Beyond the classification of meta-design catalogs for cognitive products their internal structure is of great importance because they need to cover solutions including elements of different domains. By comparing the internal structure of design catalogs from different domains, the following issues have to be considered to create holistic and unambiguous design catalogs for cognitive products that are valid independent of complexity, granularity and catalog type:

- a (formal) description of the problem/function and the solution is given
- solutions are accessible through a kind of index
- limitations of the solution space through parameters.

Next, the advantages of existing design catalogs and catalogs of design pattern are combined to create a suitable catalog structure for cognitive products. Thus, at first a universally valid index of design catalogs for cognitive products is required and proposed according the taxonomy of cognitive functions [11]. The taxonomy of cognitive functions was created by analyzing scientific publications from different domains with regard to cognitive capabilities and cognitive functions and structuring the found terms in an unambiguous way in a taxonomy. Using cognitive functions as an index for design catalogs seems appropriate because all kind of cognition can be traced back to them. Depending on the hierarchical level of the cognitive function that has to be realized technically, different links point to different solution patterns in the main part of the catalogs with different levels of abstraction. "to act", a very abstract cognitive function, points to a very generic solution pattern generally describing fundamental requirements as well as to sub-patterns providing more

Index

Primary Pattern Name	Link	Secondary Parttern Name	Link	...
to perceive	*link a*			
to learn	*link b*			
to understand	*link c*	to interpret (to construe, to see)	*link c.1*	...
		to solve (to work out, to figure out)	*link c.2*	
		to appreciate (to take account)	*link c.3*	
		...	*link c.n*	
to know	*link d*			
to think (to cogitate, to cerebrate)	*link e*	to reflect (to think over)	*link e.1*	...
		to reason	*link e.2*	
		to plan	*link e.3*	
		...	*link e.n*	
to decide	*link f*	to judge (to adjucate, to try)	*link f.1*	...
to act (to move)	*link g*	to react	*link g.1*	...
		to interact	*link g.2*	
		to move	*link g.3*	
		...	*link g.n*	
...	...	...	...	...

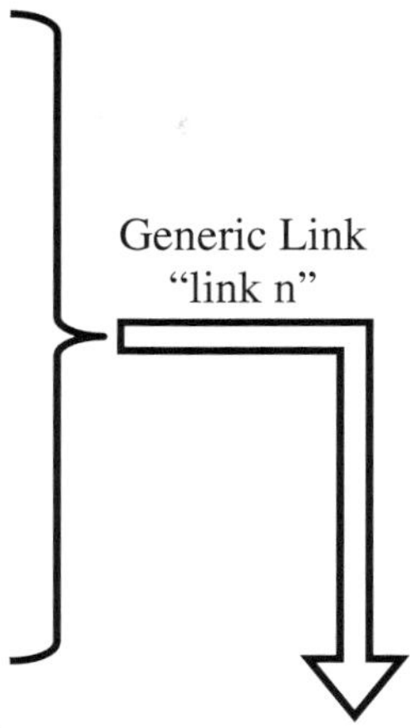

Generic Link "link n"

Main Part

Pattern Name	"name"				
Problem Description	description of the problems that pattern "name" solves				
Solution Description	Description	Sub-Pattern	Domain	Link to Domain-Specific Catalog	Description
	general description of the solution for pattern "name"	link to sub-patterns (if no further sub-pattern is available continue with domain specific solutions/patterns)	Mechanical Engineering	link me	
			Electrical Engineering	link ee	
			Computer Science	link cs	
			......		
Consequence Description	description of consequences when applying pattern "name"				

Fig. 2 Internal structure of design catalogs for cognitive functions with index and main part

specific solutions for subfunctions of "to act", e.g. for "to interact" or "to move" (Fig. 2). The main part of the catalog contains the solution patterns for the cognitive functions. Every solution pattern is first characterized by a name according to the cognitive function and a description of the problem that can be solved with it to avoid wrong applications of the pattern. Then the solution provided in the pattern is described. In addition, links to sub-patterns are included, in case the user is looking for something more detailed. Finally, the solution is broken down into domain-specific components of the solution, either objects in object catalogs, processes in process catalogs or tailored solutions in solution catalogs. A description of consequences concludes the every pattern including e.g. limitations of the pattern or possible disturbances influencing the solution.

5 Application

This section shows how solution patterns of design catalogs for cognitive products are identified, using as an example cognitive product developed by an interdisciplinary student team, and how to allocate the solution patterns to the design catalogs. Afterwards, the identified solution pattern can be used to support the search for solutions in the conceptual design phase of future cognitive products.

The cognitive product from which solution patterns are identified is a toy called "Virtual Opponent Slot Car" and was developed in a class on the theme "Cognitive Toys" by a team of four students from ME and EE. The goal of the students was to develop a slot car toy that is fun, even if no human opponent is available. To achieve human like behavior and skills it was decided that the virtual opponent learns to drive around the track from the human player. For every segment the speed of the human-driven slot car is measured and the maximum speed is stored in order to recall it during a race. Since a game with optimal performance would become boring rather quickly, a tactic function was implemented that makes the virtual opponent act according to the human driver's performance. In case the human-driven slot car is behind the virtual opponent, it drives slower depending on the distance and in case the virtual opponent is behind the human-driven slot car it drives as fast as possible. This way, the two slot cars usually stay close together and a close finish situation is generated keeping the game exciting all the time

A picture of the prototype is shown in Fig. 3 on the left and the high level cognitive function structure on the right. The virtual opponent perceives positions of the two slot cars and the average user speed for every section of the track and compares it with the known maximum speed for that section. In case the actual speed is higher than the maximum known speed it learns the new maximum speed from the user. The virtual opponent acts according to its knowledge about the maximum user speed.

A frequently reoccurring cognitive function while developing cognitive products is "to perceive". Therefore, a solution pattern for "to perceive" has been abstracted from the virtual opponent describing the general problem, the related solution and consequences exemplarily. All parts of the solution pattern are included in Fig. 4. The pattern name is *"to perceive"* according to the cognitive function and it is appropriate to solve problems related to "becoming aware of something through senses". The generic solution pattern describes how a technical system can perceive by identifying subfunctions that have been found in prototypes, e.g. the virtual opponent. The action to perceive requires the sensing of signals in the environment and internally and the processing of them according to the existing knowledge of the system and the context [15]. Sub-patterns are

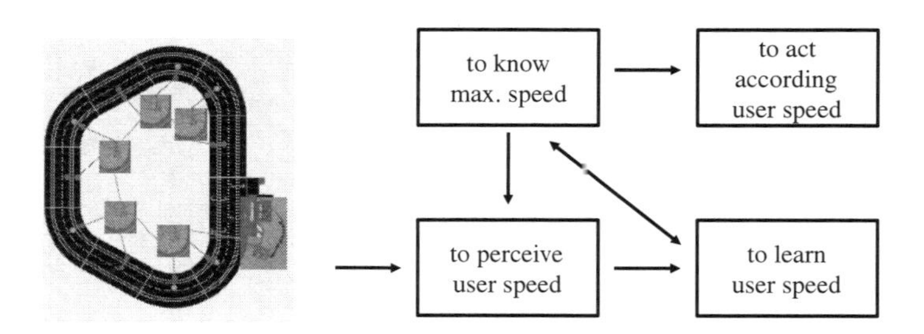

Fig. 3 "Virtual Opponent Slot Car" (*left*) and cognitive function structure of the product (*right*)

<table>
<tr><td>Pattern Name</td><td colspan="5">"to perceive"</td></tr>
<tr><td>Problem Description</td><td colspan="5">Provide a solution about how a technical system becomes aware of something through senses.
This pattern is relevant when a system is not only meant to do signal processing but has to consider context as well.</td></tr>
<tr><td rowspan="4">Solution Description</td><td>Description</td><td>Sub-Patterns</td><td>Domain</td><td>Link to Domain-Specific Catalog</td><td>Description</td></tr>
<tr><td rowspan="3">the action "to perceive" requires the sensing of signals in the environment and internally and process them according to the existing knowledge of the system and the context</td><td rowspan="3">-to know
-to sense
-to process data</td><td>Mechanical Engineering</td><td>-</td><td>-</td></tr>
<tr><td>Electrical Engineering</td><td>-sensor-catalogs
-processor catalogs</td><td>find appropriate sensor for the desired perception</td></tr>
<tr><td>Computer Science</td><td>- data processing algorithms</td><td>find appropriate algorithm for the perception task</td></tr>
<tr><td>Consequence Description</td><td colspan="5">- depending of what needs to be perceived different sensors are required
- depending on what sensor is used different software is required
- ...</td></tr>
</table>

Fig. 4 Solution pattern of "to perceive"

assumed for other functions, e.g. to sense or to process and other cognitive functions, e.g. to know. The consequences on this pattern are that, depending on what needs to be perceived, different sensors, processors and software, etc. are required. Linking this solution pattern with a domain specific pattern is partially possible, e.g. the link to object catalogs in electrical engineering containing sensors and processors is possible. The allocation of the solution pattern in the structure of meta-design catalogs for cognitive products has to consider the pattern type, complexity and granularity. The granularity in "to perceive" is low as well as the complexity. The solution pattern fits well into a solution catalog because it describes schematically how to make a technical system perceive something through senses. It neither describes an object nor a process. It is allocated to the light grey building block in the front of Fig. 1.

Because the textual description above and Fig. 4 are very abstract, an additional graphical representation of the pattern is considered helpful. This graphical representation has been developed in SysML (Fig. 5). The model includes operators and flows. The flows define the inputs and outputs of the operators and the operators conduct an activity on the flow. The combination of flow(s) and one operator is considered here as one function. In the example several functions together accumulate to the cognitive function "to perceive". Cognitive functions and flows in the activity diagram shown in Fig. 5 are further explained in [11].

In Fig. 6 components have been allocated to the subfunctions presented in Fig. 5. These components are capable of realizing the functions they are allocated to. In the case of the solution pattern "to perceive" the components belong to other domains, e.g. sensors are electronic objects and can be found in object catalogs of the electrical engineering domain.

By adding components to the solution pattern the granularity of the system model is increased but the model complexity increases likewise. For this reason it is good to start with an abstract pattern and model and successively detail it throughout the development process.

Coming back to the sample application of the virtual opponent perception is realized through the functions and components described below. The function "to sense" is realized through Hall Effect sensors that are placed around the racing

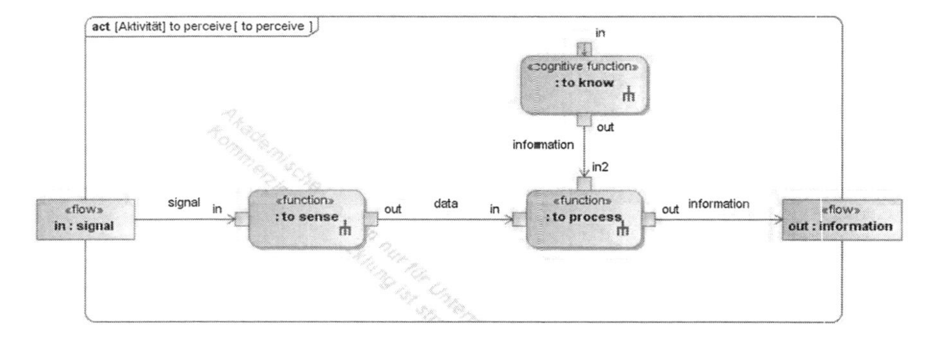

Fig. 5 Graphical representation of the solution pattern "to perceive" including functions

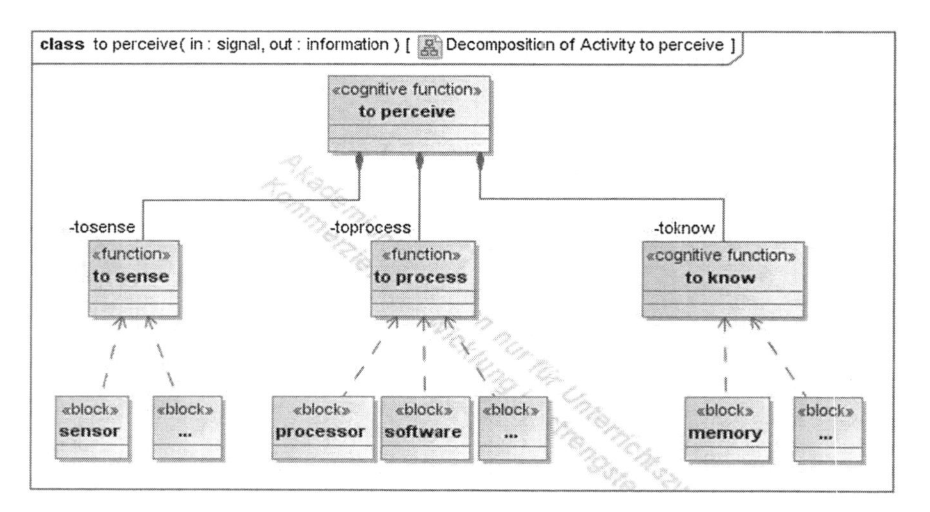

Fig. 6 Graphical representation of the solution pattern "to perceive" where components are attached to functions

track and detect magnetic fields ("signals"). They output data every time a magnetic field is within their range. A magnet is attached to the bottom to the slot car already to increase the traction. Therefore, no adaptation of the slot car was necessary. The processing of the sensor data is accomplished by a combination of the micro controller AT90USB162 mounted on an AVR-USB-162 development board from OLIMEX, a laptop and software. The software requires the sensor data and knowledge about the previous position as well as a definition about how the change in sensor data and position has to be interpreted. Therefore, knowledge about different sensor data has to be stored in the memory of the micro controller.

6 Discussion

Meta-design catalogs for cognitive products support their conceptual design phase by providing generic solution patterns that can be reused. They close the current gap of missing solution patterns tailored to cognitive functions that can not be broken down to elementary functions yet and therefore are realized using interdisciplinary solutions. By directly addressing cognitive functions a further decomposition into elementary functions is usually unnecessary.

Instead of creating meta-design catalogs for cognitive products that contain domain-specific solution patterns or even objects at component level they are interlinked with domain-specific catalogs avoiding ambiguity. That is possible because domain-specific design catalogs and catalogs of design patterns are structured similar and are partially identical to the proposed meta-design catalogs.

To date, the use of meta-design catalogs for cognitive products is limited by the number of solution patterns stored inside and the paper-based structure. The number of solution patterns must be increased significantly in order to benefit from the meta-design catalogs. This will be done in future research. As a starting point the authors are going to analyze the cognitive products they already developed, extract solution patterns and integrate them in the meta-design catalogs. It is expected that enough solution patterns will be found to do some basic evaluation of the design catalog by testing it during the development of a new cognitive product.

As proposed earlier in the paper a software implementation of the meta-design catalog structure is necessary, supporting: the systematic integration of new solution patterns, the search for solutions according to the problem description at the required level of complexity, granularity and type of catalog, and an effective representation of the solutions with links to other, domain-specific catalogs. Ideally, files containing a model of a solution pattern can be included in the meta-design catalogs, e.g. SysML models, and an interface to the modeling tool is available enabling an easy integration of the solution into the own model.

7 Conclusion

This paper presents a framework of meta-design catalogs for cognitive products supporting the conceptual development. They are structured according to catalog type, complexity and granularity from an external viewpoint and each catalog has the same internal structure consisting of an index and a main part. The main part itself consists of a pattern name, a problem description, the solution pattern and consequences arising through the application of the solution pattern. Solution patterns managed in meta-design catalogs for cognitive products are linked to other solution patterns containing cognitive functions as well as to domain specific design patterns.

Acknowledgments This research is part of the Cluster of Excellence, Cognition for Technical Systems—CoTeSys (www.cotesys.org), founded by the Deutsche Forschungsgemeinschaft (DFG).

References

1. Alexander C, Ishikawa S, Silverstein M, Jacobson M, Fiksdahl-King I, Angel S (1977) A pattern language. Oxford University Press, Oxford
2. Koller R, Kastrup N (1998) Prinziplösungen zur konstruktion technischer produkte. Springer, Berlin
3. Roth K (2001) Konstruieren mit konstruktionskatalogen—band II. Springer, Berlin
4. VDI 2222 Page 2 (1982) Design engineering methodics—setting up and use of design catalogs. Beuth
5. Mathias J, Eifler T, Engelhardt R, Kloberdanz H, Birkhofer H, Bohn A (2011) Selection of physical effects based on disturbances and robustness ratios in the early phases of robust design. Int Conf Eng Des (ICED11) 5:324–335
6. Whitney DE (1996) Why mechanical design cannot be like VLSI design. Res Eng Design 8:125–138
7. Gamma E, Helm R, Johnson R, Vlissides J (1995) Design patterns—elements of reusable object-oriented software. Addison-Wesley, Reading
8. Derhake T (1990) Methodik für das rechnerunterstützte erstellen und anwenden flexibler konstruktionskataloge. PhD Thesis, TU Braunschweig
9. Gausemeier J, Frank U, Donoth J, Kahl S (2009) Specification technique for the description of self-optimizing mechatronic systems. Res Eng Design 20:201–223
10. Metzler T, Shea K (2010) Cognitive products: definition and framework. International Design Conference—DESIGN 2010, pp. 865–874
11. Metzler T, Shea K (2011) Taxonomy of cognitive functions. Int Conf Eng Des (ICED11) 7:330–341
12. Pahl G, Beitz W, Feldhusen J, Grote K-H (2007) Engineering design: a systematic approach. Springer, Berlin
13. Rees G (2001) Seeing is not perceiving. Nat Neurosci 4–7:678–680
14. Beetz M, Buss M, Wollherr D (2007) Cognitive technical systems: what is the role of AI? Lect Notes Comput Sci 4667:19–42
15. Fellbaum C (1998) WordNet: an electronic lexical database. MIT Press, Cambridge

Extracting Product Characters Which Communicate Eco-Efficiency: Application of Product Semantics to Design Intrinsic Features of Eco-Efficient Home Appliances

Shujoy Chakraborty

Abstract So far the development of Eco-efficiency in home appliances has only concentrated on their technological attributes, overlooking the communication of Eco-efficient qualities through the aesthetical appearance of such appliances i.e. a meaning to be transmitted to the user. Technical attributes are communicated through extrinsic features i.e. labelling, branding, packing [1]. These are considered as semiotic content (not concerned with semantics) and give information regarding the product independent of meaning, and as such have no direct relation to the product appearance or character [2]. This paper will explain the application of product semantic theory to re-design the product characters of home appliances which communicate their Eco-efficient qualities through their intrinsic features within a non-instrumental product experience. The final output will be a set of design guidelines consisting of 6 product characters—*Futuristic, Feminine, Unconventional, Practical, Simple, Smart*—which appliance designers can apply. Product characters are adjectival constructs or visual metaphors [3, 4]. Design theorists have pointed out that companies which are able to communicate a specific meaning (such as Eco-efficiency) through their product appearance can achieve a competitive market advantage [5]. Athavankar [6] cited product semantics as having amongst others 2 core goals, (1) Improving user-product interaction. (2) Demystifying complex technologies. The intention of semantics as a design theory was to apply linguistic theories into a design process to develop 'readable' or 'self-evident' products through easy to apply methods [7].

Keywords Product characters · Eco-efficiency · Home appliances

S. Chakraborty (✉)
Politecnico Di Milano, 38/A Via Durando 20158 Milan, Italy
e-mail: shujoy.chakraborty@mail.polimi.it

A. Chakrabarti and R. V. Prakash (eds.), *ICoRD'13*, Lecture Notes in Mechanical Engineering, DOI: 10.1007/978-81-322-1050-4_26, © Springer India 2013

1 Introduction

Currently communication of Eco-efficiency is a mixed effort in the appliance industry, consisting of energy labelling, advertising and marketing strategies, thus giving it an ambiguous meaning confusing the user. It has been established that when the meaning of what a product has to communicate is not clear to the consumer, then, he or she will have difficulty in assessing the product and will therefore appreciate the product less [5]. Semiotic content like signage and labeling are not concerned with meaning communication [2].

The weak role of design in Eco-efficient appliance development is consequently due to the considerable ambiguity surrounding the exact meaning of Eco-efficiency in language. It is often referred to as: green design, eco design, environmental design, etc. Schmidheiny [8] says Eco-efficiency is *the production of goods and services which meet human needs while reducing environmental impacts*. This linguistic shortcoming i.e.: inaccurate usage is also reflected in the design approach towards Eco-efficiency especially within the appliance industry reflecting Krippendorff [4, 9, 10] who says meanings of artifacts are founded in language.

This research rests on 2 important design theories in order to encode the intended meaning of 'Eco-efficiency' into home appliances through their intrinsic features. First, the 'human centred' approach to Product Semantics as discussed by Krippendorff [4, 9], and second the 'product communication model' as discussed by Crilly et al. [11] both of which have overlapping concerns, and at the same time contain several unique insights into how to make products more expressive of their intended meaning.

Intrinsic product features are physical attributes—form, geometry, colour, proportion, and composition as opposed to extrinsic features which are strictly related to a manufacturer's marketing identity—packaging, branding [1]. A non-instrumental product interaction is based only on visual sensing of the artifact by the user with no mechanical or physical manipulation involved [12].

Why home appliances? Within the Europe home appliances account for the largest share of domestic energy consumption. With refrigerators and freezers accounting for 15 % of residential consumption followed by washing machines accounting for 4 %, and dishwashers, ovens, and clothes dryers accounting for around 2 % of total residential end use [13]. On its part the EU has set an efficiency guideline to increase appliance efficiency by 20 % at the end of 2020. Typically Eco-efficiency is largely looked upon as a debatable investment by important functionaries such as marketing and finance within appliance companies [14] and increasing the market performance of Eco-efficient appliances will help to address this situation.

Finally the research approach adopted in this paper is a 'Design-led' and 'Expert Mindset' approach highlighted by Liz Sanders [15] in her 'Map of Design Research'. This means that this research relies primarily on the input of design experts for attribution of product characters, and deriving visual manifestation of language based adjectives, while having an element of participatory mindset only in certain key stages of the entire process, such as capturing user feedback (i.e. adjectives) on design concepts.

1.1 Human Centred Approach to Product Semantics

Semantics as a discipline has traditionally been applied to study spoken languages and the study of meanings in languages, but its application into product design has been pioneered by Krippendorff and Butter in 1984 [16, 17]. The intention of semantics as a design theory was to apply linguistic theories into a design process to develop 'readable' or 'self-evident' products through easy to apply methods [7]. The early design theorists of semantics namely Butter [18] and Krippendorff [4, 9, 10] strove to develop it as a new approach to design [7, 19] through what they called the Human Centered view of design, a break from Luis Sullivan's "Form follows function" doctrine by instead proposing "Form follows Meaning". Unlike Crilly [11] and Krippendorff [10] warns not to confuse semantics with concerns of ergonomics which according to him is a machine centred view drawing upon objective evaluation of performance and mechanics removing human perception and faculties from consideration. The human centered approach to semantics is based on the attribution of 'characters' in products to express a meaning. The perception of visual appearances in products has a fairly rich historical precedence. Apart from semantics, other domains which overlap into this subject area are aesthetics, psychology, consumer research, sociology, marketing, semiotics, and ergonomics [16, 20].

Krippendorff defined Product Semantics as *a study of the symbolic qualities of man-made forms in the cognitive and social contexts of their use and the application of the knowledge gained to objects of industrial design*. Blaich [21] defined product semantics as *an area of inquiry or discipline concerned with the meaning of objects, their symbolic qualities, and the psychological, social, and cultural contexts of their use*. Athavankar [6] cited product semantics as having 2 core goals, (1) Improving user-product interaction, (2) Demystifying complex technologies.

The core concern of human centred design is acquiring 'second order understanding' according to Krippendorff [4]. Second order understanding entails understanding how users understand the artifacts designed by the professional. "Designers and their stakeholders merely understand differently" and since human centred design is designing for others, therefore it is the designer's job to acquire this understanding of another's understanding.

The human centred view pursued in this research is diverse from the view shared by design theorists and HCI community who propagate increasing the intuitiveness of products with the application of semantics and familiarise the user with the unfamiliar Crilly [11, 20, 22], Brown[7], [23], Evans and Thomas [24] and Norman [23]. Instead, the human centred view as Boess [19] says is an interpretation of Semantics directly derived from Semiotics, which is the study of signs, and looked at forms as proponents of language.

1.2 Meaning

A meaning is always someone's meaning [4]. Artifacts acquire meaning through use, and watching how users use an artifact can give insights into the attributed meaning. Thomas [26] points out that "meaning of form is a human production, as it is both malleable and undefined". Meaning is born in perception, thus it is essentially a human construct, and is only limited by people's capability to imagine [26]. It thus follows that any form by itself has no meaning, but is rather "a window to opportunity" according to Thomas [26]. Hekkert and Schifferstein [12] say that it is only through interaction with people that objects acquire a meaning. According to Krippendorff's human centred approach objects are open to interpretation, and language structures provide for how artifacts are perceived. Since forms trigger meanings, therefore forms can also subvert meanings [26]. The same form can trigger multiple meanings, based on the scenario in which it is present. A knife in a kitchen means entirely different than in the hands of a man in a dark alley. Thomas [26] points out that context influences the interpretation of a form, thus shaping the message it communicates.

Since meaning is born out of human perception it follows that meaning is influenced by personal past experiences of the user being a psychological construction. Past experiences and encounters influence meaning attribution. Krippendorff [4] points out that meaning is not constant and is invoked by 'sense' and so in this way related to 'perceived affordance' according to [27] theory of ecology.

Materials can have a bearing on the communication of meaning. Handcrafted wood denotes craftsmanship; metal surface is associated with precision, and plastic with cheap products [28].

1.3 Character, Characteristics, and Attributes

Both in product semantics and character theory it is a basic assumption that artefacts can be bearers of a message through the characters which they express. A character can be indicative of the external appearance, internal functioning, or behaviour of the artifact.

Demirbilek and Sener [16] say that all manufactured products make a statement through shape, form, texture, and colour. According to Janlert and Stolterman [29] a character can be ascribed with as much as a casual glance. Attributing characters makes it easier for users to anticipate the functioning of the object and also explain its behavioural patterns to the user. It thus follows that users tend to make a connection between certain appearances and the characters they attach to them i.e. transparent surfaces appear more futuristic, and organic forms appear more feminine. Based on these insights it can be said certain characters can potentially form stable relations with certain appearances and this interdependency can be an opportunity for designers.

Characters have to be "planted" into a product with great care. A complex product such as a car or an appliance is made up of several characters and even 1 character wrongly attributed can be misleading and detract from the carefully cultivated message which the product is trying to communicate. Product characters are adjectival constructs or visual metaphors and can be expressed through languages structures. [3, 4]. Thus designers have to be doubly careful with the details and features they utilise to attribute a certain character.

A character is a unity of characteristics i.e. not a simple collection, but with related characteristics integrated into a coherent whole [29]. This means all the characteristics united to send a common message to the user of the artefact.

A characteristic is interpreted as a qualifier of attributes. A characteristic is a kind of a higher order attribute, a meta attribute that applies to all attributes within a character or individual. Knowing a few of the characteristics constituting a character enables one to guess the remaining ones.

A sum of attributes which are qualified by a certain characteristic go into forming 1 characteristic, and a group of similarly constructed characteristics intending the same meaning combine to form 1 character. The hierarchy is thus pyramidal in structure.

1.4 Communication in Product Design

Based on Shannon's model of communication discussed in 'A mathematical theory of communication', Crilly et al. [11] explain how users experience the designer's intension through product forms. The designer is the source, the author of the design which seeks to communicate a meaning to the user. The product is the transmitter of this message through its intrinsic and extrinsic product features. The environment within which the product is sensed and the meaning attached is the channel, and the sensing user is the receiver of the original communication, where the sense of sight is of primary importance. The destination is the psychological response (emotional + cognitive) of the user who has sensed the product, and refers to the judgments which the user makes based on the information received by the senses. *These judgments refer to evaluation of the perceived qualities of the products being sensed* [11].

The design process discussed in the next section will aim to encode the meaning of Eco-Efficiency in washing machines based on the communication model derived below by the author (Fig. 1). This model emerges from combining the 'product communication' model as discussed by Crilly et al. [11] and the 'human centred' approach to Product Semantics as discussed by Krippendorff [4, 9]. The

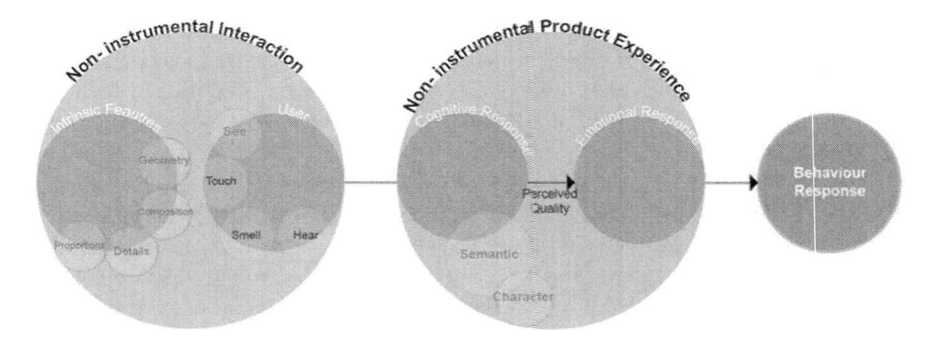

Fig. 1 Author's product communication model. *Source* Doctoral thesis

non-instrumental interaction consists of product intrinsic features (form, geometry, composition) and visual sensing of these features by the user. A non-instrumental interaction leads to a non-instrumental product experience which consists of a cognitive response being triggered (semantic interpretation) by the attributed product character (Eco-efficiency in washing machines). This cognitive response leads to an affective reaction leading to a positive behavioural outcome (approaching the product instead of ignoring the product) by the user.

2 Redesigning Product Characters Through a Design Process

The approach described here is aimed at re-designing the product characters expressing the Eco-efficiency of a washing machine, using a design process.

A washing machine was consciously selected as a test category for this study because it is an appliance category more-or-less typologically constant throughout all the major global markets. Other appliance categories, such as a cooktop or fridge for example, vary substantially in terms of typology (i.e. built-in versus free standing) across Europe, Asia, and North America which has a substantial impact on the appliance appearance. Since this design process relies heavily on capturing feedbacks from a culturally and geographically diverse set of users, therefore an appliance category had to be selected which was easily recognised. Below are the 7 steps of the design process:

1. See how product designers attribute Eco-efficiency in washing machines using a design workshop.
2. Conduct a visual questionnaire capturing adjectives through which non-expert users identify each concept.
3. Compile and classify adjectives captured through feedback from users.
4. Shortlist, rank, and group the classified adjectives into groups. Each group will represent a product character. 6 groups were constructed.

5. Build a mood board visually attributing the 6 product characters. These mood boards will act as guided stimulation for designers.
6. Conduct a second design workshop, asking designers to attribute a character set to a washing machine concept design with the help of a visual mood board representing each product character.
7. Perform a second questionnaire asking if the washing machines designed with the aid of product characters and mood boards appear Eco-efficient.

The steps 4–7 of the research process were adapted from the process suggested by Klaus Krippendorff in *The semantic turn*, 2006, itself an adaptation of the process suggested by Reinhardt Butter in *Putting Theory into Practice: An Application of Product Semantics to Transportation Design* [17] which in turn is similar to the 5-step *metaphoric links* process introduced by Athavankar [30]. Butter's *somewhat linear with clearly distinguishable phases* 8 step heuristic (experience based) process was designed specifically to apply the theory of product semantics to develop products.

2.1 The 7-Step Design Process

Step 1 *Design Workshop—IIT, Guwahati—M.Des:* The design workshop in IIT, Guwahati was planned over a period of 7 days with 8 students from the Master in Design-second year (M.Des-II) students. Since a clear trend of visually expressing Eco-efficiency has not yet emerged in the appliance industry, this workshop was designed to test how designers would approach the expression of Eco-efficiency through the intrinsic features of an appliance. The brief given to the students was: *Express the Eco-efficiency of a washing machine only through its intrinsic features affecting the appearance without disrupting too much the current technical architecture of it.*

The output of the workshop strictly maintained focus towards expressing Eco-efficiency through the form, geometry, proportions and composition of the appliances. Hekkert and Karana [28] define form as "the boundary of matter by which we distinguish (these) objects from each other and their environment" (Muller 2001). Shape is defined as that which *determines an object's boundary, abstracting it from other aspects, such as colour and material* (Chen 2005) [28].

Since at this stage of the design process, the workshop intentions were relatively general, i.e. the product character associated with Eco-efficiency were yet unknown, the concepts were delivered in a digital file format making it possible to remove a lot of background noise and colours (in the form of colour, materials, finish, and context) which might distract the user. This consideration was taken keeping in mind the insights of Hekkert and Karana [28] on how meaning attached to materials and colour can effect the user's perception of shape and form of the product. All students worked individually and delivered 8 concepts in total.

It was observed that the students were having significant difficulties in developing an aesthetic intervention for a washing machine which expressed its Eco-efficient qualities without impacting its typology. For designers, working on just the aesthetical level of a product without manipulating its internal workings on a typological level, seemed to be an inadequate level of intervention. The author realised a practical limitation of the concepts of the semantic theory proposing communication of a meaning through the product's appearance, as they were not easily accepted by the professional design community. Athavankar [6] too attests to this challenge by saying *It is a normal practice to mix issues like product function, aesthetics, technology and culture in discourses of form. Isolated discourses on form have always been seen as suspect.*

Step 2 *Visual questionnaire with non-expert users*: The objective of this step was to collect and capture a list of adjectives which potential non-expert might utilise to identify each concept. Krippendorff says *by definition, the character of an artifact consists of all adjectival constructions that a community of stakeholders in that artifact deems suitable to that artefact*. The questionnaire was displayed maintaining visual clarity and focus on only the form and geometry of the concepts, all colour, finish, material, and backgrounds were removed. All the concepts were visualised in black and white against neutral or white backgrounds.

Users were asked to attribute 3 adjectives spontaneously to each concept and rate the Eco-efficiency of each as "very Eco-efficient, some-what Eco-efficient, and not Eco-efficient". There were 30 respondents having a 50–50 male–female distribution and they were aged 23–53 with mixed nationalities. None of the respondents were from product design or allied fields.

Step 3 *Compilation and classification of captured adjectives*: Only 3 of the 8 concepts which were voted by at least 50 % of respondents as appearing Eco-efficient and were selected for evaluation of this step (Fig. 2). The objective adjectives i.e. adjectives which measure the physically measurable properties of an artifact were singled out and discarded. This was done following Butter [18, p. 55] who elaborated upon the qualitative communication which products must make. He says when generating a list of qualities which a product seeks to make to the user one must disregard the *factual attributes* which represent the objective, measurable qualities of "how something actually works" and keep only the

Fig. 2 Top 3 ranked washing machine concepts from IIT-Guwahati workshop:concept **1** (60 % voted very eco-efficient); concept **2** (50 % voted very eco-efficient); Concept **3** (50 % voted very Eco-efficient).

"expressive or semantic attributes" which represent the more communicative qualities of a product and the meaning which it communicates. All remaining adjectives coming from the 3 concepts were classified as *aesthetic adjectives, adjectives of social value, adjectives of emotions, and adjectives of interface qualities* (Krippendorff, The semantic turn, pp. 177).

Step 4 *Shortlist, rank, and group the classified adjectives* (*KJ Method*): The KJ Method is a technique which simplifies effective group decision making. It is based on the premise that when participants are informed of each other's perspectives on the same subject their decision making power can increase drastically [30]. The details of this process will not be described as they are readily available in current literature.

This process was performed as a focus group of 4 design experts within the author's research group. The objective of this activity was to rank the adjectives selected from previous step and arrange them into characteristic groups. Each group was then given a representative adjective voted from within each group which was demarcated as the product character representing that particular characteristic group (Fig. 3).

Based on the frequency of usage captured from the previous questionnaire each adjective was ranked. Thus the highest ranked adjectives became dominant characters and the lower ranked adjectives became recessive characters. The final group of 6 characters was—*futuristic, feminine, unconventional, practical, simple, smart*. The attributes within each character serve as a further indication to the designer about the qualities each character represents. From a point of view of language the attributes can also be regarded as synonyms of the character they represent. The opposite adjective (*antonym*) of each character was also indicated using a thesaurus though not visually represented. The characters will act as desirable qualities for the next stage design workshop, and their opposites will act as undesirable qualities which the designers have to definitively avoid.

Step 5 *Visual mood boards—attribution of product characters :* A visual mood board was constructed for each of the above 6 characters. If characters represent the sensory experience of the object, then mood boards are the manifestation of that experience in the form of product features depicted visually. Krippendorff [4] suggests including images of competing products, factory samples, magazine cutouts, sketches, and drawings.

Character	Futuristic (12)	Feminine (12)	Unconventional (9)	Practical (7)	Simple (5)	Smart (5)
Attributes	Modern	Light, Slender, Delicate	Unique, Different, Original, Surprising, Novel, Amazing	Convenient, Utilitarian	Easy	Brilliant, Genius, Sharp, Clever
Antonym	Traditional	Agressive	Conventional	Unpractical	Complex	Naive
	Dominant Characters			Recessive Characters		

Fig. 3 Adjectives arranged as characteristics and characters from KJ method

The mood boards were built with the participation of 4 design experts (i.e. design researchers) within a group discussion focused on which images (from mixed product categories) best manifest each product character. Although the users were the primary contributors in the previous steps of capturing adjectives and classifying the adjectives, this step was entrusted to design experts which is a reflection of the "Design-led" and "Expert mindset" proposed by Sanders [15]. There is consensus in literature [5, 18, 31] which warns that designers possess superior visualisation and attribution capabilities able to connect verbal and visual language.

Step 6 *2nd Design workshop—Poli.Design—Master in Design Engineering:* Instead of asking designers to focus on communicating Eco-efficiency, this workshop aimed to achieve the attribution of the given product characters. Each student was asked to re-design the appearance of a front loading washing machine by attributing the given character set. The 6 characters arranged in 7 combinations of 3 characters each. This was done to achieve the best fit between the number of students (14) and the number of characters (6). Each character combination setting consisted of 2 dominant characters and 1 recessive character. The design students were asked to prioritise the expression of the dominant characters with the expression of recessive character acting as a support for the overall character attribution of the washing machine. Each character set consisting of 3 characters was appointed to 2 designers (total of 14 designers were present) in order to not make the final outcome too characteristic of the capabilities of a single designer.

The character sets were as follows (the first 2 characters in each set are dominant):

(1) futuristic, feminine, smart; (2) futuristic, feminine, practical; (3) futuristic, unconventional, simple; (4) futuristic, unconventional, smart; (5) feminine, unconventional, practical; (6) feminine, unconventional, simple; (7) feminine, futuristic, simple;

Step 7 *2nd questionnaire—Capture feedback using a Semantic Differential scale :* 2 of the 14 students couldn't complete the desired objectives of the workshop therefore 12 concepts were used for the final questionnaire analysis. The students were asked to deliver their concepts against a white background, to reduce the impact of noise. It was decided to retain the presence of colour in the images, even though most of the concepts only used hues of light grey, white, and transparency to illustrate their concept. This was done to judge the impact specifically on 2 concepts which used pastel tones such as red and green and 1 concept which utilised a woodgrain finish which can be an early pointer to the appearance specifications of Eco-efficiency which designers possess and users might validate. This questionnaire was performed online utilising a semantic differential scale to evaluate each concept and consisted of 20 respondents (11 females and 8 males) of age group 19–54 with mixed nationalities. Of the 12 washing machine concepts, all but 1 (utilising a woodgrain finish) succeeded in communicating the attributed meaning of Eco-efficiency to the end users, thus potentially pointing towards woodgrain finish as a 'dead metaphor' [4] or a cliché. The criterion for success was defined as any concept which scores higher than 50 % in the user feedback study.

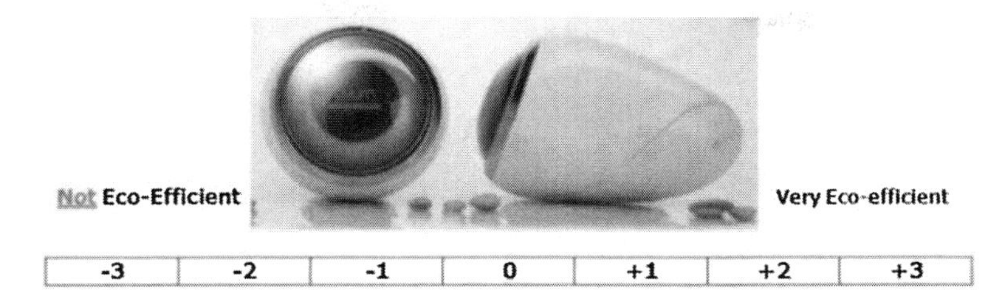

Fig. 4 Online questionnaire with semantic differential (*SD*) scale. (Figure shows concept 1: *futuristic, unconventional, simple*—score 71/120 = 59 % vote)

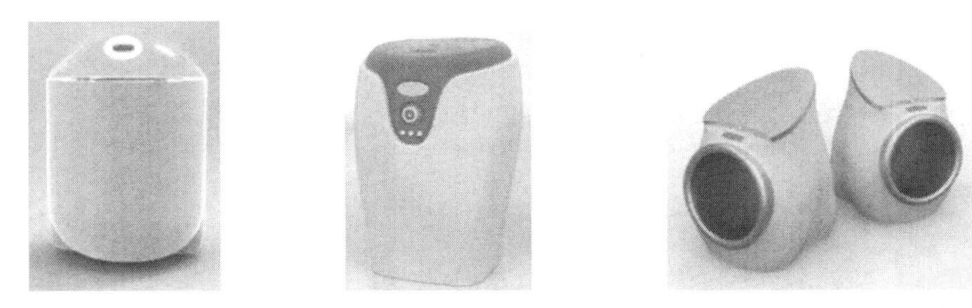

Fig. 5 Concept no. 4, concept no. 6, and concept no. 10

According to the SD scale (Fig. 4) the range of votes varied between −3 (not at all Eco-efficient) to +3 (Very Eco-efficient) for each concept, thus 'zero' being the over-all neutral median and −60 to +60 being the extremes thus achieving a range of 120 points (sum result of 20 user feedbacks).

The top 3 concepts (Fig. 5) achieved a score of 83/120, 84/120, and 86/120 on the SD scale therefore achieving a success rate of 69 % (concept 10—*futuristic, feminine, smart*), 70 % (concept 4—*futuristic, unconventional, simple*), and 72 % (concept 6—*futuristic, feminine, simple*).

3 Conclusions

It has been said more than once in existing literature that theoretical concepts relating to product semantics have been notoriously difficult by the professional design community to accept [7, 24]. Boess [19] further points out that most designers find it confusing to work with attributing meanings in product design. Keeping these observations in mind, the design students were not given an academic grounding regarding the background and theoretical constructs of Product Semantics and meaning communication, being asked only to focus on character attribution, reflecting similar observations made by Evans and Thomas [24] in their workshop exploring application of product semantics with design students.

The success of all the concepts could also be put down to the selection of the product characters (or appearance attributes) assigned to the designers, seeing as ordinary users have limited skills in reading products and even differentiating amongst them [5]. Blijlevenset al. [5] cite Simplicity, Modernity, and Playful as appearance attributes universally best recognised by non-professionals. According to them appointing these appearance attributes within design guidelines increases the likelihood of a successful communication of intended meaning to non-experts. By no means are these attributes meant to replace the expert based attributes described in literature. Among the 6 characters which the design students were asked to attribute in the final workshop at least 2 of them, i.e. Futuristic (modern) and Simple, corresponded to the universal appearance attributes cited above. This is not surprising as the product characters were user generated in the first place. Blijlevens et al. have also demonstrated these appearance attributes are universal in nature thus lending credibility to these product characters as having relevance in consumer product categories beyond home appliances.

These product characters need to be seen as a light thrown in this subject area, and would need to be tested in several appliance categories to be regarded as robust design guidelines. The results portrayed here are not immune to the influence of trends in perception of Eco-efficiency, based as they are on human centeredness and not Aesthetic Theory which prescribes normative specifications on appearances of products [4].

References

1. Lee M, Lou YC (1996) Consumer reliance on intrinsic and extrinsic cues in product evaluations: a conjoint approach. J Appl Bus Res 12(1):21–29
2. Krippendorff K, Butter R (2008) Semantics: meanings and contexts of artifacts. In: Schifferstein H, Hekkert P (eds) Product experience. doi: www.elsevier.com
3. Gorno R, Colombo S (2011). Attributing intended character to products through their formal features. Paper presented at: DPPI'11 Dppi: international conference on designing pleasurable products and interfaces, Milan, June
4. Krippendorff K (2006) The semantic turn, a new foundation for design. Taylor, Boca Raton, FL
5. Blijlevens J, Creusen MEH, Schoormans JPL (2009) How consumers perceive product appearance: the identification of three product appearance attributes. Int J Des 3(3): 27–35. Retrieved from www.ijdesign.org
6. Athavankar U (2009) From product semantics to generative methods. Paper presented at international association of societies design research Iasdr 2009: rigor and relevance in design, Seoul, Oct 2009
7. Brown C (2006) Product semantics: sophistry or success? In: Feijs L, Kyffin S, Young B (eds) In: Design and semantics of form and movement DeSForM 2006, Koninklijke: Philips Electronics N.V, pp 98–103
8. Cooper T (1999) Creating an economic infrastructure for sustainable product design. J Sustai Prod Des (8): –17. Retrieved from http://www.cfsd.org.uk

9. Krippendorff K (2008) The diversity of meanings of everyday artifacts and human-centered design. In: Feijs L, Hessler M, Kyffin S, Young B (eds) Design and semantics of form and movement DeSForM 2006 Koninklijke: Philips Electronics N.V, pp 12–19
10. Krippendorff K (1989) On the essential contexts of artifacts or on the proposition that "design is making sense (of things)". Design Issues 5(2):9–39. Retrieved from http://www.jstor.org/stable/1511512
11. Crilly N, Moultrie J, Clarkson JP (2004) Seeing things: consumer response to the visual domain in product design. Design Studies 25:547–577. Retrieved from www.elsevier.com/locate/destud
12. Hekkert P, Schifferstein HNJ (2008) Introducing product experience. In: Hekkert P, Schifferstein HN (eds) Product experience, 1st edn. Charon Tec ltd, USA, pp 1–8
13. Mills B, Schleich J (2010) What's driving energy efficient appliance label awareness and purchase propensity? Energy Policy 38(2):814–825
14. Cramer J (1997) Towards innovative, more eco-efficient product design strategies. J Sustain Prod Des (1):7–11. Retrieved from http://www.cfsd.org.uk
15. Sanders L (2008) An evolving map of design practice and design research. Interactions 15(6):13–17 doi: 10.1145/1409040.1409043
16. Demirbilek O, Sener B (2003) Product design, semantics and emotional response. Ergonomics 46(13/14):1346–1360
17. Vihma S (2007) Design semiotics- institutional experiences and an initiative for a semiotic theory of form. In: Michel R (ed) Design Research Now- Essays and selected projects. Birkhäuser, Basel, pp 219–232
18. Butter R (1989) Putting theory in practice: an application of product semantics to transportation design. Design Issues 5(2):51–67. Retrieved from http://www.jstor.org/stable/1511514
19. Boess S (2008) Meaning in product use: which terms do designers use in their work? In: Feijs L, Hessler M, Kyffin S, Young B (eds) Design and semantics of form and movement DeSForM 2006, Koninklijke: Philips Electronics N.V, pp 20–27
20. Crilly N, Moultrie J, Clarkson JP (2008) Shaping things: intended consumer response and the other determinants of product form. Des Stud 30:224–254
21. Blaich RI (1989) Philips corporate industrial design: a personal account. Des Stud 5(2):1–8. Retrieved from http://www.jstor.org/stable/1511511
22. Crilly N (2011) Do users know what designers are up to? Product experience and the inference of persuasive intentions. Int J Des 5(3): 1–15. Retrieved from http://www.ijdesign.org/ojs/index.php/IJDesign/index
23. Wikström L (1996) Methods for Evaluation of Products' Semantics, PhD Thesis, Chalmers University of Technology, Sweden
24. Evans M, Thomas P (2011) Products that tell stories: the use of semantics in the development and understanding of future products. Paper presented at City University International conference on engineering and product design education, London, September
25. Norman DA (1988) The design of everything things. Currency Doubleday, USA
26. Thomas R (2006) A new dialog. In: Feijs L, Kyffin S, Young B (eds) Design and semantics of form and movement DeSForM 2006, Koninklijke: Philips Electronics N.V, pp 10–18
27. Gibson JJ (1979) The Ecological Approach to Visual Perception, Houghton Mifflin, Boston MA
28. Karana E, Hekkert P (2010). User-material-product interrelationships in attributing meanings. Int J Des 4(3):43–53. Retrieved from www.ijdesign.org
29. Janlert LE, Stolterman E (1997) The character of things. Des Stud 18(3):297–314
31. Spool JM, (2004) The KJ technique: a group process for establishing priorities, http://www.uie.com
31. Lawson R, Storer I (2008) 'styling-in' semantics. In: Feijs L, Hessler M, Kyffin S, Young B (eds) Design and semantics of form and movement DeSForM 2006, Koninklijke: Philips Electronics N.V, pp 41–49

Indian Aesthetics in Automotive Form

Chirayu S. Shinde

Abstract Indian sense of the beautiful concerned with pure emotion applied to evolve an external appearance of a clearly defined area of automobile. Having gotten an idea of Indian aesthetics, the points and keywords that were closest to the insights gained were penned down. The process of defining Indian Aesthetics started with sketching forms that inculcated physical and emotional inspirations gained through research. Eleven primary forms were created in styrene foam, giving a tangible form to my verbal definition of ideas and emotions. A palette of contemporary Indian volumes, surfaces and lines, was hence created. It is this pallet that has been used later in coming up with an Automotive form that suggests a new style of cars for the world, the Indian Style.

Keywords India · Aesthetics · Automotive Form

1 Introduction

India has a rich culture and heritage, a result of absorption and assimilation of multitude of ideas and practises over the last 5,000 years. The advent of colonial rule and later, the focus on economic sustenance, affected the process, as the world embraced abstraction in art and automobiles. The unstructured variety of exposure post independence explains the confusion in the aesthetic preferences of people of the nation today. India's design practise tries to assimilate and adapt to two centuries of evolution that defines today's global philosophies of modernism, lightness and well-being as per Indian aesthetics and tastes [1].

C. S. Shinde (✉)
Mobility and Vehicle Design, IDC, IIT Bombay, Mumbai, India
e-mail: chirayu.shinde@gmail.com

A. Chakrabarti and R. V. Prakash (eds.), *ICoRD'13*, Lecture Notes in Mechanical Engineering, DOI: 10.1007/978-81-322-1050-4_27, © Springer India 2013

India is one of the biggest consumers of automobiles today, yet a very few true indigenous offerings have been introduced to the world. This paper suggests an approach to evolve an external appearance of a clearly defined area of an automobile with an Indian sense of beautiful concerned with pure emotion. The work hopes to come up with an automotive form that suggests a new style of cars for the world, the Indian style.

1.1 The Approach

Book reading, extensive visual examination and cursory survey was performed to garner an understanding of evolution of modern abstraction through study of art movements and their effects on automotive styling. For gaining insights into Indian Aesthetics, a study of present was done, which later went on to the study of the evolution of architecture and sculpture. The insights and interpretation were penned down and given a tangible form in the form of Styrofoam models forming the basis for further exploration and suggestion.

2 Context—Art Movements and Effects on Vehicle Design

It was important to understand how the concept of abstraction came into today's world and how it affected society. The fashion and art has changed with time, so has automotive styling. Here we will see in brief how these changes occurred from the advent of automobiles and how context was found to be relevant to design evolution.

2.1 Art Nouveau and Secessionism, Expressionism, Fauvism

Art Nouveau was started by energetic artists to create an international modern style in response to the industrial revolution. Sinuous lines were introduced. It simplified the ornate Victorian styles.

Inner moods and feelings as opposed to external appearance were incorporated through expressionist ideas using powerful colours and dynamic compositions. Similar ideals of emotions and expressions were strengthened by fauvists during the age [2].

These ideas can be seen in the vehicles, like Benz Victoria, 1893, Oldsmobile Curved Dash, world's first mass produced car, 1901 and Sunbeam Tourer, 1910 to name a few.

The whole shape of the cars of this time could be said to have lacked the appeal to the senses that the modern sculptural forms do [3].

2.2 Avant-Garde Strengthens: Cubism, Futurism, Constructivism, Art Deco, Bauhaus

Development in technology, medicine, communication, and transportation changed the social structures of the world. This was also a major shift from survival to self-actualization with theories of Sigmund Freud and Carl Jung. A struggle to break away from past art was initiated with Cubism. Dynamism, vitality and power of the Machine Age was expressed. Rhythm and Sequence was brought in to glorify the beauty of Speed. Constructivism took these ideas further with identifying and giving importance to the negative space.

Art deco's linear symmetry was a distinct departure from art nouveau representing elegance, glamour, functionality and modernity. Bauhaus tried to bring together fine arts and applied arts. The usefulness of machines was acknowledged in creating simply designed objects without excess decoration [2].

As cars shape evolved, wheels tended to become smaller. Passengers moved forwards and downwards. Cab forward design came up i.e. the windows moved forwards [3].

Voisin C11 Lumineuse, 1926, was the first effort to carry all passengers within the wheel base providing added comfort. It also provided increased side window area. It could be called as the first example of a car that emphasised on sightseeing. The design of Chrysler Airflow was imitated throughout 1930s. Examples like Fiat Balilla, 1934 and Cord 812, 1937, reflected the influences of lifestyle and the cities on vehicle styling quiet accurately.

2.3 Avant Garde Prevails: Minimalism, Conceptualism, Hyperrealism, Now

Minimalism stripped art to essentials. Focus was on Ideas behind art. Simple geometrical shapes were used reducing colours and textures. The actual final product is not as important as the process was asserted by ideals of Conceptualism [2].

Approximately by this time car bodies evolved from being desperate discrete elements into a unified form. As a result 1960s gave birth to stunning pieces of art, like the Citroen DS, Jaguar E-Type, Lamborghini Miura and Countach, Ferrari 250 GT [3].

Today limitations to one's imagination are drastically reduced due to technological advancements. Even conceptual and abstract art is brought to a stage where it is very difficult to deny the possibility of it not being real or abstract [2].

2.3.1 Observations

The world tried to detach itself from the classical theories of art and come up with art movements like art nouveau, futurism, constructivism, so that the arts and sensitivities became more abstract and emotive, rather than external/direct impressions. Advent of colonial rule in India, to be talked about later, overlapped with the birth of the discussed modern abstraction and invention of automobiles. This raises question about abstraction in India, if it has evolved as it has in the rest of the world, for over past 200 years.

2.3.2 Indian Styling for Vehicles a Need

These movements were not merely ideas towards art but also influenced other spheres like architecture, lifestyle and design. The International style in architecture created a monotony in buildings and cities all over the world. It ultimately led to growing resentment against dehumanization of modern cities and buildings all over the world. At around 1970s a critical evaluation of the modernist principles began and hence Post-Modern movement emerged in the west. This encouraged other countries to look at their heritage to create a modern architecture which respected the context and avoided the homogeneity and monotony of the International Style [1].

Cars have today evolved from being a mode of transport and a display of stature to a deeper reflection of emotions, feelings, expressions and an identity [4].

In case of vehicle styling, and the sensitivities and emotions attached with a vehicle seem to be in a state of infancy as we have been formally exposed to the industry only in 1991, i.e. after the Indian automotive industry was liberalised. India is world's sixth [5] largest producer of automobiles today. Yet a very few true indigenous offerings have been introduced to the world.

3 Indian Aesthetics

Indian Cities have been constantly "Inventing, Re-Inventing and Adjusting" themselves to their ever evolving demographics. As a result of this "Kinetic" quality, cities in contemporary India lack the legibility that they possessed at the turn of the century. A greater part of India's population has shifted to urban areas. This has produced chaotic growth and general apathy in their administration [1].

A lack of guiding principle is seen either at the Micro or Macro level. Large-scale architecture in cities is usually site specific, bound by client intentions and restricted to, in most of the cases, superficial styling [1]. Could this chaos be contagious and affect other spheres, vehicle styling being one of them?

When it comes to Indian Aesthetics, Elephants or peacocks, arches and domes, Temples and Fortresses and their sculptures, peace and spiritualism, may come into one's mind. What further raises curiosity is the thought of how much of these ideas are a result of today's work. A majority of what we see today may seem to lack meaning or purpose and merely be a replica of these [1]. We explore this looking at the major eras seen by Indian Subcontinent in this section starting from the Indus valley [1].

3.1 Indus Valley: 3300–500 BCE

The Indus civilisation produced statuettes resembling the hieratic style of contemporary Mesopotamia, while others are done in the smooth, sinuous style that is the prototype of later Indian sculpture, in which the plastic modelling reveals the animating breath of life (prana). Bronze weapons, tools, and sculptures indicate sophistication in craftsmanship rather than a major aesthetic development [6].

3.2 Mauryan Empire: 327–200 BCE

Buddhism was of great importance during this period and believed that the heart of beings is like an unopened lotus: when the virtues of the Buddha develop therein the lotus blossoms. This is why the Buddha sits on a lotus in bloom [6, 7]. Hence one can see how the elements of lotus have been used extensively in their architecture and sculpture.

The soft volume and the sinuous lines of the petals of lotus incorporated, make the sculptures and structures of this period humble and peaceful. The receptacle and the small hemispherical bumps could be an inspiration for the Stupas with a Chhatri atop. The Chaitya entrances with their repetitive arches can be thought of being derived from the structure of lotus flower [6], (Fig. 1a).

3.3 Gupta Empire and Fragmentation: 280–750 CE

The Period witnessed prolific and rigorous developments in temple architecture with extensive use of stone.

Buddhist art flourished during this period, which has often been described as a golden age. As in all periods, there is little difference in the images of the major Indian religions, Buddhist, Hindu, and Jain. Large stone figures, stone and terracotta reliefs, and large and small bronzes are made in the refined Gupta style, with extensive attention to ornamental details [6], (Fig. 1b).

3.4 Architecture and Sculpture of the Hindu Dynasties: 600–1100 CE

The Hindu Dynasties revived throughout India and a characteristic temple plan was developed. Innumerable temples were built and were so richly embellished with sculptural details that their style is called "Sculptural Architecture". The Khajuraho temples in central India (c.1,000) represent one of the high points of the Nagara buildings, and the Sun Temple at Konark (c.1,250) reveals, in its famous erotic sculptures, carvings that combine balanced mass with delicate execution. The Jain temples at Mt. Abu, constructed entirely of imported white marble and dating from the 10th and 13th century have plain exteriors but are ornately carved inside [6], (Fig. 1c).

3.5 Islamic Rule and Mughal Empire: 1206–1769 CE

A confluence of its Arabic traditions, calligraphy, inlay work and decorative traditions of India can be seen during this period. Islamic art expresses the beauty as an aspect of God through structures, designs and decoration (Fig. 1d).

Mughal art and architecture, an Indo-Islamic-Persian style combined elements of Islamic art and architecture and this resulted in Sufism (Fig. 1e). Originally, Islamic believers needed a tomb for nothing more than placing the body, but due to common people who believed in holy persons, or rulers who held attachment to the present world, erecting splendid mausoleums came about [1]. Where Islam prohibits building of mausoleums, the Taj Mahal stands today as a testimony to fusion and evolution that took place here.

3.6 The Advent of Colonial Powers: 1799–1947

A very different climate, crafts, skills, materials and technologies played a major role in the evolution of an architecture that remained British to the core but was Indian in execution. Key features would be, a dominant roof, deep verandahs all around and elaborate ventilation architecture in India. The concept of bungalows has its roots in Bengal [1], (Fig. 1f).

The concept of trade later turned to a rule and these foreign powers maintaining stronghold of their parent nations throughout their time, was found unique to this era. Little of the glorious tradition of Indian artistic achievement survived British rule [6] which has led to a constant effort to revive what is lost.

3.7 Today

There has been a prolonged distance from arts and crafts, as even after independence the artists and craftsmen were promoted to run a small scale industry which involved mere replication rather than involvement of the artist in exploring and interpreting what his clients expected [1]. Though there has been a great evolution and reinterpretation of Indian culture, it still lags behind in free thinking and expression [1].

The images of pillars of the respective eras (Fig. 1a–f) depict an aspect of absorption and assimilation discussed in this section.

Fig. 1 **a** Karla caves. **b** Ellora caves. *Source* http://asi.nic.in/images/wh_ellora/pages/024.html, 7.2.2012. **c** Sun temple. *Source* http://asi.nic.in/images/wh_konark/pages/019.html, 10.2. 2012. **d** Quwwat-ul-Islam Mosque: http://asi.nic.in/images/wh_qutb/pages/012.html, 10.2.2012. **e** Diwan-e-Khas: http://asi.nic.in/images/wh_fatehpursikri/pages/025.html, 10.2.2012. **f** Gateway of India

4 The Interpretation of Indian Aesthetics

India's design practise faces the need to assimilate and adapt to the two centuries of evolution that defines today's global philosophies of modernism, lightness and well-being as per Indian aesthetics and tastes [1]. What follows in this section are (suggestions on/interpretations of) Indian Aesthetics which are based on the study done, observations made and insights gained.

4.1 Interpretation

The forms are never skinny but voluminous. They are never stripped down to their functional elements, and if so, their functional elements are also beautified. Volumes are created with liberal, but not wasteful use of resources.

The use of Datura and Lotus flowers has played an important role in the evolution of Indian Aesthetics, as they were used by the earliest of known civilisation, the Indus Valley and the Mauryans and then later the Vedic and others that followed.

Multiple Limbs have been shown so as to hail the multiple capabilities of the divine beings. Respect to the five elements, air, water, earth, fire and space has always been given. This can be seen even today as the temples are built such that the first rays of sun fall on the idols of Gods.

Sinuous lines are a prominent feature of sculptures and motifs across all ages. A play of surface around a basic form is seen. Harmony of elements that are complete in itself, are placed around a central element gently evolving the form and giving it a meaning. A sense of celebration, happiness and joy suggests playfulness of the forms. Tranquillity and control on self should also be reflected. The form should reflect its inner beauty and external sensuousness, in subtle but direct way, like the art of dance, nothing is left to chance, as each gesture seeks to communicate ideas and each facial expression the emotions.

5 Expressions

Creation of primary forms out of styrene foam was to come up with a palette of volumes, surface and lines that defined the points of views and the key words set. It was necessary to come up with these as it was felt there exists a lack of forms that are contemporary abstractions of Indian Aesthetics. These forms are to define, and to show, a fresh take on contemporary abstractions of Indian Aesthetics that would form the basis for further explorations.

5.1 Primary Form Explorations

5.1.1 Dynamism

A voluminous form that suggests swift movement led to fusing a slender and dynamic side profile with relatively huge top profile (Fig. 2).

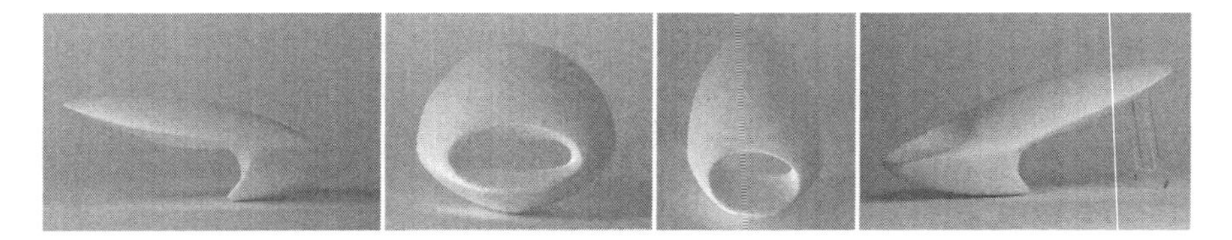

Fig. 2 Dynamism

5.1.2 Datura

Datura Flower has always been associated with Lord Shiva. This possibly explains the ribs running vertically along pillars during the Gupta period, like those on the flower (Fig. 3).

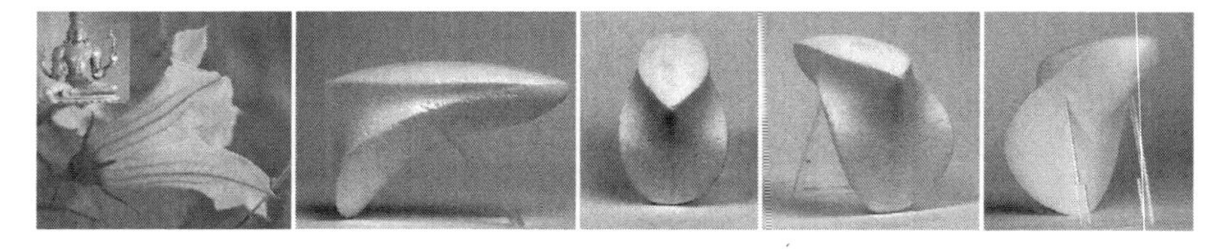

Fig. 3 Datura

The form reflects the silhouette of Datura and essence of Shiva, the destroyer and creator. The relatively flat top gives seriousness to the form. The top drooping down and the surface beneath with a broad mouth in the front narrowing down, at the rear end, gives the form a sense of power and control. The elements of the form appear emerging from behind the surfaces giving a mysterious peek into its underlying capabilities.

5.1.3 Arghya

The exact moment of water falling off one's palms as an offering to god has been captured. The Spherical bulge on the front with an element below has been thought of reflecting the generosity and faith with which offering is made (Fig. 4).

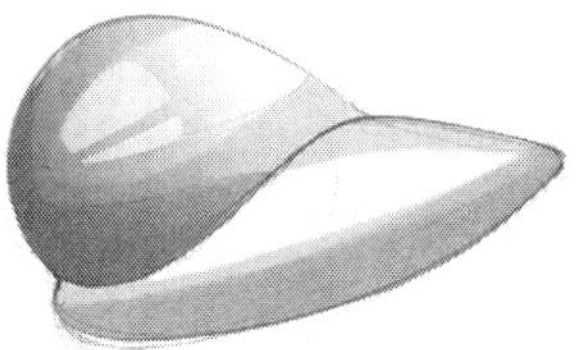

Fig. 4 Arghya. *Source* (*left first*): http://www.thesipoflife.com/wp-content/uploads/2010/04/Kumbh-Haridwar-2010-3.jpg

5.1.4 Lotus

Abode of the Buddha, Lotus is a representation of chakra or ones aura (Fig. 5).

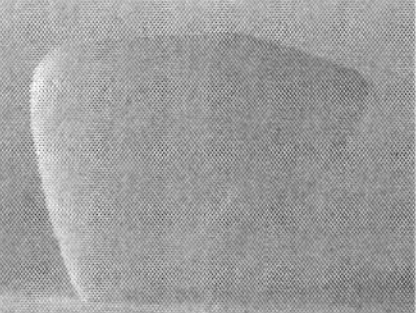

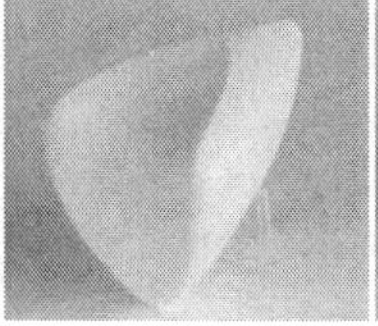

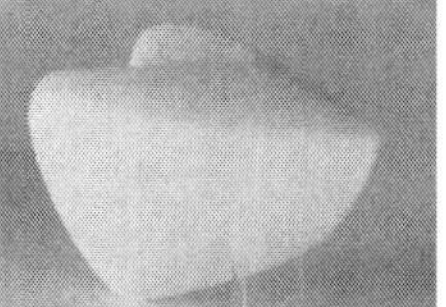

Fig. 5 Lotus

A simple, soft triangular form depicts the bud of lotus from the sides. The movements of surfaces and intersections give a mix of soft and sharp shadows. The surface on the side gently holds the core of the form. Two petals, on the top, created out negative and positive volumes depicting replication. The form raises emotions of peace and simplicity.

5.1.5 Buddha

Visualizing one meditating, sitting cross-legged, in a Chaitya that has pillars with lotus like bulges and ribs all around with a huge Stupa in the front has led to the form (Fig. 6).

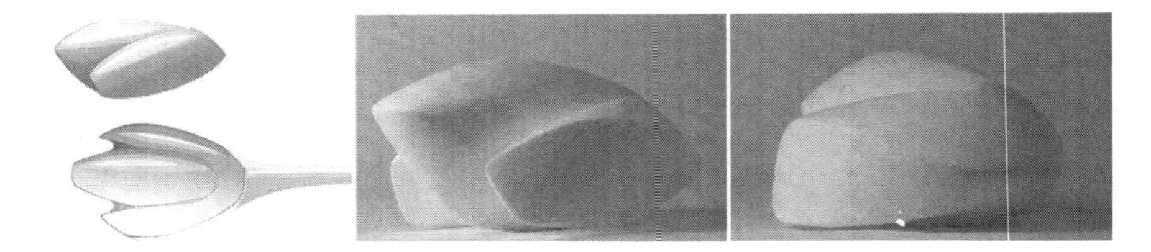

Fig. 6 Buddha

The form has two elements, one in the centre and the second around it. From the side, it comes up like enlightened being making it firm and aware, while the top surface droops down making it humble. It also mimics the stance of Buddha meditating.

5.1.6 Banyan Tree and Yogi

A perfect symbolization of eternal life due to seemingly unending expansion is the Banyan tree (Fig. 7). Often Yogis are shown meditating under it (Fig. 8).

The form consists of vast surfaces creating an immense volume that resemble the characters of a banyan tree. The bottom has been defined separately and the huge volume that grows up from it represents the shoots of the tree. The volume overhanging on the top signifies shade/closure to whatever is beneath it.

Fig. 7 Banyan tree

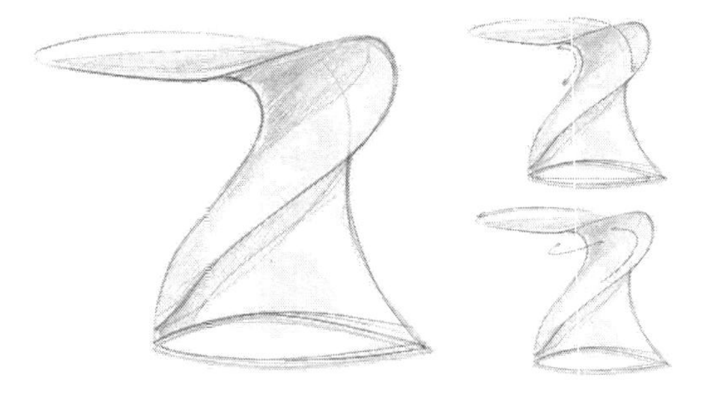

Fig. 8 Yogi

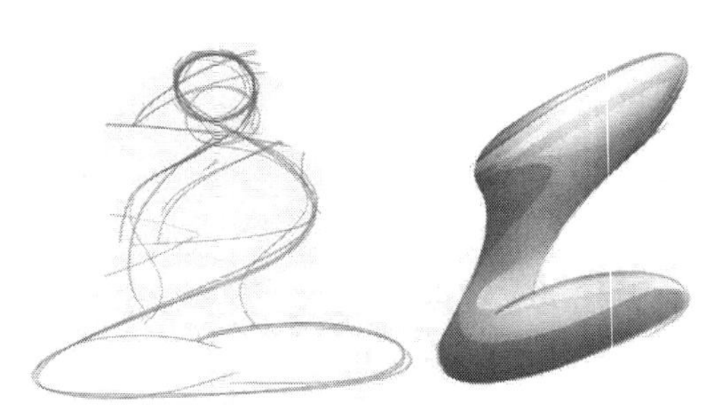

Yogi (Fig. 8) with its subtle form of a human sitting cross-legged in a meditative state shows the boldness of the enlightened saints of India.

5.1.7 Sensuous

Innumerous sensuous sculptures of Yakshis form the Mauryan period and the expressive erotic female sculptures from the temples of Khajuraho and the Sun temple of Orissa made this expression undeniable.

Fig. 9 Sensuous. *Source (left first)*: http://asi.nic.in/images/wh_khajuraho/pages/009.html

This form (Fig. 9) is an abstraction of a curvy Indian Lady wearing a sari preparing for her ablutions sitting by a water tank. The transition from the top volume to the bottom represents the slenderness of the waist of a lady as she sits down with her legs held close with the knees bent. Enough volume has been given to the thighs and the breasts as in the sculptures to accentuate the sensuality in a very subtle yet direct way.

5.1.8 Elephant and Playful Sensuous

The beloved Lord Ganesha, the care taker and his playful image, and the ride of the Kings, the grandeur of it, the infinite replications on the Indian sculptures, lead to this form of elephant.

The form (Fig. 10) is very voluminous and round giving it a presence and the playfulness. The movement to the legs and the shape given makes the huge form very nimble. The head, ears and one of the forelegs is embedded in a single volume.

Fig. 10 Elephant

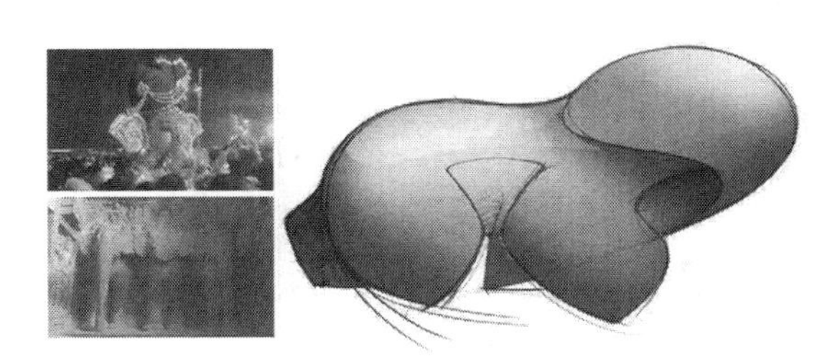

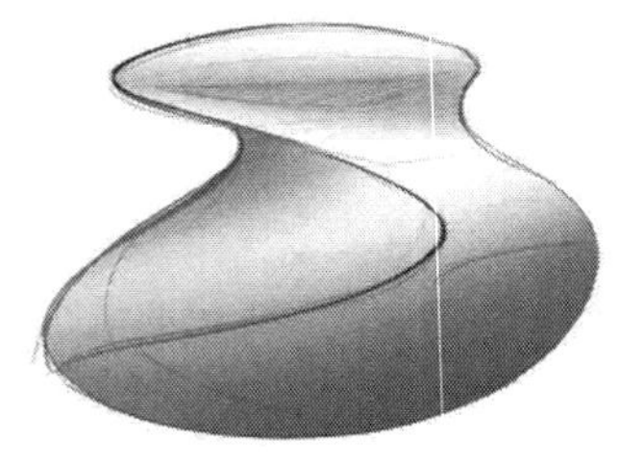

Fig. 11 Playful sensuous

Experimentation with sensuous and playfulness lead to this voluminous form with curvy and sinuous lines (Fig. 11). Opposite characters of innocence and maturity of an Indian lady can be seen in the form. The form consists of two volumes, representing the breasts and the hip of a lady, thrown along the form giving a hint of carelessness and innocence.

5.1.9 Flame

Fire has always been associated with Indian practises. The feeling of warmth and celebration attached with fire has been emulated through the soft form with crisp edges by the merging of concave and convex surfaces that give a nice play of light and shadow (Fig. 12).

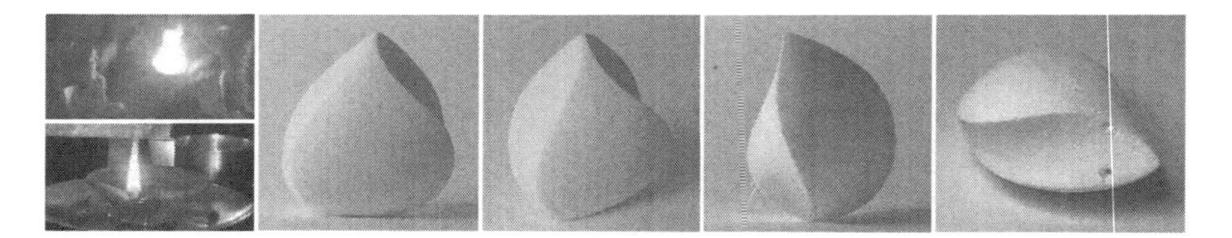

Fig. 12 Flame

Curvy edges give fluidity to the form very subtly. The volume ends in a defined way in all directions reassuring that it is a calm flame and not an erratic fire. This form shows discipline through its controlled movements.

5.1.10 Peepal (Sacred Fig)

Lord Datta, an embodiment of Brahma Vishnu and Mahesh (Shiva) hence the Lord of creation, sustenance, prosperity, well-being and destruction is believed to reside in every living being. Peepal is considered to be his dwelling (Fig. 13).

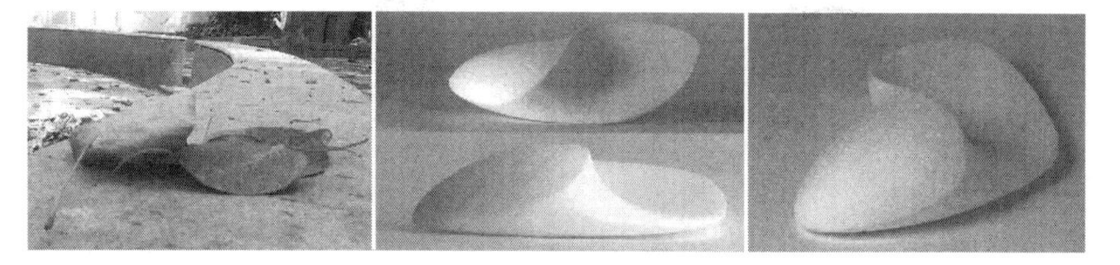

Fig. 13 Peepal

Derived from three dry peepal leaves kept together, the form is one continuous surface that folds the way a dry peepal leaf does. This surface meets at the centre giving a soothing curvy edge that emerges from the front volume. As it moves along the form, it disappears on to the surface merging with the volume. The way positive and negative space has been incorporated in the overall volume of the form makes it full and voluminous, while it appears to be so thin, raises curiosity.

5.2 Secondary Form Explorations

Having created a pallet of volumes and surfaces that defined my perception and understanding of Indian aesthetics, their usage in deriving automotive forms had to be focused on (Fig. 14).

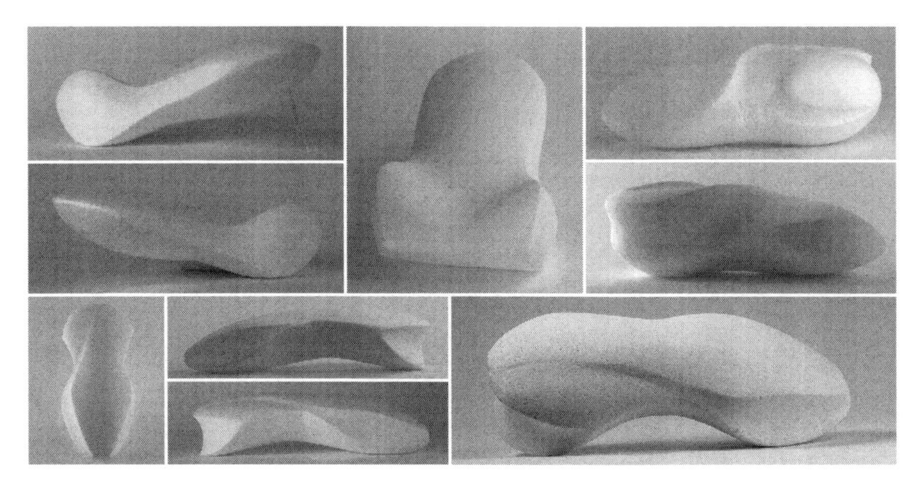

Fig. 14 The process fusion and experimentation

Hence the process of fusion of these elements was commenced through sketches at first and then a few physical models. These explorations are an experiment towards application of this pallet in creating more complex forms maintaining the feel of the primary forms. These forms are not intended to be defined as the primary forms, but an experiment of fusion of the same.

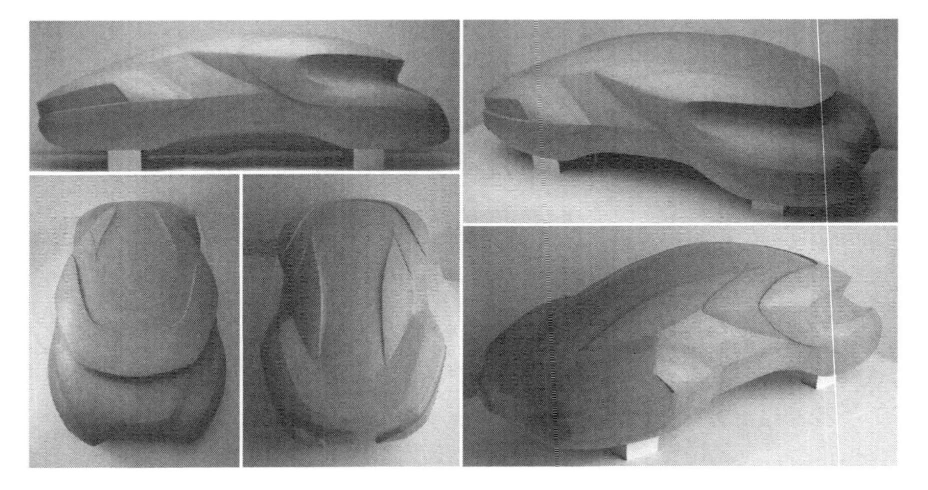

Fig. 15 Sutra

5.3 Deriving the Final Form: Sutra

The form was to appear very soft and simple from a distance, and closing and completing the form from all the sides was a move in the direction. The surfaces and elements were to adhere to the form and not appear to detached especially at the ends and hence create a soft volume. The form was to be open to deriving different configurations of vehicles out of it and an amalgamation of the definition of Indian Aesthetics perceived.

Composure and a controlled flow were to reflect from the form to resemble the Indian ideals of awareness, mental, physical and spiritual, and tranquillity. The use of long, continuous and soft lines, hence surfaces and volumes, in arriving at the form was an attempt towards the thought (Fig. 15).

The form was to involve onlookers in the play of concave and convex surfaces, the reflections and shadows created by them and the flow suggested, making it presence felt directly yet subtly raising ones curiosity to know more about it. Going closer it would reveal a play of sinuous lines flowing over the form very gently, defining each element of it one after another. The lotus petal like element that emerges out from over the rear and progresses from the side up to the centre of the form is an example in the direction. A continuous band can be seen to take the onlooker on a trip all over the form, gently arching its way along the centre and again twisting its way back to the top at the front.

Each element was handled to portray its completeness, being a part of and coherent to the form, so that even if they were to be considered individual, would still be complete. These elements then would contribute equally to the overall form of the concept. The elements were to maintain continuity, physically and visually, throughout the form.

With its contours repeating like the petals of lotus, as viewed from the top, Sutra is a work towards the use of replication of elements in automotive forms. The top surface was handled with crisp creases having a very gentle flow/waviness as they traverse the form, to render sensuousness to the form.

The form uses the pallet of volumes, surfaces and lines, created through primary and secondary forms hence suggesting the possible application of the pallet in automotive styling. Sutra is a collection of aphorisms relating to some aspect of the conduct of life [8]. The name was found closest to the purpose of the form and hence given to it.

Acknowledgments My work on "Indian Aesthetics and Automotive Forms" owes gratitude to my professors K. Munshi, K. Ramachandran and Nishant Sharma, at IDC, IIT Bombay and friends/batch-mates Keith D'souza and Mohit Gupta for their valuable inputs that helped me immensely in the ideation of the concept. I am equally thankful to anonymous reviewers who from time to time throughout the making of the model helped me with their feedback. A special mention remains to be made of the invaluable support of Anisha Malhotra, Pratik Shinde and Ellina Rath in the completion of the paper.

References

1. Pal P (2000) Reflections on the arts in India. Marg Publications, Mumbai, pp 63, 85, 87, 101, 109
2. Hodge S (2011) 50 Art ideas you really need to know. Quercus pp 92–135, 176-183, 196–203
3. Zumbrunn M, Cumberford R (2004) Auto legends classics of style and design. Merrel Publication, US
4. Andrew Noakes (text), Paul Turner and Sue Pressley (design), Philip de Ste. Croix (editor) (2011) Great Cars: a sensational collection from classics to the modern greats. Parragon, UK
5. http://oica.net/wp-content/uploads/total-2011.pdf, 8 Mar 2012
6. http://www.infoplease.com/ce6/.html. The Columbia electronic encyclopedia© 1994, 2000–2006, on Infoplease. © 2000–2007 Pearson Education, publishing as Infoplease, 28 Feb 2012. <http://www.infoplease.com/ce6/ent/A0825105.html>
7. Spirituality in art, Museum pre-visit packet, Utah museum of fine arts, pp 13 2008
8. Sutra. Online Etymology Dictionary. Douglas harper, historian. 14 Oct 2012. < Dictionary. com http://dictionary.reference.com/browse/sutra>

Bibliography

9. Taschen A, Photos, Von Schaewen D (eds) (2008) Indian style. Taschen GmbH
10. BharathRamamrutham, Ghurde M, Shah S, Von Tschurtschenthaler D (eds) (2004) Indian design. Daab GmbH
11. Kimes BR, Goodfellow WS, Furman M (2005) Speed, style and beauty: cars from the Ralph Lauren collection. MFA Publications

Understanding Emotions and Related Appraisal Pattern

Soumava Mandal and Amitoj Singh

Abstract Understanding of emotions and the appraisal patterns associated with emotions is vital in the fields such as user experience design, product design, advertising and fashion. People respond with different emotions to the same situation depending on how they interpret, or appraise the situation. It is important for a designer to understand the differentiation in emotions in order to design emotion-laden products. But assessment and mapping of emotions is always a tough task, because the topic itself is subjective and abstract. It is in this context that this paper presents an approach to differentiate emotions and to map emotions on the basis of the related appraisal patterns.

Keywords Emotion · Experience design · Abstract · Human cognition · Appraisal

1 Introduction

The world of design has been solely dominated by either usability i.e. ease of use and esthetics or combination of both. But the definition of meaningful design is slowly shifting from usability and efficiency to holistic aspects of emotional

S. Mandal (✉)
User Experience Designer, Bangalore, India
e-mail: soumava.mdes08@gmail.com

S. Mandal
IIT Delhi, New Delhi, India

A. Singh
Emotion Centered Design and Innovation Expert, New Delhi, India
e-mail: amitoj_design@yahoo.com

A. Singh
Industrial Design, IIT Delhi, New Delhi, India

A. Chakrabarti and R. V. Prakash (eds.), *ICoRD'13*, Lecture Notes in Mechanical Engineering, DOI: 10.1007/978-81-322-1050-4_28, © Springer India 2013

experiences and affect. Only in recent years, psychologists, designers, marketing gurus (Norman 2004; Demet 2002; Bagozzi et al. 1999; Jordan 2000) are trying to articulate cognitive decision making pattern of customers through emotional experience. To understand emotional aspects, design authors have been back to cognitive aspects of eliciting emotion i.e. appraisals.

1.1 Definition

According to appraisal theorists (Lazarus 1991 Oatley and Johnson-Laird 1989), emotion is a mental state of alacrity which is elicited from a diagnostic cognitive framework. This framework helps intelligent organisms to make the decision of subsequent activity. It depends on various parameters like relevance with goals and coping power of the organism according to the nature of stimuli. The coping power dictates the bipolar dimension of an emotion: positive or negative valence. Positive emotions are joy, happiness. Fun etc. and negative emotions are anger; disgust etc. emotions are varied from simple, primary which is triggered by behavior with high survival value to complex emotion which is triggered by complex multi-layered cognitive processing. Plutchik (1997), in his famous circumplex model, showed how these emotions are inter related with different intensity and combinations, and how mixing of these primary emotion can generate complex emotions with different valence like color wheel [1] (Fig. 1).

The study of positive emotions is important and interesting because designers are often prescribed to create a product experience with intension to evoke certain kind of positive emotion while interacting with users. Engagement, enticement is result of human appraisal to positive emotions. With clear goals, high level of equilibrium between individual skills and task challenges and clear immediate results user will fully engrossed in any task having high level of satisfaction-mostly leads to positive emotion [2]. If skill level or individual level of coping power is retarded with respect to challenge, it leads to anxiety.

2 Different Framework of Emotions and Their Relation Among Them

Jordan (2000), Norman (2004) and Desmet (2002) are trying to define the framework of eliciting emotion. These frameworks are outcomes of various research results done by various psychologists (Lazarus 1991; Frija 1994; Scherer 2001; Ortony 2005). Though the interpretation of these frameworks are different, but their philosophical aspects are interrelated.

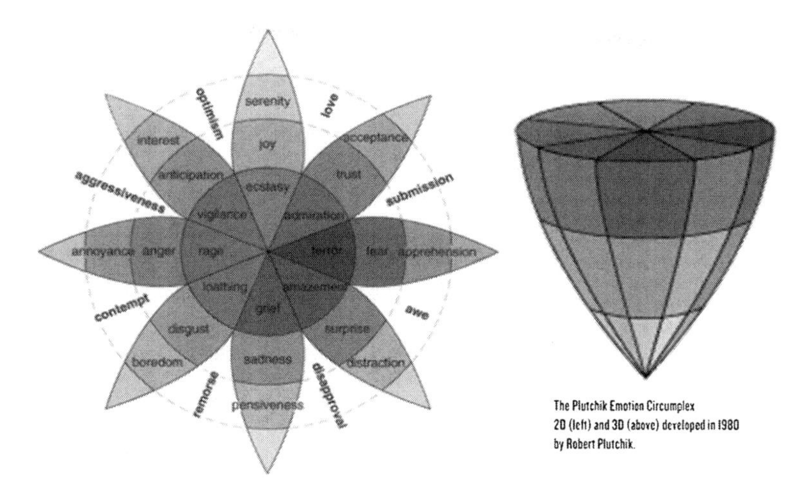

Fig. 1 Circumplex model of Plutchik. *Source* emotions and life: perspectives from psychology, biology, and evolution: Robert Plutchik

2.1 Jordan's Pleasure Theory

Foundation of famous four pleasure theory proposed by Canadian anthropologist Lionel Tiger, later popularized by Pat Jordan, lies on 'human pleasurability'. This theory is very straight forward and only implies for simple emotions, but it doesn't explain complex multi-layered emotions like thrill [3]. The 4 parameters are:

- Physio—related to the body and the senses.
- Psycho—related to the mind and the emotions.
- Socio—related to social acceptance, relationships and status.
- Ideo—related to values and beliefs.

2.2 Norman's Theory of Emotion

Neuro-biological process theory discusses how information is processed in different levels: visceral, behavioral and reflective. The automatic, prewired primary level, the visceral level: this layer deals with sensory aspects of an experience. The second level: this subconscious thinking is the Behavioral level of processing in everyday life. And the third level is the most complex advanced level: reflective layer which deals with conscious consideration and reflection learned during past experiences [4].

2.3 Desmet's Appraisal Theory

As a designer it is very important to have an in-depth knowledge on appraisal patterns of emotion so that one can use this concept while designing a product. And designer can generate a product to evoke certain emotions. People respond with different emotions to the same situation depending on how they interpret, or appraise. Appraisal theories state experiencing an unpleasant emotion requires appraising a situation as harmful to personal well-being, whereas a pleasant emotion involves appraising a situation as beneficial. In this context appraisal theorist proposed several cognitive appraisal components. These are: motive consistency component, expectation confirmation component, agency component, standards conformance component, coping potential component, certainty component, valence, goal congruency, goal conduciveness, normative significance [5].

By observing these theories, a common pattern and inter-relation can be found. For example Jordan's Physio-pleasure theory and Norman's visceral level processing manifest same interpretation in different way. Jordan's Psycho, Socio and Ideo-pleasure, Norman's behavioral and reflective level of processing and Desmet's appraisal theory demonstrate same core philosophy.

3 Research

3.1 Prolog

We have taken five positive emotions: Happiness, joy, excitement, thrill and fun. The preliminary aim of the research is to make dimensions on how each of these emotions these pleasing emotions can be differentiate.

Question we are trying to address are:

- These emotions have very similar meaning and they are abstract too. So are there any dimensions on which we can differentiate them?
- Is there any pattern?
- Which emotions are too similar for people?

3.2 Sample and Data Collection

Forty two university students (age between 24 and 30) were recruited for a volunteer research study. 15 out of 42 students were females. Half of the students have design background. An interview was designed to identify the participant's appraisal pattern. They have been asked to recollect any situation or product that they have faced or experienced last time along with a set of questionnaire and interview. Video/audio recording has been taken.

3.3 Analysis

The video/audio recordings have been transcribed. These corpuses are used as the unit of analysis. The verbatim comments are printed and cut into several corpuses, and rearrange into group with the similar corpuses. These chunks have been coded into several categories. Table 1 provides exemplary statements for all emotions. Latent content is also included into the analysis to notice silence, sighs, stammering, and postures to infer whether participants are finding difficulties to distinguish among emotions or trying hard to recollect a certain emotion. The number of times that a category repeats has been recorded in the study and analyzed. And finally mapping has been done with the co-relation among these categories and the emotions.

4 Findings

4.1 Prolog

We already conceive literary meanings of these emotions. But can the dictionary definition of these emotions exactly be mapped with actual perception of these emotions! Based on the interviews and questionnaires, 4 major types of dimensions are being address. These are Amplitude, Time, frequency of occurrence and association with social interactions. On these parameters we can map these abstract emotions.

4.2 Happiness

The dictionary meaning of emotion ‘Happy’ says:

- Feeling, showing or causing pleasure or satisfaction……Cambridge [6]
- Delighted, pleased, or glad, as over a particular thing……dictionary.com [7]
- Experiencing pleasure or joy; Satisfied; enjoying well-being and contentment……Webster [8]

The associate keywords/emotions are related to higher level of wish fulfillment, contentment and satisfaction after achieving desired goals-an activity based emotion. Most of the users say when they achieve his desired goal. The value/amplitude antecedent of this emotion is very high. But another important factor is responsible to determine the mapping of this emotion: expectancy or subjective control over achievements. For happy it is partial control over achievement i.e. the uncertainty factor is pretty high at the same time the user has high motive relevance and congruence. Users perceive that the action-control expectancy and action outcome-control and situation. For e.g. one has done a good amount preparation and he

Table 1 Example of statements referring to different parameters

Category	Sub category	Example	Participant ID	Emotion
Social interaction	With closed ones	...When I proposed her and she accepted...	Participant 1	Happy
	With others	...with my old Panjab wale friends when I booze...	Participant 6	Fun
Frequency of occurrence	Seldom	...After the interview result of Supergas where I got selected through campus placement...	Participant 1	Thrill
	Often	...when play cricket with my cousins during my vacation...	Participant 3	Thrill
	Very often	...When 1 am with my roommates and friends doing mischiefs...	Participant 1	Fun
Duration of experience	Long (lasted for days)	...Last time I was with my family members, whole week I was happy...	Participant 2	Happy
	Moderate (lasted for an hours or several hours)	...When she accepted, the whole night I felt very happy, and recollecting...	Participant 1	Happy
	Very less (lasted for a min or little more)	...when result out...	Participant 4	Excitement
Intensity of arousal	Very intense	...After the interview result of Supergas..., it was very intense...	Participant 1	Thrill
	Moderate	...hanging with my friend, it was fun but moderately intense...	Participant 2	Fun
	Less intense	...When went for marketing with my best friend, it was less intense...	Participant 3	Joy

expects that he will get through the entrance exam but at the same time he has a fear to failure as so many students are aspiring i.e. negative valence. This mixed valence creates internal predicament after the activity for the goal. Because of this uncertainty, the emotion steps up from joy level to happy level. Happy is more cherishable, flat and creates a long lasting impression (Fig. 2).

4.3 Joy

The dictionary meaning of emotion 'Joy' says:

- The emotion of great happiness; Make glad or happy......Webster [8]
- The emotion of great delight or happiness caused by something exceptionally good or satisfying; keen pleasure; elation......dictionary.com [7]
- Great happiness......Cambridge [6]

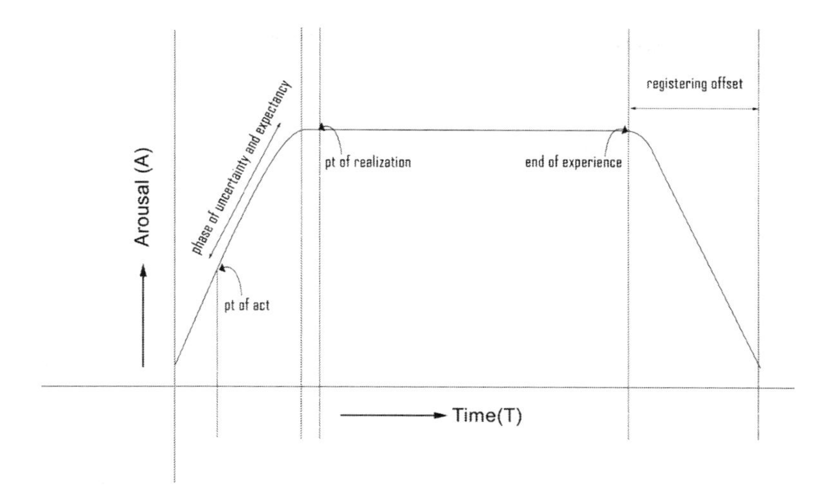

Fig. 2 Proposed journey of experience model: Happy. *Source* self-study

Joy and happy are almost synonymous. People find difficulty to distinguish between them because of the thin layer. Because like happy, joy is also a flat emotion, but the time-extent of this emotion is little less and sometimes momentary. User has less motive relevance and congruence. Being with friends, meeting with my boyfriend, riding bikes, getting stipends has low motive relevance, it's a pleasant emotion, if one is getting stipend, one will be able to spend her assistantship that money on buying things whatever he/she wants. One does not have any repulsion to feed someone else with that money. He/she is not involved with her evaluation of his/her resources and options for coping. The process of going to the level of joy is as fast as the dropping the experience line (Fig. 3).

4.4 Excitement

The dictionary meaning of emotion 'Excitement' says:

- The state of being emotionally aroused and worked up……dictionary.com [7]
- To make someone have strong feelings of happiness and enthusiasm……Cambridge [6]
- Feeling of lively and cheerful joy……Webster [8]

In excitement the amplitude of arousal is very high, but the same time the frequency of occurrence is quite low. There is low expectancy or no expectancy of outcome but which is highly consistent with the motive relevance. For e.g. first time movie watching, first time meeting with girlfriend, to see own name in the campus interview—a student who has mediocre preparation, have very low expectation of getting job, and then he suddenly sees his name in the list of successful candidate, he experiences excitement (Fig. 4).

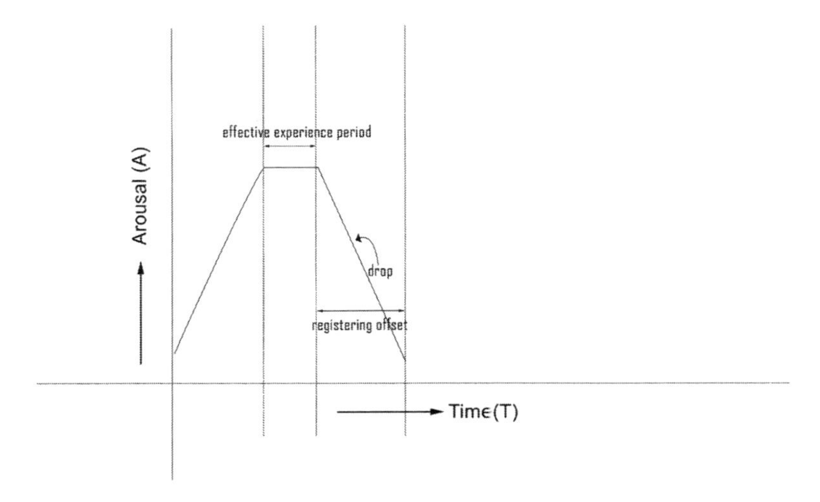

Fig. 3 Proposed journey of experience model: Joy. *Source* self-study

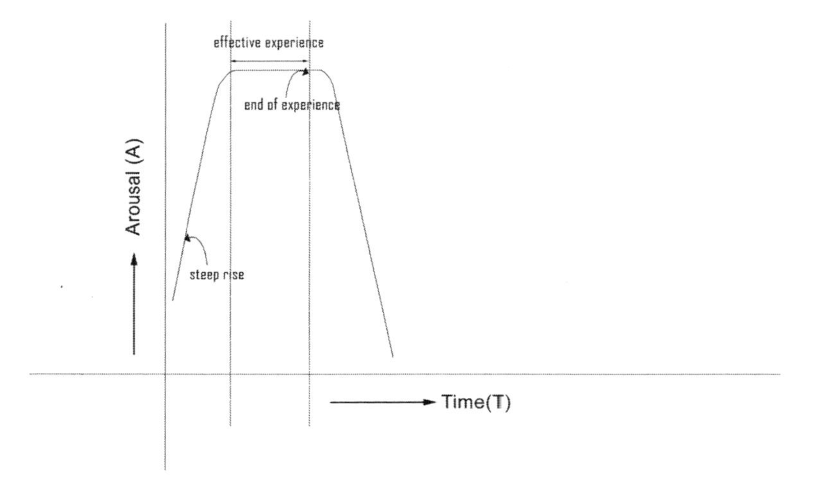

Fig. 4 Proposed journey of experience model: Excitement. *Source* self-study

4.5 Thrill

The dictionary meaning of emotion 'Thrill' says:

- The swift release of a store of affective force……Webster [8]
- A feeling of extreme excitement, usually caused by something pleasant……Cambridge [6]

Amplitude of arousal is the highest among these emotions and frequency of the occurrence is zero, most of the time it is the first time and last time for a particular event. For e.g. first time riding on rollercoaster elicits the experience of thrill, but second time experience with rollercoaster evokes excitement rather than thrill and the association with social interaction is least with this emotion. The environment is apparently adverse and it gives a perception of negative valence, but the cognition force organism to readdress as a positive valence. The arousal level is very high with physiological signs like sweating. But the cognition suggests and the event/product lies in positive valence. This emotion is ambivalence in nature (Fig. 5).

4.6 Fun

The dictionary meaning of emotion 'Fun' says:

- Providing enjoyment; pleasantly entertaining......Webster [8]
- A source of enjoyment, amusement, or pleasure......dictionary.com [7]
- Pleasure, enjoyment, amusement......Cambridge [6]

The distinct feature of this emotion is high level of social interaction but positive valence amplitude of arousal is low. For e.g. roaming with friends and the day outing with family requires high level of social interaction, and the frequency of occurrence is also very high, but the amplitude of arousal is very low. But on the other hand, the experience time-span is very high.

In the Figs. 6, 7, 8, 9, these positive emotions have been mapped with respect to different parameters. In all the graphs, parameters i.e. Social interaction, frequency of occurrence and the time-span have been plotted with respect to arousal. A particular emotion which has shown highest value on these dimension, has been taken as highest unit, and other emotions have been mapped relatively based on the content analysis.

5 Implication on Design Decision

Now the question is how these parameters can further extrapolate design decisions while designing an experience.

5.1 Synthesize Physio-Pleasure

If we analyze the experience journey of these positive emotions (refer Figs. 2, 3, 4, 5, 6), a common pattern of all these positive emotions is increase of amplitude of arousal within positive quadrant or from negative to positive valence with respect

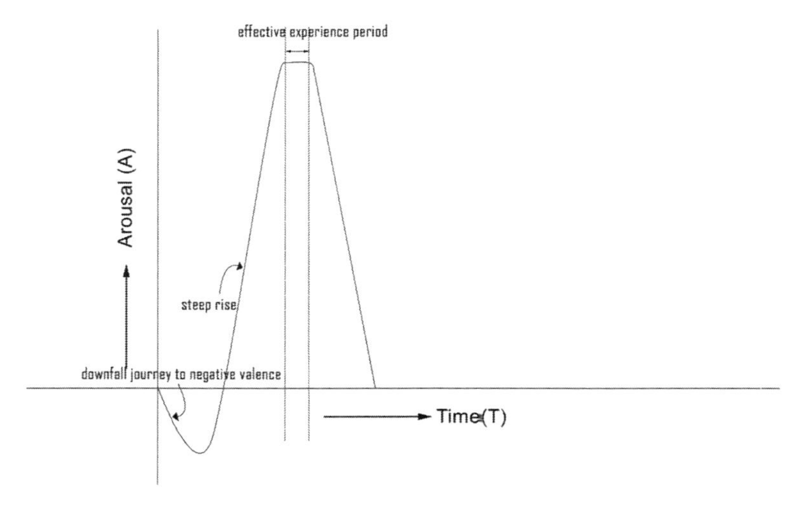

Fig. 5 Proposed journey of experience model: Thrill. *Source* self-study

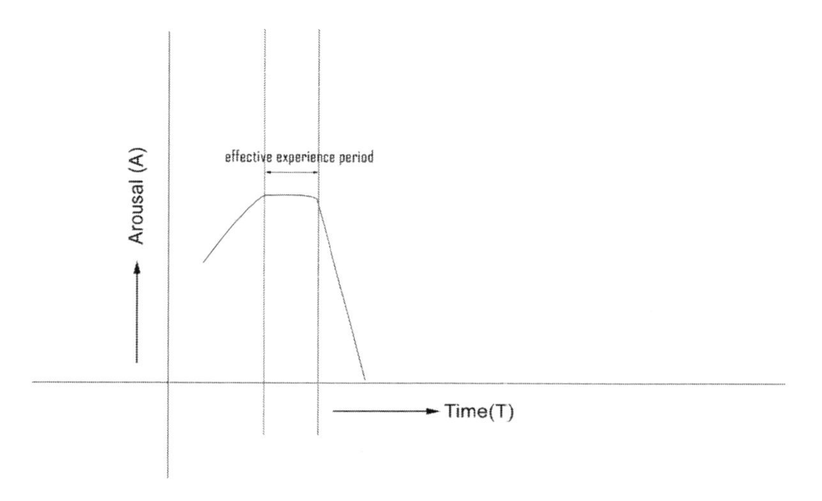

Fig. 6 Proposed journey of experience model: Fun. *Source* self-study

to time. Now question is how a designer can increase an arousal. In the research, participants often refer an experience that is associated to synthesizing physio-pleasures i.e. any five senses or combination of multiple senses. They refer to an experience of visiting Taj Mahal in the moonlight i.e. visual pleasure, listening to Bach or Mozart i.e. auditory pleasure, having dinner in a restaurant with family i.e. pleasure of taste.

The classic example of visual stimulation is experience design in stalls in Delhi auto-expo 2010: If we compare the stalls of Hero-Honda and Yamaha [9, 10]. Hero-Honda as a brand emphasizes on the fact that it is a common man's commuter where

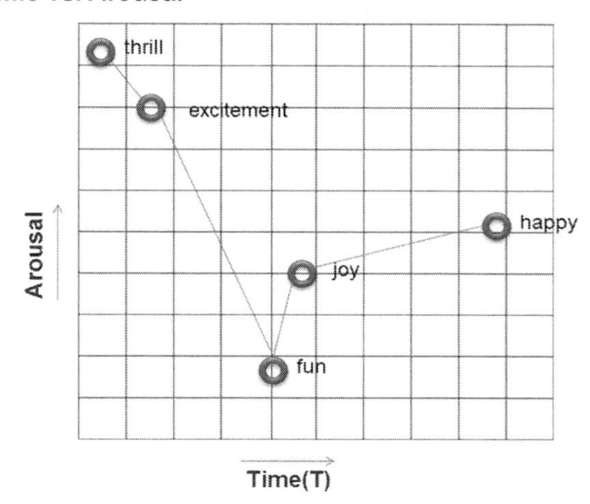

Fig. 7 Time versus amplitude of arousal. *Source* self-study

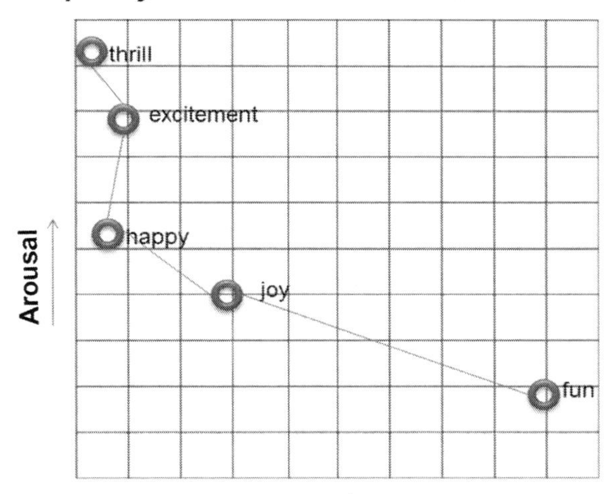

Fig. 8 Amplitude of arousal versus frequency of occurrence. *Source* self-study

joy of riding is their central philosophy, so the use of bright ambience, sounds of refreshing water-curtain creates a visual and aural pleasure which accentuates the brand identity. On the other hand, Yamaha's ambience was little dark, posing models in sensual way on top of the bikes creates a visual message of masculinity, adventurous and raunchy nature of the brand (Figs. 10, 11).

Another classic example is goosebumpspickles.com's enticing photography of pickles [11] or Tata-Nano's home page justifying their tagline "khusiyon ka chaabi-key to happiness" [12] (Figs. 12, 13).

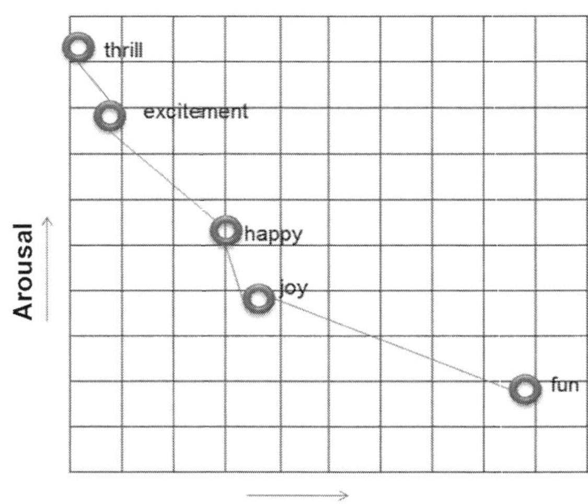

Fig. 9 Amplitude of arousal versus social interaction. *Source* self-study

Fig. 10 Hero-Honda stall auto-expo-2010. *Source* http://www.superbikesindelhi.com

5.2 Changing Persuasion with Positive Way

One of the important patterns from research is expectancy (refer to Sect. 4.2) and frequency of occurrence of events (refer to Fig. 8). Revealing unexpected positive elements in a subtle and intelligent way creates a surprise in positive way, changes the user anticipation and image towards the product experience. For example, Flipkart always promises to their customer that purchased product will be reached within 3–4 working days, but generally product comes within 1–2 days. Here users' expectation and persuasion changes because of promptness of the logistic

Fig. 11 Yamaha stall auto-expo-2010. *Source* http://www.2wheelsindia.com

Fig. 12 Homepage of Goosebumpspickle. *Source* http://www.goosebumpspickles.com/

Fig. 13 Homepage of Tata-Nano. *Source* http://tatanano.inservices.tatamotors.com/tatamotors/home.htm

and supply chain management system of the Flipkart. This transforms Flipkart to India's largest online retailer with 2.6 million users and daily revenues of Rs 2.5 crore [11] (Fig. 14).

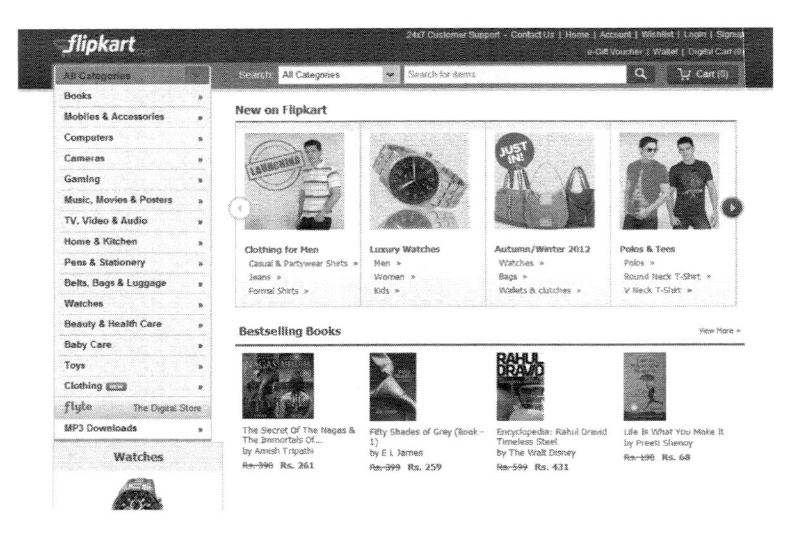

Fig. 14 Homepage of Flipkart. *Source* http://www.flipkart.com/

5.3 Stimulate Socio-Psychological Interaction

Most of the positive emotions has one common factor is high social interaction (Happy, Joy and Fun). Human beings are born with a natural instinct to find their natural place in society. For example, Facebook, top social networking website, is not a primary channel of communication or necessity like phone, but still according to Bureau of Labor Statistics report, the average Facebook user spends more than 11 h per month on Facebook [14].

5.4 Experience the Dynamics in most Usable Way

Usability is an integral part of emotional design. Though Norman says that even a product is esthetically appealing but little difficult to use, user will try to get the right path. But with excellent usability, the whole experience of easy completion of task elicits emotions and attachment with the product. Comments like "When I saw my 60 years old mother was easily able to use my gift-an iPhone, I felt so happy about it. It just took 2 days", make researcher to think about the beauty of usability.

Classic example is cleartrip.com. At the very first instance visually it's just clean website, nothing more than that, but experience during ticket booking elicits positive emotions, attachment with the website, because of efficiency and the performance of the website. The main reason behind this kind behavioral pattern is because appraisal theory says that relevance of the goal and attainment of goal in easiest possible way, elicit positive emotion (Fig. 15).

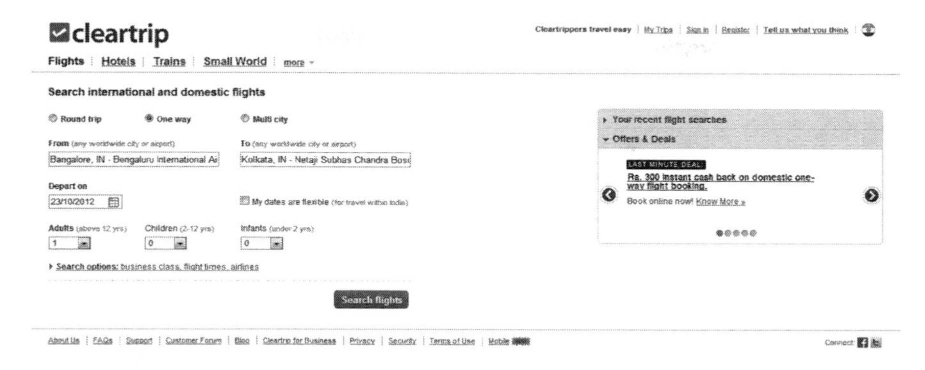

Fig. 15 Homepage of Cleartrip. *Source* http://www.cleartrip.com/

6 Conclusion

In general what we perceive or are taught about literary meanings of some of the important positive emotions are not same, as people appraise this emotion differently. There is a huge deflection in terms of articulating and understanding the dictionary meaning of these emotions. We have seen that though these emotions like happy-joy-fun or thrill-excitement are quite synonymous by nature but they have a very thin boundaries and their journey of experience are completely different with each other. But at the same time, these emotions can be mapped with certain set of parameters: Amplitude of arousal, time and social interaction and frequency of occurrence. These parameters further address the hidden cognitive meaning of these emotions and its implication on design decision. Though the appraisal of emotion is very abstract and subjective with individuals because of complex, multi-layer construct of human cognition, this paper is trying to address a framework on which these emotions can be re-articulated.

References

1. Plutchik R (2003) Emotions and life: perspectives from psychology, biology, and evolution. American Psychological Association, Washington
2. Kivikangas JM (2006) Psychophysiology of flow experience: an explorative study. Department of Psychology, University of Helsinki
3. Jordan PW (1999) Pleasure with products: human factors for body, mind and soul. In: Green WS, Jordan PW (eds) Human factors in product design: current practice and future trends. Taylor & Francis, London, pp 206–217
4. Norman D (2005) Emotional design: why we love (or hate) everyday things. Basic Books, New York
5. Demir E, Pieter MA, Desmet PMA, Hekkert P (2009) Appraisal patterns of emotions in human-product interaction. Int J Des 3(2):41–51

6. http://dictionary.cambridge.org/
7. http://dictionary.reference.com/
8. http://www.merriam-webster.com/
9. http://www.superbikesindelhi.com
10. http://www.2wheelsindia.com
11. http://www.goosebumpspickles.com/
12. http://www.cleartrip.com/
13. http://businesstoday.intoday.in/story/flipkart-online-shopping-e-commerce-order-books-online/1/20797.html
14. Bureau of Labor Statistics, June 2010 and Facebook Press Room, 2011
15. http://tatanano.inservices.tatamotors.com/tatamotors/home.htm
16. Kolb D (1984) Human factors in product design: current practice and future trends. Talyor & Francis, London, pp 206–217
17. Norman D The design of everyday things
18. Ellsworth PC,Schereï KR Appraisal processes in emotion
19. Desmet P, Hekkert P (2007) Framework of product experience. Int J Des 1(1):57–66
20. Bagozzi PR, Gopinath M, Nyer PU (1999) The role of emotions in marketing. J Acad Mark Sci 27(2):184–206

Part IV
Human Factors in Design

Force JND for Right Index Finger Using Contra Lateral Force Matching Paradigm

M. S. Raghu Prasad, Sunny Purswani and M. Manivannan

Abstract The paper aims at deriving the Just Noticeable Difference (JND) for force magnitude recognition between left and right index finger of human hand. The experiment involves establishment of an internal reference stimulus, using the left index fingers of the hand, by the subject, which is perceived and matched under contra-lateral force matching paradigm. A combination of virtual environment and a force sensor was used to derive the just noticeable difference for index-finger force application. Six voluntary healthy young adult subjects in the age group of 22–30 years were instructed to produce reference forces by left index finger and to reproduce the same amount of force by the right index finger, when the subjects were confident enough of matching same amount of force, the force values of the both the left and right index finger were recorded simultaneously for 5 s at 10 Hz. Five different trials were conducted for different force levels ranging from 2 to 5 N. The percentage real JND and absolute JND were derived for all the subjects. It was found that the Force-JND obtained was approximately 10 % across all subjects. Results also show that subjects tend to underestimate force at high force levels and overestimate at low force levels. The results obtained can be used as basic building block for the calibration of virtual reality based minimally invasive surgery related tasks and force based virtual user interfaces ranging from touch pad to assistive tools.

M. S. Raghu Prasad · S. Purswani · M. Manivannan (✉)
Haptics Lab, Biomedical Engineering Group, Department of Applied Mechanics, IIT Madras, Chennai 600036 Tamil Nadu, India
e-mail: mani@iitm.ac.in

M. S. Raghu Prasad
e-mail: raghuprasad.m.s@gmail.com

S. Purswani
e-mail: sunny.purswani89@gmail.com

A. Chakrabarti and R. V. Prakash (eds.), *ICoRD'13*, Lecture Notes in Mechanical Engineering, DOI: 10.1007/978-81-322-1050-4_29, © Springer India 2013

Keywords Just noticeable difference · Contra lateral matching · Force matching · Finger force perception · Visual cognition

1 Introduction

The Just Noticeable Difference (JND) is a measure of minimum difference between two stimuli necessary in order for human to differentiate between the two with certainty. Many researchers have focused on JND for force in human subjects but none of them have explored force JND for a single finger. Specifically the index finger force JND corresponds to the amount of differential sensation that an individual can negate while estimating the magnitude of a given stimulus.

In past, experimental results produced by various researchers have indicated the observable range of force JND to be between 5 and 10 %. Weber measured JNDs roughly 10 % in experiments involving active lifting of 907 g weights by the hand and arm [1]. Human force control and force resolution for the effective design of haptic interfaces was reported in [2] the study also emphasized the fact that humans are less sensitive to pressure changes (i.e. force changes, when contact area is fixed) when contact area is decreased. The JND was around 10 % for pinching tasks involving finger and thumb at a constant holding force [3]. A force matching experiment about the elbow, found a JND ranging between 5 and 9 % by Jones et al. [4]. Brodie and Ross [5] had obtained JNDs lying in the same range for tasks involving the active lifting of 2 oz. weight. These research works are focused either on whole limb/arm based weight estimation or finger-thumb based combinations. On the contrary, our study is primarily based on the single finger static force application of right index finger, without any involvement of any other muscle groups. Also, our study concerns the active force JND, rather than passive JND.

Various paradigms have been adopted in prior research works to obtain the force JNDs for various muscle groups. For example, Pang et al. [3] adopted one interval, two alternative forced-choice paradigms. On the other hand Jones et al. [4] utilized the method of contra-lateral force matching with generating force ranging between 15 and 85 % of the maximum voluntary contraction (MVC range: 169–482 N). An up-down transformed response rule (UDTR) paradigm, a modified form of staircase method, for both active and passive weight lifting procedures was put forth by Brodie and Ross [5]. But, most of these paradigms involve perception of force as well as movement, based on the apparatus being used for the respective experiments. Our experiment doesn't involve application of any of the conventional TSD (theory of signal detection)-based techniques due to absence of artificial external stimuli to match. The experiment involves establishment of an internal reference stimulus by the subject, which is perceived and matched by the homologous set of muscles of opposite limb, under the purview of magnitude estimation psycho-physics technique. This mechanism involves a controlled force variation and not the rate of change of the depth of skin-indentation in index-finger application [6, 7]. Moreover, the technique used here is responsive to the fact that people do not produce constant forces spontaneously unless they are artificially controlled [7, 8]. In our experiment, subjects gradually increased self-produced force to a peak value that is visually displayed and then contra laterally matched. The primary purpose of the study is to investigate the force perception of human right index finger using contra lateral matching tasks and to evaluate the % force JND for a range of forces.

1.1 Contra Lateral Force Matching Paradigm

A contra-lateral force matching paradigm is a typical method used by various researchers to study the force perception [9–13] it is a mechanism of matching forces generated by muscle group of the any of the limbs on one side of human body by using the same set of muscle group of the other side of the body. This matching action has been observed to involve CNS (Central Nervous System) and involves a small amount of lag in information exchange between the two sides when compared to the ipsilateral force matching paradigm, which involves matching action of muscle groups on the same side of human body. Contra-lateral force matching is a popular methodology for comparing force perception and control. A comparison of matching performance between ipsilateral and contra lateral finger force matching tasks and to examine the effect of handedness on finger force perception was conducted [14], the results from the experiment indicate that the absolute, rather than relative finger force is perceived and reproduced during ipsilateral and contra lateral finger force matching tasks [14].

2 Methods

2.1 Subjects

Subjects described here are 6 healthy members of the Indian Institute of Technology Madras community, age 26 ± 4.1 years, weight 69.5 ± 8.36 kg, height 172 ± 5.22 cm, hand length from the index fingertip to the distal crease of the wrist with hand extended 16.9 ± 1 cm, hand width at the metacarpophalangeal joints (MCP) level with hand extended 8.1 ± 0.6 cm. All subjects were pre-screened verbally for self-reported handedness, and history of visual, neurological, and/or motor dysfunction. All subjects gave informed consent. No subject was known to have any neurological and visual perception disorders. Five Subjects were right handed and one subject was left handed.

2.1.1 Apparatus

Each subject was comfortably seated on a chair facing a computer monitor and asked to place both of his/her upper limbs on a wooden table positioned at the same height as of the side support of the chair, thereby maintaining a correct symmetry with respect to the medial axis of the body. The angle made by the index finger with the shoulder joint was approximately 90°. Each subject was instructed to maintain a constant index-finger pressing posture during the course of the experiment. The monitor on which a visual feedback was given was placed 15° below eye level at a distance of 0.6 m away from the participant. Two Force Sensitivity Resistors (FSR) of Interlink ^{TM} make were used as force sensors.

The two FSR's one each for left and right hands were mounted on a wooden board such that symmetry was maintained with respect to the hand positions. In order to avoid fatigue precautionary care was taken in positioning the FSR's in accordance with the participant's index fingers. The two square FSRs were connected to the Analog to Digital Converter embedded in a controller over a parallel voltage divider circuit with 1 KΩ loads each, under 5 V input supply as shown in Fig. 1.

3 Contra Lateral Force Matching Procedure with Visual Feedback

During each experimental trial, subjects were asked to reach a target force bar of constant thickness (0.15 N) by pressing left-index finger over a 2″ Square FSR (Resolution—0.01 N), which were calibrated for given range of application of force [15]. The touch surfaces of both FSRs possessed same texture and were devoid from presenting any tactile cues or spatial attenuation which could result in biased force sensation. Once the subject reached the target force level the background color of display changed as shown in Fig. 2, indicating the attainment of the target force. When the subject was able to maintain the target force level over a period of 4–5 s, he/she was instructed to press the similar FSR on the right side of the arrangement using his/her right-index finger and try matching the force, without any visual feedback. No information was given to the subjects about the matched force value attained by the right index finger. Once, subject assured that he/she had attained the same force on right-index finger, data was recorded for 5 s at a sampling rate of 10 Hz and the trial was completed. Each subject was presented with 4 different force levels of 2, 3, 4 and 5 N and each level was delivered 5 times, with equal a priori probability during the course of experiment, thereby making each experiment comprise of 50 trials. A constant target force range was set across all force levels which allowed subjects to deviate from the target by a constant force of 0.15 N (i.e. 7.5 % at 2 N; 5 % at 3 N; 3.75 % at 4 N and 3 % at 5 N) this window of 0.15 N has been chosen carefully, by keeping in mind the

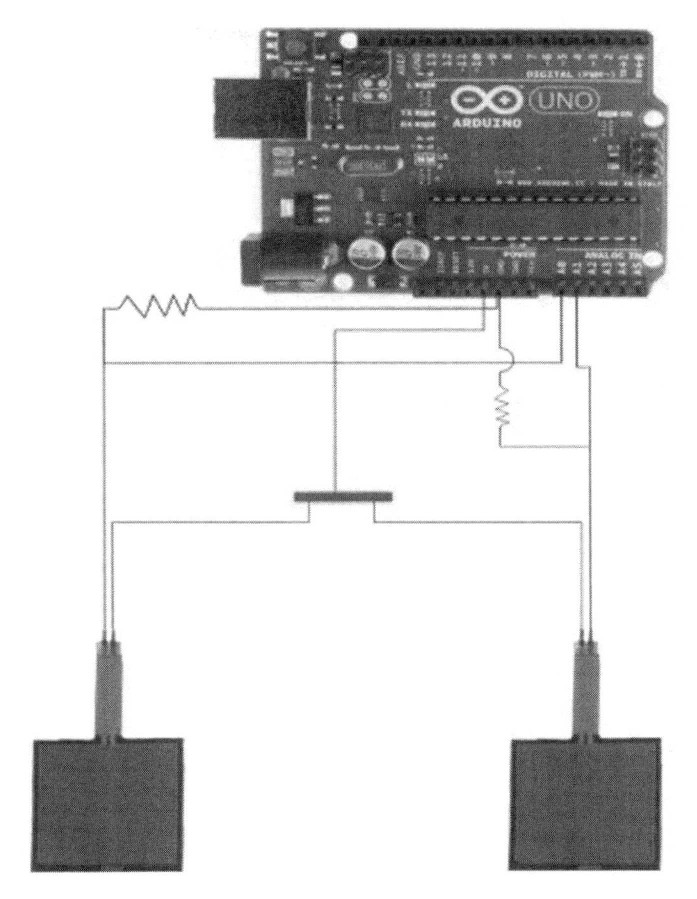

Fig. 1 Square FSRs are connected to the controller ADC

constant force error recognition results under visual feedback, as described in [8]. Subject was able to vary the force magnitude over the FSR, based on the calibration performed for electrical signal to force value conversion.

4 Data Analysis

Difference in force applied between the reference and matched value was calculated for the force-matched data across 6 subjects. The difference in real and absolute % force JND for each subject was computed separately. For a given reference force R_i and a matched force M_i the % real force JND for a single sample is obtained from Eq. (2) and the % absolute force JND is obtained from Eq. (3).

Fig. 2 Visual feedback

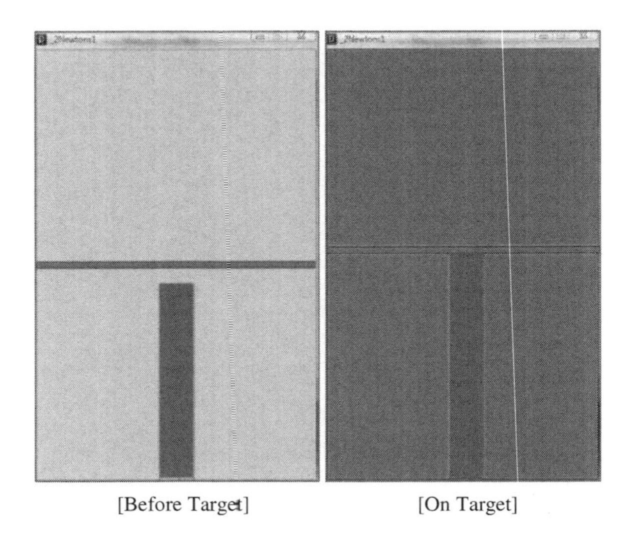

[Before Target] [On Target]

$$\% \, Force \, \mathrm{Real} \, JND = (M_i - R_i) \times 100/R_i \tag{2}$$

$$\% \, Force \, Absolute \, JND = Abs|(M_i - R_i) \times 100/R_i| \tag{3}$$

where i is the no:of of samples.

Similarly for 50 samples, average absolute % force JND and real % force JND per trial is computed using Eqs. (4) and (5).

$$\% \, Force \; Absolute \, JND = \frac{\sum_{i=1}^{50} |M_i - R_i| \times 100/R_i}{50} \tag{4}$$

$$\% \, Force \; \mathrm{Real} \, JND = \frac{\sum_{i=1}^{50} (M_i - R_i) \times 100/R_i}{50} \tag{5}$$

Parameters such as mean, standard deviation, standard error and variance were analyzed in detail to investigate the effect of change in % force JND across different subjects at different force levels ranging from 2 to 5 N. Table 1 summarizes the statistical analysis performed on the data set.

5 Results and Discussion

The absolute and real % force JND were obtained from each subject using contra lateral force matching paradigm. The distribution of % force JND values across subjects over various force levels in the graph, indicate that diversion of matched

Table 1 Mean, standard deviation (SD), standard error (SE) and variance (VAR) of the real and absolute % force JND across 6 subjects

Force level (Newton)	Mean % real force JND	Mean % abs force JND	SD % real force JND	SD % abs force JND	SE % real force JND	SE % abs force JND	VAR % real force JND	VAR% abs force JND
2	14.233	13.610	4.830	4.099	1.972	1.673	23.331	16.799
3	3.078	9.696	2.974	2.122	1.214	0.866	8.845	4.502
4	−4.996	7.856	2.418	2.825	0.987	1.153	5.846	7.980
5	−6.780	8.292	2.143	2.185	0.875	0.892	4.594	4.776

force values tend to contract as the force level increases. This indicates that subjects tend to produce similar static forces in a close range when the reference forces are high. Moreover, mean real % force JND values change sign between 3 and 4 N as illustrated in Fig. 3, which suggests that there exist certain set of force levels between this ranges, where subjects tend to match the reference forces most accurately, as per the contra-lateral force matching paradigm.

The pattern obtained from the average absolute % JNDs of each subject as shown in Fig. 4 indicate that at low force levels subjects tend to overestimate the reference forces and underestimate the reference force at higher force levels. Our

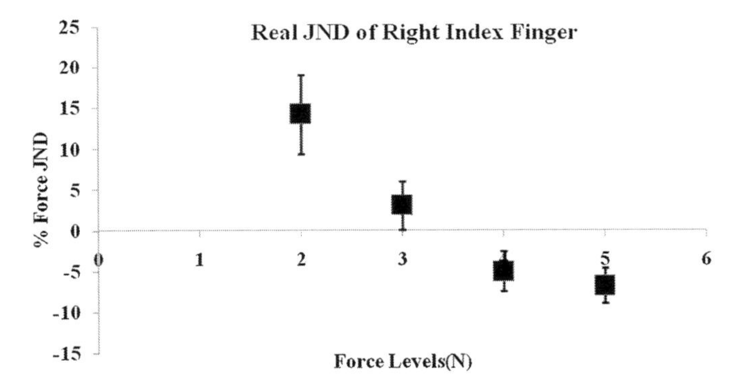

Fig. 3 Averaged real % force JND with mean and standard deviation

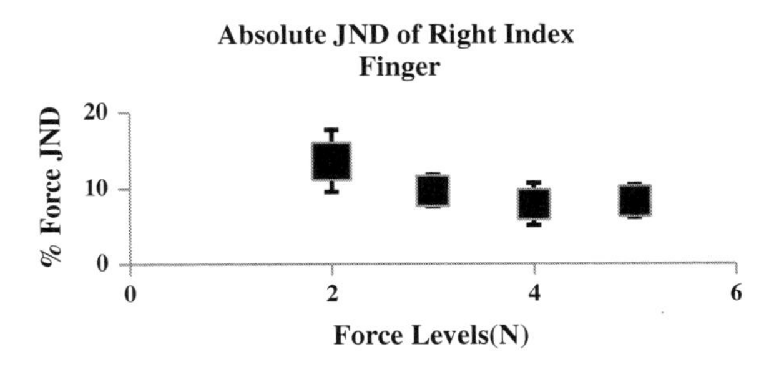

Fig. 4 Averaged absolute % force JND with mean and standard deviation

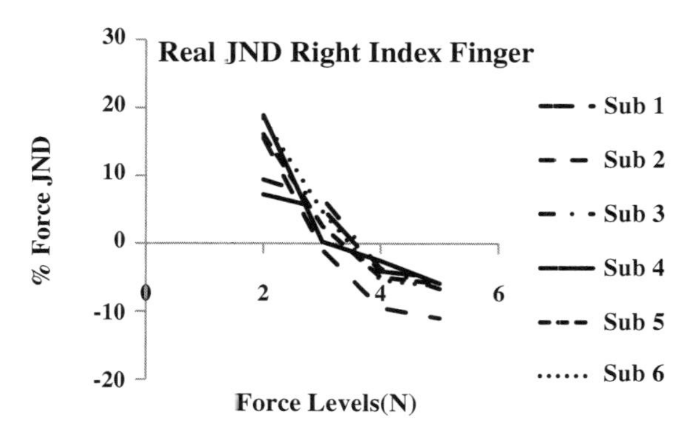

Fig. 5 Real value of % force JND of six subjects

main findings were: (1) The % force JND is larger for very small force intensity levels such as 2 N and decreases as the force stimuli increases to higher force levels such as 5 N, (2) The averaged absolute % force JND graph as illustrated in Fig. 4 follows Weber's law, (3) The decision making process of subjects took more time at lower force levels and response time was less for higher force levels.

The standard errors and standard deviations of the lower force JNDs were found to be greater than the JNDs (as shown in Fig. 4) of the higher force levels. This indicates that the subjects were more confident matching the reference force stimuli in their response to higher % force JNDs when compared to the lower force JNDs. The % force JND resolution was high at higher force levels compared to the % force JND resolution at lower force levels this indicates that the subjects were able to closely match the self generated reference stimuli by the left index finger with their right index finger at higher force levels.

Figure 5 Illustrates the real value of % force JNDs of all the subjects plotted across force levels ranging from 2 to 5 N. The graph shows that out of 6 subjects 2 subjects managed to attain lower % force JND at lower force levels. The absolute % force JND of all the subjects is shown in Fig. 6.

From our experiment we observed that absolute % force JND was roughly around 10 %, as shown in Fig. 7, when compared to JND obtained from active weight lifting

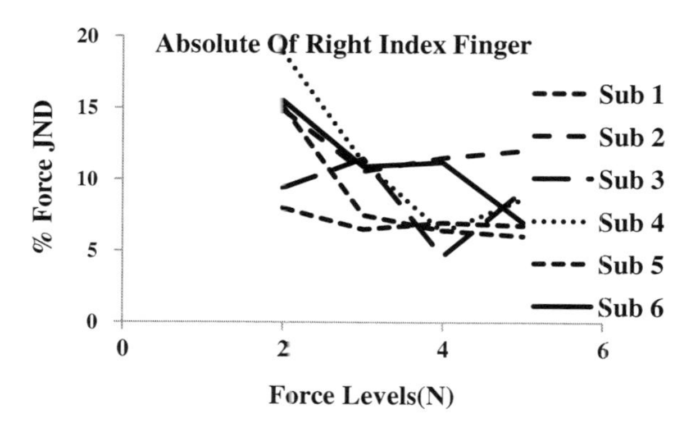

Fig. 6 Absolute value of % force JND of six subjects

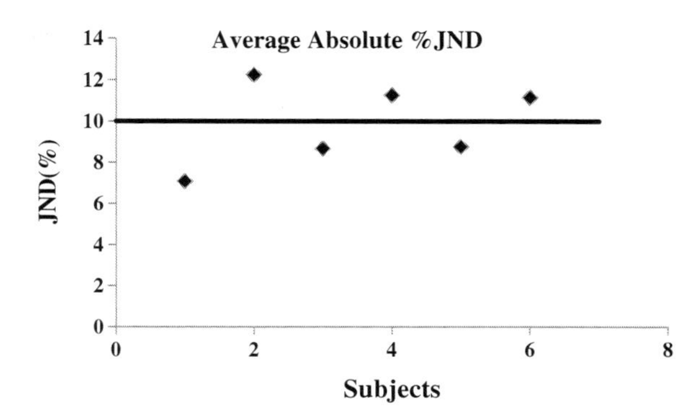

Fig. 7 Averaged % force JND of six subjects

Table 2 Absolute % force JND across all subjects

Subject	2 N (%)	3 N (%)	4 N (%)	5 N (%)	Mean (%)
One	8	6.52	6.98	6.78	7.07
Two	14.85	10.567	11.51	12.02	12.24
Three	9.38	11.45	4.63	9.237	8.68
Four	18.84	11.22	6.36	8.64	11.26
Five	15.06	7.5	6.45	6.05	8.77
Six	15.5	10.9	11.2	7.02	11.15
Mean (%)	13.6	9.7	7.85	8.3	9.86 (net)

and force matching about the elbow. A number of significant elements, however, distinguish our paradigm from previous tasks reported in literature. Table 2 presents the complete set of absolute % force JND values obtained across all 6 subjects, during course of their individual experimental runs.

6 Conclusion

% Force JND experiment produced JND values for the static force increment task averaging out 10% approximately. These JND results have closely followed the Weber's law and have fallen in the allowable range of values obtained in past research works. These experimental results prove to be pivotal in establishing a force-based virtual environment which would be operated by force based tactile interface on real-time synchronization with the system. The present paper attempts to obtain the % force JND values for static force application, specifically using index finger. A contra-lateral force matching paradigm is adopted to obtain the matching forces and corresponding force JND's. Our experiment does not consider the displacement at the point of application of force and handedness during contra lateral force matching tasks.

The experiment involves use of a force sensitive resistor (FSR), as a force transducer for obtaining the force values and related parameters. Each subject performs 20 force-matching trials for four different force levels. Each trial involved matching left-index reference force window with the right-index finger by application of force over the FSR. The forces generated by the right and left index finger and the resulting % force JND supports the notion that cutaneous feedback from the contact surface of the FSR influences the perception of force within subjects [16]. Each subject performed experiment under identical conditions without bias. It was observed that the % force JND resolution at higher force levels were better when compared to the force resolution at lower force levels, indicating that the subjects responded positively to higher force stimulus.

Also, the nature of force matching for low and high force levels was observed to be on the lines of similar past research work involving contra-lateral force matching technique. The results obtained from the experiment using the described set of attributes, opens up scope for future work involving experimental validation of the same range of force JNDs in virtual laparoscopic surgery training environment. The experiment results and data provide a frame work to build an understanding of sensory and motor deficit to the physically challenged community. Moreover, the procedures we devise here can help us explore JNDs as they change during the course of recovery. Once this has been done, the development of adaptive rehabilitative environments tailored to patients' particular sensitivities can begin. Future work could be extended to providing calibrated framework for various force-based virtual user-interfaces (UI), ranging from touchpad to assistive surgical training tools and simulators.

References

1. Weber EH (1978) The sense of touch. Academic Press for Experimental Psychology Society, London
2. Tan HZ, Mandayam BE, Srinivasan A, Cheng B (1994) Human factors for the design of force-reflecting haptic interfaces. Dyn Syst Control DSC–vol 55–1
3. Pang X, Tan HZ, Durlach NI (1991) Manual discrimination of force using active finger motion. Percept Psychophys 49(6):531–540
4. Jones LA (1989) Matching forces: constant errors and differential thresholds Perception 18(5):681–687
5. Brodie E, Ross H (1984) Sensorimotor mechanisms in weight discrimination. Attention Percept Psychophys 36(5):477–481
6. Greenspan JD (1984) A comparison of force and depth of skin indentation upon psychophysical functions of tactile intensity. Somatosens Res 2(1):3348
7. Goodwin AW, Wheat HE (1992) Magnitude estimation of contact force when objects with different shapes are applied passively to the finger pad. Somatosens Mot Res 9(4):339–344
8. Srinivasan MA, Chen JS (1993) Human performance in controlling normal forces of contact with rigid objects. Adv Robot Mechatron Haptic Interfaces DSC-vol 49, ASME
9. Gandevia SC, McCloskey DI (1977) Effects of related sensory inputs on motor performances in man studied through changes in perceived heaviness. J Physiol 272(3):653–672

10. Gandevia SC, McCloskey DI (1977) Changes in motor commands, as shown by changes in perceived heaviness, during partial curarization and peripheral anaesthesia in man. J Physiol 272:673–689
11. Cafarelli E, Bigland-Ritchie B (1979) Sensation of static force in muscles of different length. Exp Neurol 65(3):511–25
12. Kilbreath SL, Refshauge K, Gandevia SC, (1997) Differential control of the digits of the human hand: evidence from digital anaesthesia and weight matching. Exp Brain Res 117(3):507–511
13. Jones LA (2003) Perceptual constancy and the perceived magnitude of muscle forces. Exp Brain Res 151(2):197–203
14. Park W-H, Leonard CT, Li S (2008) Finger force perception during ipsilateral and contralateral force matching tasks. Exp Brain Res 189:301–310
15. Flórez1 JA, Velásque A (2010) Calibration of force sensing resistors (FSR) for static and dynamic applications. ANDESCON, IEEE, 15–17 Sept 2010
16. Jones LA, Piateski E (2006) Contribution of tactile feedback from the hand to the perception of force. Exp Brain Res 168(1–2):298–302

Modeling of Human Hand Force Based Tasks Using Fitts's Law

M. S. Raghu Prasad, Sunny Purswani and M. Manivannan

Abstract Conventional Fitts's model for human movement task finds a common application in modern day interactive computer systems and ergonomics design. According to Fitts's law the time required to rapidly move to a target area is a function of the distance to the target and the size of the target. This paper describes experimental process for prediction of minimum movement time, in a force-variation based human performance task involving right index finger. In this study we have made an attempt to extend the applicability of the conventional Fitts's model for a force based virtual movement task, without taking position into account and evaluate human performance metrics for such tasks. An experiment was conducted in which 6 healthy young adult subject's in the age group of 22–30 years performed force based movement tasks. During each trial, subjects were asked to reach an initial force bar of given thickness W Newtons, corresponding to allowable tolerance. Once the subject's had reached initial level, they were instructed to reach out the target force bar of same thickness W as quickly as possible and bring it back to the initial force level bar, thereby completing 1 iteration. Time required for 10 such iteration was noted for each subject. The results from the experiment show that the relationship between movement time and index of difficulty for force tasks are well described by Fitts's law in visual guided, force-based virtual movement task.

M. S. Raghu Prasad · S. Purswani · M. Manivannan (✉)
Haptics Lab, Biomedical Engineering Group, Department of Applied Mechanics, Indian Institute of Technology, Madras, Chennai 600036 Tamil Nadu, India
e-mail: mani@iitm.ac.in

M. S. Raghu Prasad
e-mail: raghuprasad.m.s@gmail.com

S. Purswani
e-mail: sunny.purswani89@gmail.com

A. Chakrabarti and R. V. Prakash (eds.), *ICoRD'13*, Lecture Notes in Mechanical Engineering, DOI: 10.1007/978-81-322-1050-4_30, © Springer India 2013

Keywords Fitts's law · Virtual reality · Human computer interaction · Human performance modeling · Ergonomics

1 Introduction

Display ergonomics is primarily governed by the Fitts's model [1] involving estimation of movement time involving limb movements over UIs like mouse, joystick, remote controller etc. Generic definition of the Fitts's model is based on the information capacity of the CNS to execute a physical movement task, assisted by the reflection of the same in the virtual environment over a visual display. But, there exist wide variety of modalities involving tasks which present movement in virtual environment but are restricted from significant amount of movements from the user side, and mostly involve accomplishment of the job based on application of a force—torque combination. For example, pressing and pinching task in laparoscopic grasper, z-movement using force-based touch screen etc. Hence, there was a need of establishing a relation for the estimation of movement time for such category of tasks and validate the results over one of the force-based modality.In conventional way, Fitts's law has proved itself effective in predicting movements times involving replication of human limb movements over virtual environments. In this paper we attempt to validate applicability of Fitts's law in force application-based virtual movement task, involving no human movement. In this experiment, the movement amplitude A, for a particular trial task has been replaced by difference in initial and target force, and target width W has been replaced by allowable force tolerance at both initial and target force levels. The procedure is aimed at predicting the minimum movement time, with no allowable delay, which would be set as a base equation for calibrating all sorts of possible force application based movement tasks with variable attributes.

1.1 Fitts's Law

Ergonomics of Human Computer Interface designs have been predominately driven by the Fitts's model of movement time. This model has proven its effectiveness in improving usability of UIs and optimal designing of the size and location of user interface elements. It can also be used to predict the performance of operators using a complex system, assist in allocating tasks to operators, and predict movement times for assembly line work. However it does have some disadvantages, which include un-directional movement prediction and absence of consistent technique of error detection. Various research and comparison studies have evolved around Fitts's model, some of which extend its scope beyond

1-Dimension usability others explore the possible applicability of this movement task based model in different modalities and signal forms. Fitts [1] conducted the reciprocal tapping experiment across various subjects to validate his theory stating—ability to perform a particular movement is directly characterized by the information capacity of brain and is affected by the alternate possibilities of movements. As part of his experiment, subjects were asked to tap two rectangular plates alternately with a stylus. Movement tolerance and amplitude were controlled by fixing the width of the plates and the distance between them. Subjects were instructed to take care of the accuracy by which they perform the movement task rather than speed of reaching the target points. As part of the Fitts's law expression, Index of Difficulty (ID) is specified as the amount of information required to select specific amplitude from the total range of possible movements, and is thus dependant on the amplitude of the movement (A), and the available target size/allowable tolerance to which it must be made (target width W).

$$ID = \log_2\left(\frac{A}{W} + 1\right) \quad (1)$$

Fitts's expression is closely related to the fundamental theorem of communication systems, derived by Shannon [2]. Thus, by varying ID (A and W), IP can be determined by recording MT over the various conditions. Fitts's concluded the invariability of IP over a certain range of values of ID. Using regression analysis, a linear relationship between ID and MT can be established [1, 2].

$$MT = a + b * ID \quad (2)$$

Here **a** and **b** are the information transmission coefficients. In this form, the reciprocal of coefficient b (1/b) is called the index of performance (IP), and its unit is in "bits/sec". The index of performance (IP) indicates how quickly the pointing can be done. The index of difficulty (ID) depends on the width (W) of the target and the distance (A) between the two targets. The difficulty of the task increases when the distance (A) increases or the width (W) decreases. Fitts's law has also been extended to two and three dimension tasks [3–5]. In the field of HCI, Card et al. was the first to apply Fitts' law in computer interfaces [6]. He compared the performances between a joystick and mouse. The results shown by the mouse movement task were comparable values of IP of order of 10 bits/s, to that of the tapping task perform by Fitts'. On the other hand, joystick produced an index of performance value of around 5 bits/s. The usage of Fitts's law reported in the literature [7–10] was based on position alone. After position alone and before our study the following is added. Human performance for pointing and crossing tasks depended on index of difficulty [11] and fitt's law has also been used to effectively predict movement times when steering through constrained paths and spring stiffness control [12, 13]. Our study aims at extending Fitts's law for force based tasks.

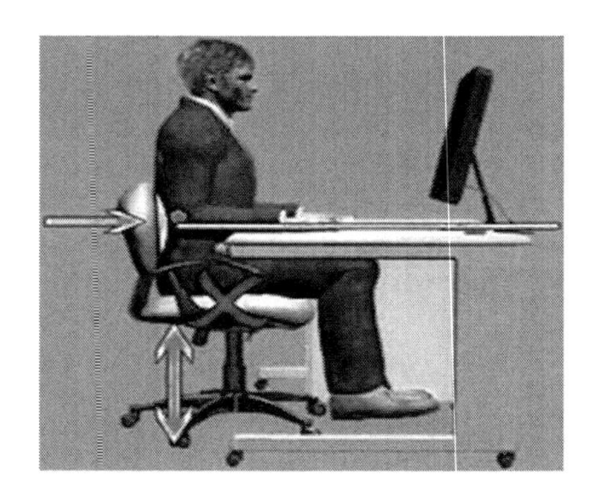

Fig. 1 Sitting posture of subject. Platform height was maintained at elbow level, providing least strained shoulder-elbow-wrist position

2 Methods

2.1 Subjects

Six healthy subjects of the IIT Madras community in the age group of 26 ± 4.1 years, weight 69.5 ± 8.36 kg, height 172 ± 5.22 cm, hand length from the index fingertip to the distal crease of the wrist with hand extended 16.9 ± 1 cm; hand width at the MCP level with hand extended 8.1 ± 0.6 cm. All subjects were pre-screened verbally for self-reported handedness, and history of visual, neurological, and/or motor dysfunction. All subjects gave informed consent. No subject was known to have any difficulty in processing proprioceptive estimation. Five Subjects were right handed and one subject was left handed.

2.2 Apparatus

Each subject was comfortably seated on a chair facing a computer monitor and asked to place both of his/her upper limbs on a wooden table positioned at the same height as of the side support of the chair as shown in Fig. 1, thereby maintaining a correct symmetry with respect to the medial axis of the body. The angle made by the index finger with the shoulder joint was approximately 90°. Each subject was instructed to maintain a constant index-finger pressing posture during the course of the experiment. The monitor on which a visual feedback was given was placed 15° below eye level at a distance of 0.6 m away from the participant. A Force Sensitivity Resistors (FSR) of Interlink™ make was used as force sensor.

An FSR for the right hand was mounted on a wooden board such that symmetry was maintained with respect to the hand position. In order to avoid fatigue precautionary care was taken in positioning the FSR in accordance with the participant's right index finger. The display was rendered using Processing 1.5, open source platform, on 160 GB–1.5 GHz, Core 2 Duo PC, running Windows Vista,

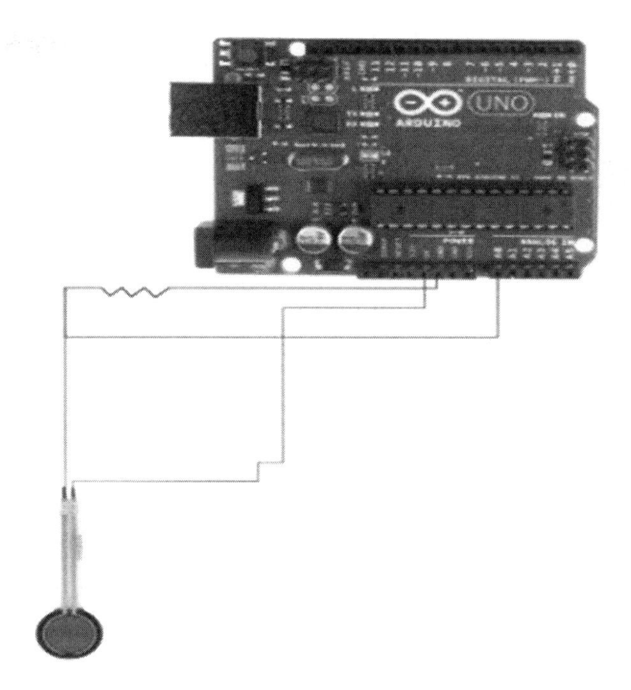

Fig. 2 Circular FSR are connected over a voltage divider circuit with 1,000 Ω loads, under 5 V supply. The input sensor values are triggered into serial port at 100 ms per iteration, with maximum information transmission capacity of 9,600 baud-rate

with 2 GB RAM. The experimental set-up as shown in Fig. 2 comprises of single FSR connected over a parallel voltage divider circuit with 1,000 Ω load under 5 V input supply. The input sensor values were triggered into serial port with a delay of 100 ms per iteration.

2.3 FSR Calibration

The FSR's were calibrated using known weights in the range of 100–1,000 g to obtain the exact values of force applied over the exposed surface of the FSR. A final relation was established between the resistance values of FSR and the applied force, Fig. 3 depicts the calibration graph of the FSR (Table 1).

3 Force Based Fitts's Protocol with Visual Feedback

During each run, subjects were asked to reach an initial force bar of given thickness W, corresponding to allowable tolerance. Once the subject had reached initial level, he was instructed to reach out the target force bar of same thickness W as quickly as possible and bring it back to the initial force level bar, thereby completing one iteration. Each subject was supposed to perform 10 such iterations per trial, where in each trial

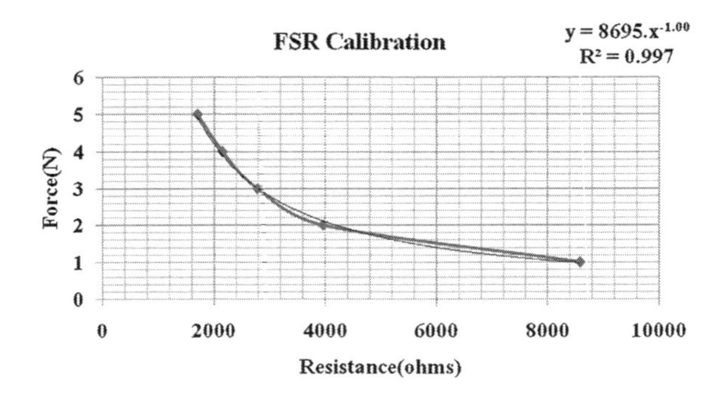

Fig. 3 FSR calibration curve with exponential fit

Table 1 FSR calibration data

Force (N)	Resistance (Ohms)
1	8,578
2	3,957
3	2,781
4	2,147
5	1,700

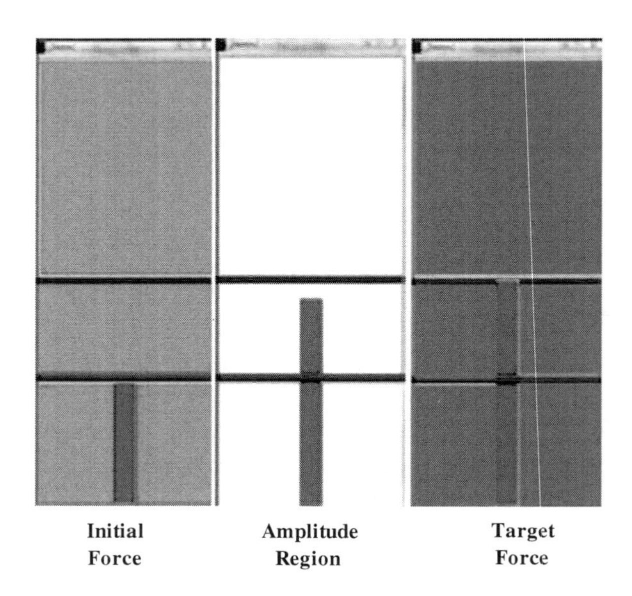

Fig. 4 Visual display

comprised of a unique set of initial force-level, target force level and thickness W. Subject was able to vary the force application by adjusting the pressure exerted by right-index finger over a 0.5″ Circular Force-Sensitive Resistor (Resolution—0.01 N).

Once the subject reached the initial force level, background color of display changed to green as shown in Fig. 4, indicating the attainment of the initial force level. Once the subject pressed over FSR and was able to move to the target force level for given thickness W, background color changed to blue, indicating attainment of target force level and instruction to come back to initial force level. Each subject underwent 50 trials as part of the experiment, force level being pair of combinations out of the (1 , 2 , 3 , 4 , 5 N) set and thickness W being values out of the set of (0.1 , 0.2 , 0.3 , 0.4 , 0.5 N). Each set of values were presented with equal a priori probability, without any bias.

4 Data Analysis

Each trial in subject-wise experiment comprised of 10 iterations, wherein sensor data values were stored at the rate of 10 samples per second. At the end of each trial, average value of all the iterations was calculated, in the following form as shown in Eq. (3).

$$MT_j = \frac{\sum_{i=1}^{10} T_i}{10} \tag{3}$$

MT_j is the average movement time for jth trial and T_i is the time taken by ith iteration. Similarly, average movement times were computed for all the 50 unique trials. Out of 50 MT values, certain set of combinations of A (Difference in force levels) and W (Allowable force tolerance) produced same A/W ration, thereby denoting same Index of Difficulty (ID). Hence, movement time for such sets of values for a given A/W was normalized to their mean.

5 Results and Discussion

Linear regression plot was performed over the final set of Index of Difficulty [ID = Log_2 (A/W + 1)] values against the respective minimum movement times (MTs). R-squared values were evaluated using least-square method for establishing the degree of correlation of the actual values with the corresponding regression plot. Process was repeated across all 6 subjects, individually. A direct

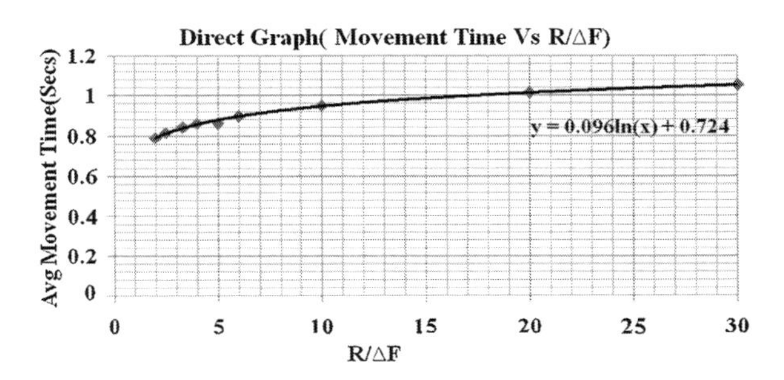

Fig. 5 A logarithmic fit to the data set of movement time and index of performance

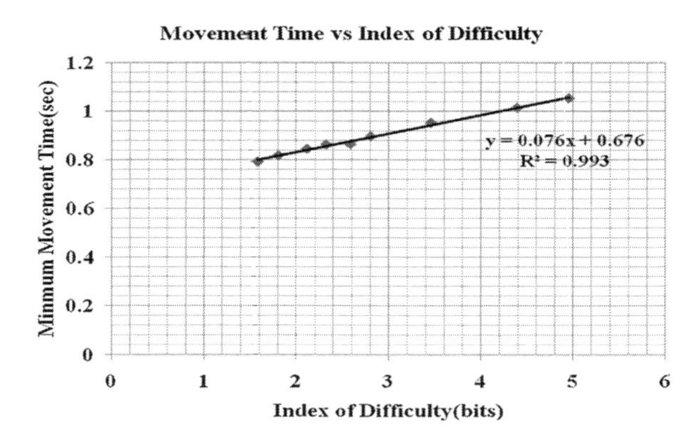

Fig. 6 A linear fit to the data of movement time versus index of difficulty

Table 2 Standard deviation and confidence values of all subjects

R/deltaF	ID	Sub1	Sub2	Sub3	Sub4	Sub5	Sub6	Avg (MT)	Std Dev	Confidence
2.00	1.58	0.71	0.73	0.89	0.69	0.83	0.90	0.79	0.09	0.08
2.50	1.81	0.75	0.73	0.98	0.72	0.88	0.85	0.82	0.10	0.08
3.33	2.11	0.74	0.73	1.08	0.80	0.87	0.84	0.84	0.13	0.10
4.00	2.32	0.83	0.79	0.73	1.10	0.86	0.86	0.86	0.13	0.10
5.00	2.58	0.93	0.80	0.79	0.89	0.86	0.91	0.86	0.06	0.05
6.00	2.81	1.10	0.69	1.05	0.75	0.79	1.00	0.90	0.17	0.14
10.00	3.46	0.87	0.76	1.18	0.90	1.06	0.94	0.95	0.15	0.12
20.00	4.39	0.96	0.78	1.23	0.91	1.15	1.05	1.01	0.16	0.13
30.00	4.95	1.03	0.79	1.28	0.93	1.20	1.08	1.05	0.18	0.14

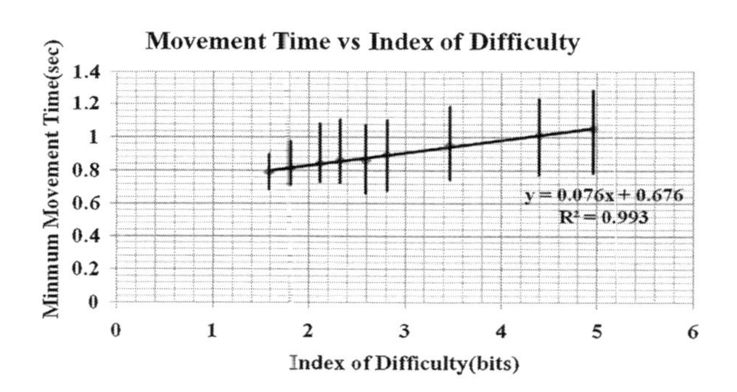

Fig. 7 Plot of movement time and index of difficulty

graph of Movement Time and index of performance is shown in Fig. 5 with a logarithmic fit where R = difference in Force levels and ΔF = allowable tolerance. The slope of the graph represents the performance of the control as for the data with visual feedback. Each subject produced R-squared values of more than 85 % accuracy corresponding to their projected linear regression plots of minimum MT versus ID, thereby validating 1D Fitts's law in force domain. Figure 6 shows the final plot of Movement time (MT) versus Index of Difficulty (ID) averaged across all 6 subjects. Plot presents an R-squared value of 0.993, denoting a very accurate representation of the Fitts's law in static force application based movement tasks over the pre-defined force range. Here final values of constants **a** and **b** are 0.676 and 0.076 respectively.

Table 2 depicts the standard deviation and confidence values across all subjects with the index of performance and index of difficulty. Figure 7 shows the graph of the ratio of difference in force levels to the allowable tolerance plotted over the data set of index of difficulty and movement time. The results indicate that Fitts's law applies to the activities in muscle space, as well as the movement and force in task space. The results from this study infer that Fitts's law applies to the activities in muscle space, as well as the movement and force in task space. This may imply that Fitts's law is determined in the level of the nervous system and is not affected by the dynamics of the limbs. The results strengthen the argument that the trade-off between speed and accuracy of the pointing movement is determined by the capacity of information transfer of the central nervous system (CNS), rather than the physical limitations of the arm, such as inertia and mechanical compliance [8]. Moreover during the course of the experiment, control rate and homing time turned out to be crucial factors in accomplishing the task under variable set of parameters [7, 14–16].

6 Conclusion

In this study we have explored the possibility of using the well established Fitts's model to force based human hand movement tasks. The final derivation of the Fitts' law expression after regression analysis over the generalized data collected across 6 subjects. R-squared value of more than 95 %, justifies the validity of Fitts's law in force-based movements in virtual environments irrespective of involvement of limb movements. The results from the experiment show that the relationship between movement time and index of difficulty are well described by Fitts's law in visual guided, force-based virtual movement task. These results, open up scope for extending this research to establishment of Fitts's law to 3-Dimensional virtual environments (e.g. laparoscopy surgical training platform) involving movements and orientations synchronized by application of multi-directional force and torque. Further the study can be used as a basic foundation in modeling and design of force based minimally invasive surgical training modules in order to assist resident doctors to hone their surgical skills.

References

1. Fitts PM (1954) The information capacity of the human motor system in controlling the amplitude of movement. J Exp Psychol
2. Shannon CE (1998) Communication in the presence of noise, In: Proceedings of the IEEE, vol 86(2). February

3. Murata A, Iwase H (2001) Extending Fitts's law to a three dimensional pointing task. Hum Mov Sci 20(6):791–805
4. Scott Mackenzie I, Buxton WAS (1992) Extending Fitts' law to two-dimensional tasks. In: Proceedings of ACM CHI conference on Human factors in computing systems
5. So RH, Chung GK, Goonetilleke RS (1999) Target-directed head movements in a head-coupled virtual environment: predicting the effects of lags using Fitts' law. Hum Factors 41(3):474–486
6. Card SK, English WK, Burr BJ (1978) Evaluation of mouse, rate-controlled isometric joystick, step keys, and text keys for text selection on a CRT. Ergonomics 21:601–613
7. Wall SA, Harwin WS (2000) Quantification of the effects of Haptic feedback during a motor skills task in a simulated environment, In: Proceedings of 2nd phantom users research symposium, Zurich, Switzerland, pp 61–69
8. Park J, Kim H, Chung W, Park S (2011) Comparison of myocontrol and force control based on Fitts' law model. Int J Precis Eng Manuf 12(2):211–217
9. Scott J, Brown LM, Molloy M (2009) Mobile device interaction with force sensing. In: Pervasive 2009, LNCS vol 5538, pp 133–150
10. Accot J, Zhai S (2003) Refining Fitts' law models for bivariate pointing. In: Proceedings of ACM CHI conference on human factors in computing systems
11. Casiez G, Vogel D, Balakrishnan R, Cockburn A (2008) The impact of control-display gain on user performance in pointing tasks. Human Comput Interac 23(3):215–250
12. Casiez G, Vogel D, Pan Q, Chaillou C (2007) Rubber edge reducing clutching by combining position and rate control with elastic feedback. In Proceedings of ACM CHI, pp 129-138
13. Dominjon L, Lécuyer A., Burkhardt J.M, Andrade- Barroso G, Richir S (2005) The bubble technique: Interacting with large virtual environments using haptic devices with limited workspace. In Proceedings of IEEE World Haptics, pp 639–640.
14. Accot J, Zhai S (2002) More than dotting the i's—foundations for crossing-based interfaces. In: Proceedings of ACM CHI 2002 conference on human factors in computing systems
15. Kattinakere RS, Grossman T, Subramanian S (2007) Modeling steering within above-the-surface interaction layers. In Proceedings of ACM CHI 2007 conference on human factors in computing systems
16. Casiez G, Vogel D (2008) The effect of spring stiffness and control gain with an elastic rate-control pointing device. In Proceedings of ACM CHI, pp 1709–1718

Self-Serving Well-Being: Designing Interactions for Desirable Social Outcomes

Soumitra Bhat

Abstract Well-being—individual and social—achieved through sustainable development is undoubtedly the overarching agenda for global public policy today. What could be the role of design in this new frame? The core question motivating this review is: How can design help us achieve well-being as individuals and society at the same time? The first part of this paper reviews the literature to frame the dilemma of social well-being. The second part reviews existing approaches to resolve social dilemmas. The third part reviews the current approaches to design for well-being. This is followed by a discussion about implications for design for achieving social well-being objectives, and an agenda, opportunities and key questions for research are outlined.

Keywords Social well-being · Interaction design · Inclusive development · Behavior change · Social dilemma

In the face of global inequality, rising consumerism, depleting natural resources and rising environmental threats, values of justice, conservation, sustainability and responsibility have become central issues in every discipline—economics, politics, social sciences, engineering, and design. This is matched by a rising awareness of the criticality of these issues and their causal links to the structural foundations of modern society in the minds of global citizens. Consequently, people are losing trust in the intention and capability of the institutions of old to look after people's best interests; they are losing faith in the free market due to corporations disproportionately valuing private profit over social equality and justice. Moreover, there is a sense that social problems represent collateral damage incurred in our apparently self-interested pursuit of personal good at the cost of the greater good—indicating a crisis

S. Bhat (✉)
UserINNOV Design Company, Mumbai, India
e-mail: soumitra@userinnov.com

A. Chakrabarti and R. V. Prakash (eds.), *ICoRD'13*, Lecture Notes in Mechanical Engineering, DOI: 10.1007/978-81-322-1050-4_31,

in personal values. Out of all this there is a rising sense of a collective responsibility to fix the state that we are in, leading us to reflect the things we've done and the ways we've gone about it. Well-being—individual and social—achieved through sustainable development is undoubtedly the overarching agenda for global public policy today [1]. What could be the role of design in this new frame?

The core question motivating this review is: How can design help us achieve well-being as individuals and society at the same time? The first part of this paper reviews the literature to frame the dilemma of social well-being. The second part reviews existing approaches to resolve social dilemmas. The third part reviews the current approaches to design for well-being. This is followed by a discussion about implications for design for achieving social well-being objectives, and an agenda, opportunities and key questions for research are outlined.

1 Individual Well-Being and Social Well-Being

Well-being has been the object of study in several disciplines including sociology, economics, psychology, biology, philosophy, etc. Well-being has been equated with welfare, capabilities, health related quality-of-life, hedonism, and the terms 'happiness', 'quality of life', 'well-being', 'life-satisfaction' have been used interchangeably, often confusingly so, in literature from these distinct disciplines. What has definitely emerged is that well-being is an umbrella concept under which several distinct types of well-being can be identified.

Veenhoven [2] presents a consolidation (Fig. 1) of the various types of well-being into a matrix schema as combinations of 'chances', 'outcomes', 'internal qualities' and 'external qualities', including a further distinction between the hedonistic and eudaemonist aspects of satisfaction with life [3]. If maximizing individual well-being is the ultimate goal, then it follows that an individual will choose to act in ways that will: (1) Improve the liveability of their environment; (2) Improve their individual ability to cope with life's problems; (3) Increase the utility of their life and make life meaningful; and (4a) get more pleasure out of life activities, (4b) Increase their level of satisfaction with individual dimensions of their life, (4c) Increase the frequency and duration of feelings of euphoria, (4d) Increase their satisfaction with their life as a whole. While some of these aims are achievable through individual action alone—mainly the individual, internal, passing well-being, most of the other aims require cooperative action amongst social members for successful achievement. Since these fruits of cooperative action are available to the entire society, a choice of action that maximizes individual well-being can therefore be seen to contribute towards social well-being as well. However, making this optimal choice is not easy. At a given instant, an individual has several known and unknown possibilities for action to increase their individual well-being. Not all actions will provide well-being to an equal extent. Some actions will increase well-being more than others while some actions will reduce well-being. Some actions will increase particular aspects of well-being

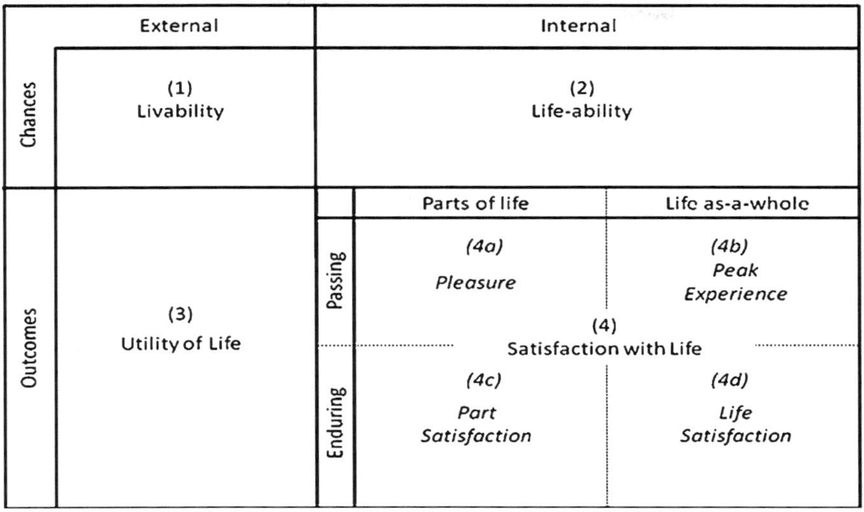

	External	Internal		
Chances	(1) Livability	(2) Life-ability		
Outcomes	(3) Utility of Life		Parts of life	Life as-a-whole
		Passing	*(4a) Pleasure*	*(4b) Peak Experience*
			(4) Satisfaction with Life	
		Enduring	*(4c) Part Satisfaction*	*(4d) Life Satisfaction*

Fig. 1 Consolidated schema of well-being. *Source* Adapted from [2]

while leaving other aspects unchanged, or even reduced. Choosing between actions leading to well-being requires judgment involving probability and value.

Studies have shown that due to limits on working memory, individuals, when faced with choices involving uncertainty and judgements of probability and value, tend to rely on highly economical heuristics, to make these judgments. These heuristics, although largely efficient and effective, lead to cognitive biases that can sometimes give rise to severe, systematic, predictable errors in decision making [4]. As the choices we face become more complicated, we are unable to always choose consistently what would be in our best interests. According to Prospect Theory [5], we often fall prey to biases where we are more averse to losing something than gaining something (loss aversion), or where we prefer gaining a certain amount now rather than gaining the same amount in the future (diminished sensitivity). Applied to choices for maximizing well-being, prospect theory would predict that people tend to favor pleasure over displeasure and, immediate satisfaction over delayed satisfaction, even in cases where this may not be the optimum choice. This bias in judgment could make an individual erroneously give more preference to short-term hedonistic outcomes over long-term social outcomes.

Given such a bias, the compatible goals of individual and social well-being come to be perceived, at times, as seemingly opposing objectives. In such cases, the problem of social well-being appears to take the form of a social dilemma—a situation where private interests are at odds with collective interests. Such social dilemmas are defined by two properties [6]: (1) Each individual receives a higher personal outcome for a non-cooperative decision no matter what the other people in the community do; (2) The entire community is better off if all or most individuals cooperate rather than act selfishly. For the collective good, cooperation needs to be sustained and defectors need to be turned into co-operators. In the context of well-being, this implies that for achieving social well-being, individuals need to overcome their short-term, hedonistic bias and consistently choose the most effective course of action that leads to individual and social well-being.

2 Theoretical Perspectives on Resolving Social Dilemmas

Social dilemmas have been the definitive characteristic of the social condition since, perhaps, the dawn of civilization. Contemporary theories and approaches to resolve social dilemmas can be categorized under three perspectives—institutional, transformational, and individual.

2.1 Institutional Perspective

From the institutional perspective, the dilemma of social well-being resembles a 'public good' or 'collective action' problem. These problems result from a lack of coordination and cooperation between individuals within society. This individual failure to coordinate and cooperate has led to the emergence of institutions to guide individual behavior and safeguard our collective interests [7]. The institutional perspective tries to resolve the coordination and cooperation problems through the design and enforcement of rule-systems that regulate individual interactions. Three types of institutional approaches can be identified: (1) centralization, (2) individualization, and (3) collectivization [8].

Centralization involves the restriction of individual access to choices that the authority believes to be antithetical to achieving collective good. There are several disadvantages to centralization as an approach: (1) individuals generally do not like impositions on the liberty and freedom to decide what is good for them, (2) the costs imposed on the non-error making individuals is high, (3) when group members have the choice of creating rules amongst themselves, they do not prefer to be ruled by a central authority, and (4) individuals prefer to be led by an elected, democratic authority that allows group members to exercise some control over the decision-making process.

Individualization involves providing individuals with a market-oriented access to all possible choices.

Individuals prefer individualized systems when compared to central authorities because the market mechanisms are considered neutral, anonymous and impersonal. Individualization works best when the outcomes of available individual choices do not affect others. However, outcomes of individual choices almost always affect others and thus individualized systems, like markets, invariably give rise to negative externalities. Another problem with the introduction of individualization is that rather than perceiving resource use as a collective problem, individual users may start to perceive it as an individual problem [9].

Collectivization involves limiting market-oriented access or limiting the choices available to members of a community through regulation by a self-organized governance regime within that community. In contrast to centralization and individualization, collectivization lays emphasis on monitoring, transparency and democratic participation in the regulation of individual choice, which leads individuals to experience the rule systems and governance regime as trustworthy, efficient and fair [10]. It gives stress on trust, reciprocity and the evolution of social norms as essential mechanisms for the emergence of cooperation in collective action. Collectivization is favorable when both, market-oriented access is desirable and negative externalities need to be managed. Collectivization is also desirable when issues faced by the community are not centrally legitimized as problems due to lack of advocacy, or lack of motivation of central policymakers to address the issues, the result of which is a lack of resources employed towards resolution of community dilemmas.

2.2 Transformational Perspective

The transformational perspective sees the social dilemma as a problem of individual choice and tries to transform an individual's perception of a specific dilemma situation by changing the decision frame and pay-off structures [11]. According to decision theory it is costly for people to make effective judgments and process full information when faced with complicated decisions especially involving uncertainty and delayed consequences [4]. Faced with a difficult choice an individual will tend to rely on heuristics rather than effortful evaluation. Transformational interventions use this inertia of individuals towards cognitive effort to nudge them towards pro-social choices by reframing decisions using strategies like setting effective defaults, expecting errors, giving feedback, structuring choices, and giving incentives [12]. These strategies are useful because of their low implementation costs, low costs on non-error making individuals, large benefits for individuals who sometimes make errors and no restriction of choices available to users. Transformational strategies have recently gained popularity as a policy tool in the form of 'Nudges'. However, there is always a possibility for abuse of paternalism by governments, firms and choice architects in positions of power. Therefore there is a need for monitoring efficacy of outcomes and transparency of implementation for asymmetric paternalism to be effective [13].

2.3 Individual Perspective

The individual perspective looks at social dilemmas as problems of individual commitment and capability. The focus of this perspective is on promoting the individual factors that motivate, facilitate and empower individuals to act pro-socially. Several theories, predominantly from social psychology, have explored person-level factors that contribute towards pro-social behavior. Whether an individual will exhibit pro-social behaviour in dilemma situations has been linked to the individual's social value orientation—or the orientation of an individual to maximize both: joint outcomes and equality in outcomes [14]. Pro-social behaviour by an individual is mediated by (1) a high level of commitment to act towards socially desirable outcomes and (2) pre-existing cooperative value orientation (rather than a competitive or an individual one) [15]. Motivation to cooperate is also mediated by needs for belongingness and identity. It is believed that human beings have a basic desire to develop and foster meaningful social relationships and, via this, build up a shared social identity. When these needs are unfulfilled, for example when people are forced to leave a social group, their mental and physical well-being suffers. This might explain why people are motivated to cooperate in groups that they feel they belong to. Interventions that develop and maintaining strong community ties, or build social capital, promote feelings of belongingness and identity [16]. Studies have also shown that individual subjective well-being and ecologically responsible behaviour were positively correlated to an intrinsic value orientation, dispositional mindfulness, and a lifestyle of voluntary simplicity [17]. However, commitment and pro-social value orientations are not enough. Commitment for doing something needs to be backed by the availability of appropriate substantive freedom and capability without which a person cannot be responsible for doing that thing [18]. This view is integrated in the theory of planned behavior which views human action to be influenced by three factors: a favorable or unfavorable evaluation of the behavior (attitude); perceived social pressure to perform or not perform the behavior (social norm); and perceived capability to perform the behavior (self-efficacy). In general the more favorable the attitude and behavioral norm, and the greater the individual's perceived control, the stronger should be the individual's intention to perform the behavior [19].

3 Design Approaches Towards Social Well-Being

It was perhaps during the 1960s that a serious discourse emerged regarding the need for the implicit power of designers, to shape society through the intended and unintended consequences of their designs, to be backed by a responsibility to use this power for the social good [20]. Since then, the idea of social good as an objective of design has gathered momentum and several recent design efforts can be noted with objectives like (design for:) happiness, socially responsible behavior, sustainable consumption, human development, and social impact. There are also several recent studies that have explicitly targeted well-being or happiness as their objective, however, these either relate to health related well-being [21], psychological well-being [22] or individual happiness [23]—and do not look at well-being as a complex whole. Both, these efforts and studies, are complemented by theoretical and methodical developments within several contemporary design approaches. Although no studies as yet describe the direct application of these approaches to social well-being problems, they are categorized here according to their focus—institutional, transformational, or individual—and potential significance to resolving social dilemmas. The institutional perspective includes approaches such as computer-supported-cooperative-work (CSCW) design [24], socio-technical-systems design [25], and participatory design [26]—aimed at increasing coordination and cooperation. The transformational perspective includes emotional design [27], experience design [28], persuasive design [29] and design with intent [30]—approaches that use the affective and meditational qualities of artifacts to reframe choices and change pay-off structures. The individual perspective includes instructional design [31] and interventions from the theory of planned behavior [32]—approaches drawing on social psychology to enhance individual commitment and capabilities. A recent review posits that the choice between these perspectives to formulate an effective strategy should be dependent upon contextual factors and the specific nature of the social dilemma [33].

Several points can be observed from a review of the current landscape of design for social good: (1) the perspectives on social good target behavior change as a strategy to resolve social problems—a shift from designing artifacts to designing interactions; (2) although social well-being is linked to social consumption and sustainable development, it is not identical. An exclusive perspective on design for social well-being is therefore lacking; (3) cases explicitly targeting well-being as an objective are few, and target a particular aspect of well-being and not well-being as a whole or social well-being; (4) the explicit perspectives on design for social good are fairly recent, with few case studies available and, therefore, the development of a specific supporting body of knowledge is still in the early stages; and (5) there are several approaches that can potentially support a design for social well-being perspective, however, specific explorations in this area are lacking.

4 Towards Designing Interactions for Self-Serving Social Well-Being

A social design approach puts interactions, rather than artifacts, at the center as the 'things to be designed'. In most social design approaches the starting point for framing the problem situation involves the notion that individual interests are *absolutely antithetical* to collective interests. This may be the case in tragedy-of-the-commons type of situations where desirable social outcomes are primarily equated with an equitable distribution of finite resources—for example, in sustainable consumption. In such cases, individuals would certainly be *materially* better off when they defect than when they cooperate. However, an exploration of the complex nature of individual well-being presents the idea that individual well-being and social well-being as intrinsically linked and positively correlated goals. Therefore, if social well-being is the desirable social outcome, then individual and collective interests are only *apparently antithetical* due to limitations of individual bounded rationality. The distinction is crucial in informing the designer's perception of social members as either helpless agents hopelessly trapped in a dilemma until saved through coercion, or as empowered agents who have the possibility—perhaps with a little education, persuasion and facilitation—to transcend their limitations and achieve individual and social well-being at the same time [34].

Although presented in this review as three separate perspectives for resolving social dilemmas—institutional, transformational and individual—the question is not one of choosing between coercion and persuasion. Any successful intervention strategy would intersect across these three perspectives. For a dilemma, apparent or absolute, to be resolved, there must be some co-operators in a group to begin with, there must be institutions that facilitate cooperation and coordination, there must be transformational solutions that reduce the temptation to defect, and finally, the numbers of co-operators must be increased by turning defectors into co-operators. In the case of social well-being, the ideal situation would occur when all the members of a collective act towards achieving individual well-being, as it is rightly understood, and thus, serve themselves with individual and social well-being. The key question then for design is: How can design support people in serving themselves with individual and social well-being?

Current case studies, under a broader topic of social design, have identified several open challenges such as: (1) defining the appropriate role, qualities and expertise of designers in promoting socially responsible behavior [33]; (2) identifying the appropriate problem frame for the successful resolution of social well-being problems [35]; (3) using dialog between designers for the effective management and resolution of dilemmas [36]; (4) facilitating coordination by building a shared understanding and shared commitment between diverse stakeholders towards resolving dilemmas [37]; and (5) taking a whole-system, long-term view and changing culturally dominant worldviews and value systems [38]. At the same time, there are potentially complimentary developments in collective intelligence and coordination theory [39] with ongoing experiments in using information technology to coordinate thousands of diverse stakeholders and lead to high quality collective decisions [40]. Fields like graph theory, evolutionary game theory [41] and behavioral game theory [42] are providing new insights into how and under what conditions bounded rational agents interact and form social networks and communities that display emergence of trust, fairness and reciprocity. Game theoretic approaches are not only interesting because of the insights they provide, but also because of the tools and methods they use for modeling complex systems, simulation and experimental design. Adopting a complex systems perspective to design for achieving well-being objectives seems to present several advantages due to its integrative nature, consideration for the autonomy of individual agents, and the inherent property of complex adaptive systems to self-organize and adapt that is desirable from a long-term, sustainability perspective [43]. Under the topic of sustainable business models, there is increasing interest in cooperative business models for their demonstrated resilience in times of crisis [44].

These issues and developments bring up concerns and possibilities that are directly relevant for design with a social well-being objective, although, systematic investigations under an explicit 'design for social well-being' perspective are lacking. Several key questions for further research can be identified: (1) How can designers build coordination and facilitate cooperation between thousands of stakeholder participants with diverse interests in an environment of fairness, transparency and trust?; (2) what are the qualities and expertise needed for a designer to be successful in designing for social well-being; and how can these be developed?; (3) what are the tools and methods that a designer can use to model dilemma situations for rapid simulation and selection between alternative intervention strategies?; (4) how can designers evaluate the short-term and long-term impact of outcomes from interventions across interdependent groups?; and (5) what are the organizational conditions, including business models, which can support design for social well-being? How can the activity of design for social well-being be organized and managed from a practical perspective?

5 Conclusion

Individual well-being is a socially interdependent construct within which individual and social goals are intrinsically linked and positively correlated. However, bounded rationality challenges individuals, while making the optimal choice between actions to maximize individual well-being, by giving the decision an appearance of a social dilemma. Derived from social psychological theories, prospect theory, and institutional theories, the individual, transformational and institutional perspectives to resolving social dilemmas outline a potential framework to support the design and analysis of interventions for achieving social well-being objectives. Although several current approaches under a generic 'design for social good' agenda can be identified, an explicit 'design for social well-being' approach is nascent and far from consolidated. There is a need for research across all formats: theoretical, case study, research through design, systematic experimentation. What has emerged so far is that an explicit focus on social well-being as an objective of design has the potential to broaden the field of design to incorporate a palette of multi-disciplinary perspectives and develop a shared vocabulary to engage in a global dialog on development. Designing for social well-being has important implications for all aspects of design: theory, expertise, methods, roles, embodiments, management and assessment. To design for social well-being, designers need to embrace complexity, share a long-term commitment towards positive change, and be aware of and manage their own bounded rational frames by actively seeking dialog and participation. In the new frame of global public policy that puts social well-being at the center of human development, design can support global efforts by supporting individuals to serve themselves with social well-being.

References

1. Stiglitz J, Sen A, Fitoussi JP (2009) The measurement of economic performance and social progress revisited. Observatoire Français des Conjonctures Economiques (OFCE). The ESPON, 2013
2. Veenhoven R (2010) Capability and happiness: conceptual difference and reality links. J Socio-Econ 39(3):344–350
3. Deci EL, Ryan RM (2008) Hedonia, eudaimonia, and well-being: an introduction. J Happiness Stud 9(1):1–11
4. Tversky A, Kahneman D (1974) Judgment under uncertainty: heuristics and biases. Science 185(4157):1124–1131
5. Tversky A, Kahneman D (1992) Advances in prospect theory: cumulative representation of uncertainty. J Risk Uncertainty 5(4):297–323
6. Dawes RM (1980) Social dilemmas. Annu Rev Psychol 31(1):169–193
7. Calvert R (1995) The rational choice theory of social institutions: cooperation, coordination, and communication. Mod Polit Econ: Old Top New Dir Polit Econ Inst Decis, Cambridge University Press 216–268
8. Van Vugt M (2002) Central, individual, or collective control? Am Behav Sci 45(5):783–800
9. Tenbrunsel AE, Messick DM (1999) Sanctioning systems, decision frames, and cooperation. Adm Sci Q 44(4):684–707
10. Ostrom E (2010) Beyond markets and states: polycentric governance of complex economic systems. Am Econ Rev 100(3):641–672
11. Tenbrunsel AE, Northcraft GB (2009) In the eye of the beholder: Payoff structures and decision frames in social dilemmas. Soc Decis Mak: Soc Dilemmas, Soc Val, Ethical Judg, Ser Organ Manag, Psychology Press 95–115
12. Camerer C, Issacharoff S, Loewenstein G, O'Donoghue T, Rabin M (2003) Regulation for conservatives: behavioral economics and the case for 'Asymmetric Paternalism'. Univ PA Law Rev 151(3):1211–1254
13. Thaler RH, Sunstein CR (2008) Nudge: improving decisions about health, wealth, and happiness. Yale University Press, USA
14. Van Lange PAM (1999) The pursuit of joint outcomes and equality in outcomes: an integrative model of social value orientation. J Pers Soc Psychol 77(2):337
15. Van Lange PAM, Agnew CR, Harinck F, Steemers GEM (1997) From game theory to real life: how social value orientation affects willingness to sacrifice in ongoing close relationships. J Pers Soc Psychol 73(6):1330
16. Forrest R, Kearns A (2001) Social cohesion, social capital and the neighbourhood. Urban Stud 38(12):2125–2143. doi:10.1080/00420980120087081
17. Brown KW, Kasser T (2005) Are psychological and ecological well-being compatible? The role of values, mindfulness, and lifestyle. Soc Indic Res 74(2):349–368
18. Sen AK (1990) Individual freedom as social commitment. India Int Centre Q 17(1):101–115
19. Ajzen I (2005) Attitudes, personality, and behavior. McGraw-Hill International, New York
20. Papanek V, Fuller RB (1972) Design for the real world. Thames and Hudson, London
21. Larsson A, Larsson T, Leifer L, Van der Loos M, Feland J (2005) Design for Wellbeing: Innovations for People. ICED 05. In: 15th International conference on engineering design: engineering design and the global economy. Engineers Australia, p 2509
22. Coyle D, Linehan C, Tang K, Lindley S (2012) Interaction design and emotional wellbeing. In: Proceedings of the 2012 ACM annual conference extended abstracts on human factors in computing systems extended abstracts, ACM, pp 2775–2778
23. Desmet PMA (2011) Design for happiness; four ingredients for designing meaningful activities. In: Proceedings of the IASDR2011, the 4th world conference on design research, vol 31

24. Bødker S, Bouvin NO, Wulf V, Ciolfi L, Lutters W (2011) ECSCW 2011. In: Proceedings of the 12th European conference on computer supported cooperative work, Springer, Aarhus Denmark, 24–28 Sept 2011
25. Whitworth B, moor AD (2003) Legitimate by design: towards trusted socio-technical systems. Behav Inf Technol 22(1):31–51
26. Bodker S, Pekkola S (2010) A short review to the past and present of participatory design. Scand J Inf Syst 22(1):45–48
27. Desmet PMA, Porcelijn R, Van Dijk MB (2007) Emotional design; application of a research-based design approach. Knowl Technol Policy 20(3):141–155
28. Hassenzahl M (2010) Experience Design: technology for all the right reasons. Synth Lect Human-Centered Inf 3(1):1–95
29. Redström J (2006) Persuasive design: fringes and foundations. In: Proceedings of the First international conference on Persuasive technology for human well-being, Springer-Verlag 112–122
30. Lockton D, Harrison D, Stanton N (2008) Design with Intent: Persuasive Technology in a Wider Context. In: Proceedings of the 3rd international conference on Persuasive Technology, Springer-Verlag pp. 274–278
31. Schunk DH, Zimmerman BJ (2012) Motivation and self-regulated learning: theory, research, and applications. Routledge, London
32. Ajzen I (2011) Design and evaluation guided by the theory of planned behavior. Soc psychol Eval, Guilford Publications, pp. 74–100
33. Tromp N, Hekkert P, Verbeek PP (2011) Design for socially responsible behavior: a classification of influence based on intended user experience. Des Issues 27(3):3–19
34. Ostrom E (1990) Governing the commons: the evolution of institutions for collective action. Cambridge University Press, Cambridge
35. Dorst K, Tietz C (2011) Design thinking and analysis a case study in design for social wellbeing. Res Des-Support Sustain Prod Dev (ICoRD'11) 104–111
36. Oak A (2012) 'You can argue it two ways': the collaborative management of a design dilemma. Design Stud 33(6):630–648. doi: 10.1016/j.destud.2012.06.006
37. Conklin J (2005) Dialogue mapping: building shared understanding of wicked problems. Wiley, New York
38. Wahl DC, Baxter S (2008) The designer's role in facilitating sustainable solutions. Des Issues 24(2):72–83
39. Malone T, Laubacher R, Dellarocas C (2009) Harnessing crowds: mapping the genome of collective intelligence. SSRN eLibrary
40. Introne J, Laubacher R, Olson G, Malone T (2011) The Climate CoLab: large scale model-based collaborative planning. IEEE 2011 international conference on collaboration technologies and systems (CTS), pp 40–47
41. Szabó G, Fáth G (2007) Evolutionary games on graphs. Phys Rep 446(4):97–216
42. Camerer CF (2011) Behavioral game theory: experiments in strategic interaction. Princeton University Press, Princeton
43. Folke C, Carpenter S, Elmqvist T, Gunderson L, Holling CS, Walker B (2002) Resilience and sustainable development: building adaptive capacity in a world of transformations. AMBIO: J Human Environ 31(5):437–440
44. Birchall J, Ketilson LH (2009) Resilience of the cooperative business model in times of crisis

Do We Really Need Traditional Usability Lab for UX Practice?

Anshuman Sharma

Abstract Many IT companies in India and across the globe have "User experience" (UX) associates to support UI and Usability. Many of these companies have a well organized UX practice with their own targets, profit and loss statements. Some of the companies boast of a having support of extensive Usability Labs and user testing infrastructure. The billing of usability testing services is much higher than the UI design and usability review services. The IT services companies have a model of offshore development for cost reduction and many usability labs are underutilized. The investment to set-up and run a usability lab is made with the objective of winning additional business and show more value to clients. There are a very few projects which really need detailed UX participation, user research and user testing through usability labs. Usability labs are expensive to establish and maintain. This brings us to the questions of ROI and whether the high cost of usability lab infrastructure is justified. This paper looks at the various aspects and business models which can help an IT company decide whether to set-up a usability lab or not and what can be an ideal model to sustain it.

Keywords HCI · User experience design · Usability practice · Usability testing · User testing · User research · Usability lab · Client lab · Usability process

A. Sharma (✉)
User Experience Design, L&T Infotech, Whitefield, Bangalore 560066, India
e-mail: anshuman.sharma@lntinfotech.com; anshusa@yahoo.com

A. Chakrabarti and R. V. Prakash (eds.), *ICoRD'13*, Lecture Notes in Mechanical Engineering, DOI: 10.1007/978-81-322-1050-4_32,

1 Introduction

User Experience Design as a practice has evolved over the last couple of decades. Businesses and clients have started realizing the value of "User Experience" (UX) practice and the contribution that a usability lab can make.

IT business management is usually reluctant to bear the cost of setting-up UX practice and majority of the expense is spent on setting up the usability lab infrastructure. The main reason could be attributed to the business model followed by IT companies of product or services businesses, where usability is still in its nascent stages [1].

The "Return On Investment" (ROI) of usability labs and what value it brings to the business and clients is still not fully visible. The penetration of usability for leading application development companies is ever increasing. One reason is the importance given to usability practice by clients. Another reason to build usability lab infrastructure by the software development companies is to show their intention to make usable products [2].

Many companies believe that usability labs are expensive and that they are a place to garner only scientifically valid data [2]. This is not entirely true as the labs evolve qualitative data as well. Usability tests conducted through Usability Lab helps the usability professionals and Product development teams to gather data about the interaction of the user with the designed application/software. This data is evolved by observing the user/participant interact with the design [3].

1.1 What is UX?

User experience in ISO FDIS 9241-210 is defined as: "A person's perceptions and responses that result from the use and/or anticipated use of a product, system or service".

Usability or user experience can be measured during or after the use of a product, system or service. The ISO definition suggests measures of user experience are similar to measures of satisfaction in usability [4]. This definition was subsequently refined to evolve the definition of Usability.

User experience or UX in short is an umbrella term coined to collectively represent visual designers, interaction designer, usability analysts and front-end engineers. The UX stream looks the software application design and its usage. The responsibility of UX team is to look at the aesthetics of the application, ease of use and its intuitiveness.

The UX team needs to work closely with the business analysts to review whether the business needs and the user needs are being met. The UX team also needs to work closely with the technology teams to get the designs and recommendations implemented. The job of a UX designer is challenging as he needs to meet business requirements, look at technological constraints and propose designs that meet the end user requirements.

1.2 How to Set-Up Usability Lab?

A Usability Lab usually consists of two rooms, the participant's test area also known as 'Participant's room' and the participant viewing area also known as 'Observation room'. Conventionally, these areas are separated by a one-way mirror (Fig. 1). Usability analysts sit in the 'Observation room' and the participants sit in the 'Participant's room'. Usability analysts can see the participants working on the application from their computer screens, while the participants cannot see or hear the usability analysts (Fig. 2).

The desktops of the usability analysts are linked to the desktops of the participants. Sound insulation and one-way mirror helps the participant work without being disturbed by the discussions of the usability analysts in the 'Observation room'. Web camera installed on the participant's desktops capture the facial expressions of the participants and the same can be viewed real time by the usability analysts. The captured video is stored for analysis.

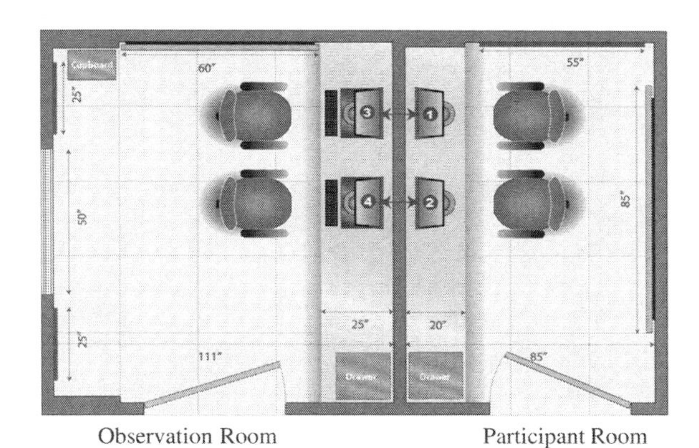

Fig. 1 Plan of usability lab set-up for testing two participants

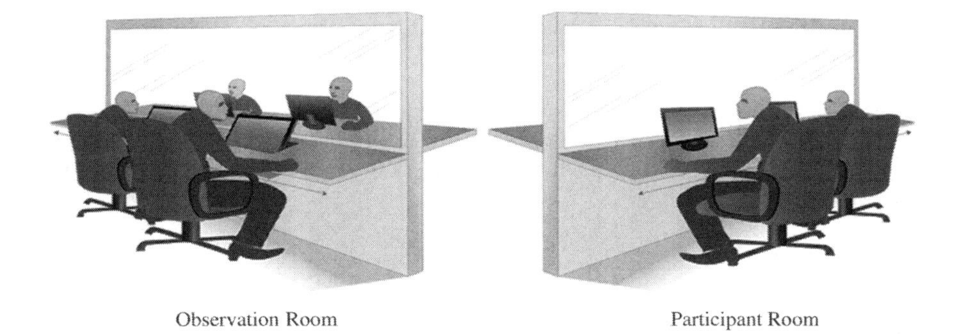

Fig. 2 Elevation of usability lab set-up for testing two participants

2 Questions to be Answered

Before setting up the Usability lab, most companies deal with numerous questions regarding its need and viability of the venture. Most usability practitioners and managers vouch for the need of a usability practice. Usability practitioners and managers would have answered questions related to usability practice and justification of having a usability lab. Most practitioners dig into research papers and articles to retrieve business justifications and ROI of usability lab infrastructure. Lets us look at some of the most frequently asked questions related to usability practice and usability lab infrastructure:

- Why do we need a usability lab?
- What are the benefits of having a usability lab?
- What do we get out of a usability lab as a deliverable?
- Which organizations should have it?
- Will client pay for usability Lab usage?
- Why we need usability lab usage when we are not a research organization?
- Are there cases when usability lab is not required?
- Are there scenarios when clients do not want the services of a usability lab?
- Can we use of some makeshift usability testing techniques and still show value to clients?

This paper will look at an IT company's business models, business focus and maturity of processes to determine if the conditions are ripe to set-up usability lab or not. One need not elaborate the benefits of a usability lab but in its absence, one needs to find ways and means to conduct usability tests and research to help UX practitioners produce user-friendly and innovative solutions. There are tools and techniques which can be used to conduct usability tests and measurements in absence of a usability lab.

Let us look at the factors which help decide the need of a Usability lab infrastructure.

3 Factors Governing Usability Lab Set-Up

There are several organizational aspects that play an important role in determining how the UX practice shapes-up and whether approval will be received to set-up usability lab infrastructure or not. Usability practice needs blessings from the senior management to stand on its own feet [1].

Lets us look at the business models and some of factors that can determine usability lab set-up within each business model.

3.1 IT Business Model: Product Companies

Most product development companies have usability labs to conduct usability tests for applications. With application customizations and new feature development, the bucket is usually full for the usability analysts. The product development companies also have the budgets and plan in place to account for usability inputs and recommendations. The prevalence of well-organized Usability labs can be seen in those product development companies where there is reasonable understanding about the importance of usability processes and its benefits.

Another important factor is the maturity of the companies with respect to having an integrated "Product Development Lifecycle" (PDLC). There are a very few companies in the Indian IT sector, where the "User Centered Design" (UCD) process has been successfully integrated with PDLC [1]. In these companies, we can see the maximum benefits of an integrated process where usability is a part of the overall product development process.

The first task for the usability practitioners is to sell usability to the senior management to get their attention and approvals to set-up usability lab. In most cases, it does not happen in the initial years of setting-up the usability or user experience design practice. One needs to engage with the existing processes and workflows for a couple of years to get noticed and get the budget to set-up a full-fledged usability practice where usability lab is an integrated part of the landscape.

But what does the usability practitioner do in the initial years of their practice without a usability lab? The usability practitioner needs to rely on their experience and the experience of the sales and marketing teams. There is a huge dependency on the sales, marketing and product management teams for approval of usability proposals. The main issue is that the usability teams do not have the tangible data like number of clicks and time savings without the usability lab set-up.

What does the usability team do in such scenarios? There are many possible ways to sell usability and get funding from the product owners. The most important step is to start engaging with clients and the senior management to influence the integration of usability as an integrated part of the PDLC [1].

In many IT companies, the usability practice needs the blessings of the senior management to flourish. Usability practice survives with a top down approach [1]. The main reason is the maturity of the UX practice in the IT company.

If the usability team is able to show the ROI of usability practice to the senior management and the clients, it is easier and faster to institutionalize it. There is always resistance to change and there are issues with sharing decision making powers from the product development teams with UX team [1]. Product development teams believe that the usability team is away from the real world and that they may not be able to provide solutions which need less time and effort to implement.

For small and mid level IT companies, usability lab is not seen as a very critical requirement. This is due to focus on specific areas of software development and delivery. However, for the companies which are working higher-up in the value chain of product development and IT services, usability is seen as a critical and game changing practice. We see a full-fledged usability lab with regular use in such companies.

3.2 IT Business Model: Service Companies

IT service companies have a different kind of problem and solutions as far as the usability practice and setting-up of usability lab is concerned. The work culture in IT services model is more volatile compared to product business.

IT services companies and their infrastructure set-up is driven by the business and client requirements. As far as the usability practice is concerned, the situation is very similar to product development companies. Usability practice and lab infrastructure needs support from senior management or it is mostly driven from client requirements.

However, there are some differences in service companies and product companies with respect to usability lab infrastructure. IT service companies need to sell capabilities to bid and win deals. Most IT service companies do not offer usability practice as service which would require specialized services of a usability lab. However, tier-1 and tier-2 IT companies do have usability labs and leave no stone unturned to show them off. But having usability lab infrastructure and having a functional and effective lab infrastructure are two different things.

3.3 Organizational Factors

Usability budget in an IT company is steadily growing. The budgets allocated for usability engineering was about 6 % of cost of the development projects in 1993. Based on visible benefits of usability methods to improve products, an IT company starts investing on usability practice and usability lab infrastructure [5].

The organizational structure of an IT company also impacts the growth and development of usability practice. Some IT companies distribute the usability resources into individual development projects, where as some other place the usability resources into a horizontal resource group which cuts across all the verticals/projects and development teams. A centralized usability practice benefits from the permanent lab infrastructure [5].

3.4 Usefulness of Discounted Usability Methods

When a company decides to improve the usability of its products, the immediate requirement is not to start building a usability lab. It is possible to get started with simpler usability methods, referred to as "discount usability engineering" [5].

In absence of usability lab, one needs to make use of methods like GOMS analysis and heuristic evaluation. Heuristic evaluation is used by individual UX practitioners to find usability issues and gaps. Heuristic evaluation is conducted based on established usability principles like learnability, efficiency, memorability, error prevention and satisfaction.

4 Cost Benefit Analysis: To Have or Not to Have

Many businesses and clients do not feel the need of having a usability lab. Many others do not have the budget and focus to engage the services of a usability lab. IT services companies rely on the off-shoring model where UX resources are budgeted and their efforts are tracked. The process followed by UX resources is usually not tracked.

In absence of usability lab, the designers and usability specialists start believing that they know what how the application is being used and what is required to be resolved. This gives them confidence and sometimes over-confidence that they know how to deal any kind of problem and can come up with the solution to meet end user needs. The designs produced out of this quick and dirty UCD process looks good but may not be effective when it comes to intuitiveness and ease of use.

Table 1 compares several parameters to evaluate the need for a Usability lab and what is its impact on business and clients. The study is based on the experience and feedback from ten UX professional over a period of last 10 years (2000–2011). The UX professionals have experience in various IT companies which range from start-ups, mid-tier and the top IT companies in India.

5 UCD Process: How Much Can We Achieve Without Usability Lab Set-Up?

The importance of usability lab for UX practice is well accepted by the academic circles and research levels, what we need to discuss specifically is how much can we achieve for a UX practice without having a usability lab in a business set-up. It can also be stressed that to evaluate the specific usability oriented data, whether it can be collected without a usability lab. Whether the data is collected through a usability lab or not, the target is to analyze the collected data and design/redesign a task in way that it becomes more effective [6].

Table 1 Usability lab: to have or not to have

Parameters	Without usability lab	With usability lab
UX services offered	80 %	100 %
Client impact	Less	High
Overall business impact	Low	High
IT product company	Low impact	High impact
IT services company	Low impact	Very high impact
ROI for delivery acceptance	Limited impact	Very high impact
ROI for business development	Limited impact	Very high impact
Customer focus	Less	High
Learning and innovation	Limited	High
Data collection and research	Limited	High
Initial set-up cost	Less	Very high
Incremental overhead user testing cost	High	Negligible cost
Running cost	Less	High
Ease of hiring and talent retention	Low contribution	High contribution
Satisfaction levels of UX team	Limited	High
Learning and customer focus for UX team	Limited	High
Learning and customer focus for product management	Limited	Very high

Various usability metrics are employed to measure usability dimensions like effectiveness, efficiency and satisfaction [7]. These dimensions are also pointed out in the ISO 9241.11 standards. The four important metrics which represent these dimensions are: task completion, error counts, task times and satisfaction scores [8].

Usability tests are devised to collect data on the usability metrics. A formal usability lab set-up is useful but not necessary. The effectiveness of the usability tests depends on the chosen tasks, the methodology and the person in-charge of the test [9].

Many researchers argue the suitability of results achieved through field study versus usability lab. The data collected through detailed usability lab set-up and through field studies differ in quality and focus. One major factor in favor of the usability labs is that the environment in usability lab is peaceful space where the participants can focus on given tasks [10]. There are interruptions, movement and noise issues in field studies [11].

The only benefit of field studies is to get understanding of real working conditions of the end user which may not be achievable in a usability lab set-up. Steve Krug has compared the cost saving benefits of "Lot-Our-Lease Testing" [12] with traditional testing. He suggests that one can achieve equivalent results with 3–4 participants in an office or conference room set-up compared to 7–8 participants in a usability lab.

Table 2 compares various usability parameters to evaluate if we can live without a usability lab and still gather all the relevant data required for understanding users and design better software. The study is based on the experience and feedback from ten UX professionals over a period of last 10 years (2001–2011).

Table 2 How much can we achieve without usability lab set-up?

Usability parameter	Without usability lab	With usability lab
User study and persona development	Yes	NA
Scenario building	Yes	NA
Competitor analysis	Yes	NA
Task analysis	Partially	Yes
Study of existing application usage and issues	Partially	Yes
User behavior and need analysis	Yes	Yes
Heuristic evaluation	Yes	NA
Iterative testing	Partially	Yes
User testing	Partially	Yes
Mouse and keyboard click analysis	Partially	Yes
Mental load analysis	Partially	Yes
Idle time analysis	Partially	Yes
Eye tracking analysis	No	Yes
Facial expression and satisfaction/ frustration levels	Partially	Yes
Time taken for click and satisfaction level analysis	Partially. Cumbersome	Yes
Preparation time for usability report	Very high	Low
Participants fear and anxiety toward hidden observation areas	Low. Since there is not partition and the observers can sit along with the participants and the participants can see what is happening during the user test.	High. One needs to spend time and effort to make them comfortable
Flexibility and portability	Can set-up a makeshift lab easily at client location	Cannot take the lab to client location
Controlling observers	We need to brief the observers not to talk and disturb the participants	It is easy to handle many observers at the same time due to partition and sound proofing
Time and effort	Less time and effort is required to quickly gauge the issues and user feedback. This is because there is less data to be analyzed.	Lot of time is required to analyze the data collected in the form of video recordings. One needs to derive the clicks, time taken and so on. This time can be reduced with specialized usability testing software

(continued)

Table 2 (continued)

Usability parameter	Without usability lab	With usability lab
Data collected	Less data is generated. Suitable for development projects	More data is generated. Suitable for research activities
Range of analysis and results	Most details like issues and pain points can be discovered	All the details and issues can be identified and analyzed
Measuring end user satisfaction	Partially	Yes
Data/video availability for reference and training	Data not readily available	Data readily available in the form of video

6 Conclusions

Usability lab set-up helps the usability team learn more about user behavior and his preferences. Usability labs are expensive to establish and maintain. Factors like maturity of usability processes and integration of UCD with PDLC in an organization, type of projects taken and delivered, user focus and budget allocation determines whether to set-up a usability lab or not.

A usability lab can be used as a common platform where various levels of researches can be carried out with representations from various industries along with academician-researchers. This arrangement can even be made for a specific context and requirement. The data collected can be used by the concerned stake holders.

With the ever changing IT working and delivery models, the relevance of a traditional usability lab is slowly fading away and giving way to a portable lab. Usability professionals have also realized over a period of time that there are other ways and means to conduct usability tests which do not demand high investment and do not take too much time in data collection and analysis. Guerrilla and discount usability methods are quite popular, but we do not need to measure usability to improve it [13].

For product development organizations, traditional usability labs do make sense. For IT service focused organizations, a make-shift or a portable usability lab will be very useful. Usability professionals are also trying out discount usability testing methods which yield similar results with minimal effort, cost and time. When a company decides to improve the usability of its products, the immediate requirement is not to start building a usability lab, but to start using 'discount usability engineering' methods [5].

Acknowledgments I would like to thank Anish Shah and Kiran Pai for enabling the UX practice and providing an environment of freedom for UX research activities. Thanks to Vinayak Hegde for the images. I would extend my thanks to Aneesha Sharma for helping me with insights into research process.

References

1. Sharma A (2011) Institutionalization of usability in banking software environment: tasks and challenges. In: International conference on research into design (ICoRD11), ISBN: 978-981-08-7721-7, 2011
2. Tara S (1999) Usability labs: our take, user interface engineering. http://www.uie.com/articles/usability_labs/. July 1999
3. Spool Jared M (2011) Bending the protocols: useful variations on usability tests. User interface engineering. http://www.uie.com/articles/bending_protocals/. Dec 2011
4. Bevan N What is the difference between the purpose of usability and user experience evaluation methods? www.nigelbevan.com
5. Nielsen J (1994) Usability laboratories: a 1994 survey. J Behav Inf Technol 13:1–2
6. Bevan N, Kirakowski J, Maissel J (1991) What is usability? 4th international conference on HCI, Stuttgart, Sept 1991
7. Ergonomic requirements for office work with visual display terminals (VDTs)—Part II: guidance on usability, ISO 9241-11:1998 (E), Geneva, 1998
8. Sauro J, Kindlund E (2005) A method to standardize usability metrics into a single score. In: CHI 2005, USA, 2–7 April 2005
9. Molich R, Ede M, Kaasgaard K, Karyukin B (2004) Comparative usability evaluation. Behav Inf Technol
10. Kaikkonen A (2005) Usability testing of mobile applications: a comparison between laboratory and field testing. J Usability Stud 1(1):4–16
11. Tamminen S, Oulasvirta A, Toiskallio K, Kankainen A (2004) Understanding mobile contexts. Spec Issue J Pers Ubiquitous Comput 8:135–143
12. Krug S (2005) Don't make me think: a common sense approach to web usability, 2nd edn. New Riders, USA
13. Nielsen J (2006) Quantitative studies: how many users to test? Jakob Nielsen's Alertbox. http://www.useit.com/alertbox/quantitative_testing.html. 26 June 2006

Muscle Computer Interface: A Review

Anirban Chowdhury, Rithvik Ramadas and Sougata Karmakar

Abstract A new kind of human computer interface using electrical activity of muscles, known as Muscle computer Interface (muCI) has been developed by researchers. With an intention of unfolding current status of muCIs, original research articles, review articles, reports, books, news etc. from authentic printed and online sources involving different search engines and libraries have been searched and critically studied by authors of present paper. This review has successfully highlighted developments of different sEMG based interfaces such as hand movement/gesture recognition interfaces, facial gesture recognition interfaces, myoelectric prosthetic arms, muscle fatigue analysis and other sort of interfaces. It has also covered the comparison between muCI and BCI, methodologies used for signal classification for muCI and the various shortcomings of the current muCIs. As muCI is still at its initial stages of development, it has been envisaged by the authors that present paper would help researchers to explore new ideas in emerging areas of muCI's.

Keywords Human computer interface · Brain computer interface · Muscle computer interface · Signals processing

A. Chowdhury · R. Ramadas · S. Karmakar (✉)
Department of Design, Indian Institute of Technology Guwahati, Assam, Guwahati 781039, India
e-mail: karmakar.sougata@gmail.com

A. Chowdhury
e-mail: chowdhuryanirban14@gmail.com

R. Ramadas
e-mail: rithvik.firebird@gmail.com

A. Chakrabarti and R. V. Prakash (eds.), *ICoRD'13*, Lecture Notes in Mechanical Engineering, DOI: 10.1007/978-81-322-1050-4_33,

1 Introduction

Many human computer interaction (HCI) technologies are now well established. Among different developmental phases of HCI, researches on brain computer interaction (BCI) are very popular. Brain computer interfaces are very much helpful to overcome many human limitations such as assisting, augmenting or repairing human cognitive or sensory-motor functions [1–3]. In recent years a new means of HCI is focused on Muscle Computer Interface (muCIs). Muscle computer interface is a user interface in which user employs electrical activity of their muscle as an input while they are executing various tasks. In other words in such interaction people may control devices using their myoelectrical signals. Though the term Muscle Computer Interface is relatively new, usage of myoelectric devices employing sEMG electrodes has long history. From the past, EMG signals have been successfully used in several fields such as medicine [4, 5], understanding neurophysiological disorders (e.g. neuropathy, myopathy etc.) [1], development of prosthetic arm [6, 7], fatigue detection devices [8–10] etc. Only few articles state information about muCI in details. On contrary, there is a lot of information available on subjects which are related to muCI like electromyography and muscle signal processing [6, 7, 11–22], HCI [1, 23], BCI [1–3, 16] etc.

In the present paper an attempt has been made to review the current status of muCI for understanding EMG signals and muCI, differences between muCI and BCI, as well as the future direction of muCI development.

2 Methodology for the Review

Authors of this paper have gone through several research articles, review papers, books and book chapters from various authentic search engines with the help of internet as well as institute library. The search engines used for this present review include Google, Google Scholar, PubMed-NCBI, ACM digital library, IEEE Xplore and other digital libraries. Following thorough study of the available literatures, analysis was made and findings were reported systematically. Apart from this, authors have tried to build connections between muCI and other related fields and have been able to give some state of the art ideas for future generation of muCI.

3 Development of muCI

Although the concept of muCI is very recent, it is actually a further improvement of HCI rather it is the general physical transducer (e.g. mouse, keyboard etc.) mediated HCI. The term muCI is relatively new to many of us but the usage of sEMG signals for developing devices is well known. Previously, EMG signals have been used for development of prosthetic arms [6, 7, 24–27]. The term muCI was first coined by Saponas et al. [28] while demonstrating the feasibility of muCI using forearm electromyography. According to them [28], muCIs are an "Interaction methodology that directly senses and decodes human muscular activity rather than relying on physical device actuation or user actions that are externally visible or audible". They were able to differentiate and classify EMG signals for change in different positions of fingers and pressure of finger presses, as well as they classified tapping and lifting gestures across all five fingers. Moreover they have concluded their work for future muCI designs [28]. Recent muCI systems include EMG sensors which are placed over muscles of interest related to a particular movement of body parts or gestures. There are varying number of EMG sensors which can be used for capturing muscle signals to develop muCI e.g. 6 sensors [10] 8 sensors [29] or 10 sensors [28] etc. Based on their previous work Saponas et al. [29] classified finger gestures on a physical surface. They introduced a bi-manual paradigm that enables use in interactive systems. Their experimental results demonstrated the classification accuracies of four-finger averaging 79 % for pinching, 85 % while holding a travel mug, and 88 % when carrying a weighted bag. Additionally, they showed feasibility of generalization across different arm postures with exploring the tradeoffs of providing real-time visual feedback. After that authors have studied and presented methods of making these interfaces more human friendly, cost effective and developing arm bands which are easy to wear without the adhesive gel. In addition, Saponas et al. reported about more practical applications of muCI in which system needed not be calibrated over and over again when removed and worn [30]. They have reported that accuracy existed in the system of about 86 % for activities like

Table 1 Different EMG signal classification methodologies and their accuracies

S.no	Methodology used	Author	% Accuracy
1	Artificial neural network (ANN)	Putnam et al. [44]	95
		Rosenberg et al. [45]	14
		Tsenov et al. [21]	98
		Jung et al. [46]	78
2	Back propogation neural network (BPNN)	Itou et al. [47]	70
		Naik et al. [17, 34], El-Daydamony et al. [12]	97
3	Log linearized gaussian mixture network (LLGMN)	Tsuji et al. [48], Fukuda et al. [49]	Not mentioned
4	Recurrent LLGMN	Tsuji et al. [50], Fukuda et al. [51]	Not mentioned
5	Fuzzy mean max neural network	Kim et al. [52]	97
6	Hidden markov model (HMM)	Wheeler [36], Chan et al. [53]	Not mentioned
7	Bayes network	Chen et al. [33], Kim et al. [15]	94

Table 2 Comparisons between BCI and muCI

S·no	BCI	S·no	muCI
1	Computer is controlled via signals sent from the brain directly. Signals are acquired as EEG signals, MRI signals etc.	1	Computer is controlled via signals received from the muscles during various activities. muCI only involves only EMG signals
2	It is generally non invasive. Electrodes used are generally specialized for EEG	2	It can be invasive as well as non invasive and EMG electrodes (with 2 to 10 channels) are used
3	Some applications of BCI include BCI based gaming interface, device interface for paralytic patients etc.	3	Some of the application of muCI includes gesture based gaming interface, prosthetics etc.

pinching across three fingers. Whatever, basic steps that need to be followed for development of muCI are: (1) Detection of EMG signals for a particular body movement; (2) classification of EMG signals and develop machine learning algorithm; and (3) use those EMG signals of a particular body movement as an input for controlling any devices. Among these steps EMG signal classification is most important.

4 EMG Signal Classification

People use EMG signal classification method for several purposes such as development of myoelectric hands [31, 32], gesture recognition interfaces [15, 17, 23, 33–37] (e.g. face and hand), blink recognition interfaces etc. Even there are various methods for EMG signal recognition process which range from ANN to Bayes Networks. All these methods have different accuracy levels. Signal classification is very crucial in the development of muCI as the machine is only able to understand/recognize right signals/signal of interest from the muscles after signal classification. Details of the various methods and their corresponding accuracy levels are furnished in Table 1.

5 Comparative Analysis Between muCI and BCI

A comparative analysis is necessary to know pre-existing and existing similar technologies for better understanding of present technological progress in the field of muCI. In this context, it is important to mention that supporting technology for Brain computer interface (BCI) is much similar to muscle computer interface technology. Brain computer interface can be defined as an interaction medium which allows the brain to directly control the computer instead of relying on direct physical interaction with computer. BCI is using electrical activity of the brain and it is independent of neuromuscular pathways. So, its application is very popular in communication of patients of neuromuscular disorders [1]. On the other hand muCI is a field where human computer interaction is based on muscular activity. Thus, muCI does not incorporate paralytic muscles and people with neuromuscular disorder(s) are unable to use muCI based communicative/interactive device. In both BCI and muCI, signals are decoded using similar methods like ANN, HMM etc. Although the approach in development of muCI and BCI is similar, still these are different in several aspects. Here, 'Table 2' is summarizing comparative analysis between BCI and muCI. This table will be helpful to have basic concepts of muCI and BCI for budding designers/researchers.

6 Present Status and Future Scopes of muCI

Muscle computer interface has a high potential for various applications to develop muscular activity based interactive devices for different purposes. How muscular signals can be used for different purposes is discussed in the subsequent sections.

6.1 Hand Gesture Recognition Interfaces Versus. Gaming Interfaces

Wheeler et al. [36] reported about neuro-electric interfaces for virtual device control. Different hand gestures were classified by them using HMM. Moreover, they conducted two experiments: one replicating a joystick and other with a keyboard, to examine their feasibility. In 2009, Xiang et al. [37] developed hand gesture based interfaces depending on American and Chinese sign languages using 6 channel electrodes for classifying various gestures. This kind of technology can further be developed as gesture based communication language tool with the help of muCI. These devices can even be interfaced with a bluetooth/wireless network for mobile applications [35] which might be helpful for differently-abled people (deaf and dumb) to communicate with ease. Kim et al. [15] has highlighted the development of an EMG based interface for hand gesture recognition with 94 % accuracy. In their experiment a single channel EMG sensor was used inside the forearm. Based on gesture recognition technology, they controlled a robot car. Similar techniques can be used to play car race. One can play such games using different gestures which would make the current games more interactive rather than the traditional joystick based control. Tang et al. [20] designed a ring like sEMG device which can be worn on the forearm easily. Their device was used for development of another device called 'Possessed Hand' where electrodes were employed for giving small harmless currents to the hand and this current helped to form various gestures. This device has been found to be helpful in training users to play string instruments like guitars. Further development of above mentioned device can replace the current device used in the 'Game Guitar Hero' which has only 5–6 buttons and one flipper. Presently 'Game Guitar Hero' can only recognize the gestures and train users to play instruments with ease. Additionally, muCI enabled device has recently been used in an application for drawing within a large scale virtual environment. Electrical changes due to muscular activity of user is acquired by the device and used by the application for controlling dimensions and colors of the brush [24].

6.2 Prosthetics and muCI

The application of surface EMG and invasive EMG in prosthetics is not new to us. Zecca et al. [22] reported traditional methods to control prosthetic hands by EMG in both clinical and research context. In their work, Chu et al. [31] used principal component analysis, self organizing feature map and multilayer perception for extracting features from the EMG signals. This was used in upper limb prostheses.

Khezri and Jahed [38] reported about current systems employed to control prosthetic hands which have limited functions or can be used to perform simple movements, they proposed to use an intelligent approach based on adaptive neuro-fuzzy inference system (ANFIS) integrated with a real time learning scheme to identify hand motion commands. The hybrid method for training the fuzzy system consists of back propagation (BP) and LMS. To design an effective system they considered the conventional scheme of EMG pattern recognition [39] system namely time domain (TD) and frequency representation. The results showed 96.7 % average accuracy and six unique hand movements were considered which include hand opening and closing; thumb pinch and flexion; wrist flexion and extension. More patterns of movements are still needed to be explored for further development of prosthetics. Emphasis should be given for detailed research in complex movements of body parts (e.g. rotation of hand), range of movement of body joints, torque analysis during movement, etc. to make the devices more accurate.

6.3 Facial Expression Recognition Based muCI

In their earlier publication Naik et al. [34] demonstrated an ICA based hand gesture recognition system for HCI using 4 s EMG sensors. Recently they have employed same method to study facial expressions. Their results reported near about 100 % accuracy excluding small errors over 360 experiments. This EMG based facial recognition system has very wide range of applications. If we can classify EMG signals for each different facial expression then we can control different devices by means of facial expression. So, it will be helpful in emotion based product design. For example, favourite music can be played by music player when got stressed. In this context input may be tensed expression based facial muscular activity. In addition, in near future one can interact with large scale virtual environment with the help of facial expression based muCI. Now, differently abled persons would be able to access mouse with the help of facial expression based muCI technology [40]. So, this facial expression based muCI technology has great potential in different application areas, researcher/designers just need to explore.

6.4 Fatigue Detection and muCI

There are several reports about muscular fatigue detection using EMG. Peper et al. [41] reported how sEMG could be used to assess several workstations to prevent repetitive strain injuries [42]. Singh et al. [9] reported about muscle fatigue detection during cycling via sEMG and this was a grand success for this methodology because it was also verified using muscle biopsy tests and blood tests. In another report, Subasi and Kiymik [10] mentioned about fatigue detection by sEMG using time frequency methods, ICA and neural network methods.

It is expected that in near future muCI based alert systems and/or health care devices would be able to detect muscular fatigue of workers who are working for long time at a stretch or repetitively in workstations. Probably this kind of alert system can be helpful to prevent musculoskeletal disorders including repetitive strain injuries.

7 Limitations of muCI

There are some limitations of muCI as it is very much dependent on sEMG signals. Locations of EMG electrodes are one of the important constraints. Hogrel et al. [43] reported how various parameters like muscular action potential, conduction velocity (CV) and mean power frequency (MPF) could vary with the electrode location, type of electrodes and their proper functioning. In addition, CV and MPF not only depend on location of electrodes but also on their initial values which change upon fatigue. This influence appears to be subject-dependent. Signal variability seems essentially to be due to the relative displacements of myotendinous and neuromuscular junctions with respect to the electrode placed. In addition to that, it is very often not possible to capture signals with the help of normal surface EMG from muscles which are very small (e.g. some facial muscles) or muscles which are located internally as placing of electrodes is itself very challenging. Researchers should take a challenge to design new kind of electrodes in this context. Even in prosthetics, current sEMG methods have proven to be inefficient and the users complained of early fatigue due to the high concentration required to executing simple tasks [7, 19, 22, 26].

However, Hargrove et al. [13] have shown that there was no significant difference when decoding the accuracy of the intramuscular EMG compared to that of surface EMG signals. Hence, proper placement of EMG electrodes, accuracy of capturing EMG signals and proper/more accurate classification of EMG signals, incorporation of Bluetooth/Wi-Fi facility for making this kind of devices more mobile [35] are still limited for better development of new muCI.

8 Summary and Conclusions

It can be summarized that muCI is a current and interesting field of HCI. It is on the verge of being a very important means of interaction in our various lively works. Present paper highlights recent researches done in the field of muCI, advances made by researchers, and also what could be the future application of such interactive devices.

Observations have been made to understand how muCI's can be used for Gesture recognition which in turn can be applied for gaming interfaces as well as interfaces for communicating with differently-abled persons (e.g. deaf and the dumb). The ways how facial expressions can be read using muCI and how this methodology can be used for controlling devices by means of different expressions, have also been stressed in the present review. The way how muCI's have been successfully used in prosthetics and how muCI's can be used to make devices for fatigue analysis of workers at workstations, have also been looked into. Lastly, some limitations of muCIs have taken into consideration to complete the knowledge-base.

Technology of muCI can be applied in various fields such as various gesture based game playing like Spiderman, Road Rash etc. design of gesture based more accurate and interactive digital drawing tool; prosthetics for more complex body movement [54], emotion based interaction with virtual world, emotion based controlling of devices, gesture based typing tool for blind persons, gesture based examination system for blind students etc. Thus, it can be concluded that present paper would be able to open new doors toward several new design ideas and using this emerging technology in future for mankind.

References

1. Ahsan MR, Ibrahimy MI, Othman O (2009) EMG signal classification for human computer interaction: a review. Eur J Sci Res 33:480–501
2. Veron S, Joshi SS (2011) Multidimensional control using a mobile phone based brain-muscle computer interface. EMBS IEEE, pp 5188–5194
3. Veron S, Joshi SS (2011) Brain-muscle computer interface: mobile phone prototype development and testing. IEEE Trans Inf Technol Biomed 15:531–538

4. Crary MA, Carnaby GD, Groher ME, Helseth E (2004) Functional benefits of dysphagia therapy using adjunctive S-EMG biofeedback. Dysphagia 19:160–164
5. Disselhorst-Klug C, Schmitz-Rode T, Rau G (2009) Surface electromyography and muscle force: limits in s-EMG-force relationship and new approaches for applications. Clin Biomech 24:225–235
6. Park E, Meek SG (1995) Adaptive filtering of the electromyographic signal for prosthetic control and force estimation. IEEE Trans Biomed Eng 42:1048–1052
7. Zecca M, Micera S, Carrozza MC, Dario P (2002) On the control of multifunctional prosthetic hands by processing the electromyographic signal. Crit Rev Biomed Eng 30:459–485
8. Gonzalez IM, Malanda A, Gorostiaga E, Izquierdo M (2012) Electromyographic models to assess muscle fatigue. J Electromyogr Kinesology 22:501–512
9. Singh VP, Kumar DK, Polus B, Fraser S (2007) Strategies to identify changes in SEMG due to muscle fatigue during cycling. J Med Eng Technol 31:144–151
10. Subasi A, Kiymik MK (2010) Muscle fatigue detection in EMG using time frequency methods, ICA and neural network. J Med Syst 34:777–785
11. Chiang J, Wang ZJ, McKeown MJ (2008) A hidden markov, multivariate autoregressive(HMM-mAR) network framework for analysis of EMG (s-EMG) data. IEEE Trans Signal Process 56:4069–4081
12. El-Daydamony EM, El-Gayar M, Abou-Chadi F (2008) A computerized system for SEMG signals analysis and classification. In: National radio science conference (NRSC 2008), pp 1–7
13. Hargrove LJ, Englehart K, Hudgins B (2007) A comparison of surface and intramuscular myoelectric signal classification. IEEE Trans Biomed Eng 54:847–853
14. Khezri M, Jahed M (2007) Real time intelligent pattern recognition algorithm for surface EMG signals. EMBS 6:45
15. Kim J, Mastnik S, André E (2008) EMG-based hand gesture recognition for real-time biosignal interfacing. In: Proceedings of the 13th international conference on intelligent user interfaces, pp 30–39
16. Morris D, Saponas TS, Tan D (2010) Emerging input technologies for always-available mobile interaction. Found Trends Hum Comput Inter 4:245–316
17. Naik GR, Kumar DK, Palaniswami M (2008) Multi run ICA and surface EMG based signal processing system for recognizing hand gestures. 8th IEEE international conference on computer and information technology (CIT 2008), pp 700–705
18. Nan B, Hamamoto T, Tsuji T, Fukuda O (2004) FPGA implementation of a probabilistic neural network for a bioelectric human interface. In: The 47th midwest symposium on circuits and systems (MWSCAS '04), vol 3, pp 29–32
19. O'Neill PA, Morin EL, Scott RN (1994) Myoelectric signal characteristics from muscles in residual upper limbs. IEEE Trans Rehabil Eng 2:266–270
20. Tang X, Liu Y, Liu C, Poon W (2011) Classification of hand motion using surface EMG signals. Biologically Inspired Robotics, CRC Press
21. Tsenov G, Zeghbib AH, Palis F, Shoylev N, Mladenov V (2006) Neural networks for online classification of hand and finger movements using surface EMG signals. In: 8th seminar on neural network applications in electrical engineering (NEUREL), pp 167–171
22. Xiong FQ, Shwedyk E (1987) Some aspects of nonstationary myoelectric signal processing. IEEE Trans Biomed Eng 34:166–172
23. Naik GR, Kumar DK, Palaniswami M (2007) Real time hand gesture identification for human computer interaction based on the ICA of surface electromyogram. In: Proceedings of the HCSNet workshop on use of vision in human-computer interaction, vol 56, pp 67–72
24. Belluco P, Bordegoni M, Cugini U (2011) A techinque based on muscular activation for interacting with virtual environments, vol 1. ASME 2011 world conference on innovative virtual reality, pp 315–324

25. Farid M, Keyvan HZ (2006) A method for online estimation of human arm dynamics. In: 28th annual international conference of the IEEE engineering in medicine and biology society (EMBS '06.), pp 2412–2416
26. Kyberd PJ, Holland OE, Chappel PH, Smith S, Tregdigo R, Bagwell PJ, Snaith M (1995) Marcus: a two degree of freedom hand prosthesis with hierarchical grip control. IEEE Trans Rehabil Eng 3:70–76
27. Scott RN, Parker PA (1988) Myoelectric prostheses: state of the art. J Med Eng Technol 12:143–151
28. Saponas TS, Tan TS, Morris D, Balakrishnan R (2008) Demonstrating the feasibility of using forearm electromyography for muscle computer interfaces. In: Proceedings of the 26th annual SIGCHI conference on human factors in computing systems, pp 515–524
29. Saponas TS, Tan DS, Morris D, Balakrishnan R, Turner J, Landay JA (2009) Enabling always available input with muCI's. In: Proceedings of the 22nd annual ACM symposium on user interface software and technology, pp 167–176
30. Saponas TS, Tan DS, Morris D, Balakrishnan R, Turner J, Landay JA (2010) Making muCI's more practical. Proceedings of the 28th international conference on human factors in computing systems, pp 851–854
31. Chu JU, Moon I, Moon MS (2005) A real time EMG pattern recognition system based on linear-non linear feature projection for a multifunctional myoelectric hand. In: Proceedings of 9th international conference on rehabilitation robotics IEEE, vol 53. pp 2232–2239
32. Farry KA, Walker ID, Baraniuk RG (1996) Myoelectric teleoperation of a complex robotic hand. IEEE Trans Rob Autom 12:775–788
33. Chen X, Zhang X, Zhao ZY, Yang JH, Lantz V, Wang KQ (2007) Hand gesture recognition research based on surface EMG sensors and 2D-accelerometers. In: 11th IEEE international symposium on wearable computers, pp 11–14
34. Naik GR, Kumar DK, Weghorn H (2007) Performance comparison of ICA algorithms for isometric hand gesture identification using surface EMG intelligent sensors. In: 3rd international conference on sensor networks and information (ISSNIP 2007), pp 613–618
35. Seregni L (2010) ERACLE wearable EMG gesture recognition system. Masters thesis, department of electrical and computer science engineering, Politecnico, Milano
36. Wheeler KR (2003) Device control using gestures sensed from EMG. In: Proceedings of the 2003 IEEE international workshop on soft computing in industrial applications, SMCia/03, pp 21–26
37. Xiang C, Lantz V, Qiao WK, Yan ZZ, Xu Z, Hai JH (2009) Feasibility of building robust surface electromyography based hand gesture interfaces. Eng Medicine Biol Soc 2983–2986
38. Khezri M, Jahed M (2007) A novel approach to recognize hand movements via sEMG patterns. EMBS, pp 4907–4910
39. Nagata K, Ando K, Magatani K, Yamada M (2007) Development of the hand motion recognition system based on surface EMG using suitable measurement channels for pattern recognition. EMBS, pp 5214–5217
40. Huan CN, Chen CH, Chung HY (2006) Application of facial electromyography in computer mouse access for people with disabilities. Disabil Rehabil 28:231–237
41. Peper E, Wilson VS, Gibney KH, Huber K, Harvey R, Shumay DM (2003) The integration of electromyography (SEMG) at the workstation: assessment, treatment, and prevention of repetitive strain injury (RSI). Appl Psychophysiol Biofeedback 28:167–182
42. Sbriccoli P, Felici F, Rosponi A, Aliotta A, Castellano V, Mazza C, Bernardi M, Marchetti M (2001) Exercise induced muscle damage and recovery assessed by means of linear and non-linear sEMG analysis and ultrasonography. J Electromyogr Kinesology 11:73–83
43. Hogrel JY, Duchêne J, Marini JF (1998) Variability of some SEMG parameter estimates with electrode location. J Electromyogr Kinesiol 8:305–315
44. Putnam W, Knapp RB (1993) Real-time computer control using pattern recognition of the electromyogram. In: Proceedings of the 15th annual international conference of the IEEE engineering in medicine and biology society, pp 1236–1237

45. Rosenberg R (1998) The biofeedback pointer: EMG control of a two dimensional pointer, wearable computers. In: Second international symposium on digest of papers, pp 162–163
46. Jung KK, Kim JW, Lee HK, Chung SB, Eom KH (2007) EMG pattern classification using spectral estimation and neural network. SICE annual conference, pp 1108–1111
47. Itou T, Terao M, Nagata J, Yoshida M (2001) Mouse cursor control system using EMG. In: Proceedings of the 23rd annual international conference of the IEEE engineering in medicine and biology society, vol 2, pp 1368–1369
48. Tsuji T, Ichinobe H, Fukuda O, Kaneko M (1995) A maximum likelihood neural network based on a log-linearized gaussian mixture model. In: Proceedings of IEEE international conference on neural networks, pp 2479–2484
49. Fukuda O, Tsuji T, Kaneko M (1999) An EMG controlled pointing device using a neural network systems. In: IEEE international conference on man, and cybernetics (IEEE SMC '99 conference proceedings), vol 4, pp 63–68
50. Tsuji T, Bu N, Fukuda O, Kaneko M (2003) A recurrent log-linearized gaussian mixture network. IEEE Trans Neural Network 14:304–316
51. Fukuda O, Arita J, Tsuji T (2004) An EMG-controlled omnidirectional pointing device using a HMM-based neural network. In: Proceedings of the international joint conference on neural networks, vol 4, pp 3195–3200
52. Kim JS, Jeong H, Son W (2004) A new means of HCI: EMG-MOUSE, systems. In: IEEE international conference on man and cybernetics, vol 1, pp 100–104
53. Chan A, Kevin B (2005) Continuous myoelectric control for powered protheses using hidden markov models. IEEE Trans Biomed Eng 52:123–134
54. Carrozza M, Micera S, Massa B, Zecca M, Lazzarini R, Canelli N, Dario P (2001) The development of a novel biomechatronic hand-ongoing research and preliminary results. Adv Intell Mechatron AIM 1:249–254

Preliminary Analysis of Low-Cost Motion Capture Techniques to Support Virtual Ergonomics

Giorgio Colombo, Daniele Regazzoni, Caterina Rizzi and Giordano De Vecchi

Abstract This paper concerns the development of a computer-aided platform to analyze workers' postures and movements and ergonomically validate the design of device a man or woman may deal with. In particular, we refer to pick and place operations of food items on the display unit shelves. The paper describes three low-cost solutions that integrate two optical motion capture techniques (one based on web-cam and another on MS Kinect sensor) and two human modeling systems (Jack and LifeMod) with the main goal of determining the suitability of operators' working conditions and, eventually, providing a feedback to the design step. The solutions have been tested considering a vertical display unit as case study. Preliminary results of the experimentation as well as main benefits and limits are presented. The results have been considered promising; however, we have planned to perform an acquisition campaign in the real environment, the supermarket.

Keywords Human modeling · Motion capture · Virtual ergonomics · Pick and place

G. Colombo
Department of Mechanics, Polytechnic of Milan, Milan, Italy
e-mail: giorgio.colombo@polimi.it

D. Regazzoni (✉) · C. Rizzi · G. De Vecchi
Department of Engineering, University of Bergamo, Dalmine, BG, Italy
e-mail: daniele.regazzoni@unibg.it

C. Rizzi
e-mail: caterina.rizzi@unibg.it

G. De Vecchi
e-mail: giordano.devecchi@unibg.it

A. Chakrabarti and R. V. Prakash (eds.), *ICoRD'13*, Lecture Notes in Mechanical Engineering, DOI: 10.1007/978-81-322-1050-4_34, © Springer India 2013

1 Introduction

Virtual ergonomics permits engineers to create and manipulate virtual humans to investigate the interactions between the worker or the consumer and the product. For example, in product design, human factors such as positioning, visibility, reaching, grasping and lifting of weights can all be evaluated by using virtual humans, providing a feedback to designers in the early steps of product development.

This paper refers to this context and addresses commercial refrigeration industry; in particular those companies specialized in display units. It describes the use of virtual ergonomics techniques in the design process of display units commonly used in supermarkets.

In previous research activities we have experienced the use of virtual manikins specifically targeted for ergonomics to evaluate postures and movements respect to requirements established by international standards to reduce health risks [1]. In this work we try to enhance that approach to provide more precise results. For example, when filling the shelves of a display unit, the workers act quite differently one from another. To face this issue we introduce the use of motion capture systems to reproduce with virtual humans the real way in which operators behave. In particular, we have evaluated the possibility to adopt and integrate Motion Capture (Mocap) and Digital Human Modeling (DHM) systems to perform ergonomic analysis relying on real movements performed by operators in everyday activities.

The paper, after a brief description of the state of the art of adopted techniques, describes three technical solutions based on low cost Mocap devices integrated with human modeling tool. Then, the case study and preliminary experimentation are presented.

2 Related Works

The implementation of the proposed virtual ergonomics platform requires the integration of Mocap and DHM systems to acquire end-users' postures and movements to determine fatigue, stress and risk for workers' health. In the following we provide an overview of both techniques.

First DHM tools for ergonomics analysis appeared in late 60s, mainly in aeronautics and automotive industries. Nowadays, in literature we can find various tools of different complexity depending on the target application. We have classified them into four main categories [2–4]: *virtual human/actors for entertainment* used to populate scenes for movies and videogames production [5, 6]; *mannequins for clothing* used to create virtual catwalks, virtual catalogues, and virtual try-on show rooms and to design garments [7, 8]; *virtual manikin for ergonomic analysis* used to define complex scenes, analyze postures, simulate tasks and optimize working environments [9–11]; and, finally, *detailed biomechanical models* [12, 13] whose applications concern ergonomics analysis, study of safety in transport, and human performance during sports activities, and medical device, etc. [14]. For our purpose, we consider DHM tools belonging to the last two categories.

Mocap techniques appeared in the 70s as a branch of imaging techniques applied to biomechanical processes and several applications have been developed for different contexts such as military, entertainment, sports, and medical applications. According to the working principle four main categories can be identified [15]: optical, mechanical, inertial, and magnetic that can be used for testing and validation also in a combined way [16].

We focus the attention on optical systems, considered the high-end technology. They extrapolate the position of body joints by triangulation between images taken from different cameras. We can have systems with active or passive markers [17] or markerless [18]. In the first case markers are placed in the remarkable points of the body while in the second case a dedicated software module recognizes the human figure.

In literature, we can also find researches that integrate DHM either with Virtual Reality (VR) and Mocap systems to improve the level of interaction and realism within the virtual environment, to drive the virtual human and facilitate the evaluation of comfort and prediction of injuries that could arise when executing a task [19, 20].

In our work we consider low cost techniques, developed for video games and entertainment, to verify their usability and performance in industrial context. Actually, such technologies benefit from a huge investment on research that leads to a rapid evolution but on the same time they keep affordable prices because of the target market they refer to. Moreover, the capillary diffusion ensure the ease of sharing of information on pros and cons, tips and unusual use researchers may find worldwide.

3 Technical Solutions

As mentioned before, main goal has been to experiment the integration of low cost Mocap system and DHM to evaluate postures and movements of workers filling the shelves of a display unit. We considered two systems, both optical and markerless: the former based on Sony Eye webcams and the latter on Microsoft Kinect sensor. We have been using them to acquire the real movements and postures of operators so that the following simulation can be based on real data. Both solutions are not expensive and can be easily moved and, with some precautions, used also outside the lab in potentially any work environment we want to acquire.

Concerning human modeling, we adopted two different tools: Siemens Jack specifically targeted for ergonomics and LifeMod to generated detailed biomechanical model. Figure 1 shows the technical solutions adopted, which are described in the detail in the following paragraphs.

3.1 Webcam Solution

This solution includes three main components: webcams, a module for data exchange and LifeMod to create the human avatar and reproduce postures and movements. Figure 2 shows the acquisition system composed by:

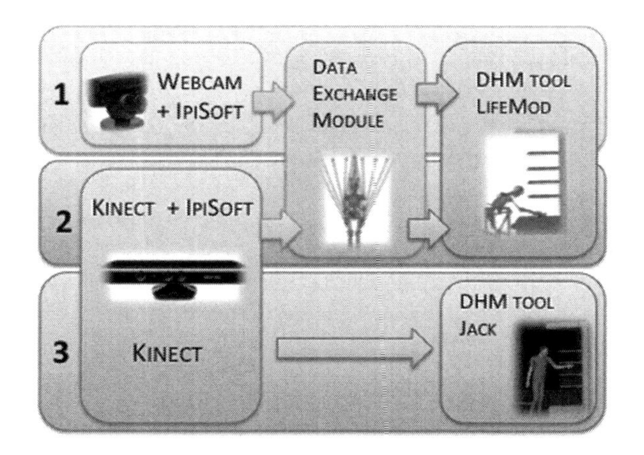

Fig. 1 Adopted technical solutions: *1*—webcam solution; *2*, *3*—Kinect solutions

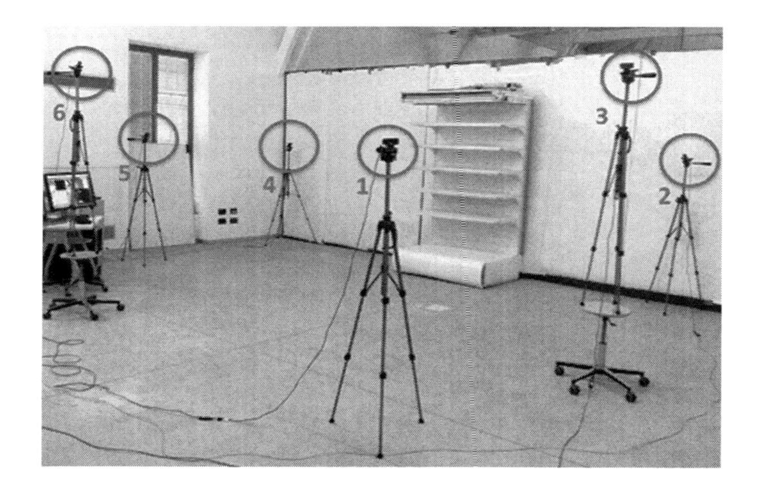

Fig. 2 Webcam solution

- Six Sony Eye webcams with a resolution of 640 × 480 pixels at 60 Hz mounted on photographic tripods.
- Portable workstation Dell Precision M6500.
- iPi Desktop Motion Capture™ (Ipi DMC) software.

iPisoft Software is a marker-less system developed to work with Sony Eyes webcams and its main features are:

- Possibility to use from 3 to 6 webcams.
- A maximum acquisition area of 7 × 7 m.
- Non real-time tracking.
- Input file format: MPEG.

The system acquires synchronized video sequences obtained with the cameras without having to apply any type of physical markers on the operator's skin. It automatically recognizes the different body segments and, then for each time step, calculates joints position and orientation. iPiDMC includes two main components: iPisoft Recorder for the motion capture phase that synchronizes images recorded from the six webcams and iPisoft Studio for the recognition and segmentation of body and tracking of movement. iPisoft Desktop Motion Capture output contains the recorded movement in Biovision Hierarchical Data (BVH) format. iPisoft adopts a skeleton made of 27 joints hierarchically organized, each characterized by proper d.o.f. and constraints. However, it has been necessary to develop a data exchange module to be integrated with the considered human modeling tool, namely LifeMod.

LifeMod is a biomechanical analysis software that permits to generate a complete biomechanical model of the human body. It is a plug-in of ADAMS software, a multibody analysis system. The creation of a model normally starts with the generation of a basic set of connected human segments based on the dimension contained in an anthropometric database; then, the joints, the muscles and the tendons are created and contact force with objects are defined. By the way, to reproduce the movement with the operators' avatar we need to exchange the data acquired from the Mocap system to the human modeling environment. This problem has been solved developing an ad-hoc conversion algorithm in Matlab. The algorithm translates the information relative to the joint hierarchy and to the motion contained in the BVH file to a SLF formatted file used by LifeMod. In order to complete the information required by a SLF file, anthropometric data are retrieved directly from user's data or from an anthropometric database (e.g., GeBOD [21], People Size [22] and US Army- Natick Database [23]).

Before proceeding in reproducing the operators' tasks using LifeMod, another issue must be faced. Actually, the conversion performed in the previous step does not take into account the fact that the information contained in the SLF file are related to the real position of the joints acquired subject while the biomechanical software use external markers placed on the skin surface. Once again to fix this problem an ad-hoc script has been created to relocate parametrically to model size the position of the markers. To reduce the number of operations necessary to run the analyzes, an automatic procedure and a CMD script have been implemented. The script is completely automatic and no user interaction is required because data relative to the markers initial position and the marker movements are produced by converting this information directly from BVH file.

Once the model is defined, simulation phase can begin. To obtain accurate simulations with the muscles and the articulations it's necessary to execute a first inverse dynamic simulation to drive the body with motion agents describing the movements to execute. Once the movements are stored a direct dynamic simulation is run to calculate the forces created by the muscles and the stresses the body is subjected to. The outcome provided by the system consists of forces and momenta acting on each joint in each time step of the analysis.

3.2 Kinect Solution #1

Similarly to the previous solution, also this one comprehends three main parts. The main difference is the Mocap system, which is composed by (Fig. 3):

- Microsoft Kinect with a resolution of 640 × 480 pixels at 30 fps, mounted on photographic tripod and connected via USB cable.
- Portable workstation Dell XPS.
- iPi Desktop Motion Capture software.

As in the previous case iPi Recorder manages the recording of images and depth videos coming from Kinect, while the iPi studio performs environment calibration and video analysis. In addition, MS Kinect SDK libraries are required.

The other two components of this solution, namely Data exchange and DHM modules, are exactly the same of the webcam solution.

3.3 Kinect Solution #2

This solution always uses a MS Kinect but, differently from the two previous one, Kinect sensor is fully integrated with Jack, a well-known human modeling system (Figs. 1, 3). It permits to define complex scenes with virtual manikins and objects, simulate many tasks and evaluate posture and ergonomics factors also using analysis tools such as Rapid Upper Limb Analysis (RULA) to investigate work-related upper limb disorders, U.S. National Institute for Occupational Safety and Health (NIOSH) lifting equations to evaluate lifting and carrying tasks, and Owako Working Posture Analysis System (OWAS) to analyze postures during work. Therefore, iPi DMC software is not necessary.

Fig. 3 Kinect solutions #1 and # 2

This solution requires:

- Mocap Toolkit, a module specifically developed by Siemens for Kinect sensor.
- MS KinectSDK v 1.0 libraries.
- SkeletalViewer sw to transfer data streaming acquired with Kinect sensor to Jack.

The skeleton used in the transition from Mocap to Jack is made of 21 joints whose positions and movements are tracked and there is not a hierarchy among them. This skeleton is less complex than Jack's one and some details cannot be taken into account (i.e., head rotation, fingers).

4 Preliminary Experimentation

In the following we describe the case study adopted for the experimentation and the preliminary results.

4.1 The Case-Study

The case study refers to a vertical refrigerator display unit typically installed in groceries or supermarket, with four or five shelves that could be placed at different heights (Fig. 4).

Different groups of people interact with the display unit: supermarket staff filling the shelves with goods, workers in charge of maintenance operations and customers picking up goods. In this paper we focus on the first category with the main goal of determining the suitability of operators' working conditions.

The task of loading a refrigerator is critical from the ergonomics point of view and may generate severe health disorders, mainly due to: repeated actions, holding and lifting loads, insufficient recovery times, uncomfortable postures, and uncomfortable environment (e.g., cold and humid).

To experiment the described solutions, we decide to focus the attention on pick and place operations of food items. We have conducted a preliminary study (interviews and direct observations) to analyze how operators really behave. This permitted to identify main operations to be reproduced in the lab and some occasional unacceptable practices such as assuming completely wrong and dangerous postures.

To acquire postures and movement we use a simplified version of the real unit (Fig. 4 right) more efficient because some elements of the complete display unit (e.g. lateral walls) may interfere with the operator during motion capture.

Fig. 4 Vertical refrigerator (image courtesy: www.costan.com) (*left*); simplified display unit (*right*)

4.2 Experimentation Results

Actors were asked to perform as much as possible as they were in the real environment and to follow a precise routine to produce comparable results with operators characterized by different anthropometric measures. The routine defines the initial and final positions of each movement to be performed. For each solution, the loading of a bottle of water in the four different shelves has been reproduced.

First the solution based on webcam has been tested. The system requires a first step of calibration that initializes the system and permits to correctly locate each camera in the space.

A semicircle disposition of the webcams at different heights around the operator is the best choice. Once calibration is done, it is possible to realize the acquisitions taking care to avoid fast movements and carefully execute the required task. It has been conducted recording ten loading routines with subjects of both genders and of different heights to evaluate if the motion capture system is affected by any problem. In particular, the conversion algorithm has been tested as well as different ways of automatically changing the position of joints depending on height and structure of subjects.

After having tested the system the real motion capture campaign has been performed and the data of the movement have been converted and analyzed with LifeMod. In Fig.5 one can see the representation of data related to the loading of the highest during the three main steps. The first one (Fig. 5a) refers to the environment in which the webcam images are captured and elaborated for each time step to gather joints positions shown in the second representation (Fig. 5b). The third representation (Fig. 5c) comes from the human modeling system where data have been converted, corrected and integrated with anthropometric databases.

The DHM tool provides different kinds of output data relative to biomechanical analysis; in particular it is possible to obtain results such as center of mass

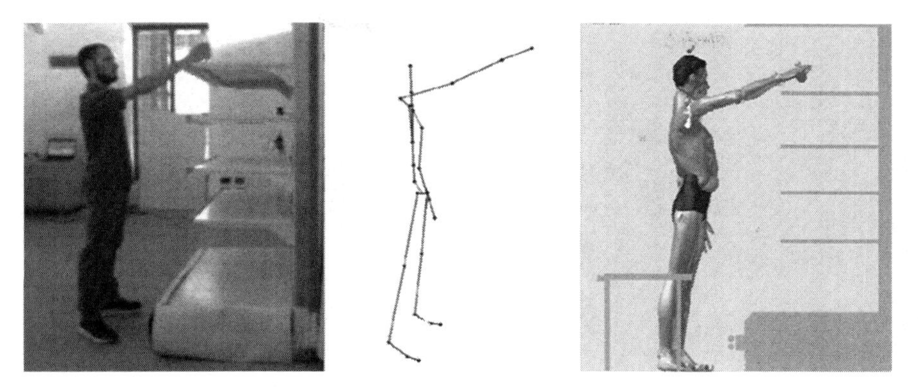

Fig. 5 Webcam solution: highest shelf loading

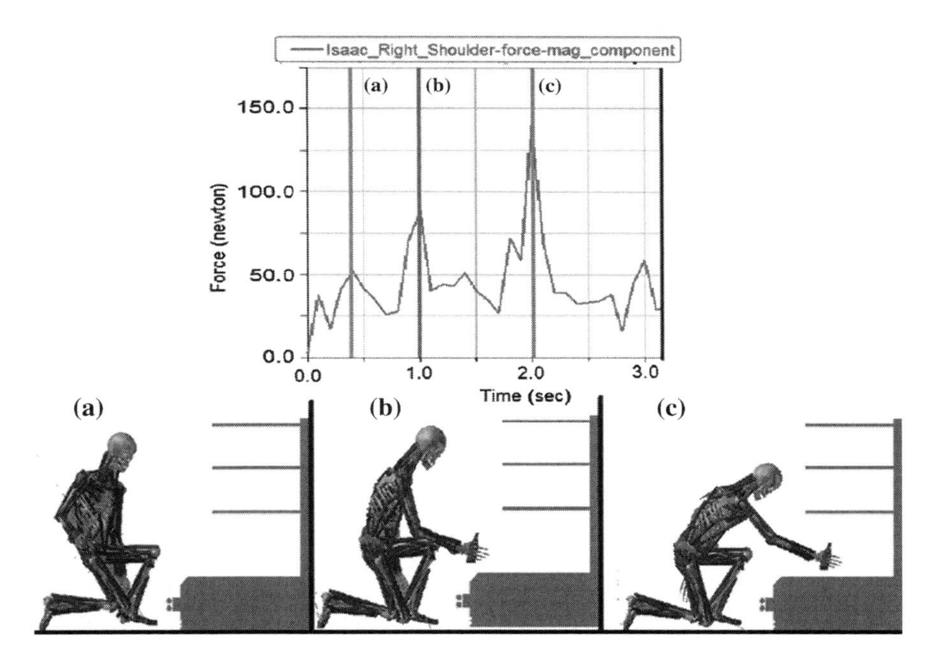

Fig. 6 Resultant force over time acting on right shoulder while loading the lowest shelf and three notable position

position, velocity and acceleration of segments, forces, torque and power for joints and additional information for soft tissues and environment interaction. Figure 6 shows an example of analysis results for the virtual human with one knee on the ground while performing the loading of a bottle on the lowest shelf. As an example the resultant force on the shoulder holding the bottle is plotted in the graphic and three screenshots corresponding to peaks are shown.

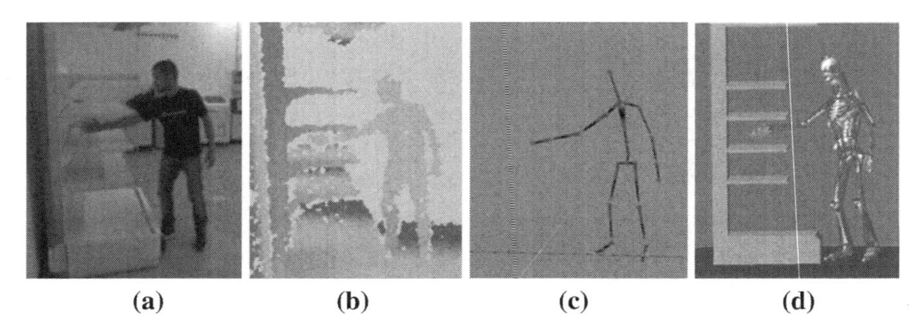

Fig. 7 Kinect solution #1: second shelf loading

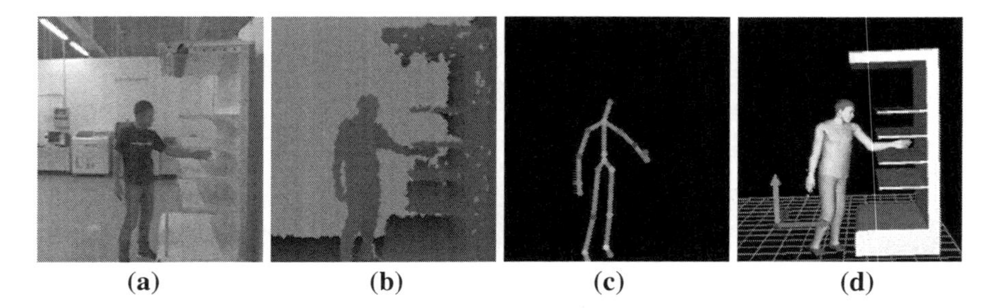

Fig. 8 Kinect solution #2: second shelf loading

Regarding Kinect solution #1 we proceeded similarly to the previous one. Figure 7 shows the results related to the loading of the second shelf. Figure 7a refers to the environment in which the Kinect images are captured, Fig. 7b shows the corresponding depth map, Fig. 7c the reconstructed skeleton and Fig. 7d the virtual avatar of the actor corrected and integrated with anthropometric databases. Results obtained were similar with those of the previous solution.

Finally we tested the Kinect solution #2, i.e. the Kinect sensor integrated with Jack. First, using Jack the virtual scene composed by the 3D model of the refrigerated unit and the operator avatar (in this case a virtual manikin 50th ile) is created; then, the Kinect plug-in and the SkeletalViewer software are launched and the acquisition session can start. Figure 8 shows the final step acquired during the loading of the second shelf. Precisely, Fig. 8a shows the image acquired, Fig. 8b the corresponding depth map, Fig. 8c the corresponding reconstructed skeleton and Fig. 8d the virtual avatar of the actor.

This solution has the advantage that the two environments, Mocap and the human modeling, are fully integrated; therefore Jack reproduces the operators' movements in real-time. For the other two solutions it is necessary to translate the data acquired with the Mocap system into a format readable by LifeMod and then reproduce the movements and postures of the actor/worker. Moreover Jack already includes dedicated modules that implement analysis tools such as RULA, NIOSH and OWAS to study and validate operator postures in order to avoid musculo-skeletal disorders.

Kinect used to track human body constitutes an interesting novelty in the instrumentation available for Mocap and virtual ergonomics at industrial level. Its low cost, broad diffusion and availability of libraries let us foresee it will be commonly used for Mocap application in near future. By the way, there are still some limitations affecting its performance when used with Jack. Actually, the skeleton does not always perfectly match the subject's posture; this is particularly true when body areas overlap. It is also recommended to acquire the entire actor so that Jack can make the posturing more robust, but Kinect limited working area, if compared to webcam solution, can constitute a limit. At the moment only one Kinect is supported for real time capture and this may cause occlusions and misunderstanding of actor position. Anyway, for the scene and tasks we consider in this paper and for most of ergonomic situations, these limitations can be acceptable and successful simulations can be carried out. Moreover, new versions of Jack that are going to be released shortly are expected to address these issues.

5 Conclusions

The goal of this work has been to verify the use of low-cost Mocap systems integrated with digital human modeling to evaluate ergonomics factors of industrial products, in our case refrigerated display units commonly used in supermarket.

Three solutions, based on two marker less Mocap systems and two different types of human modeling tools, have been preliminary tested considering the pick and place operations of a vertical display unit. The results have considered promising and interesting; however some limits have been highlighted and an acquisition campaign in a real environment, the supermarket, has been planned. In fact, starting from operators performing real tasks instead of standardized average tasks allows not only to be more precise in the final evaluation but also to consider unknown or incorrect postures or movements performed by operators in their everyday activities. In addition, data coming from acquisition campaign can be used to analyze performance variability and build a structured, domain dependant, database of real postures and movements to be applied to similar cases. This would allow benefiting from real data without the need of performing motion capture for similar cases.

The results gathered so far with the low cost Mocap techniques let us foresee a huge application in almost all industrial fields wherever a worker is asked to perform a manual operation and/or to interact with a machine. By the way, these techniques have a lower accuracy and precision of some well known and much more expensive solutions (e.g., Vicon) and this, at the moment, may limits their usage as, for instance, some medical applications.

References

1. Colombo G, De Ponti G, Rizzi C (2010) Ergonomic design of refrigerated display units. Virtual Phys Prototyping 3(5):139–152
2. Sundin A, Ortengren R (2006) Digital human modelling for CAE applications, handbook of human factors and ergonomics, 3rd edn. Wiley, New York
3. Magnenat-Thalmann N, Thalmann D (2004) Handbook of virtual humans. Wiley, Chichester
4. Rizzi C (2011) Digital human models within product development process innovation in product design. Springer, Berlin, pp 143–166
5. http://www.massivesoftware.com. Accessed July 2012
6. http://crowdit.worldofpolygons.com. Accessed July 2012
7. Wang CCL, Wang Y, Chang TKK, Yuen MMF (2003) Virtual human modeling from photographs for garment industry. Comput-Aided Des 35(6):577–589
8. Li SSM, Wang CCL, Hui Kin-Chuen (2011) Bending-invariant correspondence matching on 3D human bodies for feature point extraction. IEEE Trans Autom Sci Eng 8(4):805–814
9. Colombo G, Cugini U (2005) Virtual humans and prototypes to evaluate ergonomics and safety. J Eng Design 16(2):195–203
10. Mueller A, Maier T (2009) Vehicle layout conception considering vision requirements—a comparative study within manual assembly of automobiles. Digital human modeling for design and engineering conference and exhibition
11. Green RF, Hudson JA (2011) A method for positioning digital human models in airplane passenger seats, advances in applied digital human modeling. CRC Press, Boca Raton
12. Abdel-Malek K et al (2009) A physics-based digital human model. Int J Veh Des 51(3/4):324–340
13. http://www.lifemodeler.com. Accessed July 2012
14. Bucca G, Buzzolato A, Bruni S (2009) A mechatronic device for the rehabilitation of ankle motor function. J Biomech Eng 131(12):125001
15. Furniss M (2012) Motion capture. MIT communications forum http://web.mit.edu/ commforum/papers/furniss.html. Accessed July 2012
16. Schepers HM (2009) Ambulatory assessment of human body kinematics and kinetics. Ph. d thesis, UniversiteitTwente, The Netherlands. Available at http://www.xsens.com/images/stories/PDF/ThesisSchepers.pdf
17. http://www.vicon.com. Accessed July 2012
18. Bray J (2012) Markerless based human motion Capture: a survey. Vision and VR Group, Department of Systems Engineering, Brunel University, available at http://visicast.co.uk/members/move/Partners/Papers/MarkerlessSurvey.pdf. Accessed July 2012
19. Colombo G, De Angelis F, Formentini L (2010) Integration of virtual reality and haptics to carry out ergonomic tests on virtual control boards. Int J Prod Dev 11(1/2):47–61
20. Spada S, Sessa F, Corato F (2012) Virtual reality tools for statistical analysis for human movement simulation. Application to ergonomics optimization of work cells in the automotive industry. Work 41:6120–6126
21. Cheng H (1996) The development of the GEBOD program. In: Proceedings of the 1996 fifteenth southern biomedical engineering conference, pp 251–254
22. People size database of anthropometric measure Available at http://www.openerg.com/psz/anthropometric_dates.html. Accessed October 2012
23. NSRDEC: US Army Natick Soldier Research Development and Engineering Center, http://nsrdec.natick.army.mil/ANSURII/index.htm. Accessed October 2012

A User-Centered Design Methodology by Configurable and Parametric Mixed Prototypes for the Evaluation of Interaction

Monica Bordegoni and Umberto Cugini

Abstract The research described in this paper presents a methodology for evaluating the interaction design in new consumer products, and specifically in electronic products. The methodology is supported by mixed prototypes, i.e. prototypes made up of physical and virtual components, which are parametric and configurable, and which can be used by target users of the future product, to verify aspects of usability and preferences, and to check the satisfaction in use with respect to their needs and expectations.

Keywords User-centered design · Interaction evaluation · Mixed prototypes · User interaction

1 Introduction

Experience in product design has made clear that the design of interactive products, i.e. products that are used by the users through interactive modalities, like the majority of consumer products, require the participation and involvement of users early in the elaboration of the concept, so that the final product fully meets the users' needs and preferences. Recently, it is becoming a widespread practice in the design of interactive systems the adoption of the so-called *User-Centered* Design (*UCD*) approach, which is a method for designing and developing applications,

---

M. Bordegoni (✉) · U. Cugini
Department of Mechanical Engineering, Politecnico di Milano, Milan, Italy
e-mail: monica.bordegoni@polimi.it

U. Cugini
e-mail: umberto.cugini@polimi.it

A. Chakrabarti and R. V. Prakash (eds.), *ICoRD'13*, Lecture Notes in Mechanical Engineering, DOI: 10.1007/978-81-322-1050-4_35,

components, systems or products that take into account the views and needs of the target users [1].

The UCD is actually a process consisting of multiple tasks of design and validation, based on the iterative application of various tools of observation, test and analysis. In the context of practices and standards for interactive product design, the standard ISO-9241-210 addresses the Human-centered design processes for interactive systems [2]. Human-centered design is an approach to the development of interactive systems that focuses specifically on making systems usable. This means designing systems that meet user needs better. According to this standard, systems designed on the basis of a UCD approach have the following characteristics:

- are easier to understand and use, thus reducing training and support costs,
- improve user satisfaction and reduce discomfort and stress,
- improve the productivity of users,
- improve product quality, and appeal to the users.

The design process based on the Human-centered design standard includes four main activities. The first activity concerns the specification of the context of use, which is necessary to identify which users will use the product, what they will do with it, and how it will be used and under what conditions. The following activity regards the specification of the requirements. The requirements focus on the tasks that the users will need to complete, and on any possible business goals of the product, as the cost. The creation of design solutions follows this phase. In fact, only after the requirements have been defined, the design team can start to conceive and design the product. Various design solutions may be proposed, in the form of sketches, scenarios, and prototypes, up to a complete model. The next step of extreme importance for interactive products is that one concerning the validation of the design, made by a panel of real users through tests, interviews, questionnaires and analyzes. The involvement of users is indeed beneficial in order to verify the fulfillment of the requirements previously identified, including usability, which is a central aspect in the interaction design. Usability is defined as the effectiveness, efficiency and satisfaction with which certain users achieve certain goals in certain contexts [3]. In practice, it relates to the degree of satisfaction of users when interacting with a new product. Usability testing is a common technique used in user-centered interaction design to evaluate a product by testing it on users.

Following this ISO standard, only when the design solutions reflect the requirements of the product, the design can be released and fully implemented as market product. This approach reasonably guarantees that the designed product is appreciated by a large group of users.

The research described in this paper presents a methodology for evaluating the interaction of new consumer products, and specifically of electronic products. The methodology is supported by the use of mixed prototypes, i.e. prototypes made up of physical and virtual components [4], which are configurable and parametric, and which can be used by target users of the future product, to verify aspects of usability and preferences, and to check the satisfaction of their needs and expectations.

2 Interaction Design and Assessment

During the design of a new interactive product, several components and interaction aspects of the product are defined. These aspects have to be evaluated and assessed during their development, by using the most appropriate assessment methods.

An interactive product, such as an electronic product, consists of many components, which can be grouped as:

- outer shape (cover)
- interaction physical components
- display
- Graphical User Interface (GUI)

Many products in the consumer market are characterized by such a configuration, such as mobile phones, handheld devices, video cameras, and many others. Therefore the design of such devices comprises the creation of the overall shape and style of the device, the selection of materials and colors, the definition of the various components, as buttons, and of the layout, which includes the arrangement of the components, the definition of the graphical interface including icons, functions, etc.

In order to develop a product that is likely to be successful, these items need to be assessed at some point in time in the product development process. In case the design adopts a UCD approach, these items must be validated sooner and often with users. In the early stages of the design and development process, changes are relatively inexpensive. The longer the process has progressed and the more fully the product is defined, the more expensive is the introduction of changes [5]. It is therefore important to start the design evaluation as early as possible.

The validation of the design features, in case of interactive products, has to be performed with the target end users, with respect to a set of assessment parameters, which are specific for the type of feature. The assessment parameters relate to esthetics, ergonomics, usability, functionality, performance, reliability, etc. Many methods for the validation of product concepts are based on the use of prototyping, which can be defined as that activity aiming at testing the design solutions and that is based on the use of prototypes.

In a user-centered design approach, prototypes are not simply demonstrations to show to users a preview of the design, but are used to collect *user feedback* that is then used to drive the design process. Early in design the emphasis is on obtaining feedback that can be used to guide design, while later on -when a more complete prototype is available- it is possible to measure whether user objectives have been achieved. The expected feedback from design can be used to select the design option that best fits the functional and user requirements, and also to elicit feedback and further requirements from the users.

3 Prototyping of Interactive Devices

Prototyping includes static prototypes as sketches, scenarios and storyboards, or dynamic representations of the product made using 3D models and virtual prototypes (VP), or even through interactive prototypes, virtual or physical, that can be used by users, and which have a variable degree of realism. The various kinds of prototypes are suitable to evaluate several features of the interactive products, as shown in Table 1.

In a context of UCD, the various proposals of a new product are evaluated from the outset of the design with users, coming gradually to be reduced to a limited number of selected prototypes, which will eventually lead to having only one, which will be selected for the subsequent stages of design (Fig. 1). As said, the validation methodologies are different, with different complexity and different validation and assessment targets. For example, sketches can be used at the beginning of the design, as a simple technique to show the overall idea about the product (shape, layout, etc.), and allow us to start eliminating the solutions less pleasant and appreciated by the users.

Then one can use dynamic scenarios and storyboards, for example realized by sequences of screens or videos, to show the product components and interactive sequences.

3D models and virtual prototypes (VR) are suitable to evaluate the esthetics and functionality of the products. Finally, more complex and complete prototypes are the interactive Virtual Prototype (iVP) and physical prototypes, which can effectively be used to simulate the real use of the product [6–8].

Some aspects of the product, such as its ergonomics, can be evaluated using physical prototypes. Other aspects such as usability can be evaluated by using interactive virtual prototypes. Each of these two techniques has benefits and deficiencies. The virtual prototype does not allow a complete validation of the

Table 1 Types of prototypes used to evaluate the various product design features

	Sketch	Scenario/ storyboard	3D model/virtual prototype	Interactive virtual prototype	Physical prototype
Esthetics	X		X		X
Ergonomics					X
Usability		X		X	X
Functionality			X	X	X
Performance					X
Reliability					X

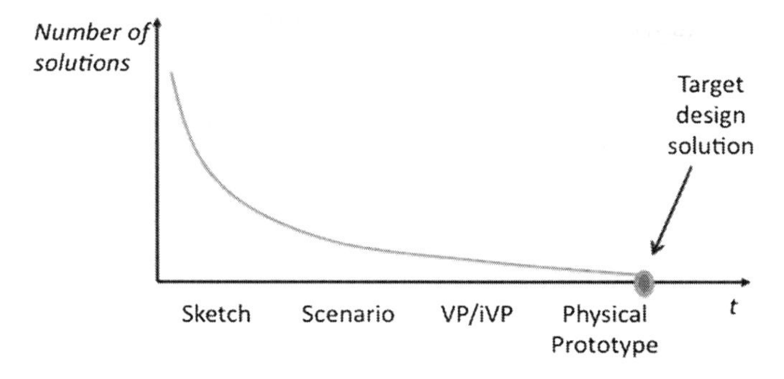

Fig. 1 Methods used to evaluate and assess the design of new products

product. Conversely, it is difficult to build a physical prototype that is very similar to the product, especially in the early stages of design; in addition the physical prototype is rigid and not easily modifiable.

A solution proven effective in other research works is based on the integration of virtual and real components [9]. Still, an open issue is that the physical components, once developed, are difficult to modify and re-submit to users for the evaluation. This research proposes the use of a mixed prototype, including virtual and physical components, where the virtual components are parametric, and the physical ones are configurable.

3.1 Classification of Prototypes

Simple prototypes are valuable at an early stage to explore alternative design solutions. Although there is benefit in making the design solutions as realistic as possible, it is important not to invest so much time, money or commitment on realistic prototypes, that there is reluctance to change the design. The effectiveness of a prototype can be evaluated in respect to several characteristics [10]. Hereafter the main ones are listed.

- *FIDELITY.* This characteristic defines the level of fidelity of the prototype in general, or of the various aspects of the product such as its visual representation, its haptic behavior, or its auditory features.
- *COMPLETENESS.* This characteristic identifies which is the degree of implementation of the product design that the prototype covers. In fact, typically a prototype implements some parts, but not all, of the product design, also in relation to the kind of evaluation and test that we intend to carry out.
- *FLEXIBILITY.* This characteristic identifies how easy and quick is to implement changes in the prototype features. For example, it indicates if it is possible to change colors and textures of the interior of a car, or if is possible to change the haptic behavior of a door handle, quickly and easily as soon as a user asks for it.

- *COMPLEXITY of REALIZATION.* This characteristic defines the complexity of developing the prototype. It depends on the technologies that are used, on the ad hoc components that need to be specifically developed for the realization of the prototype, on the software code that has to be specifically implemented, etc.

4 Mixed (Virtual and Physical) Prototypes

In order to perform an effective assessment of the design of a new interactive product, the focus is on the development of an interactive prototype that has the following characteristics: high fidelity, completeness, flexibility to configure, and easiness to implement. All these characteristics are not easily achievable in a virtual prototype.

The methodology we used for optimizing these aspects and build an effective prototype was based on the implementation of a prototype consisting of a proper mix of virtual and real components. Mixed prototyping has been proposed in some other research works [4, 11, 12]. The work presented in this paper proposes very flexible mixed prototypes where it is possible to configure both the virtual and physical components of the mixed prototype.

Therefore, it has been defined a flow of activities for the development of mixed prototypes. The definition of the activities is related to the structure of the target products. So, with reference to the typical components of products covered in this research, i.e. electronic products for the consumer market, it has been decided the hypothesis that a corresponding interactive prototype would consist of:

- virtual components
- physical components
- Graphical User Interface and interaction functional model

The initial activity plans the construction of the 3D model of the product, which comprises a set of subcomponents. It is possible to create a realistic rendering from the 3D model, which allows you to view the product as if it were real. Some parts can be developed physically for example by using fast rapid prototyping techniques, which are nowadays wide spreading rapidly [13]. For example, one can create the external cover and the interactive components, such as buttons. Typically, this type of products includes a display with which the user can interact via touch screen, buttons, interaction wheels, etc. The information is displayed on the screen through a GUI, and can be called according to sequences defined by an interaction functional model. The flow of activities for the creation of a mixed prototype is shown in Fig. 2, and includes the production of the components mentioned above (hardware and software), and a physical layout where to install the various physical components.

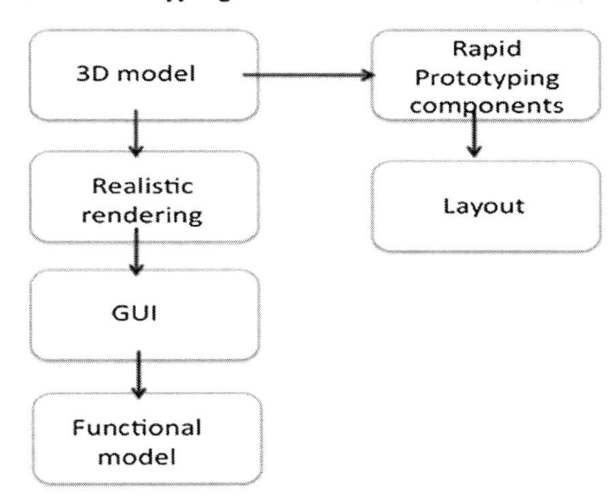

Fig. 2 Flow of activities for the development of a mixed prototype

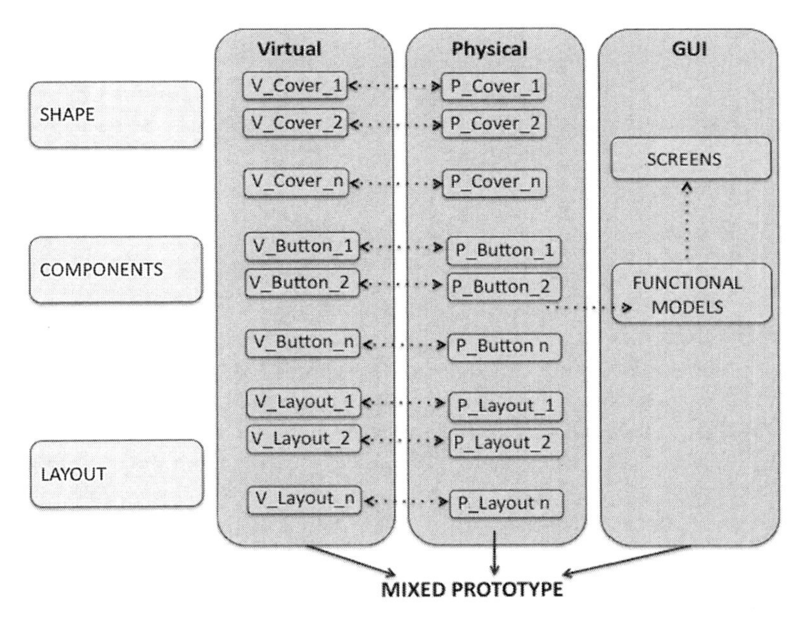

Fig. 3 Mixed prototype consisting of parametric virtual components and configurable physical components conveniently integrated

When developing a mixed prototype there is the necessity of linking and seamlessly integrating the virtual components with the physical ones (Fig. 3). Therefore, it is necessary to create a correspondence between the virtual and the physical components. In order to quickly change the virtual components, a solution can be implementing them as parametric, as well as developing the physical ones so that they are easily interchanged.

The resulting mixed prototype is highly configurable. It is possible to change the shape of the device and of the components, change the material and color, and also change the position of the interactive components as well as their behavior.

5 Case Study

One of the case studies selected to apply and test the methodology concerns a handheld device for domotics, implementing functions for the control of house equipments. The device can be placed on a holder attached on the wall, or held. After defining the device functions, we began developing the first sketches of the device, focusing on the button layout and esthetics, which were shown to a selected group of users. Then, we have developed a dynamic scenario consisting of a sequence of images (screen snapshots) that can be browsed through a hypertext structure, thus simulating the interaction with the GUI of the product. The prototype hypertext can be easily implemented by using various tools, with a limited effort, for example by using the MS PowerPoint application. These two prototypes have allowed us to select the configurations that at best suite the target users. Subsequently, we have developed a mixed prototype, as described hereafter.

5.1 Mixed Prototype Development

In order to test more accurately with users the esthetics, ergonomics, usability, and functionality of the variants of the handheld device, thus further converging to the preferred solution and configuration, we have implemented a mixed prototype. According to the methodology described in the previous section, first we have developed a 3D model of the device cover and of the interactive components (buttons), in their various shapes. Then, we have manufactured the corresponding physical components.

In order to connect the physical buttons with the functional model of the graphical user interface, and to enable user interaction with the functionality of the device, the components are connected to electronic circuitry. For that, we have used the Arduino system, which is an open-source electronics prototyping platform based on flexible, easy-to-use hardware and software [14]. Arduino is intended for designers and anyone interested in creating interactive environments. It can sense the environment by receiving input from a variety of sensors and can affect its surroundings by controlling lights, motors, and other actuators. Specifically, the mixed prototype has been built using the Arduino One. The circuitry for the control of the interface has been obtained by welding on a breadboard the various

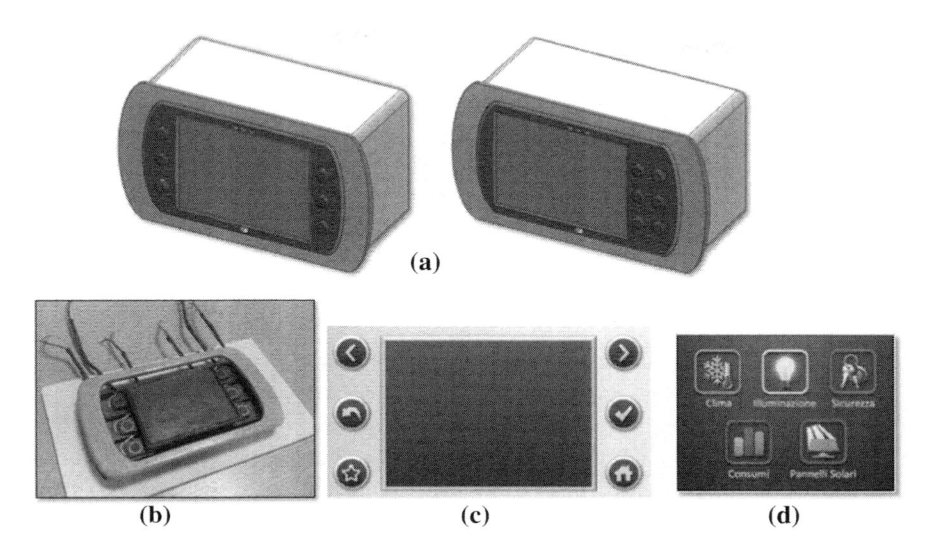

Fig. 4 Mixed prototype of the handheld device

selected electronic components, which consist of pushbuttons that detect the pressures exerted by the user, returning 5 v tension when pushed, and 0 v when released. The interaction with the buttons is controlled by a program implemented using the MenuBackend library, which allows a very flexible management of the menu, through a simple tree structure implementing the interaction functional model.

The layout has been developed by positioning some strips of velcro under the cover. Pieces of velcro has been also put under the physical buttons, so that the buttons can be conveniently repositioned. In this way it is possible to compare various variants. For example, it is possible to position six buttons on the right hand side, or three buttons on both sides. The final prototype is shown in Fig. 4. Figure 4a) shows the virtual prototypes, Fig. 4b) the mixed prototype, Fig. 4c) shows one of the functional buttons configuration, and Fig. 4d) one of the screen shots.

5.2 Evaluation of the Mixed Prototyping Methodology

After building the Mixed Prototype we have performed some tests with potential end users of the handheld device in order to validate the proposed methodology based on mixed prototyping.

The users test has been performed according to the ISO standard about ergonomics of human system interactions [2], where it is planned to consider seven characteristics that an interface between a user and an interactive system should

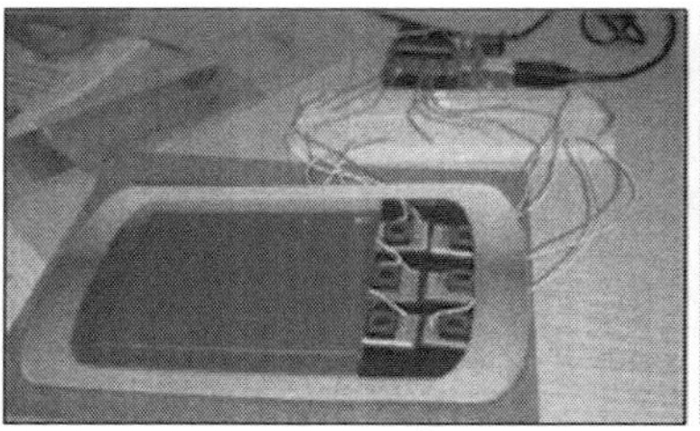

Fig. 5 Two different configurations of the device

have, for the analysis and evaluation of the functioning of the system: suitability for the task, self-descriptiveness, conformity with user expectation, suitability for learning, controllability, error-tolerance, suitability for individualization.

The objective of the user assessment was to evaluate the positioning of the interactive buttons in the layout, which have been arranged in two different configurations (Fig. 5): (1) all buttons placed on the right hand side, (2) three buttons placed on the left and three on the right side.

After a brief introduction to the functionality of the device under test, we allowed the users to use the control device for about ten minutes. The users were able to learn and test the functions of the device. The two configurations were presented to the participants in a quick sequence. In fact, the prototype set-up allowed us to quickly switch from one configuration to the other. At the end of this phase, an evaluation questionnaire has been submitted to each participant.

The questionnaire is divided into sections and provides an evaluation of the device according to the criteria of the followed ISO standard. For what concerns the section about the system usability and functionality, the users have been asked to assess the location of buttons and the functions associated with them.

At the conclusion of the test, according to the preferences expressed by the users, the preferred configuration is one in which the functions for the navigation are placed on the two sides of the device. Then, we have performed a second test for evaluating the position of the functions for browsing in the menu. The functions have been associated with the buttons, which configuration was varied as follows: (1) first buttons at the top, one on the right and one on the left, (2) two buttons on the right hand side, one below the other. The protocol used for the evaluation is similar to the previous one. At the end, again we were able to detect the user preferences concerning the functions arrangement.

Finally we have also asked the participants to evaluate the fidelity of the prototype, with respect to a real product of the same type. The perception of realism reported by the participants was quite high and satisfactory.

The tests have allowed us to evaluate the benefits of using configurable and parametric Mixed Prototyping for the evaluation of new interactive products. The tests have shown the following:

- the realism of the Mixed Prototype was considered good by the final users

- it was quick and easy to change configurations
- the configurations were switched in 2–3 min, reducing the time from one trial and the following one
- it was possible to test the esthetics of the cover and components, as well as the ergonomics and usability.

6 Conclusion

The paper has presented a methodology developed for the evaluation of interaction of new products, which is based on the use of mixed prototypes that are parametric and configurable. The methodology developed has been evaluated through the development of a case study presented and discussed in the paper.

The use of mixed prototype has allowed us to refine the evaluation previously performed with users by using other prototyping methodologies, as sketches and dynamic scenarios. The prototype has revealed to be very flexible, both for the virtual components, defined parametrically, and the physical components, which are highly configurable. The realization of the prototype includes the development of the virtual model, of the GUI and of the functional model, which can be re-used later on in the subsequent phases of the product design. The major effort has been put on the development of the correlation of the physical component with the functional model, which can rely on circuitry and modules that can be easily re-used for the development of several similar applications.

Acknowledgments The authors would like to thank Samuele Polistina, and the students Alessandro La Rosa, Luca Marseglia for their contribution to the development of the case study.

References

1. Norman DA, Draper SW (1986) User-centered system design: new perspectives on human-computer interaction. Lawrence Erlbaum Associates
2. ISO DIS 9241-210 (2008) Ergonomics of human system interaction—Part 210: human-centred design for interactive systems. International Organization for Standardization (ISO). Switzerland
3. Nielsen J (1994) Usability engineering. Academic, Cambridge
4. Bordegoni M, Polistina S, Carulli M (2010) Mixed reality prototyping for handheld products testing. Research in interactive design: proceedings of virtual concept 2010, Springer Verlag
5. Dixon JR, Poli C (1995) Engineering design and design for manufacturing, a structured approach. Field Stone Publishers, Conway
6. Kawaguchi K, Endo Y, Kanai S (2009) Database-driven grasp synthesis and ergonomic assessment for handheld product design. In: Duffy VG (ed) Digital human modeling, HCII 2009, Springer, Berlin

7. Kanai S, Higuchi T, Kikuta Y (2009) 3D digital prototyping and usability enhancement of information appliances based on UsiXML. Int J Int Des Manuf 3(3):201–222 (Springer, Paris)
8. Ferrise F, Bordegoni M, Cugini U (2012) Interactive virtual prototypes for testing the interaction with new products. Comput Aided Des Appl (in press)
9. Bordegoni M, Cugini U, Caruso G, Polistina S (2009) Mixed prototyping for product assessment: a reference framework. Int J Int Des Manuf 3(3):177–187 (IJIDeM, Springer)
10. Bordegoni M (2011) Product virtualization: an effective method for the evaluation of concept design of new products. In: Bordegoni M, Rizzi C (eds) Innovation in product design: from CAD to virtual prototyping, Springer
11. Costanza E, Kunz A, Fjeld M (2009) Mixed reality survey. In: Lalanne D, Kohlas J (eds) Human machine interaction, LNCS 5440, Springer, Berlin, pp 47–68
12. Bimber O, Raskar R (2005) Spatial augmented reality: merging real and virtual worlds. A. K. Peters, Massachusetts
13. Wright PK (2001) 21st Century Manufacturing. Prentice-Hall Inc, New Jersey
14. Arduino (2012) http://www.arduino.cc/. Accessed on Oct 2012

Study of Postural Variation, Muscle Activity and Preferences of Monitor Placement in VDT Work

Rajendra Patsute, Swati Pal Biswas, Nirdosh Rana and Gaur Ray

Abstract Various studies investigated sitting posture while working at VDT for muscle response, posture and preference, to deduce recommendations for design. The posture assumption is a dynamic activity and is often resulted out of continuous visual and physical feedback processed on cognitive level to maintain optimum comfort. The parameters that affect the work station design such as visual display terminal height directly affect the posture and comfort. This paper discusses the study of postural angles such as Head inclination, Trunk bending, Trunk inclination and sEMG muscle activity of Neck Extensors, Erector Spinae, Sternocleidomastoid and Upper Trapezius muscles when a Visual Display Terminal (VDT) user is working at a visual display terminal. The conditions were simulated on a test-rig which was developed on the basis of anthropometric data obtained. The experiment was performed on eight Indian male subjects and the VDT height was varied from 69.5 to 119.5 cm, monitor inclination was varied as 0, 30, 60°. Changes in the sitting postural angles was recorded using photogrammetry, simultaneously sEMG muscle activity of the defined muscles was recorded, also the preferred position of the VDT at each height as responded by the subjects was recorded including preferred VDT height. It was observed that the thoracic bending varied between 120 and 155° which increased with increasing VDT

R. Patsute (✉) · S. P. Biswas · G. Ray
Industrial Design Centre, Indian Institute of technology Bombay, Mumbai, India
e-mail: rajenp@iitb.ac.in; rajenpats@gmail.com

S. P. Biswas
e-mail: palswati27@gmail.com

G. Ray
e-mail: ggray@iitb.ac.in

N. Rana
Theem College of Engineering, Boisar, Mumbai, India
e-mail: ranank@rediffmail.com

A. Chakrabarti and R. V. Prakash (eds.), *ICoRD'13*, Lecture Notes in Mechanical Engineering, DOI: 10.1007/978-81-322-1050-4_36, © Springer India 2013

height. The trunk inclination with reference to horizontal varied from 96 to 65°. Similarly neck inclination was observed to change by a span of approximately 40°. The preferred monitor height was observed to be in the range of 89.5–99.5 cm. The Upper Trapezius and Sternocleidomastoid showed variation in muscle activity but not related to monitor height. The Erector Spinae and Neck extensors however responded to the variation in monitor height in exactly opposite pattern. It was observed that the point of intersection of the normalized sEMG ratio curves lies in the span of preferred VDT height responded by the subjects. The potential application of this research is in design of Sit down console, computer work stations/consoles with or without adjustability features in Indian context.

Keywords VDT height · Work station design · Sitting posture

1 Introduction

Visual display terminal height is known to influence posture and muscle during computer usage. Posture assumption is a dynamic activity and is often resulted out of continuous visual and physical feedback processed on cognitive level to maintain optimum comfort. Parameters e.g. display height directly affect the posture and comfort. The electromyographic response of the muscles is known to vary with the changes in display height. Posture is directly related to tissue loads and postural investigation associated with muscle activity reveals interesting facts. It has been the most commonly used indicator of musculoskeletal demand in studies evaluating VDT heights [1]. In design of work stations this issue of postural comfort has been partially resolved by providing flexibility of setting various work station parameters. It is but important to understand how various parameters that influence sitting posture and comfort behave in relation to each other. So that it adds to the designers knowledge base thereby affecting the outcomes. This study analyzes multiple parameters at varying heights and inclinations of VDT. The posture is studied by observing variation in defined postural angles and electromyographic response of involved muscles.

In the VDT work posture studies researchers have discussed the surface electromyographic response of the back muscles and the regions being primarily Cervical, Lumbar and Trapezoidal. The most important muscles taken into consideration are Neck extensors, Sternocleidomastoid, Upper Trapezius and Erector Spinae because of their prominent involvement in attaining the posture. The prime consideration in VDT workstation studies has been the placement of the viewing unit which is the point of information fixture around which the human body tends to manipulate and adjust itself. Since this study is about evaluation postural variation and muscle load variation in response to the changes in VDT height and inclination it is essential to take a overview of various other studies which have attempted to understand the underlying phenomenon.

It has been reported that surface EMG from electrodes located in the posterior *upper cervical region* reflected differences in VDT location, at least between extremes in location [2]. In the study carried out by Villanueva et al. [3] on ten subjects the muscle activity was recorded at c2–c3 level for the neck extensor muscles and over lateral portion of trapezius. The extensor activity normalized to maximum showed significant increase with more flexed neck and decreasing VDT height. A backward leaning trunk showed decreased trapezius muscle activity. Hamilton [4] observed that sEMG signal of Sternocleidomastoid and Neck extensor muscles changed in response to the position of the copy holder both in reading and typing task, the activity was more in the later. Turville et al. [5] compared the 15° and the 40° recommendations for VDT placement for muscle response of a set of back muscles. The 40° VDT placement or lower VDT position (LMP) showed significantly higher muscle activity for right sternocleidomastoid, right levator scapulae, right cervical erector spinae, left cervical erector spinae, right thoracic erector spinae and left thoracic erector spinae. In experiment carried out for bifocal conditioned subjects Kumar found that the trapezius and sterno-cleidomastoid sEMG was less by 30–40 % in the sunken VDT position [6]. On the other hand Aaras et al. [7] found no significant difference in trapezius EMG levels between high and low VDT position. They studied muscle load from the upper part of trapezius.

Hsio and Keyserling [8] in their pilot studies with three subjects evaluating posture behavior during seated tasks observed that postures were affected by target location, body size, and target size. Also the movement of the seat pan was seen as important for adjusting to a comfortable work posture. De Wall et al. [9] found that the position of the head in which neck muscle forces are minimal differs from the position of the head when some commonly accepted recommendations for VDU placement are used and further say that 15° above horizontal is better than the 15° below. Posture improved with increase in angle above horizontal. Villanueva et al. [10] studied the interrelationship between eye position and body posture and suggested that changes in body posture compliment the eye position in attaining a better view of the visual target. Viewing angle was mainly decided by neck and eye inclination and to a lesser extent by thoracic. Burgess-Limerick et al. [11] studied the head and neck posture influenced due to eye level and low VDT locations when subjects performed a word processing task. Lowering the VDT position did not cause changes in the posture of the neck relative to the trunk but did increase the flexion of the head relative to the neck.

Mandal [12] has displayed that with inclined seat and inclined desk a better sitting posture is achieved when the desk is inclined by 10° and seat by 15° reducing the bending that is required at the hip resulting in more comfort feeling. With horizontal chairs the same effect is achieved by users by resting only the buttocks on the seat. Fujimaki and Mitsuya [13] have discussed that backward tilt makes it difficult for the operators to keep their feet flat on the floor. Operators body posture keeps on changing and they respond in various ways like forward tilt, the reclining posture makes it difficult to keep feet flat on the floor. Operator behaviors are peculiar, in the study conducted by Ray et al. [14] it was observed that VDT operators (74 %) never adjusted their chair height even when the adjustment facility was provided, half of the users acquire forward leaning posture while working on VDT which makes the backrest practically useless.

Apart from the postural and muscular issues the visual accommodation and vergence capabilities tend to be in favor of Lower VDT placements. Since eyelids covered more region of eyeball in LMP it results in less eye dryness [10]. In quite a few studies that studied VDU placement with help of discomfort and preference the reports have been rather mixed. Kumar [6] studied the reported discomfort in three positions of VDT namely sunken, level and raised the reported rating was in the ratio of 1:2:4 respectively, informing that the discomfort was maximum in high VDT placement. In studies by Turville et al. [5] seven of the twelve subjects preferred 15° below horizontal placement whereas remaining preferred the 40° below horizontal position. Svensson [15] investigated the lowest loading height of the VDU on neck and shoulders of the operators. The study showed that to load the neck as little as possible during the VDU work, each individual should adjust the height of screen to obtain a suitable viewing angle for the eyes and one at which it is easy to keep the neck straight. The study shows that a viewing angle of +3° gave less load upon the neck than the −20° position.

Optimal VDT placement is a compromise between visual and musculoskeletal system [16]. Much research has focused on identifying the optimal screen height and there is to this date no clear consensus. So often context based studies are more useful and recommend best options for the said context. And so while recommending guidelines for design it is very essential to carry out studies simulating the contextual user parameters and derive recommendations from the studies.

The task of working at VDT involves the neck muscles, shoulder muscles and the muscles of the lumber region. The angle of head, bending of thorax and trunk inclination are prominent in deciding the sitting posture. Experimentation discussed here covers the relative behavior of the muscles and corresponding changes in posture as the VDT height is varied from a very low to high position. The study here attempts to investigate the correlation between sEMG responses of a few selected muscles from the back and the corresponding variations in the posture triggered due to input variable such as VDT height. Also it is attempted to correlate this data with the preferred position of the screen. The results are discussed in light of earlier studies and it is attempted to get a comprehensive view of the play of different variables in sitting posture and sEMG activity of involved muscles while user works at VDT to a limited extent.

2 Method

To enable variation of VDT height and inclination, a test-rig was prepared (see Fig. 1) to simulate the conditions of the VDU work station and enable controlling variables like VDT height. Slotted supporting structures were fitted on the table (Fig. 1) which enabled shifting of the VDT vertically. The VDT could also be rotated about its axis which enabled changes in VDT inclination angle (A) at

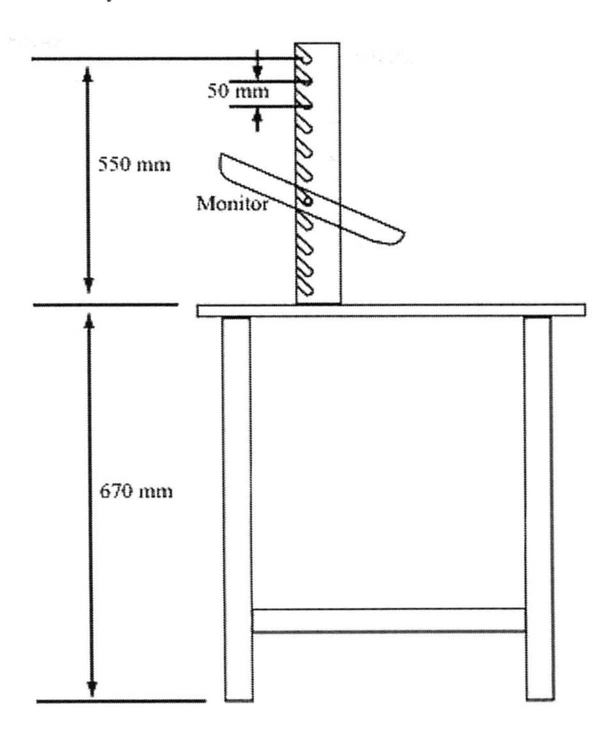

Fig. 1 Schematic of the rig developed for the experiment

various heights (H). The height of horizontal surface on which keyboard rested was fixed after studying the anthropometric dimensions of Indian male adults in sitting position equal to 670 mm. The subjects were given to sit on a seat with adjustable height which was set to the popliteal height of the subject. It has been observed that the operators while working at VDT tend to sit forward in the chairs ignoring the backrest. The eyes of the operators are strongly fixed on the screen and hands are busy operating the keys or mouse. Considering these aspects no back support or elbow-rest was provided to the subjects. And subjects were asked to adopt a comfortable working posture on the rig with thighs horizontal and feet flat on the ground. The subjects were bare bodied with minimal clothing to facilitate sticking of electrodes and markers that were used to collect postural information by photogrammetry and the electrodes captured the sEMG signal of the muscles.

The height of VDT was varied in 11 stages from 69.5 to 119.5 cm and in each stage three angles of monitor that is 0, 30 and 60° were considered. To capture information on postural changes photographs of subjects were taken in each experimental state. These photographs were then imported into AutoCAD to measure the required postural angles. White markers were put on to the subject to mark various angular positions as shown in the Fig. 2. Position of the markers on the subjects body were fixed at 7th cervical vertebra, Angulus inferior scapula (AIS), Iliac crest, Temporal canthus, In proximity of the center of the outer canal of the ipsilateral ear.

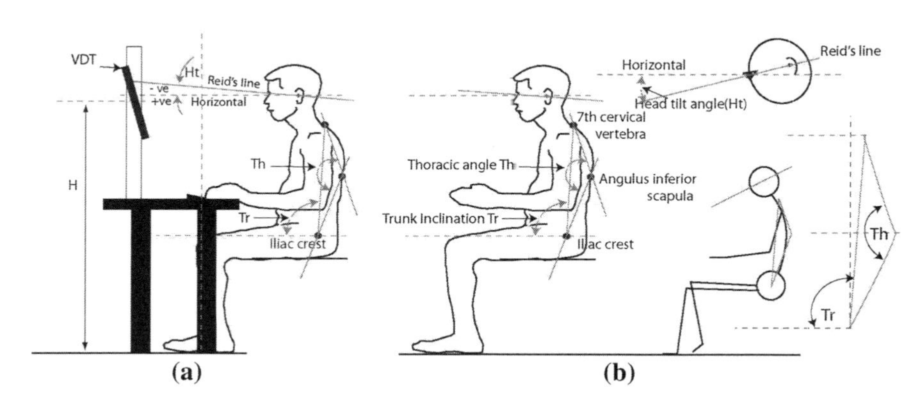

Fig. 2 Position of postural markers

Different variables considered for evaluation in the experiment were: Thoracic bending (Th) is the angle formed by the intersection of lines joining 7th cervical vertebra-AIS and AIS-Iliac crest which gave the thoracic bending of the subject. Head inclination (Ht) is the angle formed by the intersection of lines joining temporal canthus-proximity of the centre of the outer canal of ipsilateral ear and the horizontal. The angle is taken as positive when the former line is inclined and subject is looking below horizontal. Trunk inclination (Tr) is Angle formed by the intersection of lines joining Iliac crest-7th cervical vertebra and horizontal. In order to capture the sEMG signal of the neck extensor muscles electrode were placed at the level of second and third cervical vertebra, over the cervical portion of the descending part of the trapezius. To measure the activity of upper trapezius electrodes were placed at 2 cm lateral to the midpoint of the line between the 7th cervical vertebra and acromion and three cm below the processus mastoideus (Mastoid process) to measure activity of the Sternocleidomastoid. For Erector Spinae muscle the electrodes were placed at 2 cm lateral from the midline of spine at L5-S1 level.

A multichannel (16 Channel) data acquisition system MP100 from Biopac Inc was used for EMG signal recording from all the four muscles simultaneously. The electrodes used were of Ag/AgCl disc type. Before placement of the electrodes the skin was abraded to remove dead cells and was cleaned with 70 % ethanol. The data acquisition software Acq 3.7.3 was used to capture data. Figure 3 below shows the set up. The test was performed by keeping the amplifier gain 500 and sampling rate 2,000. Integrated RMS signal was obtained from the raw EMG signal for data extraction and analysis.

Eight male subjects of varying body sizes were chosen for the experiment. The subjects were asked to play an interactive game on the screen. The Sitting eye height of the subjects varied between 166 cm and 150 cm and weight varied between 49 and 81 kg. The detail statistics of subjects of subjects is shown in Table 1.

Fig. 3 Biopac MP100 data acquisition system

The task demanded the subject to keep attention continuously on the screen following the moving visual target and interact with the system via keyboard and mouse for control. Room illumination was maintained at 185 to 240 lux with partially natural and diffused artificial light. LCD VDT with display size 41.3 by 25.9 cm was used. The subjects were asked to play for approximately one and half minute in every data acquisition state the total duration therefore approximated 50 min. There were total $11 \times 3 = 33$ data acquisition states. The readings in each state were taken in the later half minute when the subject was fully engrossed in the game. The subjects were given intermittent rest after every height and was asked to relax when height of the VDT was changed. Figure 4 below shows experimental setup where subject is participating in the experiment.

Raw data was processed to be made interpretable. In every state data, average was found from the RMS signal. Normalization was carried out by dividing it by average RMS signal of reference state. In the reference state subject was asked to sit in normal posture with feet flat on the floor and hands held against trunk, bent at the elbows and palms straight (Fig. 2). This type of ratios was found for all the 33 states of a subject. These were then classified as variation of VDT height versus ratios for Upper trapezius, Sternocleidomastoid, Neck extensor and Erector spinae and the angles in respective states that is trunk inclination, thoracic bending, Head

Table 1 Subject details

Sub No	Age (years)	Weight (Kg)	Height (cm)	Eye ht (cm)	Popliteal ht (cm)	Sitting eye ht(cm)
1	34	60	165	154	48	112
2	25	57.36	168.5	155.2	47	117.5
3	28	49.5	169.5	157.8	45	118.9
4	22	63.8	164.2	150.6	43.9	112.7
5	30	56.7	170	158	46	119.8
6	25	56	164	150.7	45.7	112.7
7	25	81	173	161.4	44	122
8	22	77	176.5	166	48.5	122

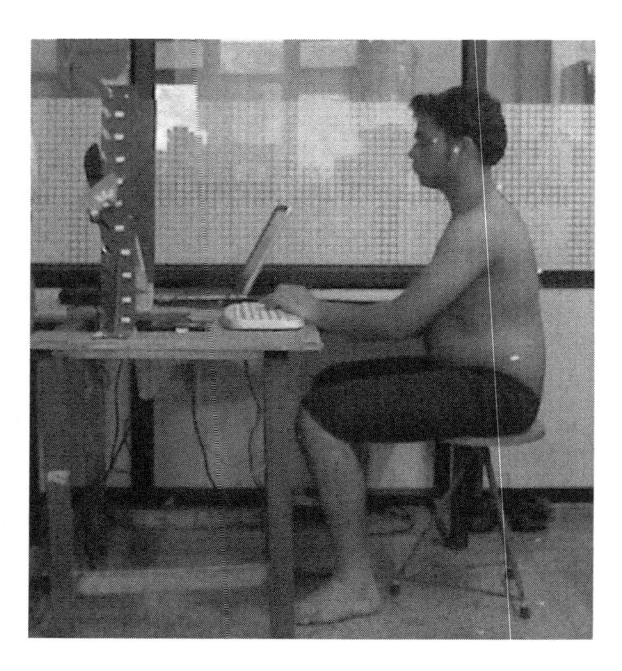

Fig. 4 Experimental set up with subject

inclination for all three states on viz 0, 30 and 60° inclination. The ratios were also classified as variation of Inclination versus Ratios for Upper trapezius, Sternocleidomastoid, Neck extensor and Erector spinae and the angles in respective states that is trunk inclination, thoracic bending, neck inclination for all states for different heights.

3 Results

The responses of posture to changes in VDT height is revealed in the variation of angular data on the subjects. The trunk angle is observed to be generally in the zone of 72–93°. Table 2 displays the summary of trunk inclination data obtained, in column heads prefix Tr is for trunk inclination and suffix 1 represents the subject number.

All subjects showed either a steady or increasing trend. Increasing trunk inclination angle means the subject is swaying away from the monitor. However this increase is not continuous and often there is steep fall in the angle indicating the subject bending more towards the screen, and may then again tilt back increasing the trunk inclination angle. The Changes in monitor inclination doesn't seem to affect trunk inclination. The average for all the subjects of the angles at various heights shows increase with increase in VDT height for all three positions: A0, A30, A60. Figure 5 displays the average plot and trend using linear regression equation for angle A0, A30 and A60.

Table 2 Summary of trunk inclination data

Sub-	Tr1	Tr 2	Tr 3	Tr 4	Tr	Tr 6	Tr 7	Tr 8
At A0								
Max	86.21	91.32	90.56	84.29	92.27	85.56	80.10	85.61
Min	81.93	82.08	74.84	80.32	84.98	82.09	72.54	74.32
Av	84.38	87.62	81.36	82.56	89.77	84.16	77.11	79.77
SD	1.34	3.40	5.43	1.38	2.90	0.99	2.52	4.03
At A30								
Max	86.19	92.60	87.42	84.06	96.27	85.25	80.67	85.43
Min	80.80	83.11	65.92	79.07	85.12	82.84	72.98	76.93
Av	84.50	87.36	79.90	82.58	90.50	84.18	77.82	80.34
SD	1.59	3.23	6.60	1.61	3.35	0.67	2.66	2.93
At A60								
Max	86.71	90.71	90.37	85.48	95.01	84.82	84.77	91.40
Min	82.73	84.38	69.12	81.15	85.64	82.43	73.25	73.21
Av	84.69	87.64	82.26	83.45	90.39	83.88	77.72	82.47
SD	1.43	2.12	6.74	1.51	3.03	0.74	3.49	4.65

The increase in trunk angle means that the subject is swaying away from the VDT screen implying that increase in height of the VDT seems to make the subject bend away from the screen. The thoracic angle varies between 125 and 155°. Increase in thoracic angle increases the height of the sitting subject and indicates straightening of the body. Though the increase may not be consistent and every now and then steep fall is seen which is compensated in the next higher stages. Generally thoracic angle is observed to increase or remain almost same with increase in VDT height indicating decrease in thoracic bending or varying around constant with increase in VDT height. Table 3 below gives summary of Thoracic angle data obtained.

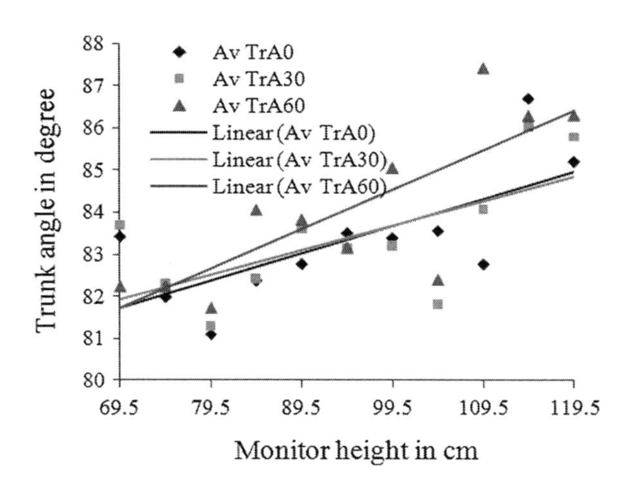

Fig. 5 Plot of Average Trunk inclination vs VDT height at VDT inclination A0, A30 and A60

Table 3 Thoracic angle variation

Sub-	Th1	Th 2	Th 3	Th 4	Th 5	Th 6	Th 7	Th 8
At A = 0								
Max	136.85	150.82	154.43	144.55	152.86	142.98	138.81	148.89
Min	127.10	123.78	124.24	141.11	133.79	138.20	131.11	139.59
Av	131.38	138.56	135.91	142.91	145.85	140.94	133.78	144.67
SD	3.26	8.62	11.76	1.02	6.44	1.70	2.24	2.59
At A = 30								
Max	135.14	149.37	152.86	145.37	151.53	142.71	136.94	147.22
Min	126.18	121.19	124.63	139.45	135.84	138.95	131.26	136.54
Av	131.00	137.50	138.85	142.96	146.29	141.03	134.03	143.80
SD	2.55	10.03	10.59	1.96	5.46	1.40	2.08	2.90
At A = 60								
Max	135.40	149.86	153.97	146.51	155.15	146.72	142.61	151.66
Min	126.76	134.12	125.26	143.22	134.23	136.27	130.71	140.20
Av	131.84	144.51	144.34	144.32	146.59	141.42	134.79	146.90
SD	2.82	4.73	11.71	1.03	6.84	2.43	3.23	3.79

The average of subjects' thoracic angles at various heights shows increase with increase in VDT height for all the three positions: A0, A30, A60. Figure 6 shows plot of Average Thoracic angle against VDT height at A0, A30 and A60 for all the subjects. The head tilt angle decreases with the increase in the VDT height, the angle below horizontal i.e. subject looking down is referred to as positive neck angle. The neck and effectively the head seems to be altered continually with changes in VDT height. The changes in VDT inclination appear to be compensated more by the eyeball movements. Following Table 4 shows summary of the head inclination variation. For all the subjects the head tilt angle is seen to distinctly vary with the monitor height. The neck bending reduces with increase in monitor

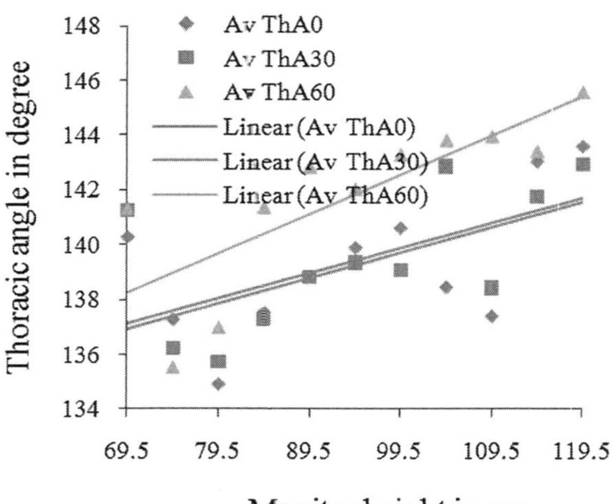

Fig. 6 Plot of average of subjects thoracic angle against VDT height at A0

Table 4 Summary of head tilt angle variation

Sub-	Ht1	Ht 2	Ht 3	Ht 4	Ht 5	Ht 6	Ht 7	Ht 8
At A = 0								
Max	0.16	22.38	3.17	−11.39	4.01	−6.62	−3.16	0.81
Min	−24.79	−13.35	−25.76	−45.21	−36.61	−20.11	−27.54	−21.73
Av	−11.52	6.25	−14.68	−24.66	−12.77	−13.18	−13.04	−10.67
SD	8.53	11.38	9.05	10.07	12.64	4.71	8.66	9.00
At A = 30								
Max	−1.14	23.41	0.18	−15.01	10.62	−4.26	−3.75	4.87
Min	−25.83	−13.49	−33.02	−40.58	−38.07	−19.20	−22.10	−33.10
Av	−12.07	6.82	−18.27	−25.63	−14.03	−13.20	−11.79	−14.69
SD	8.30	11.82	10.12	7.89	14.70	4.95	7.32	11.69
At A = 60								
Max	−2.41	17.24	−4.80	−14.32	6.47	−4.98	−1.84	4.91
Min	−22.79	−17.85	−34.03	−43.66	−36.27	−19.40	−27.93	−41.15
Av	−12.70	0.94	−17.58	−31.39	−15.75	−12.24	−15.68	−19.61
SD	7.37	11.37	10.40	8.76	13.55	5.05	9.48	13.92

height. Figure 7 below shows the plot of head tilt angle against VDT height at A0, A30 and A60. It was observed from the data that Upper trapezius activity did not show significant trend. The activity in arm and its positioning seems to introduce muscle activity which is strong but doesn't display any specific pattern.

Upper trapezius muscle was more susceptible to the movement caused by the subject in the hand arising out of adjusting to best position or better comfort and there was no specific trend observed common to all the subjects. Sternocleidomastoid also showed activity which is caused other than that due to changes in the VDT placement mostly influenced by voluntary stretching of neck muscles by the subjects and as such did not show a significant trend common to all the subjects. Table 5 below shows the summary of sEMG Ratios for Neck extensor muscles.

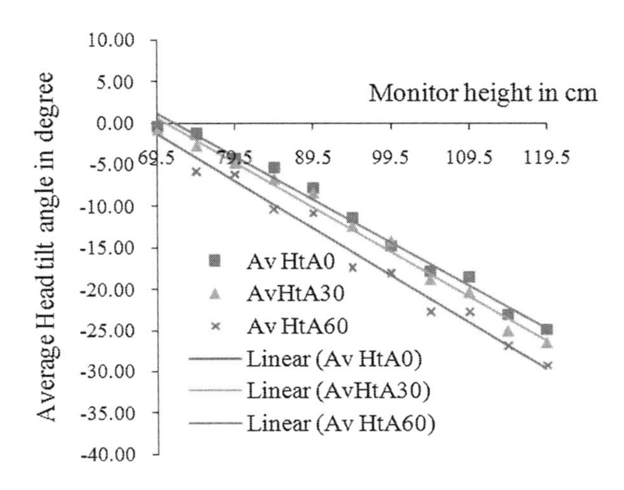

Fig. 7 Plot of average of subjects head tilt angle against VDT height at A0, A30 and A60

Table 5 Summary of results for sEMG ratios of Neck extensor

Sub-	Ne1	Ne2	Ne3	Ne4	Ne5	Ne6	Ne7	Ne8
At A = 0								
Max	1.70	1.63	1.98	1.53	1.61	1.37	1.79	2.42
Min	1.38	0.54	0.32	0.89	1.33	1.21	0.85	1.66
Ave	1.52	1.12	0.93	1.15	1.43	1.28	1.23	1.93
SD	0.11	0.33	0.46	0.18	0.10	0.05	0.39	0.24
At A = 30								
Max	1.72	1.41	1.28	1.45	1.79	1.39	1.78	2.50
Min	1.36	0.41	0.40	0.86	1.27	1.20	0.82	1.65
Ave	1.48	1.04	0.69	1.13	1.48	1.28	1.20	1.97
SD	0.11	0.36	0.24	0.17	0.19	0.06	0.36	0.25
At A = 60								
Max	1.64	1.29	0.94	1.37	1.70	1.41	1.92	2.29
Min	1.29	0.65	0.37	0.80	1.27	1.17	0.76	1.64
Ave	1.48	1.08	0.64	1.13	1.45	1.26	1.27	1.94
SD	0.11	0.22	0.21	0.19	0.13	0.08	0.44	0.23

Unlike Sternocleidomastoid and Upper trapezius the Neck extensors of all subjects tend to show a decreasing trend with increasing monitor height. Figure 8 shows plots of the average of Neck extensor values for the subjects versus the monitor height. The Av of neck extensor showed a decreasing trend with increasing VDT height.

Table 6 below shows the summary of Lumbar Erector Spinae sEMG ratios of the eight subjects. The trend lines show a pattern increasing with monitor height i.e. as the monitor height increases the Erector spinae shows greater sEMG activity. The preferred inclination of the VDT varies with height of the VDT. Table 6 gives summary of preferred monitor inclination at various heights of VDT. Following plot in the Fig. 10 shows the preferred VDT inclination at various

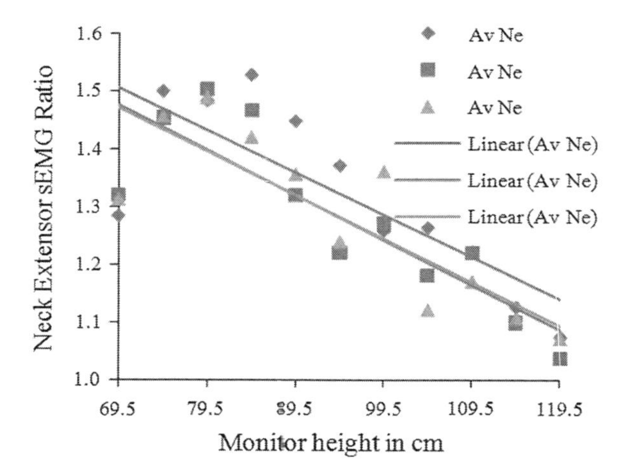

Fig. 8 Plot of average sEMG ratio of all subjects against VDT height at A0, A30 and A60 for neck extensor muscle

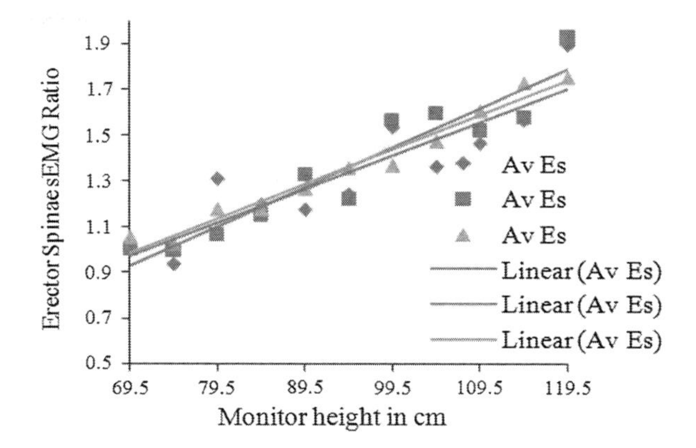

Fig. 9 Plot of sEMG ratio of different subjects against VDT height at A0, A30 and A60 for lumber Erector Spinae muscle

heights. Since the angle is measured from vertical and top edge of monitor away from subject as positive the preferred VDT inclination decreases as the height of the VDT increases (Fig. 9).

At lower heights the preferred inclination of the screen is observed to be is as high as 50°, as the height of the VDT increased the preferred inclination is seen to be decreasing and at extreme heights it has almost straightened to zero or negative values. The average of the preferred angles graph displays a steep declination as the VDT height is increased indicating that preferred VDT angle is function of the VDT height. The VDT placement preferred zone was 89.5 to 99.5 cm. Out of 8 subjects 4 preferred 94.5 cm height, 3 preferred 89.5 cm height and 1 preferred 99.5 cm height.

Table 6 Summary of results for sEMG ratios of Erector Spinae

Sub-	Es1	Es2	Es3	Es4	Es5	Es6	Es7	Es8
At A = 0								
Max	2.60	2.94	2.03	1.26	1.40	2.11	1.89	1.21
Min	0.82	1.17	0.87	0.77	0.57	0.90	0.84	0.73
Av	1.68	1.98	1.42	1.00	0.96	1.39	1.27	1.02
SD	0.59	0.63	0.40	0.16	0.27	0.38	0.30	0.14
At A = 30								
Max	2.74	2.86	2.29	1.31	1.56	2.15	1.97	1.23
Min	1.04	1.13	0.77	0.82	0.59	1.11	0.83	0.64
Av	1.82	1.99	1.39	1.00	0.94	1.40	1.30	1.04
SD	0.70	0.57	0.38	0.14	0.29	0.37	0.36	0.16
At A = 60								
Max	2.16	3.64	1.97	1.25	1.31	2.21	1.75	1.22
Min	0.64	1.28	0.91	0.84	0.59	0.97	0.88	1.00
Av	1.51	2.46	1.29	0.99	0.87	1.38	1.32	1.11
SD	0.55	0.64	0.37	0.14	0.20	0.37	0.29	0.07

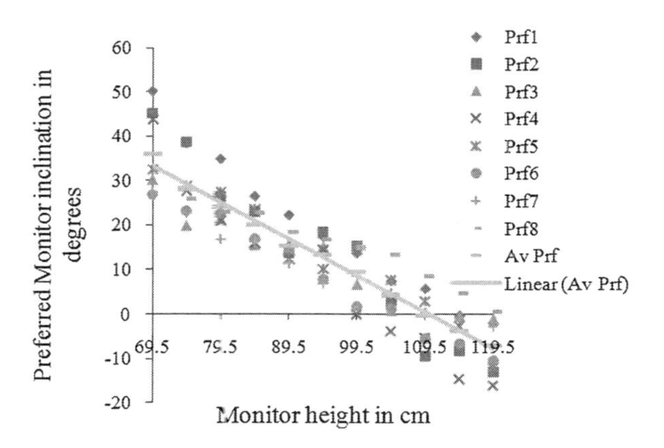

Fig. 10 Plot of preferred VDT angle against height

4 Discussion and Conclusions

Lower monitor placements increases the thoracic bending i.e. the thoracic angle reduces this coupled with thoracic inclination or the forward bend for lower monitor placements is the attained posture for viewing at lower monitor placements. The average trunk inclination angle shows increase with VDT height implying that subject bends away from VDT as height increases for viewing in higher VDT positions. The preferred VDT placement does not seem to contribute significantly to trunk inclination and subjects seem to take a suitable trunk inclination at a preferred height.

It is natural tendency to bend at the back leading to a natural preference for lower monitor placement. The average Thoracic bending angle for all three VDT inclinations show a distinctly higher thoracic angle that is less bending and more straightening of thoracic region with higher VDT positions. The variation of VDT height is distinctly reflected in head tilt however the changes in VDT inclination do not affect the same indicating that the required compensation is achieved through rotation of the eyeballs. This is supported by the observation of Villanueva et al. [10] that viewing angle was mainly decided by neck and eye inclination and to a lesser extent by thoracic inclination.

Postural changes are reflected in the sEMG activity of Erector Spinae and Neck extensors muscles. Erector Spinae was observed to show higher muscle activity with higher VDT positions. Kumar [6] also have reported higher discomfort in higher monitor placements. It may be the increase in stress in Erector Spinae that contributes due to the thoracic straightening and thereby the muscle activity.

The major response to the monitor height is by the head tilt which varies almost linearly with the varying monitor height. Lower monitor placements show a head tilt angle around horizontal and as the monitor position goes up the head moves up ward above the horizontal. The neck extensors respond to the variation by showing higher activity at lower monitor placements and it decreases as the monitor height goes up. This observation is supported by the study of Svensson and Svensson [15]

Fig. 11 Plot of average of preferred VDT angle against height

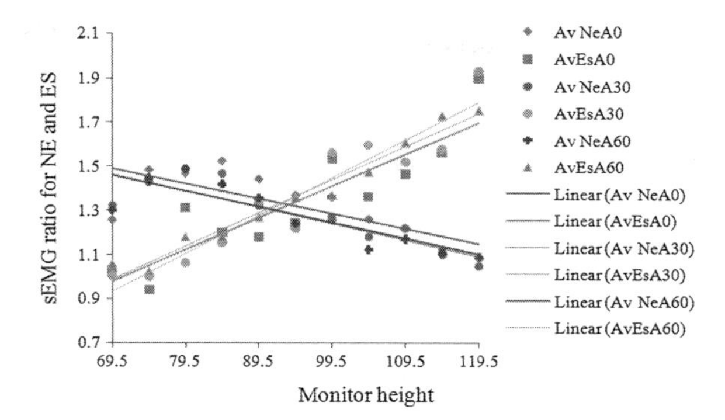

that a viewing angle of +3° gave less load upon the neck than the −20° position. Simultaneous plotting of Erector Spinae and Neck extensor ratio as in Fig. 11 shows the crossing of the two graphs as they depict two opposite trends. The crossing of the trend lines of the two graphs occurs at the H5 (equal to 94.5) which is approximate center of the preferred VDT placement zone that is 89.5 to 99.5 cm. This pattern is observed for all the three VDT inclination positions and the preferred monitor position appears to be a decision arising out of optimal muscle loading.

The plot of average preferred monitor inclination and monitor height yields the regression equation as $y = -0.8251x + 90.697$ where x is in cm and y in degrees. Using the regression equation the inclination of the monitor for a designated height of VDT can be generated. The designer may have to do finer adjustments or compensations which he may otherwise deem necessary to arrive at the final design.

The investigation attempts to answer as to what should be the position of the VDT i.e. its height and inclination. It tells the extent of influence of the VDT position on variables such as Thoracic bending and inclination which is serious concern for spine health in the long term usage. It elaborates that head tilt responds to the VDT height more directly and is reflected in the muscle activity of the Neck extensor muscles. Though the higher VDT placement tends to improve posture still it can be more fatiguing experience and cannot be adopted in design. Should we encourage lower monitor placements or to what extent the preferences should be given importance and be included in the design, how they relate to the muscular interplay has emerged.

Finally it is essential to mention that this study was carried out as part of evaluation studies of design of a sit down console that was developed for use in control rooms based on the anthropometric data of the user population in Indian context. Understanding of the preferences and muscle behavior improves the console design and increases its acceptability and creates an environment that has higher comfort levels.

References

1. Straker L, Mekhora K (2000) An evaluation of visual display unit placement by electromyography, posture, discomfort and preference. Ind Ergon 26(2000):389–398
2. Sommerich CM, Joines SM, Hermans V, Moon SD (2000) Use of surface electromyography to estimate neck and muscle activity. J Electromyogr Kinesiol 10(2000):377–398
3. Villanueva M, Jonai H, Sotoyama M, Hisanaga N, Takeuchi Y, Saito S (1997) Sitting posture and neck and shoulder muscle activities at different screen height settings of the Visual display terminal. Ind Health 35(3):330–336
4. Hamilton N (1996) Source document position as it affects head position and neck muscle extension. Ergonomics 39(4):593–610
5. Turville K, Psihogios J, Ulmer T, Mirka G (1998) The effect of video display terminal height on the operator: a comparison of the 15 deg and 40 deg recommendations. Appl Ergon 29(4):239–246
6. Kumar S (1994) A computer desk for bifocal lens wearers, with special emphasis on selected telecommunication tasks. Ergonomics 37(10):1669–1678
7. Aaras A, Fostervold K, Ro O, Thoresen M (1997) Postural work load during VDU work: a comparison between various work postures. Ergonomics 40(11):1255–1268
8. Hsio H, Keyserling MW (1991) Evaluating posture behaviour during seated tasks. Ind Ergon 8:313–334
9. DeWall M, Reil V, Aghina J, Burdorf A, Snuders C (1992) Improving the sitting posture of CAD/CAM workers by increasing VDU monitor height. Ergonomics 35(4):427–436
10. Villanueva M, Sotoyama M, Jonai H, Takeuchi Y, Saito S (1996) Adjustments of posture and viewing parameters of the eye to the changes in screen height of the visual display terminal. Ergonomics 39:933–945
11. Burgess-Limerick R, Plooy A, Fraser K, Ankrum DR (1999) The influence of computer monitor height on head and neck posture. Ind Ergon 23:171–179
12. Mandal A (1981) The seated man (Homo Sedens). Appl Ergon 12(1):19–26
13. Fujimaki G, Mitsuya R (2002) Study of the seated posture for VDT work. Displays 23:17–24
14. Ray G, Gupta N, Gupta S (2006) Experimental study to redesign visual display terminal work station for bifocal operators. J Sci Ind Res 65:31–35
15. Svensson H, Svensson O (2001) The influence of the viewing angle on neck-load during work with Video display units. J Rehab Med 33:133–136
16. Babski-Reeves K, Stanfield J, Hughes L (2005) Assessment of video display workstation set up on risk factors associated with the development of low back and neck discomfort. Ind Ergon 35:593–604

Relation-Based Posture Modeling for DHMs

Sarath Reddi and Dibakar Sen

Abstract Posture modeling for DHMs has significant effect on their usage in evaluation of product ergonomics. Direct manipulation schemes, such as joint level maneuvering for changing posture, are tedious and need more user intervention; complex scenarios are hardly simulated. This paper presents a high-level, relations based, description scheme for human postures and demonstrations for executing these descriptions using a digital human model (DHM). Here, *posture is viewed as a pattern of relations* of body segments among themselves and with the environment. These relations are then used as the criteria for the novel description based control. A few basic postures have been derived using the conventional principal planes. The basic postures and the composition rules enable description of complex postures in an easy and unambiguous way. We discussed the issues involved in the execution of descriptions and developed methods to resolve the conflicts due to link fixations. This scheme is effective for both lower and higher level control. Illustrative examples from the implementation of the concepts in our native DHM 'MayaManav' are included.

Keywords Posture control · Relations · Digital humans · Link fixations

S. Reddi · D. Sen (✉)
Centre for Product Design and Manufacturing, IISc, Bangalore, India
e-mail: dibakar@cpdm.iisc.ernet.in

S. Reddi
e-mail: sarathiisc@gmail.com

A. Chakrabarti and R. V. Prakash (eds.), *ICoRD'13*, Lecture Notes in Mechanical Engineering, DOI: 10.1007/978-81-322-1050-4_37,

1 Introduction

Digital human models (DHMs) both as avatars and agents shown in the Fig. 1 need to be controlled to make them manipulate the objects in the virtual world. DHMs are used for evaluation of product usability and identification of ergonomic issues in early stages of product design [1]. Present DHMs are technically sophisticated but difficult to use because of the low level control schemes. This in turn makes verification of DHM-based simulations in diverse platforms difficult to compare. The modeling of proper control schemes for digital humans has significant affect on their usage in the virtual environments. Several schemes are in existence to control the DHMs starting from lower level skeleton based control to higher level objective based control. The low level control schemes involve, manipulating the segments of the DHM directly using mouse or keyboard. This type of approach can be seen in commercially available modelers like Santos, Pro-E manikin [2] and DELMIA™ [3]. Though user gets the freedom to tailor the posture which satisfies the requirement, this type of control is of less use in realizing the behaviors. And also it is tedious as user has to manipulate each and every joint in a high degrees of freedom (d.o.f.) system as a DHM is. Another scheme which is being used by the above modelers involves presenting the user with a set of visuals of the posture (prototypes). In the similar lines, posture editor in the DELMIA contains predefined postures which are specified using names like stoop, squat etc. This approach is limited to situations where DHM has to be launched in the workplace in a posture, the user thinks suitable from the available predefined postures. These control schemes have limitations in realizing the postures driven by behaviors as there is little scope for autonomy. They become cumbersome in scenarios where anthropometry, scene and activity changes from the original context forcing the user intervention to readjust the configurations interactively. The idea of the predefined postures can also be seen in the scenario of hand grasp [4]. High d.o.f. of the hand poses significant challenge in its posture control. Taxonomy to simplify the description of hand posture for various everyday task scenarios can be found in [5–7]. Additionally, descriptions based type of control is seen in Jack [8] where a set of vocabulary is used to realize different kinds of behaviors for positioning the manikin and manipulating it in the virtual environment. In [9, 10],

Fig. 1 Digital human model interacting with environment

different specifications are proposed for full body gestural database and to describe the set of actions performed. To deal with higher level control in Jack [11, 12] a natural language based interface along with Parameterized Action Representation (PAR) is developed for giving instructions and connect these descriptions with actions of DHMs.

Though descriptive way of control is more efficient to the other control methods, unless a proper scheme/format is devised to construct a description, the processing cost for parsing and interpretation will be too high. And also there is no description based control scheme for posture modeling of DHMs. We developed a generic framework through which the postures are controlled by specifying only the relations between the actors (DHM and objects). Since a format is followed for descriptions; parsing and interpretation is much easy and does not involve tools to understand the intent in the description.

2 Relations Based Descriptions

There are several ways of describing the intent to be achieved. Either we can describe the actions necessary to fulfill the objective or we can describe the objective/intent itself. The second choice gives feasibility for DHM to select the actions on its own allowing it to realize the behaviors. The tasks are performed by changing the state of the objects in the virtual environment and the state of an object is usually defined with respect to other objects using its relative position, orientation and contact. Similarly, the state of the DHMs can be described using the relative position and contact between the body segments and the objects. Ability to describe these relations helps in conveying the intent and the way the tasks have to be performed. Therefore, *relations between actors* are chosen as a criterion for framing the descriptions and postures can be realized by the DHM when it satisfies the relations specified. Different cases arise pertaining to these relations between the actors (DHMS and objects); between Segment to segment, segment to object, object to object and combination of these three. As the descriptions in this scenario are expected to relate the objects, the structure of a relational description should contain the objects (segments and objects) of interest and the relation between them. It is also expected that the relation in the description should provide enough information about how to achieve the intended relation. Our intention here is to come up with a method to describe the intended relations and making the DHM act to realize the postures and also to investigate whether any finite set of descriptions (basic/canonical) are feasible combining which, will realize complex postures.

Human body comprises of segments connected with the joints. For the given number, there exist some finite combinations of relations between these segments. Suppose if the objective is to achieve a certain relation between two non adjacent segments, then that intended relation can be achieved in varieties of ways as there are many number of arrangements possible with the intermediate segments.

Therefore, numerous postures can be generated which can satisfy the intended relation between segments chosen. On the contrary if the segments are adjacent to each other, then posture attained when satisfying the intended relation between them is not influenced by the other segments as there are no intermediate segments. Therefore, the corresponding description of such a relation between two adjacent segments can be treated as basic. We consider a basic description as the one which specifies a relation that cannot be further expressed as a sequence of other basic descriptions. Basic descriptions can be used as the building elements of constructed ones following which can realize complex postures. Another important function of basic descriptions is that they can establish the links between one constructed descriptions with another.

2.1 Basic Relational Descriptions

The relations between the body segments are defined by the orientations of these joints. As the adjacent segments are finite, the relations between them can be specified using finite set of descriptions. To specify the relation we need a descriptor which relates the segments. Several terminologies are in use to describe the relative positions of the body parts, In the field of health sciences and ergonomics there is a tendency when describing a movement for it to be referred by some *dominating plane* and thus different planes and corresponding axes along with certain directions are came into existence to describe the human movements. The three planes of motion that pass through the human body are the Sagital plane, the Frontal plane and the Transverse plane. And to specify the descriptions, the literature in the field of ergonomics [13, 14] followed the convention of assuming a *reference posture* and describing only the *deviations* from this posture by specifying the magnitudes of the joint rotations between those body segment whose relations have changed. This approach is very effective in minimizing the effort needed for describing, interpreting and comparison of the postures. The idea here is to use the notions of *dominating plane* and *deviations* from a standard reference posture as the basis for specifying the basic relational descriptions so that the need of mentioning all the joint angles as is seen in the present modelers can be eliminated.

Regarding the dominant planes, these are in general defined with respect to the human body in standing posture. The sagital plane lies vertically and divides the body into right and left parts. The frontal plane also lies vertically however, divides the body into anterior and posterior parts. The transverse plane lies horizontal and divides the body into superior (above hip region) and inferior parts (below hip region), and all these planes intersect at the pelvic region. These three often termed as dominant planes as most of the human motions confines to these planes. Based on these three dominant planes, corresponding directions and axes are defined; *Anterior and posterior directions on sagital plane with reference to frontal plane; Left side and right side on frontal plane with reference to sagital*

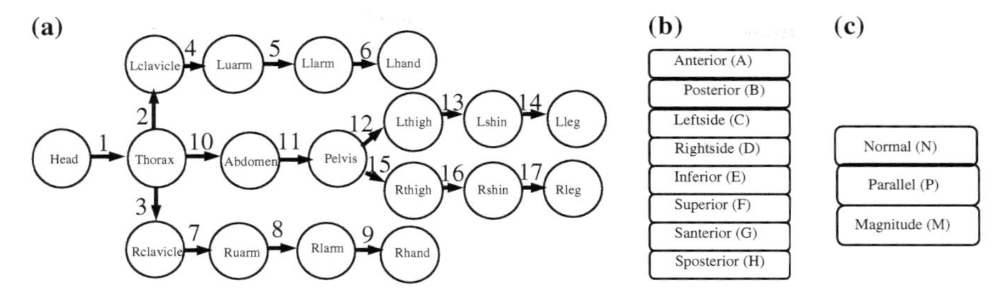

Fig. 2 **a** 17 pairs of adjacent segments. **b** 8 directions. **c** Magnitude of joint rotation. Combinations of these three elements with reference to standing posture form the basic relational descriptions

plane and Superior and inferior on traverse plane on either sides of traverse plane. The format of our descriptions is a quadruplet which consists of the objects to be related, a descriptor to specify relation and the magnitude. Following the above notations, our basic description typically consists of two adjacent segments, a direction and magnitude of the joint rotation connecting them as follows: *segment1 direction magnitude segment2*.

Though it is not uncommon to see the multi-plane movements in the daily activities of life and sports, we hardly see any naming conventions for those directions which are not in dominant planes. Our idea is to use the same dominant planes as basis to differentiate the regions such that new set of planes can be defined between the dominant planes. Plane obtained by rotating the sagittal plane about the vertical axis till it coincides with the frontal plane is referred to here as a *Sag-front plane*; a rotation $\theta < \pm 90°$ covers all possible planes. The anterior portion of a Sag-front plane is represented as Santerior-θ, and the posterior portion as Sposterior-θ. Figure 2 shows the format for basic relational descriptions. Figure 2a shows the directed graph of the adjacent segments starting from head. If the relationship between the adjacent segments is in line with the directed graph, then format is 'Adjacency INDEX'-DIRECTION-'MAGNITUDE indicator'. If it is against the directed graph, then it is 'Adjacency INDEX'-DIRECTION-'MAGNITUDE indicator'.

2.2 Postures as a Set of Basic Descriptions

In the context of relations between segments, we can understand a posture as human body segments satisfying some relation in terms of their position and orientation with respect to one another. Here the idea is to describe the postures intended using the above basic relational descriptions following the notion of describing only the deviation from reference posture [14]. The right half kneel and squat postures shown in the Fig. 3a and b are described using set of basic relational descriptions.

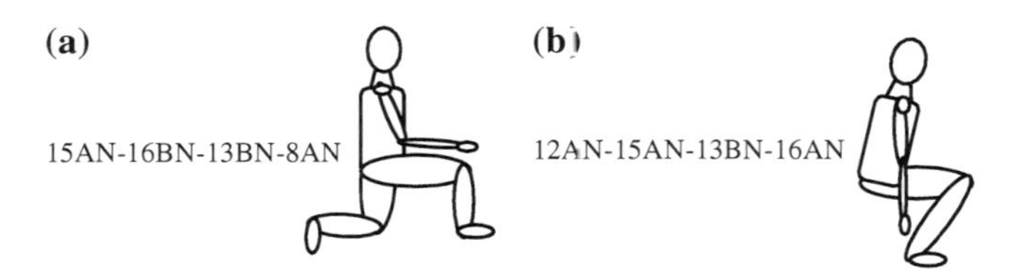

Fig. 3 **a** Right half kneel **b** squat. Describing complex postures as set of basic relations

Thus wide varieties of postures ranging from simple to complex can be described using a set of basic descriptions. The satisfaction of these relations in a DHM needs variations of direct kinematics. Advantage of representing each posture using set of basic descriptions is the transition from one posture to the other can be achieved with less effort as it is easy to compute necessary changes required in the basic relations which can bring the current posture to intended posture.

2.3 Relations Between Nonadjacent Segments

Unlike the relations between adjacent segments, the relations between non adjacent segments can be achieved in varieties of ways. Therefore, we need to specify some constraints in the description along with the segments to be related. Here the idea is to use the postures derived from set of basic relations as constraints to achieve the intended relation. If we consider the case of "hands on toes" as shown in the Fig. 4; the specified relation between hands and toes can be satisfied through many ways. Therefore, we can add a posture name as a constraint to the intended relation and in that way the relations between the intermediate segments also gets assigned. In this case the constraints can be specified using the descriptions "hands on toes by squatting" or "hands on toes by stooping" as shown in the Fig. 4a and b.

Though the exact magnitudes pertaining to the relations are specified in the basic descriptions, in this scenario they get changed to conform to the requirements posed by constraints. For example, while defining the stoop posture using the basic relational descriptions, the torso might be normal to femur; but that may not be sufficient for the hands to reach the toes. Therefore, torso has to *adjust* the angle between femur and torso as shown in the Fig. 4b.

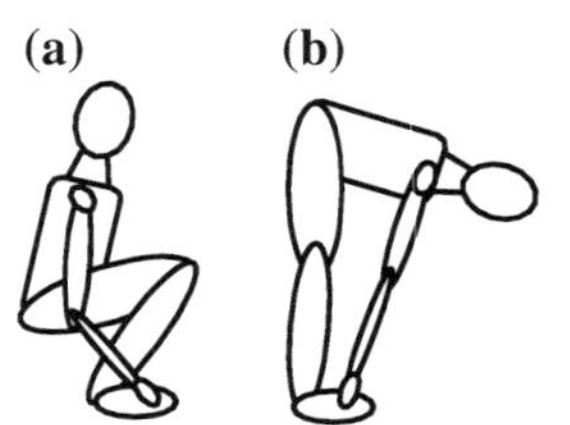

Fig. 4 **a** Reaching toes by squatting. **b** Reaching toes by stooping. Using postures as constraints to achieve relations between non adjacent segments

The format for description in this scenario is different than basic relational descriptions for it consists of a posture name to specify the constraint. The description contains the two segments involved, a relation and the posture name. The relations in this case are place prepositions such as ‘on’, ‘under’, ‘beside’ and ‘inside’ etc. These place prepositions helps in identifying the landmarks on the segments while executing the descriptions. The satisfaction of these relations in a DHM needs variations of inverse kinematics. The descriptions specify the relations using place prepositions and using inverse kinematics techniques the joint variables are determined which satisfies the relations. Thus format for describing relations between non adjacent segments: *Segment1-(place preposition)-segment2-constraint.*

2.4 Relations Between Segment–Object and Object–Object

This scenario is similar to the non adjacent segments with only difference being one of the segments is replaced by object in the environment. The description consists of a segment and object and relation between them. The relation is typically be a place preposition similar to the above scenario. Through these descriptions user can actually specify the direction to approach the object using the place prepositions. Similar to the case of relation between non adjacent segments, execution of descriptions involves employing the inverse kinematics techniques. The format is as follows: *segment-(place preposition)-object*, *object-(place preposition)-object.*

We developed the relations based control scheme with the objective of realizing behaviors. Behavior in the context of human activities can be interpreted as, out of many possible ways, choosing a particular way to accomplish a given task. Here in our scheme we are specifying only the intended relation between DHM and objects or between object and object. This approach offers lot of room to satisfy the relation governed by many aspects like human capabilities, task and initial posture etc. for example *Left hand on table*, can be accomplished in varieties of ways posing constraints on elbow, torso and orientation of hand. And also user can specify the constraints based on his understanding of the task by extending the description similar to the description of non adjacent segments. The above format in the scenario of constraints is modified as *Segment-(place preposition)-object-constraint.*

3 Resolving the Conflicts During Execution of Descriptions

The descriptions specified above are based on intended relations between segments and objects without concerning whether those relations are possible to achieve or not in presence of prevailing environmental constraints. There is no indication of how to overcome the effect of these constraints. Therefore, the responsibility of checking the feasibility and resolving the conflicts lies on the DHM itself.

The main constraint for the execution of the intended relations comes from the present configuration of the DHM. It influences the way the descriptions are executed as it determines the feasibility. The configuration of the DHM is not only a set of relations between body segments but it is defined based on the fixation of the links (links that are grounded). This aspect brings difference in the execution of relations for two different link fixations. Consider the case where the DHM is initially in standing posture, fixing its legs on the ground, and the objective is to achieve a relation between tibia and femur.

Following the format proposed above, it can be described as "16′ AN" (*right tibia posterior normal to right femur*). But as the legs are fixed, the movement of tibia is arrested and therefore the intended relation in the description cannot be satisfied. Only the relation "16AN" (*right femur posterior normal to right tibia*) as shown in the Fig. 5a can be realized with the given fixation. To satisfy the intended relation "16′AN", the fixation has to be changed to the pelvis or segments above pelvis as shown in the Fig. 5b. Therefore, when intended relations are described, first we need to check the feasibility conditions; whether the relation specified can be achieved in the existing posture with a given link fixation. If these conditions are satisfied then descriptions can be executed. The issues that arise due to change in link fixations like change in the global posture are resolved through our novel scheme developed to build the kinematic structure [15]. Extending the above discussion from basic descriptions to posture descriptions; we have seen that the postures can be described as a set of basic relational descriptions. From the above discussion we have also seen that for the same set of relations the way postures are realized is different (joints involved in moving segments) for different link fixations. Consider the case shown in Fig. 6. where attaining half right kneel posture from standing posture is the objective. Depending on the link fixations the sets of relations necessary are achieved with different joint rotations. When the right leg is fixed, the right knee rotates anticlockwise, torso rotates clockwise and

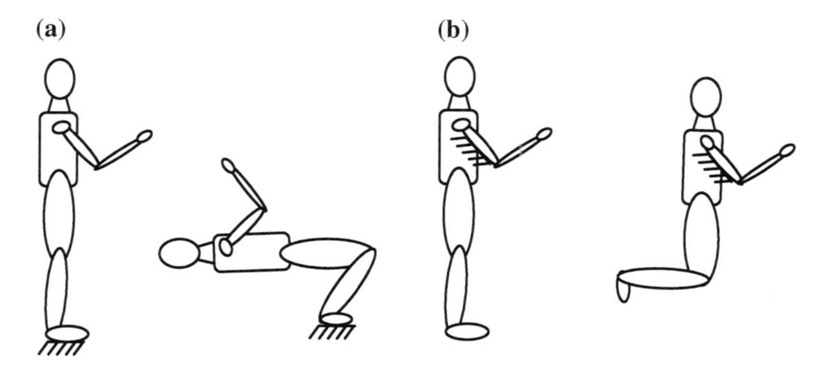

Fig. 5 Choosing the joint rotations based on link fixations. **a** Satisfying the relations between femur and shin when legs are fixed. **b** Satisfying the relations when upper torso is fixed

Fig. 6 Achieving the half kneel with different leg fixations

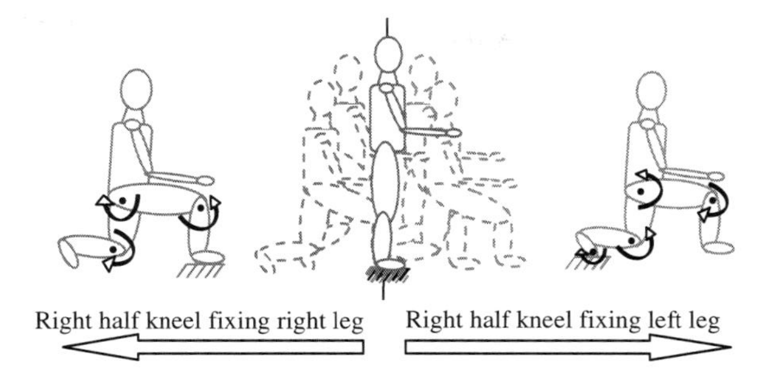

left knee rotates clockwise directions and to realize the same posture, the left toe in clockwise, left knee in anticlockwise, right femur in anticlockwise and right tibia in clockwise, are rotated fixing the left toe. Thus though the relations achieved are same, the execution differs considerably with different link fixations. The objective in this scenario of change in link fixations is, making the DHM compute the joint level solutions when it needs to switch over between two postures autonomously. Here two cases arise; one is achieving transition from one posture to another with same fixation and the other is posture transition between different fixations. Transition from squat (12AN-15AN-13BN-16AN) to stoop (15′AN) can be an example for the first case as legs got fixed for both the postures. The transition in this case is straight forward as there is no change in the link fixation. Each posture is a set of relations and only involves changing these relations by actuating respective joints.

The descriptions specified are actually the deviations from standing posture (base posture) where the principle planes and directions are described. The transitions involve first comparing the deviations in the two postures in the light of base posture and then bring the relations between segments to base posture which are not required while executing the new relations in the destination posture. In the case of "squat to stoop", the existing relations (12AN-15AN-13BN-16AN) are not required for stoop. Therefore, their relations can brought down to that of base posture and the new relation between torso and femur is executed to obtain stoop (15′ AN) posture. Regarding the second case where the link fixations are different between existing and destination postures, the link fixation needs to be changed while transition takes place. Consider the case of "halfkneel (15AN-16BN-13BN) to squat (12AN-15AN-13BN-16AN)" where one of the legs are fixed in existing posture while two legs are fixed in the destination posture. The procedure in this case is not straight forward and varies with leg fixation (right leg and left leg are fixed in the destination posture). Unlike the above case, the comparison is not only between the basic relations involved but also involves comparing the existing link fixation with that of destination posture. *The execution of transition starts from the existing fixed link by bringing the non relevant relations to the base posture and executing the new relations starting from the segment attached to the fixed link.* The satisfaction of relations in intended posture is carried out by first prioritizing the segments in the set of basic descriptions that constitutes the posture description with respect to the fixed link. As the link fixation of the intended posture determines the feasibility conditions, along with relationships, the link fixation also should be embedded in the description.

4 Demonstration Using DHM

The prerequisite for executing the descriptions by the DHM includes identification and naming of landmarks on the segments and the objects in the virtual environment. The first step in the procedure for reconstructive approach includes processing the scanned data to identify the location of the joints and body segments. This step is carried out interactively and different regions are rendered on the scanned data to identify the body segments and landmarks as shown in the Fig. 7a. Rendering is required to associate the segments with their respective joints. Segment lengths are then can be computed from the joint locations. Second step includes importing the scanned data (in the form of vertices and indices) and reconstructing the geometry to make DHM appear as a human. The computed link lengths are passed on to the kinematic structure module so that it can construct the skeleton from the scanned data. The vertices identified for each segment are stored separately to manipulate them along with the skeleton defined on the geometry. The third step involves applying the transformations to the segments obtained from the kinematic structure. Similarly the landmarks on the objects are rendered interactively based on their features of interaction as shown in the Fig. 7c. Each landmark can be named which is identified on the object for describing the relationship with respect to DHM. The naming can be done using the place prepositions so that if the description specifies a particular preposition, DHM will interact with the landmark associated with it.

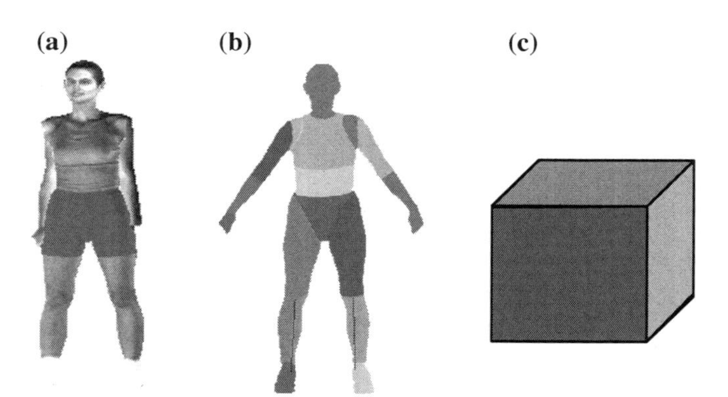

Fig. 7 **a** Scanned model of a human. **b** Body segments rendered to identify the corresponding vertices of the joints. **c** Object rendered to identify the landmarks

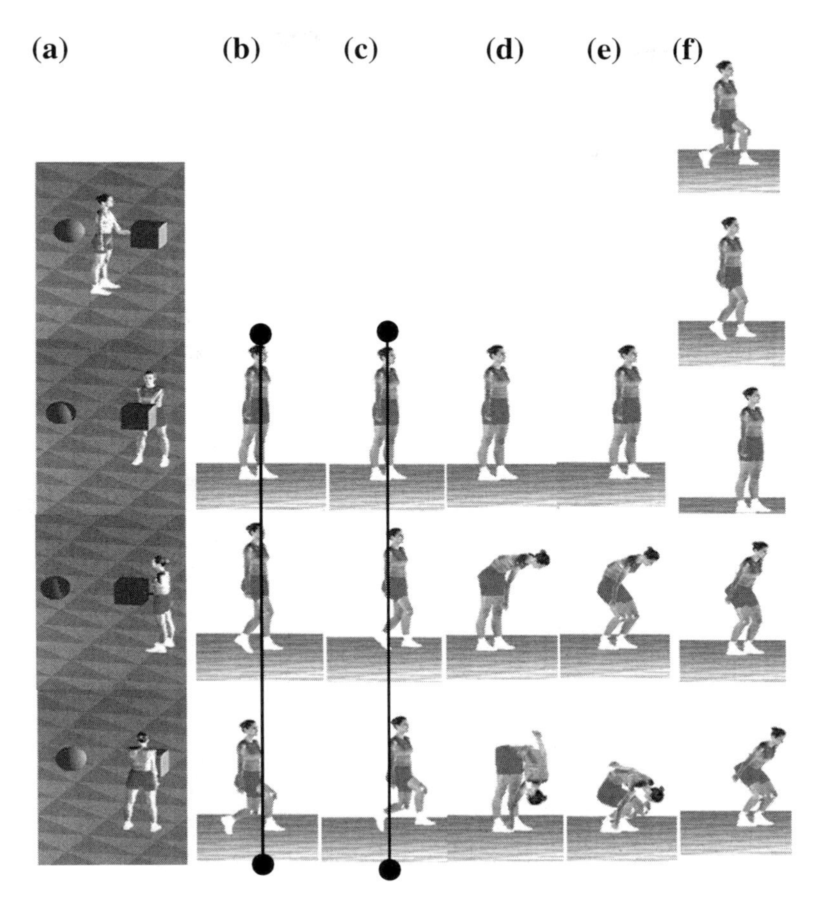

Fig. 8 **a** DHM satisfying relation "lefthand place preposition box". According to the place preposition DHM approaching the box in different directions. Prepositions used are, front side, backside, left and right side. **b** DHM realizing the half kneel posture fixing right leg. It moves backward from its initial position achieve the posture. **c** Realizing the same half kneel posture with left toe getting fixed. In this case DHM moves forward and joint angles are different from the previous scenario. **d** Depicting the relations between non adjacent segments. "right palm on right toe by stooping". Stooping is the constraint to achieve the specified relation. **e** "right palm on right toe by squatting". The squat posture is used as constrain to achieve the relation. **f** Posture transition from half kneel to squat by changing the link fixations from right toe to left and right ankles

The second step in the implementation of the descriptions is developing tools to make the DHM act to satisfy the intended relations. The execution of basic descriptions is a direct kinematics problem as it involves only rotating the joint to a magnitude and in the direction specified in the description. The execution of descriptions which involves non adjacent segments and segment-objects require inverse kinematics models to satisfy the relations between segments and objects specified. To execute the direct kinematics, kinematic structure is built [15] which can handle the posture alterations due to change in link fixations. Regarding inverse kinematics, a novel geometric based framework is developed to find the joint angles so that DHM can satisfy the relations (Fig. 8).

5 Conclusion

The descriptions to specify the relations and postures as a set of relations involve simple triplets and quadruplets. As the descriptions are to be specified according to the format devised, unlike existing modelers the parsing and interpretation aspects will not involve referring to complex grammars and meanings. As the descriptions constitute only the relations between actors, incorporating different behavior models while executing the relations is feasible through this framework. Along with the constraints imposed through behavior models, new constraints can be added using the postures defined as a set of relations. This framework gives the user feasibility to specify the relations intended in any order suitable to a task scenario. Both lower and higher level aspects of DHM can be controlled through this framework.

References

1. Chaffin DB (2005) Improving digital human modelling for proactive ergonomics in design. Ergnomics 48(5):478–491
2. Creo Manikin Extension http://www.ptc.com/products/creo-elements-pro/manikin/whats-new.htm
3. DELMIAErgonomics Evaluations http://www.3ds.com/products/delmia/portfolio/delmia-v6/v6-portfolio/d/digital-manufacturing-and-production/s/program-control-engineering/p/ergonomics-evaluation/?cHash=be2ef83cc230451ef0f39a010e68afb9
4. Cutkosky MR (1989) On grasp choice, grasp models, and the design of hands for manufacturing tasks. Robot Autom IEEE Trans 5(3):269–279
5. Sandra JE, Donna JB, Jenna DM (2002) Developmental and functional hand grasps, 1st edn. Slack Incorporated, Thorofare (ISBN-10: 1556425449)
6. Elliott JM, Connolly KJ (1984) A classification of manipulative hand movements. Dev Med Child Neurol 26(3):283–296
7. Iberall T (1997) Human prehension and dexterous robot hands. Int J Robot Res 16(3):285–299
8. Badler NI, Phillips CB, Webber BL (1993) Simulating humans: computer graphics, animation, and control. Oxford University Press, Oxford
9. Ianni JD (1893) A specification for human action representation. In: SAE technical paper series, 1999-01-1893
10. Hwang BW, Kim S, Lee SW (2006) A full-body gesture database for automatic gesture recognition. In: 7th international conference on automatic face and gesture recognition, pp. 243–248, Apr 2006

11. Bindiganavale R, Schuler W, Allbeck J, Badler N, Joshi A, Palmer M (2000) Dynamically altering agent behaviors using natural language instructions. In: Autonomous agents 2000, pp 293–300, June 2000
12. Badler N, Bindiganavale R, Allbeck J, Schuler W, Zhao L, Palmer M (2000) Parameterized action representation for virtual human agents. In: Cassell J, Sullivan J, Prevost S, Churchill E (eds) Embodied conversational agents, MIT Press, pp 256–284
13. Norkin CC, White JD (1995) Measurement of joint motion: a guide to goniometry, 2nd edn. FA Davis Co, Philadelphia (ISBN-10: 0803609728)
14. Cailliet R (1992) Soft tissue pain and disability, 2nd edn. FA Davis Co, Philadelphia (ISBN-10: 0803601107)
15. Reddi S, Sen D (2011) A novel scheme for modeling kinematic structures of DHMS with mutable fixations. In: International conference on trends in product life cycle modeling, simulation and synthesis

How People View Abstract Art: An Eye Movement Study to Assess Information Processing and Viewing Strategy

Susmita Sharma Y. and B. K. Chakravarthy

Abstract Perception of a form is primarily an individualistic experience. Viewing a product may involve perception of its function or an inherent coding of an idea in its form. Perception to an object or an artifact is like having a visual dialog with the form. The literature suggests that, the visual experience of a work of art, such as painting, is constructed in the same way as the experience of any aspect of the everyday world. Therefore, it is extremely difficult to analyze every aspect of the viewer perception. But, if the viewing aspect itself is observed, it may offer interesting insights, on how the visual dialog to an artifact or designed object is formed by the observer. In this research paper, we aim to explore some of the viewing aspects of viewers, observed through the eye movement research (EMR). The study has been done with 17 participants to establish an understanding in viewing strategy and information processing, while viewing select abstract paintings. Six paintings were used in this study. Selection of the paintings was done as a combination that included Abstract expressionistic paintings by Jackson Pollock and Willem DeKooning, which are spatially spread compositions. Mary Abbot and W. Kandinsky's directional and vigorous-movement stroke oriented paintings, and Piet Mondrian's neo-plastic, pure geometrical painting. Images were shown to the participants to investigate viewing strategy through visual attention and exploratory behavior: diverse or specific; in order to understand how abstract art is viewed. It was observed that spatially spread, uniform paintings offered maximum components of information for the viewer to process while pure abstraction evoked high visual search. We further discuss which paintings evoked

S. Sharma Y. (✉) · B. K. Chakravarthy
Industrial Design Centre, Indian Institute of Technology Bombay, Mumbai, India
e-mail: sharma.susmita@gmail.com

B. K. Chakravarthy
e-mail: chakku@iitb.ac.in

A. Chakrabarti and R. V. Prakash (eds.), *ICoRD'13*, Lecture Notes in Mechanical Engineering, DOI: 10.1007/978-81-322-1050-4_38,

high attention and discuss qualitatively the possible reasoning. One significant finding is that existence of high information processing may not necessitate a high diverse exploratory behavior.

Keywords Eye movement · Visual perception · Viewing strategy

1 Introduction

Visual response is one of the means to gauge the visual perception of an observer. Eye movement behavior (EMR), since the first works of Yarbus [1], has been used in various fields from reading, usability evaluation, cognitive psychology, artificial intelligence to object perception. Though extensive researches have been carried out in eye movement, the understanding is primarily based on viewer's response to usability, where s/he approaches with a destination or action in mind with response to time. On the contrary, viewing an art object, or a designed form, entails esthetic experience and personal involvement that may not aim to elicit an action on the part of the viewer. Therefore, it is also important to reinstate and re-contextualize the eye movement measures used for web-assessment to be seen in a new light and create meaning for visual assessment of art and form esthetics. Present Study with EMR aims to throw light on some of the viewing aspects of viewer strategies as observed in the eye movement behavior to establish an understanding of viewer viewing strategies and information processing trends.

A few areas have been studied in art, but mostly for studying the painterly way of thinking [2] analyzing representational or semi-representational art [3] and viewers in an art-Museum [4] for comparative study of diverse styles. Zeki compares artists with neuroscientists in their capacity to manipulate brain enabling the esthetic experience [5]. Exploration in displacement and slow revelation of pictorial element to enhance esthetic pleasure in art is also reported, [6] as well as preference and its relationship with typicality in paintings is explored [7].

Interesting insights are reported in comparison of Surrealist (Dali) and Baroque art (Caravaggio) for esthetic rating proposing that style of the painting also affects the viewing response [8]. Their results find that after the initial global exploration of the painting, specific visual scrutiny follows, and the style was observed to be an effecting factor for the viewing response. The study explored attention on compositional features like figures and other identifiable items in the figurative style selection. Also the effect of additional information to the paintings was sought, in consistence with previous studies [9]. In the studies, emphasis was laid on the effect of providing titles of the paintings as added information to visual experience [10]. This idea is based on the model that for the esthetic experience to be enhanced and engaging, it is important that the viewer understands the painting, in term of its projected meaning through its title.

Present study on the contrary, contextualizes the visual perception on non-identifiable components in the artwork. We developed the current approach to study visual response on Abstract art, where no added information is provided. The approach was to observe viewer inclination toward abstraction in art that propounds the viewer to create their own references and emergent meanings by unidentifiable compositional sources in the compositional space of a painting. This perception of art can be equated with first hand encounter of any art form or designed object that evokes and retains attention, as the esthetic evaluation of product form perception [11].

A viewer might just choose to see how a painting would have been painted, its compositional elements, layout, identifiable triggers or establishing meaning with their own unique backgrounds and exposures. The study explores some of the viewing aspects to provide an understanding of trends of visual attention through information processing and visual strategy as affected by style and genre of abstract paintings.

2 Experiment Design

Objective of the experiment is to explore viewer's visual attention and visual exploration strategies while viewing abstract paintings. Selection of the paintings was carefully done as a combination that included Abstract expressionistic paintings by Jackson Pollock and Willem Dekooning, which are spatially spread abstract expressionist compositions. Mary Abbot and W. Kandinsky's directional and vigorous-movement stroke and line oriented paintings, and Piet Mondrian's neo-plasticism geometrical painting (Fig. 1).

2.1 Equipment and Participants

Seventeen subjects took part in the study (12 m, 5 F, mean age 28). The subjects were post graduate students who volunteered for the experiment. All had normal or

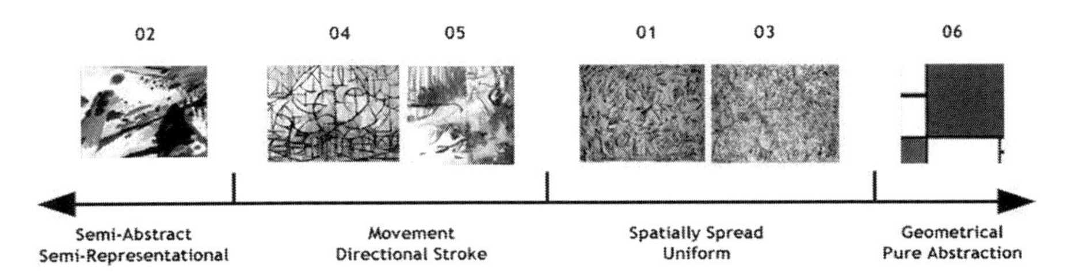

Fig. 1 Paintings grouped based on their treatment and morphology in style in a range from semi-abstract to pure abstract geometrical paintings

corrected to normal vision. An iView X system with RED III camera was used for eye-tracking, which is a dark pupil tracking system that uses infrared illumination and computer-based image processing. A pentium V, PC computer at 60 Hz speed, 96 dpi 17″ LCD monitor with a 1024 × 768 screen was used to display abstract painting images. The subject- screen distance was 720 mm and a chin rest was used to minimize head movements.

2.2 Stimulus Design and Procedure

The six select paintings were used in this study and participants were asked to view paintings at their own timing. The paintings were set in black border. A five point calibration was performed at the beginning of each viewing.

No two similar styles were put in order of viewing as shown in Fig. 2, to avoid the eye getting accustomed to a particular style. A gap of blank neutral gray slide was inserted between each stimulus to prevent any visual biases and to provide a visual break.

3 Rationale for Analysis

The Eye movement data was recorded and analyzed in Bgaze software. Recorded scan path of subjects visually appeared in data as shown in Fig. 3, where circles denote fixations. A Fixation is the moment when the eye is relatively still and focused on a target, indicative of attention, information processing and visual comprehension on that point of gaze. Radius corresponds to the duration of a fixation. Connecting lines to the circles represent saccades; these are rapid directive movements of the eye that help locating the target of attention [11, 12].

The EMR analysis considering the Fixation time and duration signifies the preference with respect to attention [12, 13], also indicating that a form that evokes maximum attention may have more interest areas [14]. Total viewing time, fixation-saccade ratio and Average Fixation Duration (AFD) were compared to assess information processing and search trends on each painting. The total fixation duration is divided by the number of individual fixations to reveal the average

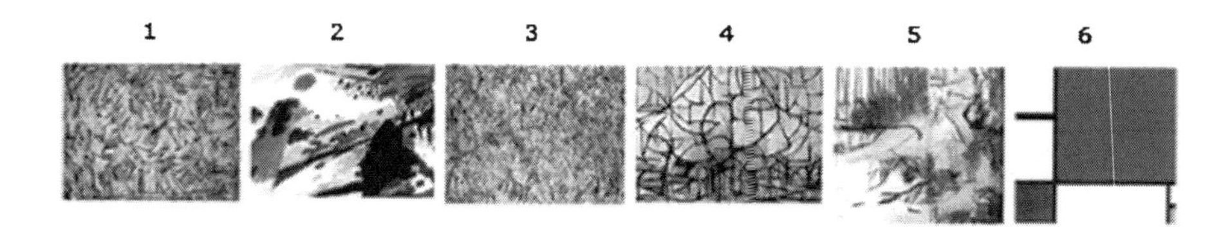

Fig. 2 Viewing order of the stimulus paintings

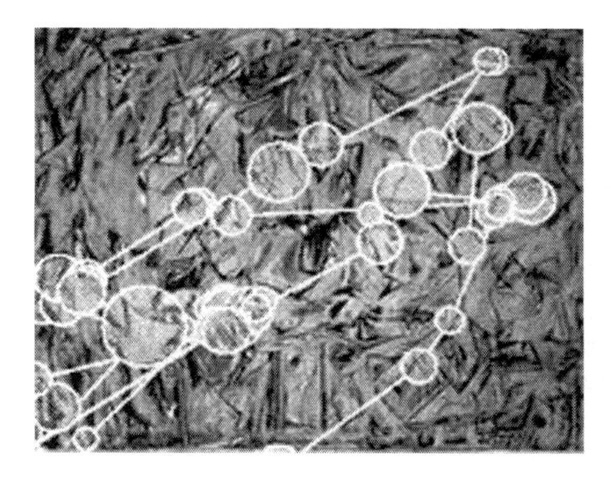

Fig. 3 A scan path of the eye movement consisting of fixations and saccades

duration (AFD in milliseconds). A fixation average length can be between 200–300 ms, an increase is indicative of higher attention, higher information processing and visual comprehension.

Visual exploration trend was sought by comparing long and short clusters of fixations. Short gazes reflect global surveying of compositional design related to diverse exploration and long gazes reflect focal scrutinizing of local pictorial features related to specific exploration [15]. Comparative analysis of these measures could indicate if higher information processing is congruent to the exploratory behavior. Further to the analysis, zonal preferences shall be observed on maximum attention areas to analyze attention density patterns.

4 Results and Analysis

4.1 Attention on Each Painting

Average fixation as shown in Figs. 4 and 5; indicate that paintings 1 and 3 evoked high attention (Fig. 5). Both are the uniformly-spatially spread paintings that evoked high information processing providing maximum number of information components. The AFD also recorded an increased score (2.6 and 2.61 s respectively).

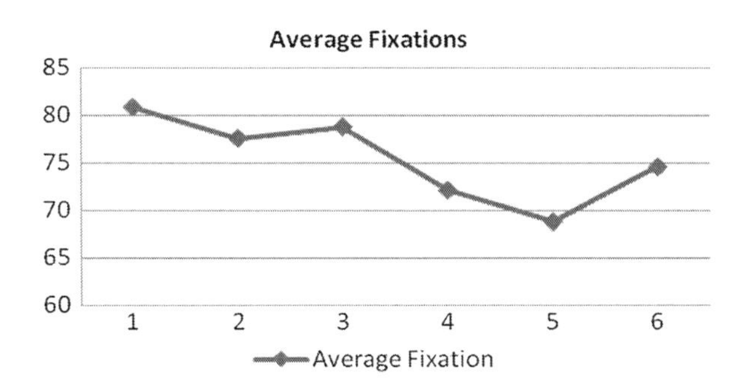

Fig. 4 Average fixation scores indicating attention on each painting

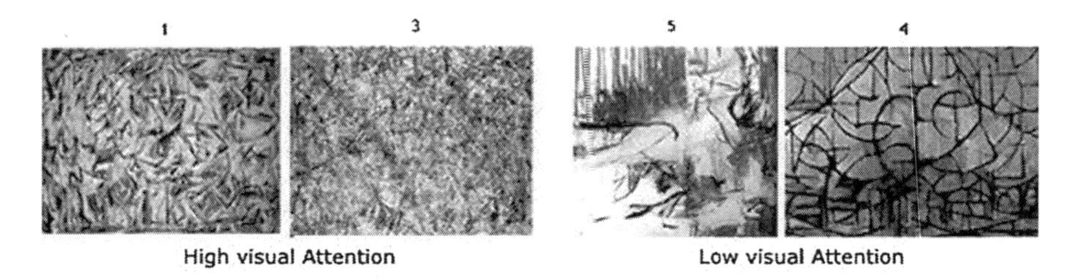

Fig. 5 Spatially spread paintings that received high attention 1 and 3. Stroke and line oriented paintings that received minimum attention 5 and 4

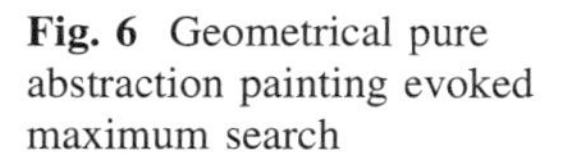

Fig. 6 Geometrical pure abstraction painting evoked maximum search

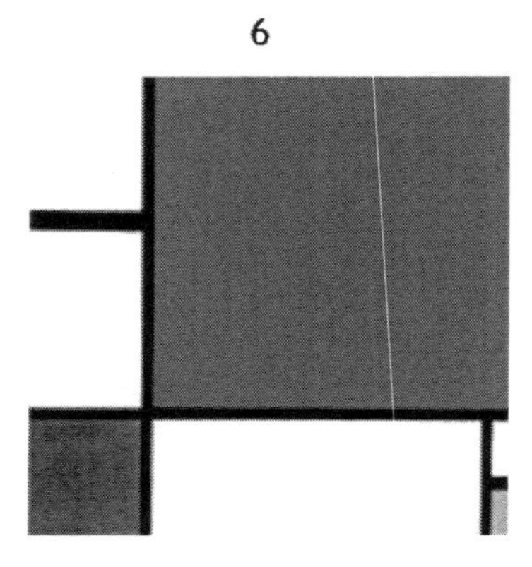

Painting 5 evoked minimum attention followed by painting 4, Figs. 4 and 5. (AFD also recorded a decreased score 2.1 and 2.2 s respectively.) These are both stroke and strong line oriented paintings that recorded low information processing.

Painting 6, (Fig. 6), evoked more saccades than other paintings showing that search activity was more on this pure abstraction geometrical style (Average Fixations 74.59 and saccades 15.20).

4.2 Visual Exploration

Short gazes reflect diverse and global viewing strategy on compositional aspects and long gaze reflect focal viewing strategy indicating specific exploration. These are the two visual exploration strategies analyzed.

Painting 1 (spatially spread, abstract expressionist) offers maximum components of information for the viewer to process. It also shows maximum diverse exploratory trend. However painting 6 (pure geometrical) is observed to be having less diverse exploration trend compared to high information processing as shown in Fig. 7.

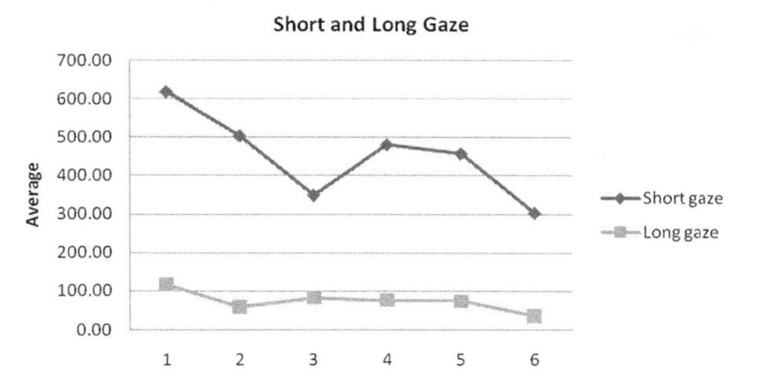

Fig. 7 Short and long gaze

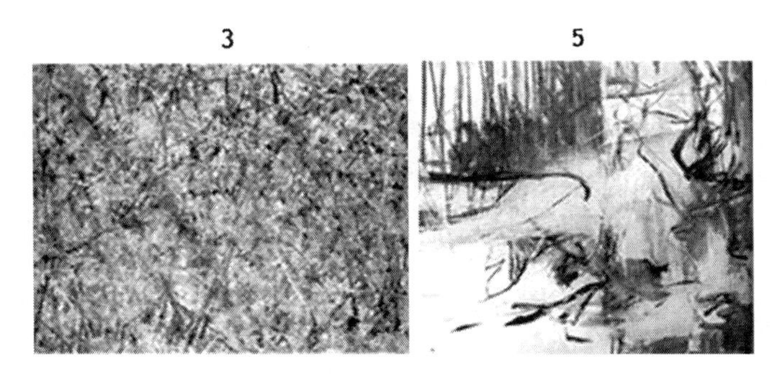

Fig. 8 Painting 3 (spatially spread) evokes high information processing but low at diverse exploratory behavior. Painting 5 (stroke and line oriented) evokes less information processing but shows high diverse exploration trend

Comparing the information processing trend by analyzing attention, and viewing strategies for the paintings, we observed that Painting 3 and 6, offers high information processing and comparatively less diverse exploration; but for 4 and 5 inversely, high diverse exploration is evoked with a decreased information processing trend (Fig. 8). Therefore, indicating that existence of higher information processing may not necessitate a high diverse exploratory behavior.

No significant difference is observed in focal scrutiny, while 1 is marginally high, 6 recorded lowest (Fig. 7). The average Fixation-Saccade ratio overall is 86 and 14 %; the higher ratio indicates more processing and less search activity overall for viewing paintings.

4.3 Attention Density on Zones

Regarding zonal attention, the maximum density of fixations is observed at the center of the painting 1, with a unique rigorous open stroke combination of red and

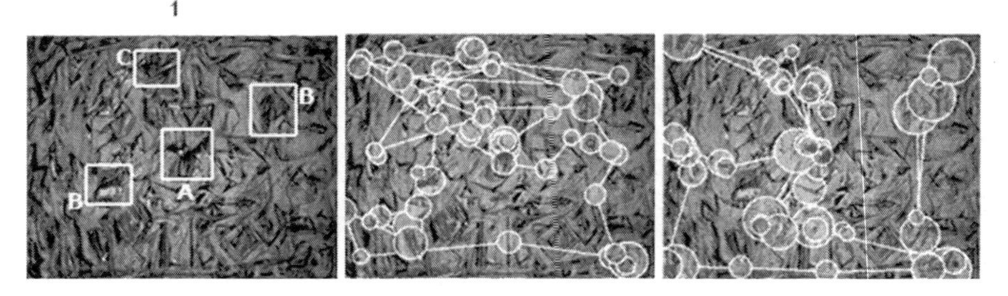

Fig. 9 Attention zones for the painting 1

blue (A); (Fig. 9). The second highest density is shared by two diagonally opposite strokes in red (B). Redundancy in pattern and stroke-style of the painting might have caused an attempt to fixate on a top red stroke clubbed with a star-shaped stroke in black (C). The viewer seems to be assembling these color codes and directional strokes in form, to establish a relationship between diversely located clues for visual mapping to develop their understanding.

For painting 2, the maximum density area is the center. The horse shapes overlapping (A) (Fig. 10). The second highest density area is shared by the foliage on right and the sun (B and C). Though the sun is less detailed than other areas, it appears as a point of reference and reconfirmation junction for visual perception.

For 3, the high point is positioned a little above the center (A). There are very few long gazes observed in this painting. Primarily the surface has been covered with very short and medium gazes implying a discursive route by overall composition to establish visual perception.

For painting 4, the maximum attention is on the off-centered left area with a significantly visible and prominent black stroke and the form in shape of an arc (A) (Fig. 11). It is observed that second highest attention area at the bottom left where there is a play of color glazes in ocher and blue (B). The curvature form is echoed at various places but most predominantly on right in a more perfect circle-stroke which is given the third highest attention (C). It is noticeable that though there is a deliberate effort to delegate interest toward right with continual inclined strokes and geometry, on the contrary the bottom left area evokes visual attention as there are seemingly complex, asymmetric stroke combinations. There are sweeping lines that are ignored at the right side. Organic and surprising form and spatial contrast takes most of viewer's attention in this painting than more geometrical and symmetric elements.

To sum up the results, the spatially spread abstract expressionist paintings evoked high information processing also providing maximum number of information components, while stroke and strong line oriented paintings offered lesser information processing. The pure abstract geometrical painting by Piet Mondrian evoked more search activity (Fig. 12).

While both spatially spread paintings evoke high visual processing, they are explored differently when it comes to diverse visual exploration. Similarly the

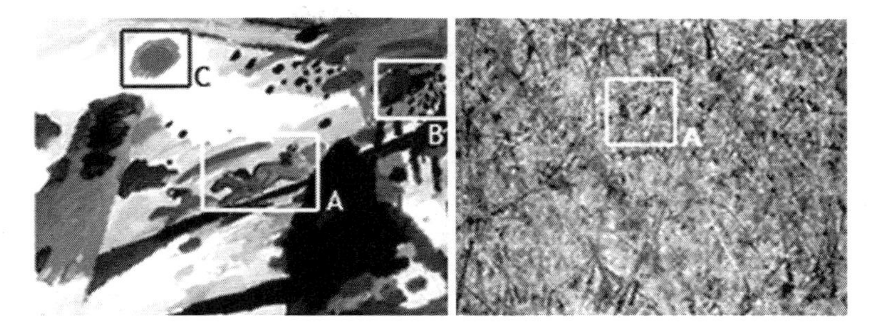

Fig. 10 Attention zones for the painting 2, 3

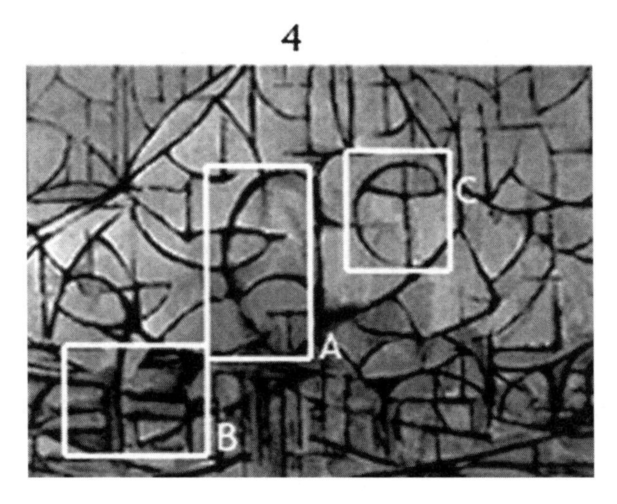

Fig. 11 Attention zones for the painting 4

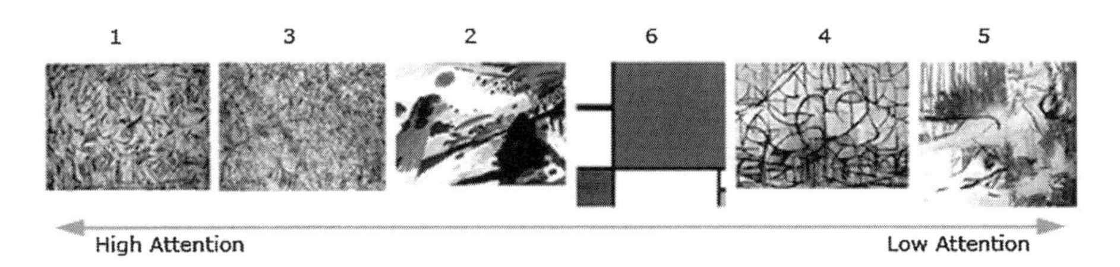

Fig. 12 Attention (information processing) trend for the paintings

stroke and line oriented paintings receive less information processing, but comparatively high diverse exploratory behavior (Fig. 13). The comparison of Information processing on components with the visual exploration trend shows that the two are distinct traits of visual response. Indicating that existence of higher information processing may not necessitate a high diverse exploratory behavior.

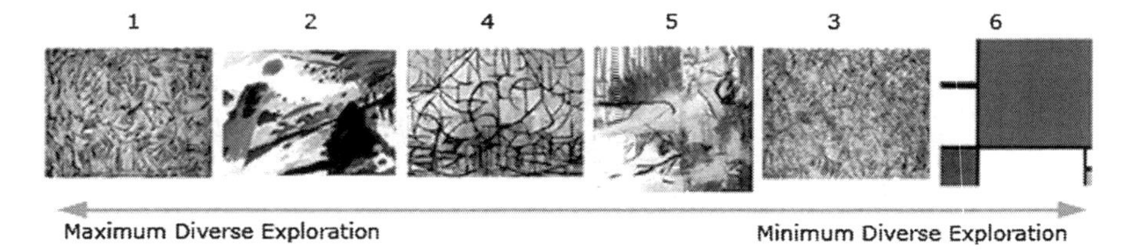

Fig. 13 Diverse visual exploration (short gaze) trend for the paintings

5 Conclusion

Viewer attention and visual exploration are probed specifically in the eye movement study, to understand viewer information processing through visual attention and viewer perception aims through diverse and focal visual strategies. We found that style of the painting affects the viewing response, which is in line with previous studies [9]. Further we form an understanding that, existence of higher information processing may not necessitate a high diverse exploratory behavior. These could be considered as two distinct traits of visual strategies which need to be explored further in various genres, to further the understanding.

This account of visual response as an area to study visual perception is an effective means of exploring the formal and in-built composition of art and design objects with visual perception [11]. A study of viewing strategies not only facilitates understanding of esthetical appreciation, but helps develop a rationale for understanding role of spatio-compositional use of elemental features within the morphology of a form. These features facilitate the visual communication and trigger the visual dialog with objects while abstracting through visual attention [16].

The number of participants, their age, and the limited number of stimuli are the limitations of the study. A large number of paintings of different genres with diversely variated subject groups may offer deeper insights and substantiate current learning.

References

1. Yarbus AL (1967) Eye movements and vision. Plenum Press, New York
2. Berlyne DE (1975) Studies in the new experimental aesthetics. Am J Psychol 88(3):520–523
3. Mial RC, Tchalenko J (2001) Eye movement in portrait drawing. Leonardo 34(1):35–40
4. Stark CM (2005) Algorithms for defining visual regions of interest: comparison with eye movement. IEEE
5. Zeki S (2001) Artistic creativity and the brain. Science 293:51–52
6. Privitera CM, Stark LW, Zangemeister WH (2007) Bonnard's representation of the perception of substance. J Eye Mov Res 1(3):1–6
7. Purcell AT (1993) Relations between preference and typicality in the experience of paintings. Leonardo 26(3):235–241

8. Hristova E, Grinberg M (2011) Time course of eye movements during painting perception. European perspectives on cognitive science
9. Leder H, Carbon CC (2006) Entitling art: influence of title information on understanding and appreciation of paintings. Acta Psychol 121:176–198
10. Leder H, Belke B (2004) A model of aesthetic appreciation and aesthetic judgments. Br J Psychol 95:489–508
11. Sharma SY, Chakravarthy BK (2011) Perception of form: a peep on the eye (ICoRD'11). In: International conference on research into design, Bangalore, pp 818–825
12. Duchowski AT (2002) A Breadth: first survey of eye tracking applications. Behavior and Research Methods instruments and Computers BRMIC
13. D Salvucci, Goldberg J (2000) Identifying fixations and saccades in eye tracking protocols. In: Proceedings of the eye tracking research and applications ETRA symposium, pp 71–78
14. Goldberg JH, Kotval XP (1999) Computer interface evaluation using eye movements: methods and constructs. Int J Ind Ergon 24:631–645
15. Nodine CF, Locher PJ (1993) The role of formal art training on perception and aesthetic judgment of art compositions. Leonardo 26(3):219–227
16. Sharma SY, Chakravarthy BK (2009) Abstraction: investigating the phenomena to facilitate form creation in design. In: International association of societies of design research (IASDR'09), - COEX, Seoul

Stimuli painting details

Painting title and style	Artist and year	Other information
1. Excavation	Willem De Kooning	Medium: oil on canvas. Size: 6 ft 8 in × 8 ft 4 in.
2. Paisaje romantico	Wassily Kandinsky	Medium: oil on canvas Size not known
3. Lavender mist	Jackson Pollock	Medium: oil and other media Size: 7 ft 2 1/2 in × 9 ft 11 in.
4. Alberi in fiore	Piet Mondrian	Medium: Oil on canvas Size: 86 × 66 cm.
5. Antioch (part of the painting)	Mary Abbot	Medium: oil and oil crayon on canvas. Size: 49 × 85 in.
6. Composition with red ,blue and yellow	Piet Mondrian	Medium: oil on canvas Size: 37 × 27 7/8 in.

Web source for paintings
1. http://www.artic.edu/aic/collections/artwork/76244
2. http://www.royal-painting.com/htmllarge/large-22917.html
3. http://www.ibiblio.org/wm/paint/auth/pollock/lavender-mist/
4. http://oseculoprodigioso.blogspot.in/2007/03/mondrian-piet-neo-plasticismo.html
5. http://raggedclothcafe.com/2007/04/22/mary-abbott-abstract-expressionist-by-clairan-ferrono/
6. http://www.wikipaintings.org/en/piet-mondrian/composition-with-red-blue-and-yellow-1930#close

Part V
Eco-Design, Sustainable Manufacturing, Design for Sustainability

Sustainability and Research into Interactions

Suman Devadula and Amaresh Chakrabarti

Abstract Sustainability is an ambitious interdisciplinary research agenda. The required knowledge, tools, methods and competencies being spread across wide-ranging areas pose challenges for researchers in sustainability who often specialize in one discipline. The efforts of researchers to understand sustainability comprehensively and contribute will be benefited if research outcomes are presented against an integrating framework for sustainability knowledge. Though general systems theory has this agenda, it targets consilience and not sustainability in particular as in sustainable development. However, systems concepts provide for a structure to imbibe aspects of sustainability. We propose a nested structure for organizing relevant research across the various scales of concerns that characterize sustainability. As understanding sustainability fundamentally requires understanding the interactions between natural and human systems, we discuss this in the context of the proposed structure and research into interactions.

Keywords Sustainability research · Nestedness · Interactions · Systems coherence

1 Introduction

Sustainability science is an ambitious agenda comparable to the Copernican revolution [1] and aspiring to integrate theory, applied science and policy, making it relevant for development globally and generating a new interdisciplinary

S. Devadula (✉) · A. Chakrabarti
Centre for Product Design and Manufacturing, Indian Institute of Science,
Bangalore 560012, India
e-mail: devadula@cpdm.iisc.ernet.in

A. Chakrabarti and R. V. Prakash (eds.), *ICoRD'13*, Lecture Notes in Mechanical Engineering, DOI: 10.1007/978-81-322-1050-4_39, © Springer India 2013

synthesis across fields [2]. It emphasizes management of the human, social and ecological systems from an engineering and policy perspective at earth scale. A systemic conception of earth comprises four spheres i.e., atmosphere, biosphere, lithosphere, hydrosphere, and the interactions between them. To make a further distinction, researchers in climate change have added the *Anthroposphere* or *technosphere* as separate from biosphere in comprising anything anthropogenic i.e., the effects of human and social systems in terms of the emissions off the first-world industrial revolution [3], resource over-use [4], etc. On the other hand, skeptics opine that progress of any civilization, both cultural and economic, is afforded by the provisions of the environment, and that when environmental conditions are themselves dependent on other cycles, periods of rise and extinction of species human or otherwise, become a consequence of these cycles. The frequency, amplitude and the coupling of these cycles can lead to periods that afford life or prove detrimental to it [5], relegating questions of sustainability to happenstance.

The questioning of current development trajectories and the future that the burgeoning third-world should take, leaves little scope for chances to be taken. Hence, addressing unsustainability at the required scale and intensity requires a systemic understanding of interactions between human, social and ecological systems for making meaningful inferences and consequent action. While there are dangers of conclusive inferences out of trials to force simple reductionist models onto a diverse set of world situations [6], approaches that rely on a sub-set of potential variables of socio-environmental systems (SESs) and propose abstract cure-alls for solving complex SES problems prove detrimental too [7]. Sustainability of SESs requires us to build a coherent understanding of how systems are progressively linked to ever larger systems and how upward and downward causation linkages occur within an SES as well as across diverse sectors and scales. This is a prescription for sustainability research involving interactions across scales. The varied nature of reading material, knowledge of worthy disciplinary contributions to sustainability requires a framework for structuring disciplinary knowledge in the broader context of sustainability. In this paper, we propose a structure for supporting systemic understanding of interactions at various scales that can also be used for organizing literature on sustainability.

2 Interactions

For understanding complex systems, it is believed that a focus on interactions rather than the entities within a system opens up vistas. Interactions are ontologically equivalent to entities in *entitification*. The process of *entitification* (Fig. 1) identifies entities. The existence of a differential simultaneously provides an ontological basis for entities and interactions. Interactions are self-liquidating as the differential gradient deteriorates with interaction time. Interactions are inferred through changes in energy, material, information or entropy generically.

The dynamic of interacting entities, as real world systems, is provided by modern thermodynamics. The second law of thermodynamics gives a direction to progression of systems. Isolated systems progressively increase their disorderliness toward thermodynamic equilibrium. Contrarily, open systems, when exposed to a sufficient differential, tend to spontaneously decrease their disorderliness while increasing that of their environment. Behaving in this way, these systems evade thermodynamic equilibrium by self-organizing and increasing in complexity. Such systems are referred to as far from equilibrium self-organizing dissipative structures (FFESODS), and by nature of their behavior including information, there is much debate on such interactions to be characteristic of organisms and life. Contrasting these two types of systems, which together comprise parts and the whole of the universe, shows that while the whole tends towards disorganization and equality, parts further organization and distinction. Such systemic behavior, of the particulars as well as the whole, fundamentally describes interactions among them. This is illustrated as a juxtaposition of two cycles (Fig. 1): one, in bold arrows, represents natural cycles where sustainability is not a question as this 'is'; and two, in outlined arrows, is the anthropogenic cycle, in which, out of knowledge of natural laws, the impression of differentials onto existing entities changes the availability of opportunity (resources) to natural cycles and hence potentially, their course e.g. the construction of dams changing downstream natural cycles, accumulated GHG's changing the intensity of monsoons etc. As the services humans get out of natural systems are irreplaceable, the magnitude of change they initiate affecting natural cycles to their own detriment is the subject matter of sustainability. Note that both these cycles ever abide by natural laws, the difference being the magnitudes of action of the anthropogenic cycle resulting in magnitudes of reaction of the natural cycle. The management of consequences arising due to differences of these magnitudes is the interest of sustainability, as a praxeological prescription of what should be the case for the human use of earth. Note that the above systemic notions used to define interactions assume as entities characterized by their boundaries, when in reality it is a continuum of their properties (or attributes). This leads to the concept of boundary uncertainty influencing interactions at different scales. These scales are nested by nature. As sustainability requires a reconsideration of many prevailing ways of interaction, Fig. 2 is conceived to systemically consider interactions across nested scales of reality. Though the concept of nestedness exists in ecological, AI and social sciences, the consideration of the unifying force across scales provides a context for explicitly considering interactions. The annular areas represent scales of reality; the concentric circles demarcate scales, the crossing of which marks interactions relevant between scales from sub-atomic (part) reality to the universe (whole). The entities that comprise a scale are alone not sufficient to describe that which results in the next level. Figure 2 should be read as any annular area being a scale of reality comprising of the physicality at the scale lower to it along with the interactions relevant between these scales as indicated in the interaction column to the right, e.g. particles along with nuclear forces comprises the atom, atoms along with chemical forces comprise molecules, etc. Proceeding from the organism ring either

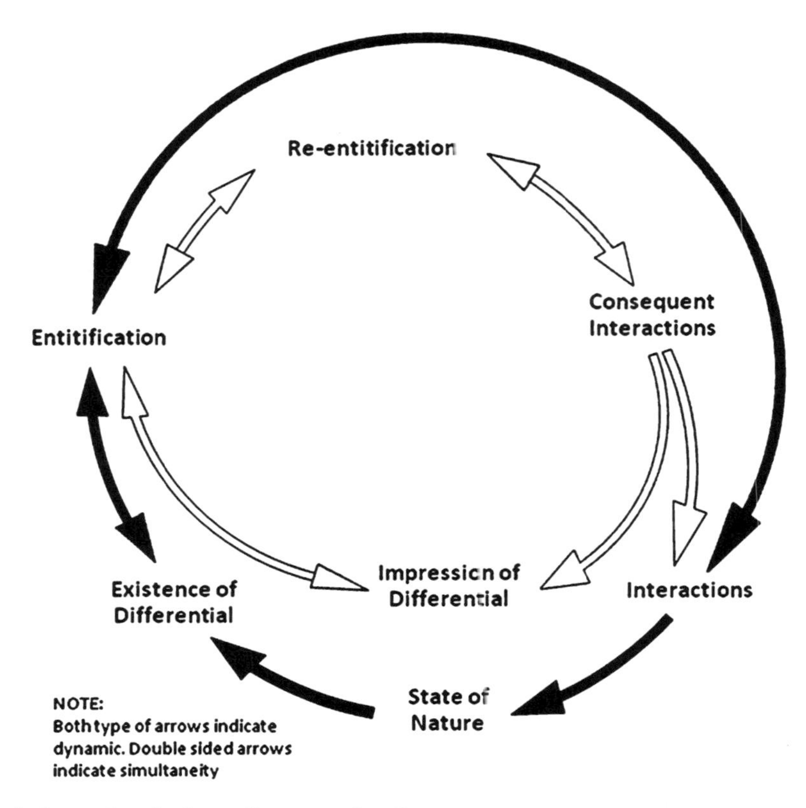

Fig. 1 Interactions in the cyclic nature of reality

ways, the dotted circles indicate the uncertainty in knowledge associated at the micro or macro scales. The organism ring is demarcated with full lines as the notion of certainty through identity, either in the form of cell wall for the cell or territory for the animal, is explicitly reinforced. However, proceeding either ways leads one to realize that identity, and in extension reality, is actually punctured. This realization of reality, the other inference of the dotted circle in Fig. 2, paradoxically leads to being nothing (represented by zero, at the centre) and everything (represented by infinity, at the outset). The implications of findings at the fundamental particle scale are applicable at the highest scale. Thus, at the scale of constituent atoms we are indisputably creatures derived from the cosmos [8]. However, the implications at the intermediary scales in terms of human-environment interactions are unclear, makes research inquiry challenging given the indeterministic, normative nature prevalent at these scales [9]. Given the fact that sustainability is characterized by requiring a transition from the self-obliterating state of affairs, the knowledge required for informing such a transition is felt at various scales culminating in a worldview which is an individual's or community's

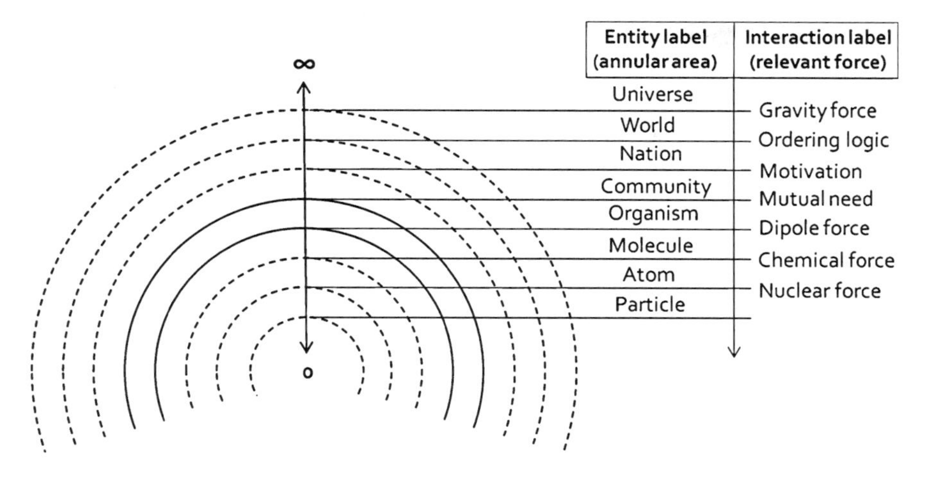

Fig. 2 A structure of reality for ordering knowledge of interactions for sustainability

conception of reality. We present issues and questions ordered according to the scales they address as relevant for a transition to making development sustainable.

2.1 Universe

The universe is so grand to our sense of time and space that we conceive it as one infinite i.e., 'uni'. Gravitational force, acting by virtue of the planetary mass and relative distance, is the unifying force of entities of the lesser scale. The grand scale of space, time and matter that the universe spans make it the bearer of evidence for answering questions of origin of matter and life within cosmic phenomena. The fact that certain life-sustaining physical constants have just come right in this earthly corner of the universe is attributed to chance [10]. We are yet to have a cosmic neighbor to share our mundane experiences of the planet and possibly learn from theirs. A view, provided by the universe beyond the earthly limits, that can establish any fact of planetary life elsewhere affords pan-earth notions of life and hence a provision for a cosmic-praxeological framework to hang our individual planetary behavior in comparison of our earthly responsibility. An example theorizing life's origin is *pan-spermia* which states that a great percentage of interstellar dust is microbial and that striking comets or meteors can be potential vehicles for these microbes to prime evolution [8] on habitable planets like earth. Another example is the notion of 'self-realization of the cosmos' [11], which, carrying forward Spinoza's idea of substance, states that the essence of universal substance is to seek plenitude. The metaphors linking this scale of reality to ours are *spaceship-earth, cosmic speck of dust, pale blue dot, cosmic sea, planetary stewardship* etc. What matters to our sustainability at this scale is the prompt knowledge of stellar events involving the earth and the ability to handle their effect e.g., a comet about to intersect the earth's revolutionary path, radiation from novae, the influence of the sun, etc. At the grandest scale the universe affords being considered as an isolated system. Consequently its ever increasing entropy makes chaos, irrespective of life's activity otherwise on earth, increasingly more probable. Though the time-scale of this fact is way beyond what can matter in a life-time, it provides an eventuality within which we may strive to achieve a goal that fits our human condition and is commonly agreed upon for its worth.

2.2 World

The world comprises nations and people in them. Nations, as protected sovereign territories by governments authorized through the consent of their people in a democratic election, come together out of common interest based on their respective foreign policy based on ordering logic [12]. The metaphors relevant at this scale of reality are mother earth, *Gaia, only home, earth democracy, earth system governance* etc. This scale and the two lower ones, nation and community, are partly or wholly human constructs, and are included as real scales in Fig. 1 as they are the sources of our highest institutions. As we have grown to be dependent upon our institutions, they serve as the instruments with which we organize effort in the economy, both for routine conduct of affairs of the state and for working out sustainable transitions required. The institutions at this scale are supranationalist e.g., the UN, ILO, IMF, WB etc. These were mostly products of first-world flagship and the necessity was/is felt particularly at times of challenges that span national borders e.g. League of Nations for stopping WW2, ILO for internationally protecting worker rights, UN for stopping WW3 and working for world peace and security, G5 nations comprising the superpowers having veto power in the UN etc. Consequently, interactions at this scale are determined by how states conduct in the international society of states based on the agreements they sign or ratify.

Sustainability is a problem irrespective of national divides and hence may be said to be situated at this scale of reality. One conception of the earth is as Gaia, indicating that the earth is a self-correcting system and that humans should consider consuming its resources within its spring-back. Though the climate crisis has proven that sustainability challenges need unqualified planetary scale efforts, nations continue to be divided on how to share responsibility for the causes of damage (CBDR argument of the Third-World), their repair, and the monetary and intellectual investment necessary for working out a global agenda. The failure of the recently held Rio+20 conference in relation to any of its preceding summits is testimonial to this fact. Leading environmentalists have observed that the call for sustainability is of earth scale, recommending 'earth democracy' from a preservation perspective. Others have called upon 'earth governance' in an effort to free individuals from the hold of their nationalist identities that offer no protection in times of natural calamity proving detrimental to humanity. Sustainability needs sincere, committed effort from the nations to invest in the required intellectual and technological capability towards helping all span the transition to a better world in which everyone realize their full potential.

2.3 Nation

The continental shelves along with geographic surface features provide for the political demarcations between countries that are otherwise explicitly erected and surveyed continuously. Beyond this the idea of a nation is constructed by two forces. One, the vesting of authority in the government by the will of the people participating in free and fair elections of representatives, and two, the patriotic feeling driven by the resourcefulness of the country necessary for sustaining its population. The metaphors in use e.g., *mother, homeland, motherland, fatherland* etc., are testimonial of the perspective of a provider of nourishment. The idea of lack of resources elsewhere accentuates the second feeling, more of which may seed fundamentalism [13]. Consequently, the matter of interest for sustainability at this scale is the preservation and distribution of national resources to its citizens through drafting effective policy [14–16] and appropriately realized and represented systems of governance and administration, the well-being indices by which national progress is measured, the human rights treaties to which nations are internationally signatory that serve to check its domestic policy against its own citizens as humans first, the transparency and independence of its judicial system to civilians, institutional provisions for recognition of civil societies etc.

2.4 Community

Environment provides necessary resources for the survival of plants and animals. Man is unique among these animals in having abilities to communicate using language [17, 18]. This enabled him to think, extend his cognition, gather groups, organize effort, enterprise and effect a change in the resources to produce tools and artifacts that in turn better equipped him for survival. Consequently his capabilities had multiplied beyond his rather frail abilities; however, this development has been non-uniform over the planet wherever people thrived into communities and civilizations. Each progressed at their own rate based on the limitations of locally available resources and their own physical and mental limitations in effecting change for their advantage. This may have led to exchange/trade across settlements and civilizations for mutual benefit driving a curiosity for and interest in exotic resources. Unlike the previous scales, the smaller size of this scale stands greater chance of suiting the physical and cognitive limitations of more of its members and hence results in more individuals knowing most about their communities. The first notions of community among simple bacteria that grow in local environs can be called 'cultures' while mutualism, commensalism, amensalism etc. are fitting labels for animals interacting within communities for satisfying shared needs. Ecologists, identifying the scales at which a phenomenon is applicable [19], have extensively researched establishing community structure within which they are now able to predict interactions with some certainty. Consequently, the scales at which such communities form, limited by available natural resources and mutual necessity of its members, seem to be closer to the natural cycle than higher scale constructs that are positively maintained to cater to the masses by the exploitation of resources elsewhere. This implies that communities at this scale are generally self-sustained within their knowledge-base of the surroundings and of collective action necessary for corrective measures to be taken in an event of disturbance [15]. Prevalent institutions here derive their authority either through democratically elected representatives in free and fair elections or are vested in members practicing traditional occupations that emerged in mutual necessity. Other communities that form are institutions e.g., welfare associations, manufacturing institutions, and virtual communities e.g., social networking, chat groups, etc. The role of these institutions for sustainability is commensurate with their resource use. One example is the corporate sustainability performance initiatives that have become important while emphasizing the role of the corporation in the affairs of the state and of people. Hence it is of interest to sustainability at this scale to understand how such institutions can be steered towards meeting goals of sustainability amidst tighter constraints. On another note, the community an individual is part of partly provides for the construction of his identity [20] and this is essential in framing his interactions with the other members and his contribution to the society as a whole.

2.5 Organism

From unicellular organisms to plants and humans, this level consists of all entities that are capable of self-maintenance and self-replication. This scale consists of all biota of the planet listed by the entities of the trophic levels in ecology. The fundamental unit of this scale is the 'cell' in its capability to maintain and replicate itself. Consequently of interest at this scale of reality are the capabilities of these organisms amidst changing contexts. The changing contexts are primarily of the environment requiring the organism to adapt to it or perish. However, contexts also change due to organismic activity, e.g., decreasing availability of resources as organismic populations increase. It is argued that the need, for formation of cell wall is necessitated by the competition for resources amidst increasing organismic populations in the *primordial soup*. This situates the problem for the organism's sustainability amidst organisms similarly driven.

Homo Sapiens are the first to alter natural courses at planet scale and also be aware of this fact [6]. Though this generally occurs post facto, we only have instantial knowledge of doing otherwise under less severe situations like avoiding ozone depletion (Montreal Protocol). This gives rise to metaphors like *technological adolescence, earth-scale stupidity, geo-engineering, techno-fascination, earth-worthiness, megabuck science* etc. The capacity of humans residing in this scale to be aware of their bio-physical structure as well as their possible realization of entailing concepts of uncertainty and pervasiveness, make this a scale in which conceptions, however rudimentary or refined, of all the other scales become possible. One such concept is anthropocentrism that provides a basis for rights and hence sustainable development too. Objective validation of hypotheses, framed at various scales of reality through appropriate experiment and method may support the anthropocentric notion of sustainability. It is stated that to proceed "...from bacterium to people is less of a step than to go from a mixture of amino acids (molecules in Fig. 2) to that bacterium (organism in Fig. 2)" [21]. This scale of reality is also where notions of identification with oneself as an entity on the general basis of sustenance forms the pivot of all arguments in the process of evolution by natural selection. At the extremes, and exemplary in this context, are the creationist and naturalist [22] explanations to life. Consequently, the matters of concern at this scale are behavior, conduct, intentionality, responsibility, rights, duties, truth, purpose, agency etc. Correspondingly, the fields of interest are axiology and praxeology as ends, and ontology and epistemology [23] as means. In short, the worldview of humans is the fundamental concern as this [24] influences our conception of reality, including the earth and its use, in designing [25] to meet our requirements. Motivation for human action at various levels of satisfaction is otherwise provided by a hierarchy of human needs [26], though the influence of a worldview is beyond these motivations.

Worldview is a conceptual system by which we order reality, irrespective of its scale, and hence seem to find our way comfortably in it. Worldview orders interactions and the plurality of worldviews implicates sustainability of the whole. While ontological inquiry provides insight into what is and what being is, epistemological inquiry focuses on the process of acquiring this knowledge, particularly on tools and methods, and their limitations for such acquisition. Language of discourse, scientific method, deduction and inference are some examples of such tools and methods [27]. Axiology explores criteria for evaluation of reality, meaningful life worth living and what should one strive for. Sustainability, as a concept, with its notion of human centrality takes some answers for such axiological questions, granted. That human life is worth striving at the cost of anything else for by the individual himself and the community so that every individual realizes his full potential is one such answer. Even in phrases like 'environmental sustainability' or 'X sustainability', it is the human whose well-being has to be preserved, for which an environment's or x's contribution is required and hence to be ensured. However, if the epistemological basis for human centrality is ill-founded, the concept of sustainability needs to re-work its priorities as might be the case under a paradigm of 'life-centrism or ecological naturalism' [9]. Ontology and epistemology need provide the means for answering axiological and praxeological questions related to sustainability. The choices we make depend on the worldview we inherit or learn to adopt. Until we become conscious of our worldview and plan to override its influence, we may not be acting sustainably [28]. In this connection, leaders of religion and nations on different occasions have urged mankind to inculcate 'universal consciousness' that is fitting for world peace, security and sustainability.

2.6 The Lower Three Scales

These scales comprise of all a biota of the planet. Nomenclature is extensively developed in these fields, though development of integrative frameworks for organizing knowledge for interdisciplinary research has been an afterthought [29]. Implications of research findings at this level are fundamental to at all scales of reality, though within the assumed anthropocentrism, of relevance to sustainability is research that has direct implications to humans alone at whatever cost to the rest. Concepts like uncertainty encountered at these scales have far-reaching implications to our conception of reality as it could change our paradigms towards that necessary for sustainability.

3 Discussion

Sustainability science is a relatively new, inter-disciplinary, and fast developing field of inquiry, which lacks coherent reference material and textbooks [30]. Sustainability, as humanity's concern and at the scale at which it is required renders national, political and personal boundaries porous. Consequently, economies must invest in and drive the sustainability agenda domestically while co-operating to follow the sustainability roadmaps laid out by neutral, supranationalist organizations. In this regard, though the Rio Summit of 1992 was promising, the recent Rio+20 conference of the UN was a failure, described as the longest suicidal note by activist groups. Henceforth it is personal conviction in sustainability that should motivate people to commit time and effort towards influencing the short-term interests of their domestic governments towards making and leaving the planet a better place to lead life. Individuals value different scales of reality differently. These differences set up arenas for people to field opinion and discuss perspective comparatively. Sustainability, as a human ability, is predominantly a social argument, in requiring us to be able to mutually support successfully informing ourselves of realizing purposive action determined by the limits of the earth. Hence, within the natural tendency of systems to deteriorate, sustainability is a human ability to communicate action to come to grips with a deteriorating environment individually while limiting its voluntary deterioration in attending to needs alone [31]. With this basis to the argument, it is appropriate to refer to interventions as 'tackling unsustainability', in a Sisyphean sense [32], that is more real in involving more time and effort rather than a vainglorious phrase of 'achieving sustainability' which is only momentary.

The outlined inner cycle of Fig. 1 represents the design cycle of making, reflecting and modifying. From the perspective of interactions and the scales of reality across which they can occur, designers are implicated to control the amount of change that the product/process life-cycles have on the natural cycles. Philosophically this requires design to be negative rather than its conventional understanding of being positive. Negative design is about design being reactive, initiated only under circumstances where it is widely indicated that "...life has gone wrong" [33]. The stimulus for negative design is not an imagined future but a real problem in the form of a changed context similar to evolution fitting a new environmental context. Negative design intends only as much as necessary to bring the state of affairs back to being in accord with the new context. While positive design imagines a future and designs for it, negative design reacts to real world [34] problems and designs to resolve them. This contrast of design philosophy may be likened to the thermodynamic context of sustainability, i.e., as the contexts to which we need to self-organize as complex biological and social systems change naturally, personally furthering their magnitude and frequency of change is uncalled for as it may demand organization of a scale that we may not be able to match, in scale and in time. Though LCA assessments provide for point-estimates of environmental impacts, designers should be sensitive to the dynamic concern of sustainability, and assess the life-cycles of products and processes for their consequences on vital natural cycles, before going forth with product development locally or globally.

4 Conclusions

Understanding sustainability necessarily requires knowledge of aspects of multiple scales of reality and their interactions. The entailing requirements for data are so huge that sustainability research is limited to only few aspects of reality thereby falling short of making any holistic claims about sustainability. The integrative research agenda spanning scales of reality of sustainability science implicates science to coherently structure its disciplinary findings within a framework. Addressing this, a structure of reality is proposed to order knowledge of sustainability. Within the organism scale of this structure, the difficulty of addressing normative aspects within the methods of science and discussions on other approaches to conceive and understand reality is accommodated. Sustainability springs from a human rights and dignity core. As inviolable rights claiming their bearings in natural rights derived from natural law, the concept of sustainability needs to be founded naturally and thereby ground anthropocentrism, it relies on. The proposed description of interactions based on thermodynamics aids understanding the dynamics of interactions across all scales of real world systems. This, along with the description of the scales of reality proposed, indicate a probable theory of interaction that could order inquiry across disciplinary borders accommodating relevant normative aspects. Towards this, we have systemically structured exemplary sustainability literature, presenting aspects of it in context.

Acknowledgments The authors thank Ramani K for valuable comment on the draft and The Boeing Company for financial support under contract PC36018 at SID, IISc.

References

1. Schellnhuber H (1999) Earth system analysis and the second Copernican revolution. Nature 402(2):C19–C23
2. Wiek A, Withycombe L, Redman CL, Ferrer-Balas D (eds) (2011) Key competencies in sustainability: a reference framework for academic program development. 30 Mar 2011
3. Zalasiewicz J et al (2010) The new world of the anthropocene. Environ Sci Technol 44:2228–2231
4. Vitousek PM et al (1997) Human domination of earth's ecosystems. Science 277:494–499
5. Yu JF et al (2008) Recognition of Milankovitch cycles in the stratigraphic record: application of the CWT and the FFT to well-log data. J China Univ Min Technol 18(4):594–598
6. Mörner NA, Burdyuzha V (eds) (2006) The danger of ruling models in a world of natural changes and shifts: the future of life and the future of our civilization. Springer, Berlin, pp 105–114
7. Ostrom E, Janssen MA, Anderies JM (2007) Going beyond panaceas. Proc Natl Acad Sci 104(39):15176–15178
8. Wickramasinghe C, Burdyuzha V (2006) The spread of life throughout the cosmos: the future of life and the future of our civilization. Springer, Berlin, pp 3–21
9. Wang LS (2011) Causal efficacy and the normative notion of sustainability science, vol 7(2) Proquest Fall, NY. http://sspp.proquest.com
10. Sagan C (1961) On the origin and planetary distribution of life. Radiat Res 15:174–192
11. Mathews F (1999) The ecological self. Routledge, Great Britain. ISBN 0-203-00974-6
12. RP Claude, BH Weston (1992) Human rights in the world community: issues and action. In: Falk RA (ed) Theoretical foundations of human rights, pp 31–41
13. Ackoff RL (2001) Syst Pract Act Res 14(1):3–10
14. Hardin G (1968) The tragedy of the commons. Science 162:1243–1248
15. Ostrom E (1990) Governing the commons, ISBN 0 521 40599 8
16. Poteete AR, Ostrom E (2008) Fifteen years of empirical research on collective action in NRM: struggling to build large-N databases based on qualitative research. World Dev 36(1):176–195
17. Clark A (2004) Is language special? some remarks on control, coding and co-ordination. Lang Sci 26(6):717–726
18. Wheeler M (2004) Is language the ultimate artefact? Lang Sci 26(2004):693–715
19. Levin SA (1995) Scale and Sustainability: a population and community perspective. Munasinghe M, Shearer W (eds) Defining and measuring sustainability: the bio-geophysical foundations. The UN University, New York and The World Bank, USA, pp 103–114
20. Goffman E (1959) Presentation of self in everyday life. Doubleday Anchor Books, New York
21. Margulis L (1996) End of Science. Horgan J (ed) Addison Wesley, MA (Chapter 5)
22. Hume D (1994) Dialogues concerning natural religion: the english philosophers from bacon to mill. In: Burtt EA (ed) Random house modern library, New York. p 1779
23. Gallopin GC (2004) Sustainable development: epistemological challenges to science and technology. ECLAC. Chile
24. Galle P (2008) Candidate worldviews for a design theory. Des Stud 29(3):267–303
25. Galle P (1999) Design as intentional action: a conceptual analysis. Des Stud 20(1):57–81

26. Maslow AH (1943) A theory of human motivation. Psychol Rev 50:370–396
27. Aerts D et al (2007) Worldviews: From Fragmentation to Integration. In: Vidal C, Riegler A (eds) Internet edition, Originally published in 1994 by VUB Press
28. van Egmond ND, de Vries HJM (2011) Sustainability: the search for the integral worldview. Futures 43:853–867
29. Boulding KE (2004) General systems theory: the skeleton of science. E:CO vol 6 no 1/2, pp 127–139. First published in management science vol 2, no 3, Apr 1956
30. Chichilnisky G (1977) Economic development and efficiency criteria in the satisfaction of basic needs. Appl Math Model 1(6):290–297
31. Kates RW (ed) (2010) Readings in sustainability science and technology. CID working paper no. 213. Center for International Development, Harvard University. Cambridge, Dec 2010. http://www.hks.harvard.edu/centers/cid/publications/faculty-working-papers/cid-working-paperno.-213
32. Camus A (1955) The myth of sisyphus
33. Jones JC (1980) Thoughts about the context of designing: in the dimension of time. IPC Business Press, Design studies, pp 172–176
34. Papanek V (2005) Design for the real world: human ecology and social change. Academy Chicago Publishers, vol 2, p 394. First Published in (1972), ISBN13: 9780897331531

Residential Buildings Use-Phase Memory for Better Consumption Monitoring of Users and Design Improvement

Lucile Picon, Bernard Yannou and Stéphanie Minel

Abstract Residents' usages and behaviour are inadequately known and understood, as well as being highly variable. However, they play a determining role in the variability of both the energy consumption and environmental impact of residential buildings during their use-phase. This paper proposes a use-phase memory model for residential buildings, which stores energy and resource consumption, and usage patterns. Useful information is further extracted by data crossing and visual data representation. Building experts refer to it for two specific use-cases, namely designing a new sustainable building and renovating an existing one. This information helps them to understand energy and resource consumption and, real users' behaviour and activities. Building's users obtain different kinds of service in return for their collaboration and contribution. Our model is presently being deployed on a residential building, based on beneficial services for each building's stakeholder, thus introducing a sustainable relationship between designers, the residential building and its users.

Keywords Use-phase memory · Environmental impact · Use behaviour · Design tools

---

L. Picon · B. Yannou (✉)
Laboratoire Genie Industriel, Ecole Centrale Paris,
Grande Voie des Vignes, 92290 Chatenay-Malabry, France
e-mail: bernard.yannou@ecp.fr

L. Picon
Design Industriel, Ecole Nationale Superieure de Creation Industrielle,
48 Rue Saint-Sabin, 75011 Paris, France
e-mail: lucile.picon@gmail.com

S. Minel
Technopole Izarbel, Ecole Supérieure Des Technologies Industrielles Avancées,
64210 Bidart, France
e-mail: s.minel@estia.fr

A. Chakrabarti and R. V. Prakash (eds.), *ICoRD'13*, Lecture Notes in Mechanical Engineering, DOI: 10.1007/978-81-322-1050-4_40, © Springer India 2013

1 Introduction

Announced by the ADEME energy supervision agency [1] in 2011, the building sector consumes 43 % of total energy consumption and ejects more than 24 % of the national CO_2 emissions every year in France. This sector consumes and pollutes more than heavy industry and transport. That is why energy and environmental experts generate standards and regulations, such as the "Thermal Regulations 2012" in France. Molle et al. [2] explain that the "TR 2012" recommends the limitation of energy consumption and needs to five specific end-uses: heating, sanitary hot water, lighting, heat-ventilation-air-conditioning (HVAC) and auxiliaries using renewable energies, thermal insulation, building orientation, etc. However according to Yu et al. [3] total building energy and resource consumption is influenced by a building's environment, building's characteristics, as well as by residents' behaviour and activities.

The issue is complex because resident usages and behaviour vary widely, are far from being completely known and understood, and different settings interact. The influence of appliances on building energy consumption and its environmental impacts is another factor to be considered because it varies according to their energy-efficiency, technology type, kind of usage and user's behaviour. For instance, National Resources Defense Council [4] in 2007 has indicated that a new high-definition set-top box and its attached television consume more energy per year than a refrigerator. In addition, according to Yu et al. [3] "current simulation tools can only imitate human behaviour patterns in a rigid way" and Hoes et al. [5] explain that the algorithms used, most of the time and at best, focus on the manual opening of windows and lighting control. As a consequence, results of computer-aided design simulations are far from being representative of reality because of the number of influencing factors that are ignored and ignorance of residents' needs, usages and behaviour. Building experts, for instance engineers and architects, need to know more about the way people transform, renew or degrade their equipments and appliances, which also in turn influence the resulting energy eco-efficiency, equipment life duration and environmental impacts. They would benefit from better knowing users needs. This knowledge would be useful in both situations of redesigning a sustainable construction and renovating an existing one. Concurrently, building users, such as residents, building administrator and employees, would benefit in getting feedbacks to improve their usages, tasks and quality of daily life. The establishment of links between building users, building experts and residential building's lifetime is essential for more sustainable building designs and renovations.

That is why we propose in this paper a use-phase memory model for application to residential buildings, which can store during a building's lifetime any information that can be useful for both construction companies and residential building's stakeholders. In Sect. 1, we study the literature concerning impacts and understanding of usages, transmission of knowledge and relationship between companies and the end-users of their products or/and services. Then, in Sect. 2, we present our model of use-phase memory of residential building, its main aim and its beneficial services, which establish a sustainable relationship between building experts, the residential building and its users. We conclude by demonstrating the potential of use-phase memory for building experts. We illustrate by usage scenarios how useful information can be obtained in these two use-cases for the purpose of the design of a new sustainable building or the renovation of an existing one.

2 Background and Related Work

Although several studies have been carried out to analyse the influence of usages and behaviour on the variability of product or system's environmental impacts, and to understand their reasons, few of them have established sustainable link and dialog between users and designers.

Telenko et al. [6] underline that for many products "the environmental impacts during use can be more significant than manufacturing and end-of-life impacts". Studying the combination of usage and domestic lighting appliances on energy consumption, Zaraket et al. [7] conclude that energy consumption is influenced by three factors: the presence of the occupant at home, the domestic activities and corresponding usages, and the types of lighting appliances. And Borg et al. [8] demonstrate the effect of energy efficiency on domestic electric appliances and they underline stand-by consumption as an increasing problem. In 2002, Enertech [9] has also highlighted this problem in a report on the assessment of potential electricity savings in European households. Energy consumption and environmental impacts of products are deeply influenced by characteristics of product utilization and, beyond, by the purchasing choice people make more or less consciously. That is why, it is essential to understand usages and users' needs.

User Centered Research approach often assists understanding of usages. Lilley et al. [10] define this approach as "the process of gaining information about practices, habits or behaviours in order to inform the design of a product, service or system". Lofthouse et al. [11] present a range of techniques, based on UCR approach, which can help designers to better understand the end-users of their products or/and services, their misuses and ways in which they adapt products to better satisfy their needs. However a residential building is a very complex system, more complex than a mobile phone. In addition, it is unsuitable tool and method for gathering, storage and reuse data and information about usages on a long period.

As expressed by Escalfoni et al. [12], the knowledge of an organization and its sharing are important for developing innovation, which is a way to improve the quality of products and services, identifying new activities and business opportunities. Gibbert et al. [13] distinguish Customer Relationship Management from Customer Knowledge Management. Knowing customers needs and tastes is a good way to satisfy them. However understanding customers, their needs, their ways of using and transforming a product or/and service, is a strategic key to innovate.

The influence of usages and behaviour on the variability of product's environmental impacts is admitted and, organizations and designers are aware of users' experience worth. However, few sustainable relationships including at the same time, user satisfaction, problems encountered, and the ways people use, handle, and/or adapt a product or service, are established between these different stakeholders. Shen et al. [14] introduce a communication tool between building users and designers. Developing an interactive platform by means of a realistic 3D representation of the building, it allows building users to improve their understanding on the design, help them to specify their activities in the new building and increase their involvement in communication with designers. In that way, designers can realize pre-occupancy evaluation and collect building users feedback on the design.

3 Model of Use-Phase Memory of Residential Building for Future Sustainable Constructions and Renovations

We develop below a model of use-phase memory for residential buildings, illustrated in Fig. 1.

The system gathers and stores useful information during a residential building's lifetime, which are energy and resource consumption, recounted events and encountered problems and, noticed residents' usages and behaviour. Useful information is further extracted by data crossing and visual data representation. Building experts refer to it, automatically or at their request, to be assisted for two specific use-cases, namely designing a new sustainable building and renovating an existing one. This information helps them to understand energy and resource consumption and, real users' behaviour and activities. Building's users obtain different kinds of service in return for their collaboration and contribution. Firstly, residents monthly get an invoice giving detailed energy consumption, helping them to identify potential savings. Secondly, if they need and are interested, they can be backed by a customized diagnostic, allowing understanding their usage, needs, appliance consumption so as to inventory consumption savings. The building administrator can control the residential building's state, assisting him to identify recurrent and costly problems (technical, deterioration and damage). Our model is presently being deployed on a residential building, based on beneficial services for each building's stakeholder, thus introducing a sustainable relationship between designers, the residential building and its users.

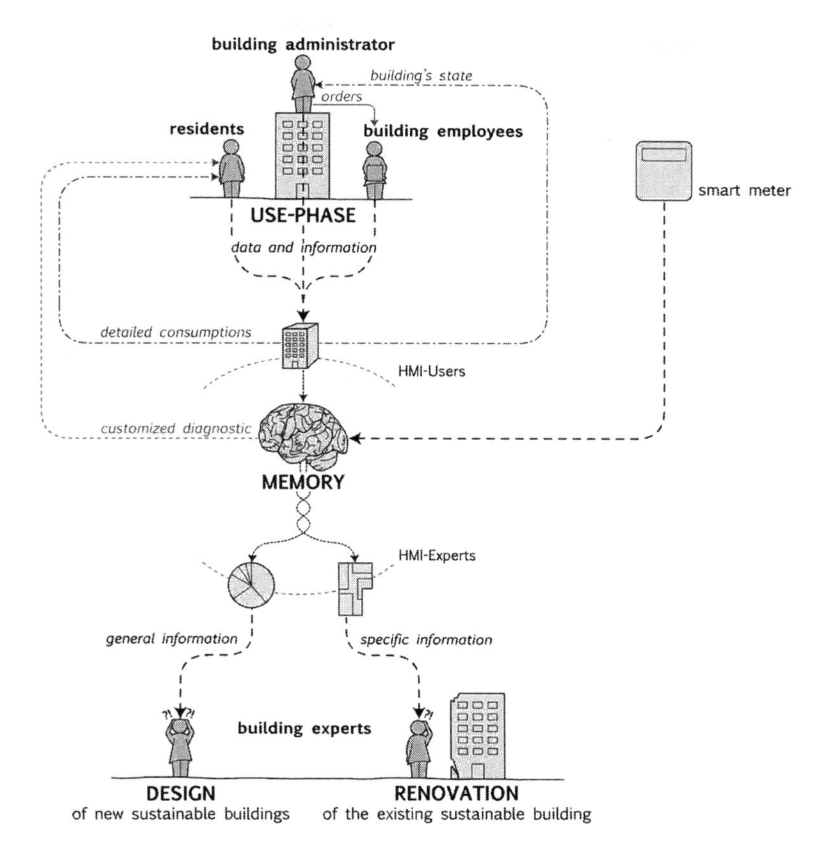

Fig. 1 Model of use-phase memory of residential building

3.1 Methodology

Dieng et al. [15] and Matta et al. [16] inspired our design of the use-phase memory and helped us to structure it. Of the three steps—gathering, storage and reuse—we focused on the input stage (gathering) and on the output stage (reuse). Several issues need to be dealt with: what data to collect, how to collect the data, what tools should be used for the input stage, how this information should be consulted, and how useful information should be extracted.

3.1.1 Data Gathering

The main issues are what data to collect, and how to collect these data.

On the one side automatically and quantitatively energy and resource consumption, which are consumed quantities of electricity, gas and wastewater, is measured thanks to automatic means. A principle of non-disclosure agreement with data providers is integrated into the agreements between the building administrator, residents and the use-phase memory of the residential building. For this study, we collaborate with a startup company that commercializes an electrical signal analysis box, further called *smart meter*. Used algorithm is able to assign each elementary signal to an electrical consuming category like: a washing machine, a toaster, lighting for a certain kind of electrical power and bulb technology. At the moment, existing installed gas and water meters being are used without detailed and automatic analysis. Consequently, residents or the building administrator (depending on the meters' location) must read the meters and transmit these data each month. And on the other side, usage patterns, noticed events and encountered problems are voluntarily and qualitatively gathered by building users.

3.1.2 Means of Obtaining Access to the Memory

Two aims and specific modes allowing access to the building memory are identified.

The feeding with gathered data and information by building users is done through an oriented Human Machine Interface user (HMI-User). Inspired by the platform described by Shen et al. [14] the HMI-User is connected to the use-phase memory of the residential building and allows the users to enter data and information directly on the 3D representation of the building. The feeding with electricity consumption is automatic thanks to the *smart meter*.

The consultation by building users is also done through the HMI-User. It displays the residential building's state through the 3D representation of building. The building experts use the HMI-Expert. Allowing them to display useful information for the two uses-cases. The next section details kinds of displayed information.

3.1.3 Useful Information and Knowledge Retrieval

The main aim of the use-phase memory for a residential building is to get building experts and building users useful information. Thanks to data crossing and different means of visualizing data, we extract from the collected data, information and knowledge for the building experts, as well as for the building administrator and residents in return for their collaboration.

First of all, the building experts need general information to design a new sustainable building, thus the HMI-Expert displays, by automatic analysis of the building's memory, statistical representations about energy and resource consumption. Renovating an existing building, they look for some specific information on the given building, therefore HMI-Expert firstly automatically displays the tagged building maps by most costly facts and problems. Then, for these two uses-cases, they can ask for specific information from the use-phase memory thanks to the request mode based on Structured Query Language.

The building administrator gets the building's state regarding energy and resource consumption, costly facts, and successes from the use-phase memory of the residential building through the HMI-User. Data gathered allow a first reference consumption resident to be defined on the scale of a building. The residents obtain the building's state and also their own consumption levels, which may be compared to the "reference consuming resident" previously evoked, this comparison being a source of energy saving questions and insights. This point is an important motivation for people to be challenged comparatively.

4 Demonstration of Useful Information Retrieval From Use-Phase Memory of Residential Building for Building Experts

In this section, we present the potential of our use-phase memory of residential building in terms of consultation of useful information and knowledge by building experts in the two use-cases of the design of a new sustainable building and renovation. We show one example of a possible scenario for each use-case.

4.1 Usage Scenario During the Design of a New Sustainable Building

Building experts quickly obtain global and useful information on energy and resource consumption. Then, they can browse and consult more detailed information on electricity consumption by end-uses and appliances such as shown by Figs. 2 and 3. Crossing data by automatic electrical consumption measurement and information passed on by residents, it is possible to identify which end-uses and appliances consume more energy, and their causes such as reduced energy-efficiency of an appliance or misuse.

They can get information on specific subjects from the use-phase memory through the request mode, which is detailed in the next section (Sect. 4.2) through the example of usage scenario by request mode during the renovation of an existing sustainable building.

4.2 Usage Scenario During the Renovation of an Existing Sustainable Building

Building experts automatically display specific information on most costly problems of the residential building. The ten most costly problems are located on the maps of the residential building, their financial importance being represented by the size of the bubble. Building experts can consult details of the problems thanks

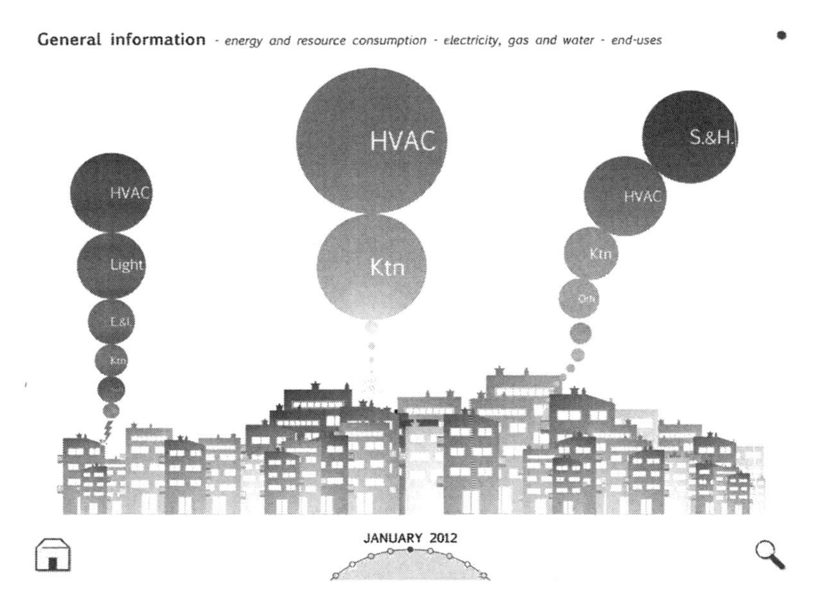

Fig. 2 Automatic display of electricity and gas consumption, and wastewater according to kind of end-uses

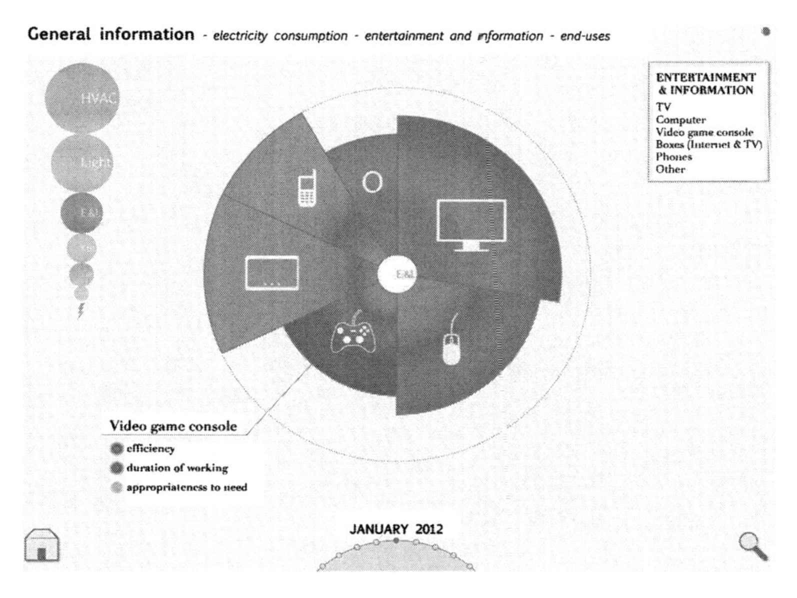

Fig. 3 Consultation of electricity consumption details according to kind of end-use and appliance

to reported information by the building administrator, residents and building employees. They can also browse in the different surface areas of the building. Information on costly problems is updated according to the scale of residential building such as block of flats, and accommodation and service spaces.

Using the request mode, building experts must specify the subject and the chosen setting(s). The request mode, based on Structured Query Language, is made up of two categories of information: subject and settings. The subject defines the type of information being researched. Two kinds of subject are proposed, namely "Consumption" and "Activities and needs", and they are divided into several sub-categories. The settings allow building experts to specify their request. We selected 10 different settings, which are duration, frequency, period, cost, surface area, equivalent resident number, kind of household, level of satisfaction or dissatisfaction, kind of end-uses (Heating, Ventilation and Air-Conditioning: HVAC, kitchen, lighting, entertainment and information, housework and sanitary or other) and kind of appliances. Then, they display located and represented information on the maps of residential building. Figure 4 presents information for "problems: technical problems, damage, deterioration and other" according to "cost: costly" and "level of satisfaction/dissatisfaction: all expressed dissatisfaction". For instance, it is possible to notice that for a costly problem the level of dissatisfaction is also very high and understand the reasons why. Of course, cost may not be the immediate problem for a resident: it could be problem's duration, frequency or level of inconvenience. However, these details allow building experts to understand the problem, its reasons and also its consequences for residents.

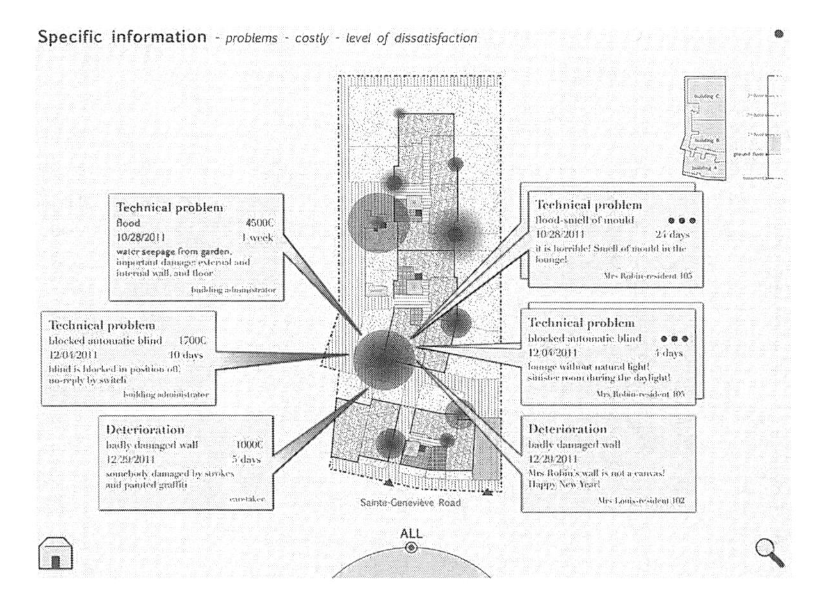

Fig. 4 Displayed result for "problems: technical problems, damage, deterioration and other, according to cost: costly, and level of satisfaction/dissatisfaction: all expressed dissatisfaction"

5 Conclusion

In this paper, we carry out a literature review on the impacts and understanding of usages, transmission of knowledge, existing concepts of memory and relationship between companies and the end-users of their products or/and services. Then, we detail the main aims of our proposed model for use-phase memory of residential buildings and explain which services and relationships are developed between residential building, building experts, building users. We end by demonstrating the potential of our proposition during the two use-cases, namely the design and the renovation processes of a new sustainable building.

Our proposition helps to better gather the usages and patterns of behaviour within a residential building thanks to gathered data and information on energy and resource consumptions, and also the events, successes and problems noticed and reported by building administrator, building employees and residents. For building experts, it is a means to obtain insights to develop or to improve solutions that will reduce the building's energy consumption and its environmental impacts. Use-phase memory also allows building experts to gain time during the two use-cases because much information and data are centralized internally. Our model is presently being deployed on a residential building. This stage of test and experimentation will allow us to develop, to improve and to validate the model of use-phase memory of residential buildings. Residents have just started to participate in the electricity/gas/water consumption bill feedback service. For electricity, data are automatically transferred to a central data treatment and warehouse; for gas and water, it requires a monthly personal declaration by Internet. We need at least two years to draw firm conclusions about how people behave both when they face their detailed consumption bill compared to others and for providing relevant and useful storytelling.

Some issues need to be studied and detailed. It is important to carry on HMI development, in terms of interface for users and experts, but also in defining technical aspects to feed, to share, to consult and to question the use-phase memory. And the development of data representation may be really useful to increase the creation of information and knowledge for building experts. Nowadays much research is being conducted on this question, such as David McCandless [17]. Other issues must be questioned, in particular the existing design and renovation processes, and the integration of use-phase memory into them. Building experts work with specific processes according to the project, its context and its aims. In our context of residential buildings, we notice many steps dividing up tasks, responsibilities and also building experts' dialogue. It may be necessary to change these approaches to designing and to renovating a sustainable building, or to invite other experts such as sociologists and lifecycle experts for instance. This study developing a model of use-phase memory of residential building places the focus on residents. They influence the variability of a building's energy consumptions and environmental impacts, but also they hold much useful knowledge allowing innovation and the development of more sustainable buildings. So one of the first tasks in carrying out this study is the detailed analysis of design and renovation processes and definition of changes needed to allow residents and their usages to be taken into account.

References

1. ADEME, www.ademe.fr
2. Molle D, Patry P-M (2011) RT 2012 et RT existant: réglementation thermique et efficacité énergétique pour construction et rénovation. Éditions Eyrolles 180. ISBN 978-2-2121-2979-3
3. Yu Z, Fung BCM, Haghighat F, Yoshino H, Morofsky E (2011) A systematic procedure to study the influence of occupant behaviour on building energy consumption. Energy Buildings 43(6):1409–1417
4. National Resources Defense Council (2007) Tuning into energy efficiency: prospects for energy savings in TV set-top boxes. p 4
5. Hoes P, Hensen JLM, Loomans MGLC et al (2009) User behaviour in whole building simulation. Energy Buildings 41(3):295–302
6. Telenko C, Seepersad C (2012) Probabilistic graphical models as tools for evaluating the impact of usage-context on the environmental performance. In: IDETC 2012: international design engineering technical conferences, Chicago, 12–15 Aug 2012
7. Zaraket T, Yannou B, Leroy Y, Minel S, Chapotot E (2012) An experimental approach to assess the disparities in the usage trends of domestic electric lighting. In: IDETC 2012: international design engineering technical conferences, Chicago, 12–15 August 2012
8. Borg SP, Kelly NJ (2011) The effect of appliance energy efficiency improvements on domestic electric loads in European households. Energy Buildings 43(9):2240–2250
9. ENERTECH (2002) End-use metering campaign in 400 households of the European community-assessment of the potential electricity savings. Commission of the European Communities, pp 321–339
10. Lilley D, Lofthouse V Bhamra T (2005) Investigating product driven sustainable use. Sustainable innovation '05, 24–25 October, Farnham Castle Internal Briefing and Conference Centre, p 14
11. Lofthouse VA Lilley D (2006) What they really, want: user centered research methods for design. In: International design conference—design 2006, Dubrovnik, Croatia, May 15–18, p 9
12. Escalfoni R, Braganholo V, Borges MRS (2011) A method for capturing innovation features using group storytelling. Expert Syst Appl 38(2):1148–1159
13. Gibbert M, Leibold M, Probst G (2002) Five styles of customer knowledge management. And how smart companies put them into action. Research report, UG-HEC-CR—02-009, p 16
14. Shen W, Shen Q, Sun Q (2012) Building information modeling-based user activity simulation and evaluation method for improving designer–user communications. Autom Constr 21:148–160
15. Dieng R, Corby O, Giboin A et al. (1998) Methods and tools for corporate knowledge management. Research report, n°3485. INRIA, p 44
16. Matta N, Ribiere M Corby O (1999) Définition d'un modèle de mémoire de projet. Rapport de recherche, n°3720. INRIA, p 45
17. McCandless D (2011) Datavision. Robert Laffont. p 224. ISBN 978-2221126752

Developing Sustainable Products: An Interdisciplinary Challenge

Kai Lindow, Robert Woll and Rainer Stark

Abstract This paper presents an interdisciplinary method that allows engineers to map their view, that is usually focused on product parameters and characteristics, to the view of sustainability assessment experts, whose view is on a higher abstraction level and mainly focused on officially accepted sustainability indicators. The presented approach proposes to bridge the conceptual gap between the different views of both disciplines by focusing on a product's life-cycle processes such as manufacturing, distribution and end-of-life processes. Details about these processes need to be provided by respective experts that thus must be included into the interdisciplinary design process. The presented method builds on the House of Quality method and uses the existing Sustainability Dashboard software tool for a visual comparison of different design alternatives.

Keywords Sustainability · Interdisciplinary challenge · House of sustainability · Life cycle sustainability assessment · Sustainable engineering

K. Lindow (✉)
Department of Machine Tools and Factory Management, School of Mechanical Engineering and Transportation Systems, Technische Universität Berlin, Berlin, Germany
e-mail: kai.lindow@tu-berlin.de

R. Woll · R. Stark
Fraunhofer Institute Production Systems and Design Technology, Virtual Product Creation, Berlin, Germany
e-mail: robert.woll@ipk.fraunhofer.de

R. Stark
e-mail: rainer.stark@ipk.fraunhofer.de

A. Chakrabarti and R. V. Prakash (eds.), *ICoRD'13*, Lecture Notes in Mechanical Engineering, DOI: 10.1007/978-81-322-1050-4_41, © Springer India 2013

1 Introduction

The life cycle of a product/system roughly comprises design, manufacturing planning, manufacturing, use, maintenance, remanufacturing and the end of life situation respectively. Since design engineers basically do not have sufficient knowledge to evaluate the sustainability impacts of a product regarding the entire life cycle they have to rely on expertise from other disciplines, such as environmental and manufacturing engineering. Hence, evaluating and assessing a product/system's sustainability impact requires an interdisciplinary and cooperative approach, already during the design phase.

The paper presents on the one hand the results of the research project "Methodological sustainability assessment of machine components in the development process". The purpose of the project was to develop a method that allows a design engineer to estimate the impact of design decisions on the sustainability of a product/system already in the early design phase. To make this possible, design decisions have to be associated with the evaluation criteria for sustainability assessment. This in turn necessitates deep knowledge about the product/system's process, such as manufacturing, logistics, end-of-life treatment.

Apart from the new approach that was developed—a methodology that combines the House of Quality (HoQ) [1] approach with an integrated Life Cycle Sustainability Assessment (LCSA) covering (environmental) Life Cycle Assessment (LCA), Life Cycle Costing (LCC) and Social Life Cycle Assessment (SLCA)—the greater challenge occurred while developing and testing the new approach. The approach necessities the expertise and knowledge of different engineering disciplines. Due to that fact, design, manufacturing process and environmental engineers had to work jointly. During the research, several workshops among the different disciplines were carried out. It revealed that the most important challenge is to ensure a meaningful linking between the technical properties and characteristics (e.g. dimensions, fatigue behavior) of a product/system, the characteristics of manufacturing processes based on the manufacturing technology (e.g. cleaning technology) and indicators of LCSA (e.g. electrical energy consumption) [2].

The research has shown that it is very difficult to estimate in which context certain technical parameters and characteristics of the products/systems are directly associated with specific sustainability indicators (e.g. relation between diameter of a rotor and the consumption of electrical energy) [3]. It has shown that the relationship between technical parameters and sustainability indicators can only be conducted via its processes. For that reason, the design of a product/system has a significant influence on the processes in the following life cycle phases.

2 Objectives and Challenges of Sustainable Product Development

Nowadays, environmental pollution and shortage of natural resources are anchored in people's mind that also highlights the importance of the development of sustainable products. The term "sustainability" comprises—in addition to environmental challenges—economical and social challenges. Lately, the social aspect has been stressed making it even more difficult to overcome the challenges of a sustainable product development [4]. Due to that, questions arise like: How much workload and what kind of a working environment can be considered healthy for a human being or how is it possible to prepare employees for a dynamically changing labor market? It becomes obvious that the term sustainability alone raises versatile and important views.

It has not been answered yet which implications the unspecified demand for sustainable product development does have. Companies are trying to develop products that make the customer/the consumer feel that they are using a sustainable product during their utilization phase [5].

The question of how a product is developed, manufactured or recycled/remanufactured is usually not on the focus of the user. Nevertheless, a supposedly sustainable product can be expected to fulfill all the requirements of sustainability in its entire lifecycle [6]. For example, a fuel-efficient vehicle should also be manufactured in a healthy and harmless working environment by using as less resources as possible. Examining the sustainability along the entire product lifecycle makes the goal of having a sustainable product a rather complex and difficult mission.

Considering a product along its entire lifecycle is especially a challenge from the perspective of engineering and design. The reason for that is that the decisions made during the design phase significantly influence how a product is going to be manufactured, used, recycled or re-manufactured [7]. Important effects of decisions made during the early stages of design are mostly not well-known. Topics like manufacturing processes, environmental guidelines and recycling processes are important for the development of a sustainable product but it can hardly be expected of a design engineer to be an expert in all these fields. Considering this fact, the development of a sustainable product also becomes a challenge regarding the qualification of people who participate in this process. A team of experts with different qualifications is needed to be able to evaluate and determine the lifecycle of a product regarding sustainability.

Materials, supplies or auxiliary materials used along the product lifecycle should also be sustainable. A product which can only be re-used if it is remanufactured manually by using harmful chemicals cannot be considered sustainable since it may harm the health of the workers.

It can be summarized as follows: A product can only be considered to be sustainable if it fulfills social conditions along its lifecycle while its negative influence on the environment is as little as possible [7]. Engineering is only sustainable if it contributes to the creation of a socially acceptable (justifiable), eco-friendly and economically successful product. Considering the fact that the properties and characteristics of a product are determined during the early stages of design it is possible to argue that late stages of the lifecycle are predetermined. This implies that engineering has a decisive impact on the sustainability of a product along its entire life cycle.

3 Different Languages and Perspectives of Different Disciplines

Within the research project "Methodological sustainability assessment of machine components in the development process" a method was developed that allows a design engineer to estimate the impact of design decisions on the sustainability of a product already in the early design phase. In order to make this possible, design decisions have to be associated with evaluation criteria for sustainability. The central challenge is: While a design engineer makes specific choices in terms of technical parameters and characteristics (e.g. dimensions, applicable load) of the product, the sustainability of a product is measured at a different, much more abstract level (e.g. Global Warming, Child labor), cp. Figs. 1 and 2. For design engineers it is very difficult to estimate in which context certain technical parameters and characteristics of the product are linked/associated with specific indicators of sustainability. To fill this conceptual gap, a way must be found on how these different aspects can be accommodated in an understandable manner and in a correct context. Regarding the project a specific example of an alternator that has to be re-designed, in order to make it suited for sustainable remanufacturing, was chosen (cp. Fig. 1). To limit the complexity of the research case, the focus has been limited to the process of disassembly, cleaning and re-assembly (cp. Fig. 2). The different design alternatives were evaluated with respect to sustainability issues. The evaluation did not allow a reliable assessment of sustainability throughout the entire life cycle since only a fragment of the last life cycle phase was considered—though it provided evidence on how to proceed in order to establish a relationship between technical parameters and sustainability indicators.

The most important finding was: The relationship between technical parameters and sustainability indicators can be drawn via processes. The design of a product has a significant influence on the processes in the life cycle phases. Experts can assess quite accurately how the process at a particular stage of life will behave, depending on characteristics of the designed product. Thus, an expert can estimate different manufacturing processes and pathways which will be necessary in order to manufacture a product with certain characteristics. For the assessment of sustainability on the other hand, information on exactly these processes are needed in order to determine measurable results for sustainability indicators. In the selected

Part	Material	Reman - Scenario 1	Reman - Scenario 2
Stator	Steel	Material recycling	Reuse
Rotor coil	Copper C10100	Material recycling	Reuse
Rotor	Iron Cast G25	Disposal (landfill)	Reuse
Drive shaft	Steel	Material recycling	Material recycling
Belt fitting	Alumi	ig	
Fan	Steel		
Spacer	Alumi		
Bearing	Rollec	ig	
Slip ring N	Coppe	ig	
Slip ring S	Coppe	ig	

X_1: Height of rotor coil

X_2: Radius of rotor

X_1: Height of stator

Fig. 1 Engineer's view on different remanufacturing scenarios of an alternator (product's properties and characteristics)

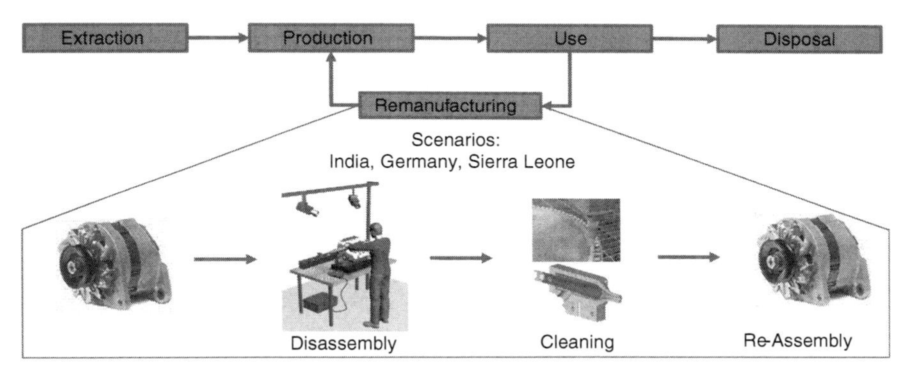

Fig. 2 Sustainability expert's view on different remanufacturing scenarios of an alternator (assessment of processes and location regarding Global Warming, child labor etc.)

research case, the disassembly and assembly process were examined using the method of Life Cycle Sustainability Assessment (LCSA) [8].

Figure 3 illustrates the link between the language of the design engineer and the language of the sustainability expert. To connect between technical product parameters, process characteristics and sustainability indicators, an expert on remanufacturing and an LCSA expert have been involved. The sustainability expert, together with a process expert, created a list of the process properties that are relevant for the determination of the values of sustainability indicators. Afterward design engineers and process experts identified together the influence of the relevant technical parameters for the design on the previously created elements in the list.

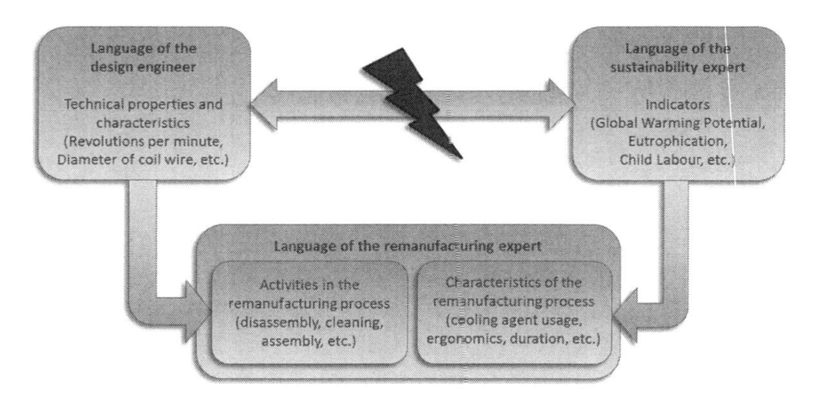

Fig. 3 Bridging the gap between design engineer's and sustainability expert's language

4 An Interdisciplinary Design Approach: The "House of Sustainability"

Based on the findings described in the previous sections the methodology "House of Sustainability" (HoS) was developed. It represents a combination of the "House of Quality" (HoQ) approach combined with the LCSA approach. The HoS approach consists of different steps that are partly supported by software. Initially, the relationships between the product parameters and manufacturing process characteristics (in this case those of the assembly and disassembly process for remanufacturing) are specified together by a team of design engineers, manufacturing process experts and sustainability experts. Subsequently, the design engineers determine design alternatives in the form of different configurations of properties and characteristics. At the same time, sustainability experts draw different remanufacturing scenarios based on the processes at different remanufacturing locations. Afterward, the different design alternatives are evaluated with respect to their remanufacturing scenario. Eventually, the interpretation of assessment results is done together by design engineers and sustainability experts.

The aim is to figure out which design alternative suits best to which remanufacturing scenario—and which scenario has the best performance on sustainability.

A detailed process flow chart was developed using the Business Process Modeling Notation (BPMN) (Fig. 4). The first half of the process ensures the creation of a data base for the different design alternatives. After that indicators for the sustainability assessment are identified. By using the HoQ the relationships between customer requirements and product parameters can be modeled. This allows for ranking the parameters depending on their importance for the fulfillment of customer requirements.

At the beginning of the HoS approach (shown in Fig. 5) a sustainability expert and a process expert jointly determine the quantifiable properties of a particular production process that are relevant to certain sustainability requirements.

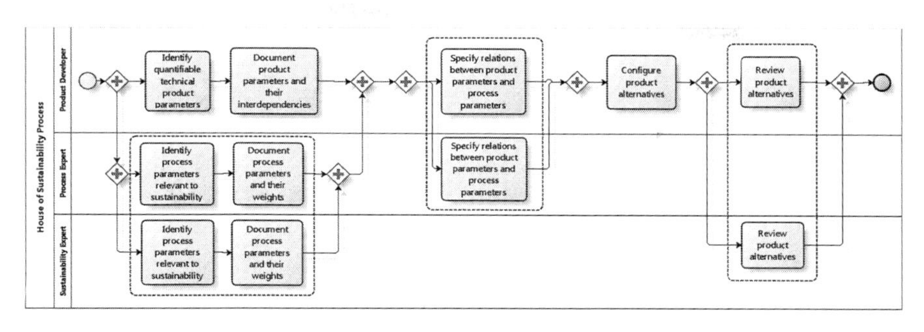

Fig. 4 Process chart for creating a "house of sustainability"

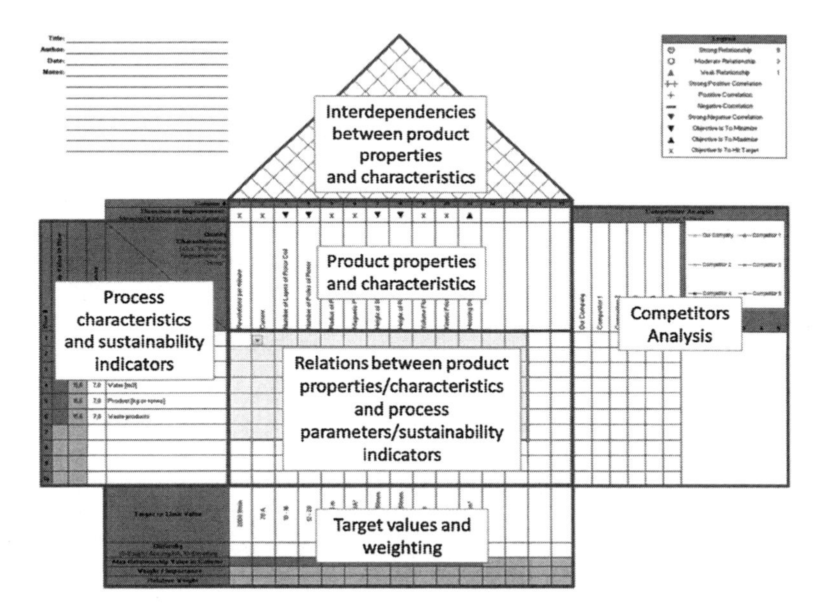

Fig. 5 House of sustainability as a modified "house of quality" editor

The properties identified in this process are listed in the left section of the HoS (Fig. 5). In parallel, the design engineer can already gather the product parameters in order to estimate the interactions between them and thus determines the top of the HoS. Subsequently, the process expert and the design engineer estimate together what relationships exist between the product parameters and characteristics and enter them in the middle of the HoS.

This approach differs from the typical use of a HoQ in such that process characteristics are listed in the left section (where usually customer requirements would be listed). Additionally, the design engineer usually estimates the dependencies between customer requirements and product parameters whereas in this case relations are specified between process characteristics and product parameters. Process characteristics are usually out of scope in the HoQ. The two previously mentioned sections of a typical HoQ, the competitor analysis (right in Fig. 5) and the weighting of the technical product parameters can optionally be filled.

Once the HoS is completed the design engineer configures three different design alternatives. The configuration of design alternatives is done by setting the technical parameters for each alternative. At this time only the technical parameters for which there exists no fixed target value can be varied (e.g. power output is fixed, diameter and length can vary). For parameters with a predetermined interval low (near the lower limit of target value), medium and high (near the upper limit of target value) levels can be set. The user interface only displays those parameters that do have at least one relationship with a process property. Additionally, the parameters of the product are sorted that the properties with the highest influence on the process are listed on the top. Since those parameters which have fixed target values are also shown, the design engineer consciously confines only those parameters that are relevant for a sustainable outcome.

When the configuration of design alternatives is completed the system performs an automatic assessment and evaluation based on the information provided by the HoS. This assessment is based on a simple computation that takes as input the degrees (and direction) in which the different product properties influence the process characteristics (this information is specified by the users in the central matrix of the HoS) and the settings of the technical parameters for the different design alternatives.

For example, if the technical parameter 'diameter' affects the energy consumption of the development process (and a larger diameter means more energy consumption) then the alternative with the largest diameter will have the most negative effect on energy consumption. And since energy consumption is an environmentally relevant sustainability indicator this alternative will have the worst relative score related to the environmental dimension. The degrees at which a technical parameter influences a process can have three different values: low (weight 1), medium (weight 3) and strong (weight 9). Hence, if multiple parameters influence the same process and one of them has a strong influence it has as much impact as 3 medium-scored parameters or 9 low-scored parameters. For simplicity it is assumed that different settings for a parameter (please note: this is not the 'influence degree' from the HoS but a specific value for a design parameter that is specified when the design alternatives are configured) have a linear proportional relation with the influence degree. That means that if the diameter has three alternative settings (small, medium, large) and the influence degree of it on energy consumption is medium (i.e. weight 3) then the small version has a negative effect of 0 (0 * 3), the medium version 3 (1 * 3) and the large version 6 (2 * 3).

For each design alternative the algorithm computes the negative effects of all parameters on all process characteristics and sums them up for each process characteristic. Once the results are computed they are visualized in the "Sustainability Dashboard". The Sustainability Dashboard is used for rapid visual analysis of assessment results and uses an intuitive color coding (traffic lights) to compare the three coexisting product alternatives. For each alternative it is shown how they were individually evaluated in terms of their economic, environmental and social sustainability. For simplicity it is assumed that for each process characteristic one alternative can only perform either well, medium or bad, i.e. if the cumulated negative effect of all parameters of alternative A on energy consumption is 9, the one for alternative B is 10 and the one for alternative C is 40, then alternative a is considered good (green color code), alternative B mediocre (yellow color code), and alternative C (red color code) bad. The actual relative difference between their ratings is omitted.

This way every single aspect of sustainability can be considered in detail, in which case the specific sustainability indicators are displayed and how each alternative was evaluated. The sustainability expert can explain the exact meaning of these indicators to the design engineer.

Based on the sustainability assessment the design engineer can choose an alternative or configure new alternatives. In this context it is important to note that the reliability of the results is directly depending on how well the HoS was elaborated. A reliable assessment of sustainability is only possible if all relevant process characteristics have been identified, and if the correlations were estimated realistically between the product parameters and process characteristics.

5 Complexity, Reliability and Effort: Challenges for the Future

The paper shows the evidence on how to proceed in order to establish a relationship between design parameters and sustainability indicators against the background of measuring the sustainability impact of a product/system. Moreover, the paper discusses the complexity, reliability and effort about the presented approach. Eventually, future challenges of interdisciplinary sustainability assessment are discussed.

During the development of the House of Sustainability approach and during the final evaluation with expert teams of three different research departments of the Technische Universität Berlin a number of basic challenges regarding the sustainability evaluation of products were identified. The first challenge has been mentioned in the previous section implicitly: The information that forms the basis of assessment results is estimated and it is usually not verifiable on the basis of detailed measurements. Poorly specified correlations between product properties and process characteristics can lead to assessment results that do not reflect reality sufficiently well. In order to avoid—or at least to minimize—disputable design decisions, a more detailed analysis of the processes has to be approached. The effect of a detailed analysis on the reliability of the evaluation should be investigated further. Likewise the costs for analysis are increasing with the degree of detail. An investigation of different business cases should be subject of further research.

A key challenge in carrying out a combined LCSA is allocating the large amount of information that must be procured to perform a full assessment. In the presented research project solely the process of assembly and disassembly in remanufacturing was investigated. The workshop in which three different experts provide the necessary product, process and sustainability information took more than a week already. Since sustainability is assessed along the entire life cycle with all its processes therein, it can be concluded that similar workshops have to be conducted. The creation of a comprehensive data base for a complete assessment of sustainability using the House of Sustainability approach is a task that could thus take several persons months until completion. Hence, the question of the cost-benefit ratio arises. However, there is already a multitude of companies that carry out similar efforts for performing LCAs on their products—at a point of time in the project when no design changes are realistically possible. Though, the overall aim should be to develop a sustainable product from the early phases of design instead of checking retrospectively if the product is sustainable. Therefore, another research question would be whether the efforts in the acquisition of information during product design can lead to a reduction of expenses for future LCSA.

Another important finding is the difficulty of involving social aspects in product design. Generally, design relations with social aspects of ergonomics and workplace safety can be identified even though a number of important other social sustainability indicators exists, such as working hours or wages. Social aspects that have no relation to other life cycle processes cannot directly be affected by product design. Nonetheless, they can be influenced by the choice of (re-) manufacturing locations, manufacturing equipment and business partners.

The research has identified a very practical problem that has already been indicated: Technical properties and characteristics of a product have no range of target values but a fixed target value at the time a LCSA is carried out. Such parameters cannot be changed in design alternatives and their impact on sustainability may thus not be optimized. The effects of design decisions on the sustainability of the product have to be examined already in the early stages of the design process.

Future research has to focus on approaches that deal with supporting the engineer's work in setting-up the foundation for sustainable products. Therefore, life cycle knowledge, expert know-how and experiences from other disciplines are needed and have to be merged in early phases of product development. In particular, solutions have to be developed for modeling the relationships between design decisions and the sustainability impact on the entire life cycle. Finally, all these information must be presented in a way that a design engineer is able to make sustainable decisions.

Acknowledgments We thank Kilian Müller and the sustainability team involving Erwin Schau, Marzia Traverso and Matthias Finkbeiner at the Department of Environmental Technology, Technische Universität Berlin. This research project is funded by the German Research Foundation DFG as part of the university wide research project (DFG PAK 475/1) "Sustainable value creation exemplified on mini-factories for remanufacturing".

References

1. Akao Y (2009) Development history of quality function deployment. The customer driven approach to quality planning and deployment. Asian Productivity Organization, Minato, Tokyo 107 Japan, p 339
2. Schau EM, Traverso M, Lehmann A, Finkbeiner M (2011) Life cycle costing in sustainability assessment—a case study of remanufactured alternators. Sustainability 3(11):2268–2288
3. Birkhofer H, Wäldele M (2008) Properties and characteristics and attributes and... —an approach on structuring the description of technical systems, AEDS 2008 Workshop, Pilsen—Czech Republic
4. Hanusch D, Birkhofer H (2010) Creating socially sustainable products—examinating influence and responsibility of engineering designers. In: Marjanović H, Štorga D, Pavković M, Bojčetić N (eds) Proceedings of the 11th international design conference DESIGN 2010, Dubrovnik, Croatia, pp 771–778
5. Kerga E, Taisch M, Terzi S (2011) Integration of sustainability in NPD process: Italian experiences. In: Proceedings of the international conference on product lifecycle management (PLM2011), Eindhoven
6. Niemann J, Tichkiewitch S, Westkämper E (2009) Design of sustainable product life cycles. Springer, Berlin
7. Devanathan S, Ramanujan D, Bernstein WZ, Zhao F, Ramani K (2010) Integration of sustainability into early design through the function impact matrix. J Mech Design 132
8. ISO 14040 (2006) Environmental management—life cycle assessment—principles and framework. International Organisation for Standardisation (ISO), Geneva

Life Cycle Assessment of Sustainable Products Leveraging Low Carbon, Energy Efficiency and Renewable Energy Options

S. S. Krishnan, P. Shyam Sunder, V. Venkatesh and N. Balasubramanian

Abstract Design of sustainable products is a critical process with enormous implications on the energy consumption and Green House Gas (GHG) emissions. The impact on energy and emissions can be measured by an analysis of specific performance indicators at each stage of the product lifecycle. The analysis can provide useful insights to the design process. In this paper, Life Cycle Costs (LCC), Life Cycle Energy (LCE) and GHG emissions over the use phase of common appliances such as refrigerators, air conditioners, distribution transformers, street lights and irrigation pumps are examined. This study presents the current energy ratings and analyses the impact of energy efficiency, renewable energy, and low carbon material options on the overall energy and emissions at the individual unit and national aggregate levels.

Keywords Life cycle costs · Appliances · Energy efficiency · Product design · Sustainable manufacturing

S. S. Krishnan (✉) · P. Shyam Sunder · V. Venkatesh · N. Balasubramanian
Center for Study of Science, Technology and Policy, Bangalore, India
e-mail: ssk@cstep.in

P. Shyam Sunder
e-mail: shyam@cstep.in

V. Venkatesh
e-mail: venkat@cstep.in

N. Balasubramanian
e-mail: nbalu23@cstep.in

A. Chakrabarti and R. V. Prakash (eds.), *ICoRD'13*, Lecture Notes in Mechanical Engineering, DOI: 10.1007/978-81-322-1050-4_42, © Springer India 2013

1 Introduction

The deployment of sustainable products is gaining importance and has huge demand across the globe. Designing such products needs technological update and management of life cycle energy and associated emissions. Managing energy for products over their life cycle phases starts from materials extraction, manufacturing and end use. The associated emissions are directly or indirectly involved with the environment. India is witnessing a huge demand for energy to manage all the sectors and particularly has huge challenge for delivering reliable energy to residential, commercial and agricultural sectors. These sectors are considered to be unorganized and volume of appliances is outnumbered. The sector wise energy consumption is as shown in Fig. 1. According to IEA the residential sector consumed around 168 Mtoe (37 %) out of India's total energy consumption of 449 Mtoe in 2009 [1] and GHG emissions from residential sector have contributed to about 138 Mt CO2 in 2007 [2].

Several policy measures have been undertaken across the world to address energy related aspects within the product life cycle.

The energy star programme by the U.S. Department of Energy (DoE) first introduced the energy efficiency concept amongst consumer durables to help consumers save money and to protect the environment through energy efficient products and practices [3]. In India, the Bureau of Energy Efficiency (BEE) is implementing the star labelling system for appliances under the National Mission for Enhanced Energy Efficiency (NMEEE) [4].

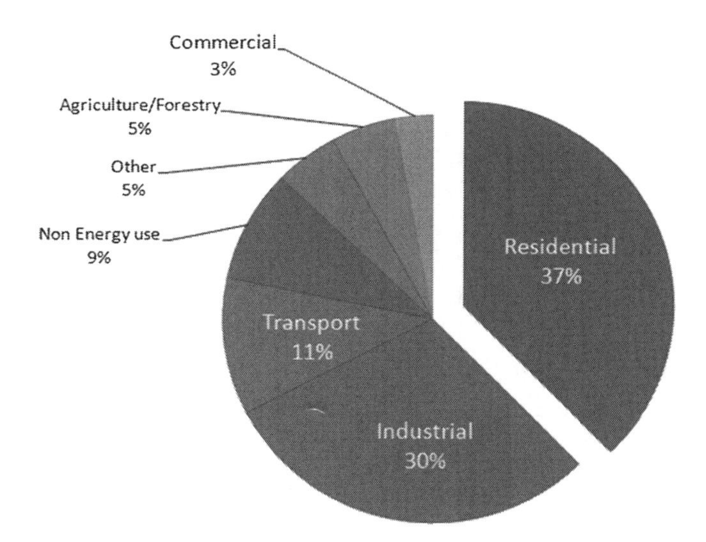

Fig. 1 Sector-wise energy consumption in 2009 (includes non-energy use—the petrochemical feed stocks)

2 Sustainable Design and Manufacture

Typically a manufacturing system considers materials, energy consumed and released accompanied by the corresponding wastes released. The life cycle system is broadly classified into various stages such as, raw material extraction, manufacturing, the use phase of the product and finally end-of-life phase. This is a far broader view of manufacturing than simply looking at consumption, wastes and pollutants occurring at the factory. It has become clear that integrating manufacturing into a sustainable society requires the broader systems view [5]. A unique Sustainable Manufacturing (SM) process model detailing the sequence of processes that occur in the life cycle of manufacturing was developed by researchers at the Centre for Study of Science and Technology (CSTEP) [6]. This model shown in Fig. 2 illustrates the major activities in the product life cycle and their interaction with the environment directly and indirectly.

Activities such as raw material mining, energy production, manufacturing, use phase, recycling and others have a direct interaction with the environment. The interactions are in the form of accessing raw materials, energy to waste disposal and emissions. Other activities such as the design process, maintenance and end-of-life analysis have indirect interactions, and also have the potential to alter the activities direct impact to the environment. For example design stage is more theoretical and involves technical brainstorming activities. An efficient or

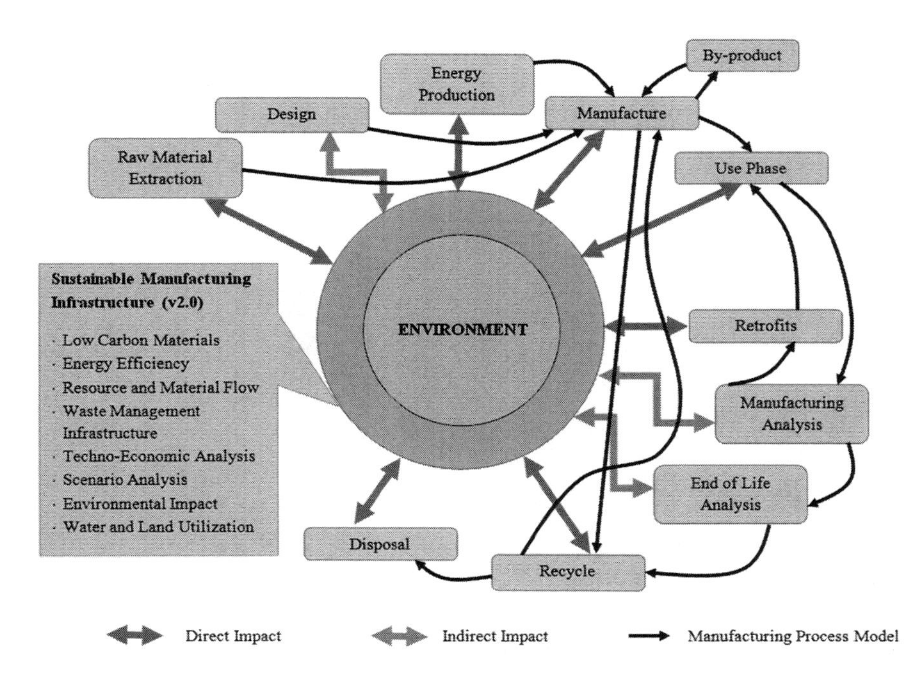

Fig. 2 Sustainable manufacturing process model

outstanding design can foresee the best and economical way to manufacture sustainable products provided several manufacturing intricacies, best manufacturing process, selection of materials, reduction of emissions and many others are assessed in this stage, This may reduce the actual material consumption, lead to energy efficient manufacturing, and also improve sustainable life of the product. Figure 2 describes the sustainable manufacturing process model. One can observe the different components and the inter component interactions as depicted by black arrows. These show that processes not only have direct and indirect interaction but must also follow the path to disposal after passing the efficient use phase, recycle, reuse of parts, and retrofits. The model also highlights that all the processes must interact with the environment through the sustainable infrastructure layer.

The component view of sustainable manufacturing infrastructure shows the involvement of various stake holders. Each stakeholder in SM infrastructure forms an aggregate relationship in unified modelling language (UML) terminology. The report states that sustainability has to be integrated in all the activities that comprise the current economic processes of human endeavour. Thus, sustainability analysis has to be incorporated in different components shown in the SM infrastructure model. Figure 3 shows the component view in the sustainable manufacturing infrastructure.

Fig. 3 Sustainable manufacturing infrastructure: component view

Each activity of the SM process model has to perform its entire repertoire of sub-activities whilst treating sustainability consideration as an additional factor. This may be treated in various formulations by different components as an optimisation function, a hard constraint, a soft constraint, a policy option, a policy mechanism guideline, a compliance target parameter or in other ways a modification of societal preferences, value systems and demands. However, the fact remains that in a systems view of the SM process, sustainability needs to be integrated into the current set of activities.

3 Life Cycle Energy Efficiency

Table 1 shows life cycle energy consumption and savings over 5, 10 years for a typical refrigerator, and air conditioners Each appliance is assessed based on efficiency level class 'A, B and C', where 'A' being the most efficient and C being least efficient.

For example, if an average rated refrigerator with annual energy consumption of 663.05 kWh, is replaced by an efficient unit with annual energy consumption of 397.50 kWh, it could lead to life cycle energy savings of 315.19 GWh over 5 years and 895.91 GWh over 10 years for the population of units rated under each efficiency levels. The estimated volume of units under each efficiency levels is based on number of units rated by BEE in the energy labelling system in 2010 [7]. A 13 % growth rate is taken in estimating the energy consumption over 5 and 10 years respectively. The volume of appliances used in the calculation under each category of appliances is derived from BEE.

In the case of irrigation pumps, most of the units sold were with most efficient rating. About 67,518 units were rated under 5 star, 252 units of 3 star and only 4 units of least efficient were sold in 2010.

The savings are estimated based on actual energy efficiency rated under 5-star to 1-star. In the agricultural sector, a submersible pump of 10 stages, 90 m pump head which is widely used in irrigation with 3 efficiency levels also shows potential savings of 0.12 and 2.25 GWh from least and average rated units respectively if the corresponding units are replaced with efficient units. Alternatively, if a portion of energy itself is supported from renewable energy resources, there are even greater potential of energy conservation in the domestic sector. Figure 4 shows the comparative conventional energy consumption and with renewable energy mix (75 % Conventional, 25 % Solar).

The study also examined distributional transformer (DT). DTs are widely used to step-up or step-down the electrical voltage. The assessment of its efficiency is based on losses at 100 and 50 % load over the years. BEE has rated a variety of DT models ranging from 16–200 kVA in a similar format.

The Losses from a 63 kVA distribution transformer at 50 and 100 % load are 490 and 1,404 W respectively. These losses are from least efficient type transformers. The estimated avoidable losses at 50 % load if a least efficient DT was

Table 1 Life cycle energy consumption (use phase only) of appliances and savings

Appliances		Annual per unit energy consumption (kWh)	Life cycle energy consumption		Life cycle energy savings	
			in 5 years (GWh)	in 10 years (GWh)	in 5 years (GWh)	in 10 years (GWh)
Refrigerators	B	663.05	786.98	2,236.95	315.19	895.91
	A	397.50	1,666.73	4,737.56		
Air conditioners	C	7,777.39	18,229.69	51,816.71	7,032.23	19,988.67
	B	5,627.43	25,252.91	71,779.75	3,815.33	10,844.83
	A	4,777.21	7,325.62	20,822.60		

Calculation is based on data available from Bureau of energy efficiency (BEE)

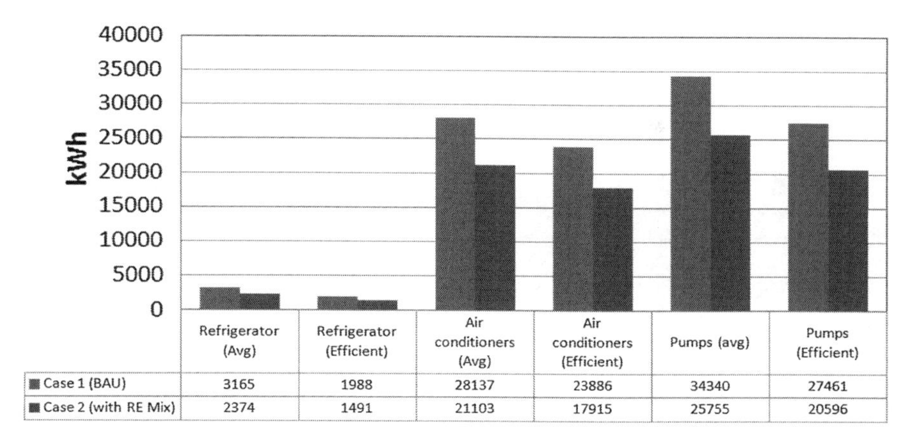

Fig. 4 Per unit energy consumption scenarios with and without renewable energy options

replaced by the most efficient one for 5 and 10 years were 8,322 and 16,644 kWh respectively (Table 2).

Lastly, street lighting is considered to be the more by numbers and energy intensive in the appliances sector. According to USAID and BEE, lighting alone accounts for 10–38 % of total energy bills in typical cities worldwide [8]. Energy efficient technologies and designs can cut street lighting costs dramatically (often by 25–60 %); these savings can eliminate or reduce the need for new generating plants and provide the capital for alternative energy solutions for population in remote areas. The BEE, based on Central Electricity Authority (CEA) statistics, has estimated gross energy consumption for public lighting to be 6,131 million kWh in India for the years 2007–2008 [8], Table 3 shows the typical lamp technology and life of units.

Table 2 Life cycle energy losses in the use phase of 63 kVA distribution transformers

Appliance	Class	Losses at 50 % load (W)	Life cycle energy losses (kWh)		Avoidable energy losses (kWh)	
Distribution transformers			5 years	10 years	5 years	10 years
(63 kVA)	C	490	21,462	42,924	8,322	16,644
	B	368.8	16,153	32,306	3,013	6,026
	A	300	13,140	26,280		

Calculation is based on data available from Bureau of energy efficiency (BEE), ‘A’ most efficient, and ‘B’, average efficient and ‘C’ least efficient

Table 3 Street lights technology and efficiency (data from USAID-BEE) [8]

Type of lamp	Luminous efficacy (lm/W)	Lamp life hours
High pressure mercury vapour (MV)	35–65	5,000
Metal halide (MH)	70–130	8,000
High pressure sodium vapour (HPSV)	50–150	15,000
Low pressure sodium vapour (LPSV)	100–190	15,000
Low pressure mercury fluorescent tubular lamp (T12 and T8)	30–90	5,000
Energy-efficient fluorescent tubular lamp (T5)	100–120	5,000
Light emitting diode (LED)	70–160	50,000

4 Life Cycle Cost Assessment

This section discusses the life cycle assessment of appliances over 5–10 years. The life cycle cost is calculated with a discount rate of 12 % and also assumes the electricity tariff of Rs 5/kWh for residential appliances and Rs 3/kWh for irrigation pumps. Table 4 provides the life cycle cost and payback period for the investment incurred on shifting to efficient refrigerators, air conditioners and irrigation pumps. In the case of refrigerators, most units were rated average or above average, as per BEE star labelling scheme. The estimation of cost is for BEE rated units in operation during 2010.

4.1 Distribution Transformers

The life cycle cost assessment of DTs has been computed for least efficient and most efficient units at 50 and 100 % load. From Table 5, we observe that the life

Table 4 Life cycle cost of appliances over 5, 10 years of life of product

Appliances	Category Class	Life cycle cost (use phase, million Rs)		Additional investment (Rs)	Pay back period for additional investments (years)
		5 years	10 years		
Refrigerators	B	2,188.9	3,430.9	13,000	9.79
	A	4,635.7	7,266.2		
Air conditioners	C	50,703.1	79,473.4	10,000	0.67
	B	70,237.1	110,091.5	5,000	1.18
	A	20,375.1	31,936.5		
Irrigation pumps	C	0.7	1.1	25,000	1.08
	B	31.0	48.6	10,000	1.46
	A	6,648.4	10,420.9		

*Estimation corresponds to the existing products

Table 5 Life cycle energy and operating cost of 63kVA DT

Annual energy costs	(@ 50 %)	(@ 100 %)
Annual operating cost (63 kVA) 1 star (Rs)	21,497	61,608
Annual operating cost (63 kVA) 4 star (Rs)	14,916	48,807
Savings (Rs)	6,580	12,800
Savings (kWh)	1,314	2,555
Life cycle operating cost over years	5 years	10 years
Least efficient at 50 % load (Rs)	107,486	214,972
Most efficient at 50 % load (Rs)	74,582	149,164
Least efficient at 100 % load (Rs)	3,080,426	616,085
Most efficient at 100 % load (Rs)	2,440,380	>488,076
Total life cycle savings 50 % load (Rs)	32,904	65,808
Total life cycle savings 100 % load (Rs)	74,582	149,164
Payback		Years
Pay back on incremental cost (@ 50 % load)		4.56
Pay back on incremental cost (@ 100 % load)		2.34

cycle operating cost increases with higher load and further, the payback period by switching to efficient units is less at 100 % load operation,

4.2 *Street Lights*

In street lights, the total annual energy cost is less for high pressure sodium vapour lamps. By replacing all high pressure mercury vapour lamp with high pressure sodium vapour lamps with slightly lower wattage, our analysis finds that savings of 20–25 % can be achieved [8] (Table 6).

Table 6 Annual costs of street lighting (data from USAID-BEE) [8]

Type of lamp	Installed cost (only lamp + luminaries supply) (Rs)	Annual energy cost (Rs)	Annual operating cost (Rs)	Total annualized cost (energy cost + operating cost) (Rs)
(MV)	465,800	805,920	43,625	849,545
(MH)	2,449,615	464,954	77,703	542,657
(HPSV)	1,750,286	345,394	10,512	355,906
(LPSV)	1,370,400	394,200	119,837	514,037
(T12 and T8)	390,857	550,629	36,041	586,670
(T5)	510,000	474,500	105,120	579,620
(LED)	6,000,000	372,300	–	372,300

Assuming 7.5 m wide, dual carriageway type, 1 km long road

5 Integration of Low Carbon Material Assessment

We have developed a scenario based analysis on estimating the impact of substitution of low carbon materials in place of energy intensive materials. This indicative analysis gives an insight during materials selection in the design stage of product life cycle. In this section we examined low carbon based design scenario for refrigerator, air conditioners and distribution transformers which are the most widely used appliances. The proportion and kinds of material used in this model are arrived by using the Gabi software model. Moreover the selection of materials was based on the functionalities of the replaced material i.e. Strength, durability albeit with less carbon content. Figure 5, shows the raw materials composition of a 2.1 kW single duct room air conditioner with a cooling capacity of 0.68 ton. The total weight of the unit is 28.96 kg, with metals 67.05 % (includes the steel, copper, iron and aluminium) and 30 % plastic materials (including rubber). The estimated emissions from single Air conditioners are 88.258 kg CO_2. Steel is extensively used material (about 10.15 kg) which involves energy intensive process. If 25 % (2.539 kg) of steel is replaced with low carbon materials with close approximate properties imitating similar strength and durability, the per unit emissions from the manufacturing stage alone can reduce by 6.24 % (Table 7).

The selected DT [11] for our analysis is a 315 kVA—Primary 11 kV and Secondary 433 V. An estimated 533 kg of steel and 200 kg of aluminium are used in manufacturing this DT; respective per unit embodied emissions are 1,321.73 and 1,955 kg CO_2 (combined Al sheets and wire). Table 8 provided similar analysis, showing LC materials inclusion in distribution transformers manufacturing. Case 1 depicts the sample conventional transformers materials analysis, Case 2, an LC based design replacing 25 % of Steel.

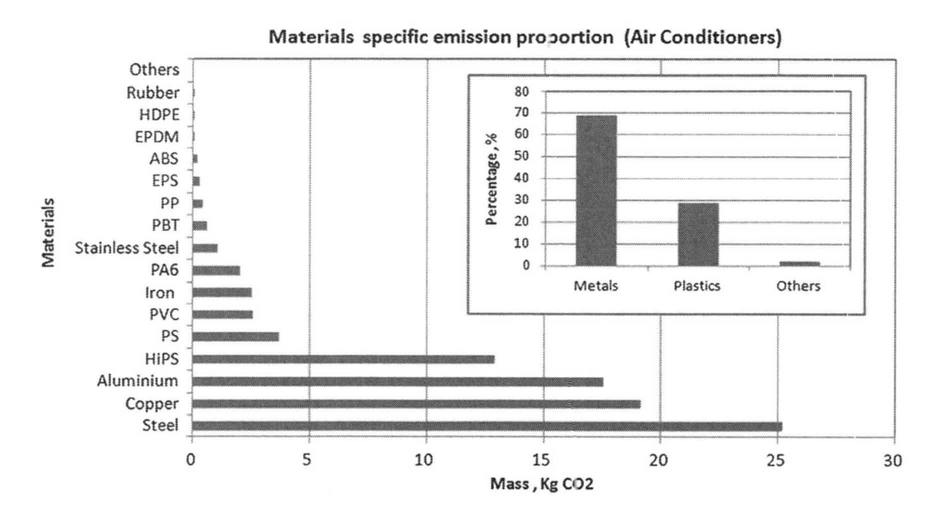

Fig. 5 CO_2 emissions from sample air conditioners

Table 7 Low carbon assessment of 2.1 kW air conditioner

Materials [9]	kg	%	CO_2 emission from each materials ($kg-CO_2$)	kg	CO_2 from each materials ($kg-CO_2$) [10]
	Case 1			Case 2	
Aluminium	1.795	6.22	17.567	1.795	17.567
Copper	4.918	17.04	19.134	4.918	19.134
Iron	2.062	7.14	2.526	2.062	2.526
Stainless steel	0.424	1.47	1.051	0.424	1.051
Steel	10.155	35.19	25.182	7.616	18.887
HDPE	0.021	0.07	0.033	0.021	0.033
PP	0.237	0.82	0.417	0.237	0.417
PS	1.894	6.56	3.702	1.894	3.702
EPS	0.114	0.39	0.294	0.114	0.294
Hi-PS	4.678	16.21	12.903	4.678	12.903
PVC	1.17	4.05	2.554	1.170	2.554
PA6	0.366	1.27	1.999	0.366	1.999
PBT	0.174	0.60	0.606	0.174	0.606
ABS	0.062	0.21	0.189	0.062	0.189
Rubber	0.006	0.02	0.019	0.006	0.019
EPDM	0.043	0.15	0.082	0.043	0.082
Others	0.742	2.57		0.742	
Low carbon materials	–	–	–	2.539	4.722
Total	28.861	100	88.258	28.861	86.684

25 % of steel is only replaced with LC material; percentage of rest of the material is same as existing product

Table 8 CO_2 emissions for sample transformer

Materials [11]	kg	%	CO_2 emission from each material ($kg-CO_2$)	kg	CO_2 emission from ach material ($kg-CO_2$)
	Case 1			Case 2	
Core steel	533	36.09	1,321.73	399.75	991.30
Transformer oil	340	23.02	1,14.92	340	114.92
Steel tank	324	21.94	803.46	324	803.46
Al wire	113.51	7.69	1,110.90	113.51	1,110.90
Al sheet	86.3	5.84	844.60	86.3	844.60
Insulation	59.9	4.06	194.82	59.9	194.82
Porcelain	11	0.74	3.36	11	3.36
LC material	–	–	–	133.25	247.83
Other	9	0.61		9	
Total	1,476.71	100	4,393.78	1,476.7	4,311.18

25 % of steel is only replaced with LC material with the same functionalities; percentage of rest of the material is same as existing product

Table 9 Impact on emissions for low carbon designed refrigerator

Refrigerator materials [12]	Case 1 Mass (kg)	Case 2	Case 1 Emissions ($kg\text{-}CO_2$)	Case 2
Aluminium	2.321	2.321	22.715	22.715
Polystyrene expandable granulate (EPS)	6.875	6.875	17.711	17.711
Acrylonitrile–butadiene–styrene granulate (ABS)	5.577	5.577	17.020	17.020
Copper	2.97	2.97	11.555	11.555
Glass	3.256	3.256	7.254	7.254
Cast iron	5.016	5.016	6.144	6.144
Brass	0.781	0.781	3.539	3.539
PVC	1.111	1.111	2.481	2.481
Styrene–butadiene rubber mix (SBR)	0.187	0.187	0.580	0.580
Steel	52.305	39.22	56.68	42.51
Low carbon materials	–	13.07	–	10.62

The results show that metals such as steel and aluminium emit high quantity of CO_2, 56.68 and 22.71 kg CO_2 per unit respectively and plastics materials like ABS and EPS together were estimated to emit 34 kg CO_2 as shown in Table 9 illustrate similar analysis, showing LC materials inclusion in refrigerator manufacturing. Case 1 depicts the materials analysis of a sample refrigerator, Case 2, an LC based design replacing 25 % of Steel. Additionally, from the Gabi software analysis also estimate the particulate matter (SPM) in the atmosphere from typical refrigerator units were contributed by aluminium, glass and other metals. Collectively from metals, about 100–110 g of SPM is displaced into the atmosphere and overall, from all the selected materials, approximately 170 g of SPM is displaced in the air from a single refrigerator unit.

6 Conclusion

This work depicts some of the valuable indicators in the life cycle of sustainable product design. Managing energy for a populous country like India in near future is a challenging task. Though current policy by BEE assists in the penetration of energy efficient appliances in the market, but, considering the population growth rate coupled with increase in income levels of the population, the demand for energy and appliances are increasing at an alarming rate. The study identifies several factors including replacement of conventional materials by their low carbon counterparts which can be integrated in products at the design stage itself and this can result in large energy and emissions savings over the entire product life cycle. These factors need to be integrated into the framework for sustainable product design for a low carbon economy. Further, the proposed methods need to become standard practices for increasing the competitiveness of products given the societal and political aspersion for sustainable practices and stricter environmental norms. We are pursuing the above work with the development of specific design options in the context of Indian manufacturing.

References

1. IEA (Online) (2010) Energy balance for India. http://www.iea.org. Accessed on 30 June 2012
2. Ministry of Environment and Forest (2010) India: Greenhouse Gas Emissions 2007
3. Environment Protection Agency (EPA) (2012) Energy Star. ENERGY STAR Qualified Products (Online). http://www.energystar.gov. Visited on 30 June 2012
4. Bureau of Energy Efficiency (BEE) (2012) Standard and labeling program. BEE (Online). http://www.beeindia.in/. Visited on 15 June 2012
5. Gutowski TG (Panel Chair), Murphy CF (Panel Co-chair) et al (2001) WTEC panel report on Environmentally Benign Manufacturing. International Technology Research Institute, 1 April 2001
6. Krishnan SS, Balasubramanian N, Subrahmanian E, Arun Kumar V, Ramakrishna G, Murali Ramakrishnan A, Krishnamurthy A (2009) Machine level energy efficiency analysis in discrete manufacturing for a sustainable energy infrastructure. Second annual international conference on infrastructure systems and services, Chennai, India, December 2009
7. Bureau of Energy Efficiency (BEE) (2010) Verified energy saving activities of Bureau of Energy Efficiency (BEE) for the year 2009–10. National Productivity Council (NPC), September 2010
8. USAID-BEE (2010) Guidelines—energy efficient street lighting. Bureau of Energy Efficiency, New Delhi
9. Boustani A, Sahni S, Gutowski TG, Graves SC (2010) Appliances remanufacturing and energy savings. Environmentally Benign Manufacturing Laboratory, January 2010
10. PE international GmBH (2009) GaBi sustainable assessment tool and database
11. ABB, Environmental product declaration. Distribution Transformers' Darra, QLD 4076
12. Yuhta Alan H (2004) Life cycle optimization of household refrigerator and freezer replacement. Center for Sustainable Systems, University of Michigan

Inverse Reliability Analysis for Possibility Distribution of Design Variables

A. S. Balu and B. N. Rao

Abstract Reliability analysis is one of the major concerns at the design stage since the occurrence of failures in engineering systems may lead to catastrophic consequences. Therefore, the expectation of higher reliability and lower environmental impact has become imperative. Hence the inverse reliability problem arises when one is seeking to determine the unknown design parameters such that prescribed reliability indices are attained. The inverse reliability problems with implicit response functions require the evaluation of the derivatives of the response functions with respect to the random variables. When these functions are implicit functions of the random variables, derivatives of these response functions are not readily available. Moreover in many engineering systems, due to unavailability of sufficient statistical information, some uncertain variables cannot be modelled as random variables. This paper presents a computationally efficient method to estimate the design parameters in the presence of mixed uncertain variables.

Keywords Fuzzy variables · High dimensional model representation · Inverse reliability analysis · Random variables

A. S. Balu (✉) · B. N. Rao
Structural Engineering Division, Department of Civil Engineering,
Indian Institute of Technology Madras, Chennai 600036, India
e-mail: arunsbalu@gmail.com

B. N. Rao
e-mail: bnrao@iitm.ac.in

A. Chakrabarti and R. V. Prakash (eds.), *ICoRD'13*, Lecture Notes in Mechanical Engineering, DOI: 10.1007/978-81-322-1050-4_43,

1 Introduction

The solution procedure for inverse reliability problems is required to determine the unknown design parameters such that prescribed reliability indices are attained. One way to solve the inverse reliability problem is through trial and error procedure, using a forward reliability method like first-order reliability method (FORM) and varying the design parameters until the achieved reliability index matches the required target [1]. However, the trial and error procedures are inefficient and involve difficulties resulting from repetitive forward reliability analysis. As a result there is considerable interest in developing an efficient and more direct approach to determine the design parameters for a specified target reliability level. An inverse first-order reliability method (inverse FORM) was developed for the estimation of design loads associated with specified target reliability levels in offshore structures, later it was extended to general limit state functions [2]. The inverse reliability methods were extended to design the wind turbines against ultimate limit states [3]. To overcome the drawbacks of the inverse FORM, artificial neural network (ANN)-based inverse FORM [4] as well as polynomial-based response surface method [5] were developed. The traditional performance measure approach (PMA) was modified to improve the efficiency and accuracy [6]. The various inverse measures and their usage for Reliability-Based Design Optimization (RBDO) were described to establish the relationship between Probabilistic Performance Measure (PPM) and Probabilistic Sufficiency Factor (PSF) [7]. The most probable point (MPP) based dimension reduction method was developed for RBDO of nonlinear and multi-dimensional systems [8].

Traditionally, inverse reliability methods require complete statistical information of uncertainties. These uncertainties are treated stochastically and assumed to follow certain probability distributions. However, in many practical engineering applications, the distributions of some random variables may not be precisely known or uncertainties may not be appropriately represented with probability distributions. In the design problem with both statistical random variables and fuzzy variables, if the random variables are converted into fuzzy variables by generating membership functions, the method may yield a design that is too conservative because treating the random variables as fuzzy variables loses accuracy of the uncertainties. On the other hand, treating fuzzy variables as random variables by adopting approximate probability distributions may lead to an unreliable optimum design. Therefore, in this paper a novel solution procedure for inverse reliability problems with implicit response functions without requiring the derivatives of the response functions with respect to the uncertain variables is proposed to determine the unknown design parameters such that prescribed reliability indices are attained in the presence of mixed uncertain variables.

2 High Dimensional Model Representation

High Dimensional Model Representation (HDMR) [9, 10] is a general set of quantitative model assessment and analysis tools for capturing the high-dimensional relationships between sets of input and output model variables. It is a very efficient formulation of the system response, if higher order variable correlations are weak, allowing the physical model to be captured by the first few lower order terms. Practically for most well-defined physical systems, only relatively low order correlations of the input variables are expected to have a significant effect on the overall response. HDMR expansion utilizes this property to present an accurate hierarchical representation of the physical system. Let the N-dimensional vector $\mathbf{x} = \{x_1, x_2, \ldots, x_N\}$ represent the input variables of the model under consideration, and the response function as $g(\mathbf{x})$. Since the influence of the input variables on the response function can be independent and/or cooperative, HDMR expresses the response $g(\mathbf{x})$ as a hierarchical correlated function expansion in terms of the input variables as

$$
\begin{aligned}
g(\mathbf{x}) = g_0 + \sum_{i=1}^{N} g_i(x_i) + \sum_{1 \le i_1 < i_2 \le N} g_{i_1 i_2}(x_{i_1}, x_{i_2}) + \cdots \\
+ \sum_{1 \le i_1 < \ldots < i_l \le N} g_{i_1 i_2 \ldots i_l}(x_{i_1}, x_{i_2}, \ldots, x_{i_l}) \quad + \cdots + g_{12\ldots N}(x_1, x_2, \ldots, x_N)
\end{aligned}
\tag{1}
$$

where g_0 is a constant term representing the zeroth-order component function or the mean response of $g(\mathbf{x})$. The function $g_i(x_i)$ is a first-order term expressing the effect of variable x_i acting alone, although generally nonlinearly, upon the output $g(\mathbf{x})$. The function $g_{i_1 i_2}(x_{i_1}, x_{i_2})$ is a second-order term which describes the cooperative effects of the variables x_{i_1} and x_{i_2} upon the output $g(\mathbf{x})$. The higher order terms give the cooperative effects of increasing numbers of input variables acting together to influence the output $g(\mathbf{x})$. The last term $g_{12,\cdots,N}(x_1, x_2, \cdots, x_N)$ contains any residual dependence of all the input variables locked together in a cooperative way to influence the output $g(\mathbf{x})$. Once all the relevant component functions in Eq. 1 are determined and suitably represented, then the component functions constitute HDMR, thereby replacing the original computationally expensive method of calculating $g(\mathbf{x})$ by the computationally efficient model. The expansion functions are determined by evaluating the input–output responses of the system relative to the defined reference point $\mathbf{c}$ along associated lines, surfaces, subvolumes, etc. in the input variable space. This process reduces to the following relationship for the component functions in Eq. 1.

$$g_0 = g(\mathbf{c}) \tag{2}$$

$$g_i(x_i) = g\left(x_i, \mathbf{c}^i\right) - g_0 \tag{3}$$

$$g_{i_1 i_2}(x_{i_1}, x_{i_2}) = g(x_{i_1}, x_{i_2}, \mathbf{c}^{i_1 i_2}) - g_{i_1}(x_{i_1}) - g_{i_2}(x_{i_2}) - g_0 \tag{4}$$

where the notation $g(x_i, \mathbf{c}^i) = g(c_1, c_2, \ldots, c_{i-1}, x_i, c_{i+1}, \ldots, c_N)$ denotes that all the input variables are at their reference point values except x_i. The g_0 term is the output response of the system evaluated at the reference point $\mathbf{c}$. The higher order terms are evaluated as cuts in the input variable space through the reference point. Therefore, each first-order term $g_i(x_i)$ is evaluated along its variable axis through the reference point. Each second-order term $g_{i_1 i_2}(x_{i_1}, x_{i_2})$ is evaluated in a plane defined by the binary set of input variables x_{i_1} and x_{i_2} through the reference point, etc. Considering terms up to first-order in Eq. 1 yields

$$g(\mathbf{x}) = g_0 + \sum_{i=1}^{N} g_i(x_i) + \mathcal{R}_2 \tag{5}$$

Substituting Eqs. 2 and 3 into Eq. 5 leads to

$$g(\mathbf{x}) = \sum_{i=1}^{N} g(c_1, \ldots, c_{i-1}, x_i, c_{i+1}, \ldots, c_N) - (N-1)g(\mathbf{c}) + \mathcal{R}_2 \tag{6}$$

Now consider first-order approximation of $g(\mathbf{x})$, denoted by

$$\tilde{g}(\mathbf{x}) \equiv g(x_1, x_2, \ldots, x_N) = \sum_{i=1}^{N} g(c_1, \ldots, c_{i-1}, x_i, c_{i+1}, \ldots, c_N) - (N-1)g(\mathbf{c}) \tag{7}$$

Comparison of Eqs. 6 and 7 indicates that the first-order approximation leads to the residual error $g(\mathbf{x}) - \tilde{g}(\mathbf{x}) = \mathcal{R}_2$, which includes contributions from terms of two and higher order component functions. The notion of 0th, 1st, etc. in HDMR expansion should not be confused with the terminology used either in the Taylor series or in the conventional least-squares based regression model. It can be shown that, the first order component function $g_i(x_i)$ is the sum of all the Taylor series terms which contain and only contain variable x_i. Hence first-order HDMR approximations should not be viewed as first-order Taylor series expansions nor do they limit the nonlinearity of $g(\mathbf{x})$. Furthermore, the approximations contain contributions from all input variables. Thus, the infinite number of terms in the Taylor series is partitioned into finite different groups and each group corresponds to one cut-HDMR component function. Therefore, any truncated cut-HDMR expansion provides a better approximation and convergent solution of $g(\mathbf{x})$ than any truncated Taylor series because the latter only contains a finite number of terms of Taylor series. Furthermore, the coefficients associated with higher dimensional terms are usually much smaller than that with one-dimensional terms. As such, the impact of higher dimensional terms on the function is less, and therefore, can be neglected. Compared with the FORM and SORM which retain only linear and quadratic terms, respectively, first-order HDMR approximation $\tilde{g}(\mathbf{x})$ provides more accurate representation of the original implicit limit state function $g(\mathbf{x})$.

3 Inverse Structural Reliability Analysis Using HDMR and FFT

The objective of the inverse reliability analysis using HDMR and FFT [11, 12] is to find a new MPP, denoted by $\mathbf{x}^*_{\mathrm{HDMR}}$, which will be then used in the subsequent iteration of analysis. The proposed computational procedure involves the following three steps: estimation of failure probability in presence of mixed uncertain variables, reliability index update, and MPP update.

3.1 Estimation of Failure Probability in Presence of Mixed Uncertain Variables

Let the N-dimensional input variables vector $\mathbf{x} = \{x_1, x_2, \ldots, x_N\}$, which comprises of r number of random variables and f number of fuzzy variables be divided as, $\mathbf{x} = \{x_1, x_2, \ldots, x_r, x_{r+1}, x_{r+2}, \ldots, x_{r+f}\}$ where the subvectors $\{x_1, x_2, \ldots, x_r\}$ and $\{x_{r+1}, x_{r+2}, \ldots, x_{r+f}\}$ respectively group the random variables and the fuzzy variables, with $N = r + f$. Then the first-order approximation of $\tilde{g}(\mathbf{x})$ in Eq. 7 can be divided into three parts, the first part with only the random variables, the second part with only the fuzzy variables and the third part is a constant which is the output response of the system evaluated at the reference point $\mathbf{c}$, as follows

$$\tilde{g}(\mathbf{x}) = \sum_{i=1}^{r} g\left(x_i, \mathbf{c}^i\right) + \sum_{i=r+1}^{N} g\left(x_i, \mathbf{c}^i\right) - (N-1)g(\mathbf{c}) \tag{8}$$

The joint membership function of the fuzzy variables part is obtained using suitable transformation of the variables $\{x_{r+1}, x_{r+2}, \ldots, x_N\}$ and interval arithmetic algorithm. Using this approach, the minimum and maximum values of the fuzzy variables part are obtained at each α-cut. Using the bounds of the fuzzy variables part at each α-cut along with the constant part and the random variables part in Eq. 8, the joint density functions are obtained by performing the convolution using FFT in the rotated Gaussian space at the MPP, which upon integration yields the bounds of the failure probability.

3.2 Transformation of Interval Variables

Optimization techniques are required to obtain the minimum and maximum values of a nonlinear response within the bounds of the interval variables. This procedure is computationally expensive for problems with implicit limit state functions, as optimization requires the function value and gradient information at several points in the iterative process. But, if the function is expressed as a linear combination of interval variables, then the bounds of the response can be expressed as the summation of the bounds of the individual variables. Therefore, fuzzy variables part of the nonlinear limit state function in Eq. 8 is expressed as a linear combination of intervening variables by the use of first-order HDMR approximation in order to apply an interval arithmetic algorithm, as follows

$$\sum_{i=r+1}^{N} g\left(x_i, \mathbf{c}^i\right) = z_1 + z_2 + \ldots + z_f \tag{9}$$

where $z_i = (\beta_i x_i + \gamma_i)^\kappa$ is the relation between the intervening and the original variables with κ being order of approximation taking values $\kappa = 1$ for linear approximation, $\kappa = 2$ for quadratic approximation, $\kappa = 3$ for cubic approximation, and so on. The bounds of the intervening variables can be determined using transformations [13]. If the membership functions of the intervening variables are available, then at each α-cut, interval arithmetic techniques can be used to estimate the response bounds at that level.

3.3 Estimation of Failure Probability using FFT

Concept of FFT can be applied to the problem if the limit state function is in the form of a linear combination of independent variables and when either the marginal density or the characteristic function of each basic random variable is known. In the present study HDMR concepts are used to express the random variables part along with the values of the constant part and the fuzzy variables part at each α-cut as a linear combination of lower order component functions. The steps involved in the proposed method for failure probability estimation as follows:

1. If $\mathbf{u} = \{u_1, u_2, \ldots, u_r\}^T \in \Re^r$ is the standard Gaussian variable, let $\mathbf{u}^* = \{u_1^*, u_2^*, \ldots, u_r^*\}^T$ be the MPP or design point, determined by a standard nonlinear constrained optimization. The MPP has a distance β_{HL}, which is commonly referred to as the Hasofer–Lind reliability index. Construct an orthogonal matrix $\mathbf{R} \in \Re^{r \times r}$ whose rth column is $\boldsymbol{\alpha}^* = \mathbf{u}^*/\beta_{HL}$, i.e., $\mathbf{R} = [\mathbf{R}_1 | \boldsymbol{\alpha}^*]$ where $\mathbf{R}_1 \in \Re^{r \times r-1}$ satisfies $\boldsymbol{\alpha}^{*T}\mathbf{R}_1 = 0 \in \Re^{1 \times r-1}$. The matrix R can be obtained, for example, by Gram–Schmidt orthogonalization. For an orthogonal transformation $\mathbf{u} = \mathbf{R}\,\mathbf{v}$. Let $\mathbf{v} = \{v_1, v_2, \ldots, v_r\}^T \in \Re^r$ be the rotated Gaussian space with the associated MPP $\mathbf{v}^* = \{v_1^*, v_2^*, \ldots, v_r^*\}^T$. The transformed limit state function $g(\mathbf{v})$ therefore maps the variables into rotated Gaussian space $\mathbf{v}$. First-order HDMR approximation of $g(\mathbf{v})$ in rotated Gaussian space $\mathbf{v}$ with $\mathbf{v}^* = \{v_1^*, v_2^*, \ldots, v_r^*\}^T$ as reference point can be represented as follows:

$$\tilde{g}(\mathbf{v}) \equiv g(v_1, v_2, \ldots, v_r) = \sum_{i=1}^{r} g\left(v_1^*, \ldots, v_{i-1}^*, v_i, v_{i+1}^*, \ldots, v_r^*\right) - (r-1)g(\mathbf{v}^*) \tag{10}$$

2. In addition to the MPP as the chosen reference point, the accuracy of first-order HDMR approximation in Eq. 10 may depend on the orientation of the first $r-1$ axes. In the present work, the orientation is defined by the matrix $\mathbf{R}$. In Eq. 10, the terms $g(v_1^*, \ldots, v_{i-1}^*, v_i, v_{i+1}^*, \ldots, v_r^*)$ are the individual component functions and are independent of each other. Equation 10 can be rewritten as

$$\tilde{g}(\mathbf{v}) = a + \sum_{i=1}^{r} g\left(v_i, \mathbf{v}^{*i}\right) \tag{11}$$

where $a = -(r-1)g(\mathbf{v}^*)$.

3. New intermediate variables are defined as

$$y_i = g\left(v_i, \mathbf{v}^{*i}\right) \tag{12}$$

The purpose of these new variables is to transform the approximate function into the following form

$$\tilde{g}(\mathbf{v}) = a + y_1 + y_2 + \cdots + y_r \tag{13}$$

4. Due to rotational transformation in $\boldsymbol{v}$-space, component functions y_i in Eq. 13 are expected to be linear or weakly nonlinear function of random variables v_i. In this work both linear and quadratic approximations of y_i are considered. Let $y_i = b_i + c_i\, v_i$ and $y_i = b_i + c_i\, v_i + e_i\, v_i^2$ be the linear and quadratic approximations, where coefficients $b_i \in \Re$, $c_i \in \Re$ and $e_i \in \Re$ (non-zero) are obtained by least-squares approximation from exact or numerically simulated conditional responses $\left\{g(v_i^1, \mathbf{v}^{*i}), g(v_i^2, \mathbf{v}^{*i}), \ldots, g(v_i^n, \mathbf{v}^{*i})\right\}^T$ at n sample points along the variable axis v_i. Then Eq. (13) results in

$$\tilde{g}(\mathbf{v}) \equiv a + y_1 + y_2 + \cdots + y_r = a + \sum_{i=1}^{r} (b_i + c_i\, v_i) \tag{14}$$

$$\tilde{g}(\mathbf{v}) \equiv a + y_1 + y_2 + \cdots + y_r = a + \sum_{i=1}^{r} \left(b_i + c_i\, v_i + e_i\, v_i^2\right) \tag{15}$$

The least-squares approximation is chosen over interpolation, because the former minimizes the error when $n > 2$ for linear and $n > 3$ for quadratic approximations.

5. Since v_i follows standard Gaussian distribution, marginal density of the intermediate variables y_i can be easily obtained by simple transformation.

$$p_{Y_i}(y_i) = p_{V_i}(v_i)\left|\frac{1}{dy_i/dv_i}\right| \quad (16)$$

6. Now the approximation is a linear combination of the intermediate variables y_i. Therefore, the joint density of $\tilde{g}(\mathbf{v})$, which is the convolution of the individual marginal density of the intervening variables y_i, can be expressed as follows:

$$p_{\tilde{G}}(\tilde{g}) = p_{Y_1}(y_1) * p_{Y_2}(y_2) * \ldots * p_{Y_r}(y_r) \quad (17)$$

where $p_{\tilde{G}}(\tilde{g})$ represents joint density of the transformed limit state function.

7. Applying FFT on both sides of Equation 17, leads to

$$FFT\left[p_{\tilde{G}}(\tilde{g})\right] = FFT[p_{Y_1}(y_1)]FFT[p_{Y_2}(y_2)]\ldots FFT[p_{Y_r}(y_r)] \quad (18)$$

8. By applying inverse FFT on both side of Eq. 18, joint density of the limit state function $\tilde{g}(\mathbf{v})$ is obtained.
9. The probability of failure is given by the following equation

$$P_F^{\text{HDMR}} = \int_{-\infty}^{0} P_{\tilde{G}}(\tilde{g})d\tilde{g} \quad (19)$$

After computing the probability of failure P_F^{HDMR} using coupled HDMR-FFT technique, the corresponding reliability index β_{HDMR} can be obtained by

$$\beta_{\text{HDMR}} = -\Phi^{-1}\left(P_F^{\text{HDMR}}\right) \quad (20)$$

where $\Phi(\bullet)$ is the cumulative distribution function of a standard Gaussian random variable.

3.4 Reliability Index and MPP Update Procedure

As expected it is very likely that the β_{HDMR} (computed using Eq. 20) is not the same as the target reliability index $\beta_t = -\Phi^{-1}(P_F^{\text{Tar}})$, and hence, using the difference between these two reliability indices, a recursive formula is obtained as

$$\beta^{(k+1)} \cong \beta^{(k)} - (\beta_{\text{HDMR}} - \beta_t) \quad (21)$$

where $\beta^{(k)}$ is the reliability index at the current step, with $\beta^{(0)} = \beta_t$ at the initial step.
The updated MPP is approximated as

$$\mathbf{u}^*_{k+1} \cong \frac{\beta^{(k+1)}}{\beta^{(k)}}\mathbf{u}^*_k \quad \text{or} \quad \mathbf{v}^*_{k+1} \cong \frac{\beta^{(k+1)}}{\beta^{(k)}}\mathbf{v}^*_k \tag{22}$$

The updated MPP obtained through Eq. 22 is called the coupled HDMR-FFT based MPP, denoted by $\mathbf{u}^*_{\mathrm{HDMR}}$ in **U**-space or $\mathbf{x}^*_{\mathrm{HDMR}}$ in **X**-space.

4 Numerical Examples

To obtain the approximation of the HDMR component functions of the nonlinear limit state function in Eq. 12, n sample points $\mu_i - (n-1)\sigma_i/2$, $\mu_i - (n-3)\sigma_i/2$, ..., μ_i, ..., $\mu_i + (n-3)\sigma_i/2$, $\mu_i + (n-1)\sigma_i/2$ are deployed along axis of each of variable x_i. If N and n respectively denote the number of uncertain variables, the number of sample points taken along each of the variable axis, then using first-order HDMR approximation the total cost of original function evaluation entails a maximum of $N \times (n-1) + 1$ by the proposed method. The efficiency and robustness of the proposed method is expected to increase with increase in the complexity of the structure, number of uncertain variables.

4.1 Hypothetical Limit State Function

This example considers a four dimensional quadratic function of the following form:

$$g(\mathbf{x}) = -x_1^2 - x_2^2 - x_3^2 - x_4^2 + 9x_1 + 11x_2 + 11x_3 + 11x_4 - 95.5 \tag{23}$$

where $x_1 - x_3$ are assumed to be independent normal variables with $N(5, 0.4)$. The variable x_4 is assumed as fuzzy represented by the triplet [4.6, 5.0, 5.4]. The objective is to find the membership functions of $x_1^* - x_3^*$ values, such that the target reliability index $\beta_t = 1.645$ (which corresponds to a failure probability $P_F = 0.05$) is achieved. The presence of mixed uncertain (both random and fuzzy) variables, leads to the membership function of MPP ($x_1^* - x_3^*$) instead of having a unique value at the target reliability index. Figure 1a–c respectively show the membership functions of $x_1^* - x_3^*$ values at the target reliability index estimated by the proposed method using linear and quadratic approximations. The effect of number of sample points is studied by varying n from 3 to 9. In Fig. 1a–c it can be observed that the

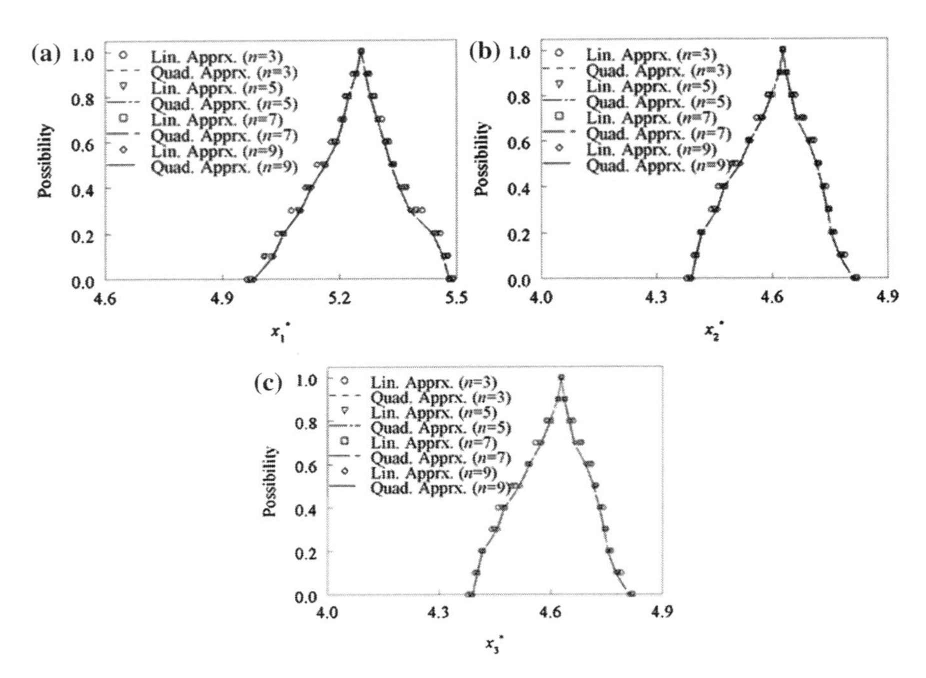

Fig. 1 Membership function of design variables **a** x_1^*, **b** x_2^*, and **c** x_3^*

membership functions of $x_1^* - x_3^*$ values estimated by the proposed method using $n = 7$ and 9 are overlapping each other.

4.2 Single Story Linear Frame Structure

In this example a linear frame structure of one story and one bay as shown in Fig. 2a is considered. The cross sectional areas A_1 and A_2 are assumed to be log-normally distributed random variables with mean values of 0.36 and 0.18, and standard deviation values of 0.036 and 0.018 respectively. The horizontal load P is treated as fuzzy with a triplet of [15, 20, 25]. The sectional moments of inertia are expressed as $I_i = \alpha_i A_i^2$ ($i = 1, 2;\ \alpha_1 = 0.08333,\ \alpha_2 = 0.16670$). The Young's modulus E is treated as deterministic. $E = 2.0 \times 10^6$ kN/m^2. In this study, the functional relationship to define the horizontal displacement at the top of the frame is:

$$g(A_1, A_2, P) = \Delta_{\text{lim}} - u_h \tag{24}$$

where Δ_{lim} is taken as 10 mm. Our interest is to find A_1^* and A_2^*, such that the target reliability index $\beta_t = 2.831$ (which corresponds to a failure probability $P_F = 2.32 \times 10^{-3}$) is achieved.

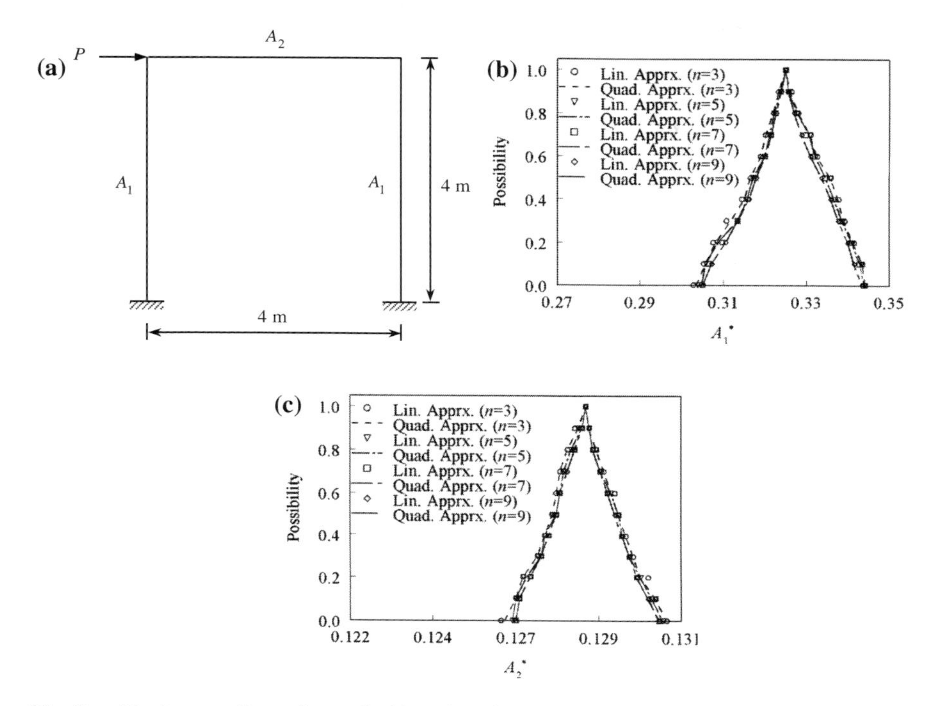

Fig. 2 **a** Single story linear frame, **b** A_1^*, and **c** A_2^*

The implicit limit state function given in Eq. 24 is approximated using first-order HDMR by deploying n sample points along each of the variable axis. The approximated limit state function is divided into two parts, one with only the random variables along with the value of the constant part, and the other with the fuzzy variables. The joint membership function of the fuzzy part of approximated limit state function is obtained using suitable transformation of the fuzzy variables. Using the proposed inverse reliability method in conjunction with linear and quadratic approximations, the membership functions of A_1^* and A_2^* values at the target reliability index are obtained, and shown in Fig. 2b and c. The effect of number of sample points is studied by varying n from 3 to 9 and observed that the membership functions of A_1^* and A_2^* values estimated by the proposed method using $n = 7$ and 9 are overlapping each other.

5 Summary and Conclusions

An efficient, accurate, robust solution procedure alternative to existing inverse reliability methods is proposed for nonlinear problems with implicit response functions, that can be used to determine multiple unknown design parameters such that prescribed reliability indices are attained in the presence of mixed uncertain (both random and fuzzy) variables. The proposed method avoids the requirement of the derivatives of the response functions with respect to the uncertain variables. The proposed computational procedure involves three steps: (1) probability of failure calculation using High Dimensional Model Representation (HDMR) for the limit state function approximation, transformation technique to obtain the contribution of the fuzzy variables to the convolution integral, and fast Fourier transform for solving the convolution integral, (2) reliability index update, and (3) most probable point (MPP) update. The limit state function approximation is obtained by linear and quadratic approximations of the first-order HDMR component functions at MPP. The methodology developed is versatile, hence can be applied to highly nonlinear or multi-parameter problems applicable involving any number of fuzzy variables and random variables with any kind of distribution. The accuracy and efficiency of the proposed method is demonstrated through six numerical examples.

In addition, a parametric study is conducted with respect to the number of sample points used in approximation of HDMR component functions and its effect on the estimated solution is investigated. An optimum number of sample points must be chosen in approximating HDMR component functions. Very small number of sample points should be avoided as approximation may not capture the nonlinearity outside the domain of sample points and thereby affecting the estimated solution.

References

1. Li H, Foschi RO (1998) An inverse reliability method and its application. Struct Saf 20:257–270
2. Der Kiureghian A, Zhang Y, Li CC (1994) Inverse reliability problem. ASCE J Eng Mech 120:1154–1159
3. Saranyasoontorn K, Manuel L (2004) Efficient models for wind turbine extreme loads using inverse reliability. J Wind Eng Ind Aerodyn 92:789–804
4. Cheng J, Zhang J, Cai CS, Xiao RC (2007) A new approach for solving inverse reliability problems with implicit response functions. Eng Struct 29:71–79
5. Cheng J, Li QS (2009) Application of the response surface methods to solve inverse reliability problems with implicit response functions. Comput Mech 43:451–459
6. Du X, Sudijianto A, Chen W (2004) An integrated framework for optimization under uncertainty using inverse reliability strategy. ASME J Mech Des 126(4):561–764
7. Ramu P, Qu X, Youn BD, Choi KK (2006) Inverse reliability measures and reliability-based design optimization. Int J Reliab Saf 1(1, 2):187–205
8. Lee I, Choi KK, Du L, Gorsich D (2008) Inverse analysis method using MPP-based dimension reduction for reliability-based design optimization of nonlinear and multi-dimensional systems. Comput Methods Appl Mech Eng 198:14–27
9. Rabitz H, Alis OF (1999) General foundations of high dimensional model representations. J Math Chem 25:197–233
10. Sobol IM (2003) Theorems and examples on high dimensional model representations. Reliab Eng Syst Saf 79(2):187–193

11. Rao BN, Chowdhury R (2008) Probabilistic analysis using high dimensional model representation and fast Fourier transform. Int J Comput Methods Eng Sci Mech 9:342–357
12. Sakamoto J, Mori Y, Sekioka T (1997) Probability analysis method using fast Fourier transform and its application. Struct Saf 19(1):21–36
13. Adduri PR, Penmetsa RC (2008) Confidence bounds on component reliability in the presence of mixed uncertain variables. Int J Mech Sci 50:481–489

Analyzing Conflicts Between Product Assembly and Disassembly for Achieving Sustainability

S. Harivardhini and Amaresh Chakrabarti

Abstract Environmental performance of a product could be increased throughout its life cycle by incorporating design requirements which consider Design for Disassembly (DfD) from a life cycle perspective by aiding ease of disassembly of the product across its life cycle. These design requirements, including DfD for different life cycle phases, should be made compatible with Design for Assembly (DfA) requirements within an integrated framework. Using such an integrated framework should reduce various layers of complexity introduced into design and should help designers to develop products that are easy to both assemble and disassemble, without compromising the product's functionality. Prerequisites to developing the integrated framework are to: understand the requirements for DfD and DfA, identify if they are in conflict with one another, understand the underlying causes, and develop means to resolve these. To determine whether DfD and DfA requirements conflict one another, various existing products are analyzed, for conflicts among their assembly and disassembly processes. Various conflicts are found to be present among these processes. These conflicts are outlined, and possible causes for these are identified.

Keyword: Disassembly · DfD · DfA · Conflicts

S. Harivardhini (✉) · A. Chakrabarti
IDeaS Lab, CPDM, Indian Institute of Science, Bengaluru 560012, India
e-mail: vardhini@cpdm.iisc.ernet.in

A. Chakrabarti
e-mail: ac123@cpdm.iisc.ernet.in

A. Chakrabarti and R. V. Prakash (eds.), *ICoRD'13*, Lecture Notes in Mechanical Engineering, DOI: 10.1007/978-81-322-1050-4_44, © Springer India 2013

1 Introduction

Sustainability is a growing concern among all countries. This is due to various factors, such as the following: manufacturers continue to introduce huge quantities of products without considering future reuse during their development, and customers often get dissatisfied with their products even through these are in good working condition. Factors such as these lead to a wide variety of products being disposed rather than being recovered and reused.

Society has started realizing the likely environmental threats that might result from disposal; for instance, governments are enacting strict legislations for disposal of products in an environmental friendly manner. One impact of these regulations is increased responsibility on the part of manufacturers in the End-of-Life (EoL) phase of their products. This scenario makes manufacturers rethink about the decisions taken during their product design and manufacturing stages. Decisions taken during the design stage are critical because it is during this stage that most product attributes are decided [1] and most of the considerations that have the potential to resolve environmental issues are incorporated. Designing products with reduced impact on environment in their EoL phase is a far better option than products that are not designed for this purpose, and hence destined to end in disposal. However, in order to achieve environmental sustainability, it is not enough to focus only on the EoL phase of the product. Impacts caused by the product in its other life cycle phases also add to its environmental consequences. Therefore, it is crucial to improve the environmental performance of a product throughout its life cycle [2]. The product should be developed in such a way that its likely impact on the environment is minimized in each of its life cycle phases.

One way to improve the environmental performance of a product throughout its life cycle is to design the product such that it aids ease of disassembly in all its life cycle phases. Design for (ease of) Disassembly (DfD) is one of the strategies to improve disassemblability of the product. It is an approach in which disassembly considerations are incorporated into the product at the design stage itself [3], thereby increasing the product's ease of disassembly. While doing this, DfD should be balanced against other design considerations [4] such as DfA in order to avoid new problems being introduced into the design; evidence for this can be found [5], who observed that introducing Design for X guidelines for one aspect (e.g. assembly) without considering other often led to new problems.

This paper emphasizes the need for developing an integrated framework for DfD. Prerequisites to this are: to understand the requirements for DfD and DfA, to identify whether or not they are conflicting, identify the reasons behind the conflicts, if any, and develop means to resolve the conflicts. Existing products were analyzed for conflicts in their assembly and disassembly processes to understand DfD and DfA compatibility. We found that various conflicts exist among assembly and disassembly processes of these products. In this paper, these conflicts are outlined, and possible causes are discussed.

Section 2 elaborates the importance of carrying out disassembly in each life cycle phase; Sect. 3 explains what DfD is and how DfD needs to be different for each life cycle phase; Sect. 4 reviews existing literature on the relationship between DfD and DfA, and establishes the need for an integrated framework for DfD; Sect. 5 reports on a pilot study carried out to determine whether conflicts exist among ease of assembly and disassembly processes of existing products and what cause these conflicts; conclusions and future work are presented in Sect. 6.

2 Disassembly

Brennan et al. [6] defined disassembly as ''the processes of systematic removal of desirable constitute parts from an assembly while ensuring that there is no impairment of the parts due to the process"

"Ease of disassembly" is one of the requirements for achieving easy transportation, easy service and maintenance [7], easy recovery of parts at EoL [7, 8]. We argue, therefore, that disassembly has the potential to improve environmental performance of a product throughout its life cycle by improving efficiencies of the operations carried out during all the life cycle phases of the product. Henceforth, we refer to this objective as "disassembly for all life cycle phases".

2.1 Disassembly for Various Life Cycle Phases

2.1.1 Disassembly for Production

During assembly, the possibility of parts ending up in a wrong fitting is high when parts have similar geometric structure and multiple possible ways of being fitted. In such cases, disassembly would be necessary for removal of those parts for reassembly in the same product, thus preventing the use of new parts. Another reason is that testing may reveal issues with functioning of the product; to resolve these, product must be disassembled.

2.1.2 Disassembly for Distribution

Some complex products are difficult to distribute, because their product architecture will not allow their components to get separated during transportation and later reassembled for use. Disassembly of such products could improve the distribution efficiency by making products occupy less storage space during transportation.

2.1.3 Disassembly for Use

Disassembly enables maintenance and enhances serviceability [7]. Thus it increases the life of a product [1].

2.1.4 Disassembly for EoL

The recovery processes are often economically unviable if the products are originally designed with no consideration to their future reuse. So, very often, disposal is the only option for such products. To resolve this issue, products should be variously remanufactured, reused or recycled, so as to maximally recover its sub-assemblies, components or materials, from used products, in order to make these available for new products. Disassembly is necessary in carrying out these recovery processes.

3 Design for Disassembly

According to Giudice et al. [23], DfD is a design approach with the objective of optimizing the architecture and all other constructional characteristics of a product in relation to the following main requirements: limiting the time and costs of disassembly; simple and rapid separability of parts to be serviced or recovered. DfD can also be defined as "the consideration of the ease of disassembly during the design process" [3].

3.1 DfD for various life cycle phases

Production phase

The objective of DfD for production is to design such that the parts having similar geometric structure and ambiguous fitting possibilities are easily accessed, disassembled and assembled in order to rectify the assembly, if applicable.

Distribution Phase

The objective of DfD for distribution is to design the product with high modularity, i.e., easy access, disassembly and reassembly of all modules, with all functional requirements satisfied after reassembly.

Use phase

The objective of DfD for service and maintenance is to make design choices that most efficiently ease accessibility, disassembly and reassembly of certain predetermined components that require servicing intervention.

EoL phase

The objective of DfD for EoL is to design a product such that its subassemblies, parts and materials, at the end of its useful life, are easily accessible and separable (and in some cases re-assemblable) from their adjacent subassemblies, parts and materials, so as to make them amenable to appropriate EoL treatments e.g. remanufacturing, reuse, or recycling.

4 Compatibility Between DfD and DfA

4.1 Literature on Relationships Between DfD and DfA

Boothroyd and Alting [9], Jovane et al. [10], Penev and De Ron [11] and Gupta and McLean [12] have studied DfA methods and discussed research opportunities in DfD. Shu and Flowers [13] showed that joints designed for ease of assembly and recycling may not facilitate remanufacturing. One problem with disassembly of existing products, reported by Alting and Legarth [2], is that it requires a large number of steps to take products apart as joining techniques are directed towards assembly and not disassembly.

Harjula et al. [14] identified that though DfA redesigns could be beneficial in simplifying disassembly, additional design changes have to be incorporated for simplifying removal of critical items. Several differences between assembly and disassembly have been identified, such as (1) irreversible operations like welding, riveting or breakage of components [15], (2) selective disassembly [16]. Based on the implications of these differences, Srinivasan et al. [17] concluded that the most economical assembly sequence need not be the most economical disassembly sequence. Kroll et al. [8] pointed out that DfD and Design for Manufacture and Assembly (DfMA) may seem similar in intent, but are often quite different in practice. They reported that many products designed for assembly are very hard to disassemble, e.g., those with certain types of snap-fit joints.

Westkamper et al. [18] have compared assembly and disassembly for different EoL options (including repair), based on following criteria: productivity, quality, lead time, time to delivery, process time, and flexibility. Their study highlights how to integrate assembly and disassembly given that logistics, systems, technical installations, flexible automation, management of product life cycle data were to be made common for both assembly and disassembly. While this work focuses on integrating assembly and disassembly systems for existing products, the focus of our work lies in integrating assembly and disassembly requirements at the design stage.

Nof and Chen [19] argued that Design for assembly and disassembly (DFAD) involves integrating the specific domain knowledge of manufacturing, design, and decision-making. They have developed an approach called Cooperation Requirement Planning (CRP), the output of which is analyzed for conflicts among task assignments and assembly planning in CRP. The focus of our work is distinct from this work by resolving conflicts among DfA and DfD requirements to achieve sustainability rather than resolving conflicts among task assignments and assembly planning in CRP to achieve optimum utilization of cooperation among robots.

Also our definition of DfAD is to design products such that it enables easy assembly and easy disassembly. But Nof and Chen considers disassembly as reverse of assembly. DfAD is approximately equal to DfA in their work.

Motevallian et al. [1] have modified the DfMA process, and incorporated DfD into a framework; however, this integrates DFMA and DfD in a serial manner, allowing possibility for conflicts among these to remain in the final product. Gkeleri and Tourassis [4] pointed out that disassembly concerns must be balanced against other design considerations. They also mentioned that industrial firms complained about the increasing layers of complexity imposed upon the product design process. Integrating various DfX concepts into a single framework is required [5].

4.2 Need for an Integrated Framework for DfD

As discussed in Sect. 2, a major means for improving environmental performance should be to support "disassembly for all life cycle phases". From literature (Sect. 4.1), it can be argued that substantial differences can exist between requirements for DfD and DfA; design requirements that enable easy assembly can be different from those that enable easy disassembly. There is a need to balance disassembly concerns with other design considerations, and a need to integrate various DfX concepts into a single framework. The overall objective is therefore to develop an integrated framework that supports consideration of design requirements for ease of disassembly for all phases of product life cycle while being compatible with requirements for ease of assembly in these life cycle phases.

To achieve this, the following steps are necessary:

1. Understand the requirements for DfD and DfA,
2. Identify whether they conflict one another,
3. Understand the underlying causes for conflicts or their absence, and
4. Develop means to resolve or learn from these

To carry out Steps 1–3, a series of existing products are taken, and their assembly and disassembly processes are analyzed to answer the following research question: Are there any conflicts among the assembly & disassembly processes for the same product, if yes, what are the conflicts, and why do they occur?

5 Conflicts Among Assembly & Disassembly Processes

As a preliminary investigation to answer the research question in Sect. 4.2, two studies were undertaken. The first was a literature based study, where existing cases in literature that report conflicts among assembly and disassembly processes of a product are analyzed to identify the underlying causes. The second is a pilot

Table 1 Results of literature based study—conflicts in assembly and disassembly processes

Products (Mechanical assemblies)	Assembly process	Disassembly process	Conflicts	Causes of conflicts
Rivet in Aircraft structure (from literature [20])	Rivet is passed through the holes in the parts and then forming (upsetting) a second head in the pin on the opposite side The deforming operation can be performed hot or cold and by hammering or steady pressing	Support the structure to prevent distortion and permanent damage to the remainder of the structure Undercut rivet heads by drilling Drilling must be exactly centered and to the base of the head only After drilling, break off the head with a pin punch and carefully drive out the shank Inspect rivet joints adjacent to damaged structure for partial failure	More effort and longer time to disassemble than to assemble	Use of many disassembly tools (bucking bar, drill, pin punch) Fastener design did not consider disassembly. (deforming the second head)
Retaining ring in Gear assembly (from literature Ref 21)	Installation is manually carried out using hammer	These rings with no lug holes are impossible to remove without either destroying the ring or warping it out of specified tolerances Once installed, the rings become tamper proof and make it difficult to be removed.	More effort to the extent of destruction is required in disassembly unlike in assembly.	Fastener design did not consider disassembly (e.g. without lug holes). Becomes tamper proof
Shrink fit in Shaft hub assembly (from literature Ref 22)	The external part is heated to enlarge by thermal expansion, and the internal part either remains at room temperature or is cooled to contract its size The parts are then assembled and brought back to room temp so that external part shrinks and internal part expands to form a strong interference fit	If evenly distributed heat is used to remove parts from shafts. This will increase the time cycle and create heat buildup in the shaft that can result in both parts expanding thus causing difficulty in removal. In this case, it is often best to shock that particular component with a rapid heat. This should be done carefully to prevent expansion of both the parts	More effort to disassemble than to assemble	Evenly distributed heat leading to expansion of both parts Accessibility and visibility influence the shock given since shock needs to be given only for a particular component

Table 2 Results of questionnaire based study—conflicts in assembly and disassembly processes

Products	Assembly process	Disassembly process	Conflicts	Causes of conflicts
Welding in Levers (from questionnaire)	Connection between parts to be welded is established using an agent that together with the material of the parts undergoes phase transition	Usually destructive disassembly is used to separate welded joints	More effort and longer time to disassemble than to assemble	Difficult to access the parts Low clearance for tooling
Cotter and nut in Bicycle pedal crank (from questionnaire)	Slide in pedal crank into axle	The process of removal of the cotter from the pedal crank requires reversal of the tight fit between them	More effort and longer time to disassemble than to assemble	Corrosion
	Align cotter into the pedal and hammer it for tight fit	Since they had rusted and joined up with each other, power drill was used to drill out		
	Screw in the nut on to the cotter from other end of the cotter head	Cotter and pedal got damaged while trying to separate them		
Snap fit in Laptop Key (from questionnaire)	Used fingers to carefully engage the projection in one part to other	Used nail to force open the snap fit. However, it was difficult to take apart	More effort to disassemble than to assemble.	Parts were hidden. Joints were invisible Structure was delicate to handle
Snap fit in Wrist watch (from questionnaire)	A small hammer was used to establish snap fit between back cover and dial of the watch	A strong blade was to disassemble	More effort to disassemble than to assemble	Low clearance for tooling Fit design has little. consideration to disassembly (back cover and dial are almost jammed)

study that was conducted with data collected using a semi-structured questionnaire (with both open and close-ended questions) among Masters and PhD students with formal engineering training at Indian Institute of Science and in some cases with industrial experience; the students were asked about products known to them that have conflicts among assembly and disassembly processes, and according to the subjects, what the causes might have been. Three products (each of them a mechanical assembly) from literature and associated information about their assembly and disassembly processes, were selected for analysis. A further four products from participants in the questionnaire survey, feedback on these from 12 participants on 12 questions (answered over a period of 15 days) are also analyzed. The results from these seven products are shown in Tables 1 and 2, respectively.

5.1 Results and Discussion

The above studies showed the existence of conflicts among assembly and disassembly processes of existing products. In all cases, conflicts seemed to occur in the amount of effort and/or the time required to carry out assembly and disassembly processes. Causes behind the effort and/or time to disassemble are interpreted to be among the following:

Design Issues

1. **Fasteners** The fastener or fit design did not consider (or considered little) the disassembly requirements. Additional conflicting requirements (e.g. tamper proof) forced the design to be difficult to disassemble.
2. **Product architecture** Accessibility and visibility of parts and joints were low, or structure of parts was delicate to handle during disassembly.
3. **Materials** Rust formed due to corrosion and made disassembly difficult.

Other Issues

1. **Tooling** Use of many tools, and/or low clearance for tooling made disassembly difficult.

6 Conclusions and Future Work

Existing products were analyzed for conflicts among assembly and disassembly processes using two studies. It seemed that conflicts occurred in the amounts of effort and/or time required to carry out the assembly and disassembly processes. Various causes for these conflicts were identified: these came either from the product (parts, interfaces and their materials), or from joining elements or associated tools. The study indicates that conflicts exist among assembly and disassembly processes, and all of the causes identified could be addressed during the design process. However, this is only a pilot study, with relatively few subjects and products, and with subjects who are not assembly/disassembly professionals. The goal is to expand this study into a comprehensive study involving professional engineers and assemblers from industry.

Acknowledgments I would like to acknowledge the research students of IDeaS Lab and Masters students of VDS Lab, CPDM, IISc for participating in the study conducted using the questionnaire.

References

1. Motevallian B, Abhary K, Luong L, Marian RM (2007) Integration and optimisation of product design for ease of disassembly. In: Dudas L (ed) Engineering the future, Sciyo, pp 317–340. ISBN 978-953-307-210-4. doi:10.5772/291
2. Alting L, Legarth JB (1995) Life cycle engineering and design. Annals CIRP 44(2):569–580
3. Veerakamolmal P, Gupta SM (2000) Design for disassembly, reuse and recycling. Environmentally responsible engineering. Butterworth-Heinemann, Oxford, pp 69–82
4. Gkeleri VP, Tourassis VD (2008) A concise framework for disassemblability metrics. In: IEEE, 2008
5. Chiu MC, Kremer GEO (2011) Investigation of the applicability of design for X tools during design concept evolution: a literature review. Int J Prod Develop 13(2):132–167
6. Brennan L, Gupta SM, Taleb KN (1994) Operations planning issues in an assembly/ disassembly environment. Int J Oper Prod Manage 14(9):57–67
7. Desai A, Mital A (2003) Evaluation of disassemblability to enable design for disassembly in mass production. Int J Ind Ergon 32:265–281
8. Kroll E, Hanft TA (1998) Quantitative evaluation of product disassembly for recycling. Res Eng Des 10:1–14
9. Boothroyd G, Alting L (1992) Design for assembly and disassembly. Annals of CIRP 41:625–636
10. Jovane F, Alting L, Eversheim W, Feldmann K, Seliger G, Roth N (1993) A key issue in product life cycle: disassembly. Annals CIRP 42(2):651–658
11. Penev kD, De Ron AJ (1996) Determination of a disassembly strategy. Int J Prod Res 34(2):495–506
12. Gupta SM, McLean CR (1996) Disassembly of products. Comput Ind Eng 31:225–228
13. Shu LH, Flowers WC (1995) Considering remanufacture and other end-of-life options in selection of fastening and joining methods. In: IEEE, 1995
14. Harjula T, Rapoza B, Knight WA, Boothroyd G (1996) Design for disassembly and the environment. Annals CIRP 45(7):109
15. Lee K, Gadh R (1996) Destructive disassembly to support virtual prototyping. IIE J Des Manuf 30:359–72
16. Srinivasan H, Gadh R (1997) Virtual selective disassembly: a geometric tool to achieve net positive environmental value. In: 30th ISATA, Florida, Italy
17. Srinivasan H, Shyamsundar N, Gadh R (1997) A Virtual Disassembly Tool to support environmentally conscious product design. In: IEEE, 0-7803-808-1
18. Westkamper E, Feldmann K, Reinhart G, Seliger G (1999) Integrated development of assembly and disassembly. Annals CIRP 48(2):557–585
19. Nof SY, Chen J (2003) Assembly and disassembly: an overview and framework for cooperation requirement planning with conflict resolution J Intell Rob Syst 37:307–320

20. AFS-640 (1998) Acceptable methods, techniques, and practices aircraft inspection and repair, Title 14 of the code of federal regulations (14 CFR) guidance material : Advisory Circular 43.13-1B /U S Department of Transportation; Federal Aviation Administration. pp 4–12
21. Ref 1: Retaining ring from Wiki - http://en.wikipedia.org/wiki/Retaining_ring
22. Ref 2: Induction shrink fitting from Wiki - http://en.wikipedia.org/wiki/Induction_shrink_fitting
23. Giudice F, La Rosa G, Risitano A (2006) Product design for the environment: a life cycle approach. CRC/Taylor & Francis, Boca Raton, p 348

A Conceptual Platform to View Environmental Performance of a Product and Its Usage in Co-Design

Srinivas Kota, Daniel Brissaud and Peggy Zwolinski

Abstract All the stakeholders in product life cycle needs to work together to achieve best possible sustainable solution. User perspectives need to be considered in design, for products to be sustainable in use. Literature review and empirical studies helped in identifying requirements to include user perspectives in design. An activity model is developed after thoroughly studying the usage of electric kettle. In this paper we propose a computer aided conceptual platform to visualize and interact with the product in virtual environment by the user for performing basic activities in use of that product. The platform also supports designer to create product and its usage scenarios based on requirements' from users. It also captures and stores the data generated while user performs the activities virtually for assessment. This is achieved with the help of 3D stereo display, motion capture devices and visualization tool kits. A questionnaire is planned to obtain designer and user feedback on the platform to evaluate the support.

Keywords Eco-design · Co-design · User centered design · 3D visualization · Use · Activity

S. Kota (✉) · D. Brissaud · P. Zwolinski
Grenoble-INP/UJF-Grenoble 1/CNRS, UMR5272 G-SCOP Laboratory, Grenoble 38031, France
e-mail: srinivas.kota@g-scop.inpg.fr

D. Brissaud
e-mail: Daniel.Brissaud@g-scop.inpg.fr

P. Zwolinski
e-mail: Peggy.Zwolinski@g-scop.inpg.fr

A. Chakrabarti and R. V. Prakash (eds.), *ICoRD'13*, Lecture Notes in Mechanical Engineering, DOI: 10.1007/978-81-322-1050-4_45, © Springer India 2013

1 Introduction

Eco-friendly characteristics of the product and the stake holders throughout the life cycle need to be considered from the requirements through detail design to develop eco-friendly products. This calls for different stakeholders with different perceptions work simultaneously. Pre-use and post-use phases in product lifecycle are influenced by use phase via user and usage; hence this research focuses on those. Designer has to take into account user and usage in design, as products are designed for use. User has to provide usage requirement scenarios and subsequently help in test and feedback for improvement. Designer and user needs to work together to achieve best possible solution, this needs new methods to support different perspectives of different stakeholders in design.

Life Cycle Assessment (LCA) [1] is a proven approach for determining the environmental impact and needs experts for interpretation of assessment results. It needs detailed information about processes due to which impact is generated. As usage determines the information about the processes in use it needs to be observed. Required information can be captured based on this observation. Virtual Reality especially 3D visualization and interaction can help in this regard. When the environmental impact results are shown to users and designers, these impacts need to be filtered as the users and designers have varied interests and skills. For example, if environmental impact is conveyed in points, which is the measurement in Eco-indicator'99 which is one of the methodologies used in LCA for calculating environmental impact to user it won't make any sense as the user is not aware of the methodology.

Designers develop products based on user requirements. Users perform activities using products that results in environmental impact during use phase. Actual usage environmental impact varies a lot from designed usage as the products are designed for ideal use and in reality it is not so. It will be good to see the impact based on actual usage in design so that impact can be minimized by changing design.

The following sections report the research carried out on understanding the requirements of stakeholders in use phase and proposing a conceptual support to fulfill the requirements. Literature review (Sect. 2) helped in general understanding and establishing the need for a support in use, empirical studies on electric kettle helped in understanding the multiple stakeholders' requirements in use phase (Sect. 3). Based on these requirements, a platform is proposed in Sect. 4 and discussion about evaluation is detailed in (Sect. 5).

2 State of the Art

Lifecycle thinking helps in thinking holistically about the whole product lifecycle and its stakeholders. The use phase of energy consuming products is very important towards reducing environmental impact, as 80 % of the environmental impact occurs in that phase [2]. Studies highlight the importance of the reporting of valid alternative usage phase scenarios; the lack of rigor and justification for this step can indeed be significantly detrimental to the validity of LCA results [3]. Most of the authors and practitioners point to the need for methods to define usage scenarios; which are scarcely found in LCA studies. This is an obstacle to the truth of results of some LCA.

In practice, users have their own way of using products and it results in varied impacts [4, 5]. The data used in assessment were often extremely simplified and defined arbitrarily due to complexity and diversity involved in use phase. The data is generalized and solutions are generated for generic requirements which have wide scope in determining impact in usage. Normally requirements are considered while developing solutions, but the solutions depend on many factors including context, actual actions etc. The impact is going to change based on variety of factors which needs to be considered while developing solutions [6]. It is important to instrument designers with tools that model the user characteristics and use phase scenarios of the product.

Research towards studying and influencing user behaviour through design measures is reported in literature [7]. Studies done to see how some strategies like on product information (OPI) and explicit ecological instruction affect the usage [8]. But to conduct these kinds of studies, we need multiple physical prototypes with respective strategies. In actual practice these studies are limited as they are costly, time consuming and require lots of effort. The situation can be improved by providing the product and strategies in virtual environment so that a number of concepts and use scenarios can be explored with little effort, time and money to achieve those. Various representations of users with highlighted behavioral characteristics that influence the environmental performance of products in use phase need to be identified by different exercises using users. This will help in producing, consuming and disposing products in a sustainable way.

Virtual reality [9] gives experience to several senses without the presence of real object. The environment can be non-immersive (3D display) or semi-immersive (HMD) or fully immersive (CAVE) with visual [10], tactile and audio technologies. The capabilities of the systems depend on resolution, interference, and lag. The decision to opt for any of these systems should consider set up time, ease of use, and cost.

Virtual reality is being used successfully in different domains: manufacturing [11], architecture, automobile, aerospace, and retail [12] for different purposes. Virtual reality is also successfully used in making 3D annotations in collaborative environments [13] for design. Methodologies like VRID [14] and [15] developed to guide the creation of virtual reality interfaces. Work is also reported on successfully combining conventional CAD and Virtual reality [16].

Co-design can help development of environmentally friendly use by supporting generation and evaluation of concepts, user preferences, and use scenarios with user inputs. This requires supporting multiple views, interactions of product and user in usage at the same time. Virtual reality can support co-design efficiently by showing multiple views of different aspects of real world complex situations [17]

Fig. 1 Generic kettle parts *Source* http://karisimby.wordpress.com/2009/08/09/parts-of-a-kettle/

to different stakeholders for analysis at the same time. Computerized tools ease design and assessment by helping in capture, manipulation and communication of the required data iteratively. Advanced visualization and interactive equipment are needed to achieve these objectives. Successful use of Kinect, an advanced skeletal motion capture device is reported in [18] for querying and visualization of complex datasets. Next section details the empirical studies done for identifying requirements that need to be supported.

3 Empirical Studies

Empirical studies were performed to understand better, the use phase in terms of its constituents, characteristics of its constituents, and interactions [19] among its constituents. A device which has impact in use is studied for requirements to be satisfied, functions which fulfill the requirements, behaviour of different constituents and required structure to achieve those.

Generic electrical kettle for producing hot water for making tea is used in this study. The kettle and its use are studied thoroughly. Figure 1 shows the individual components of the structure of the kettle. The kettle consists of power cable (1), base (2), water container (3), handle with lid top and on/off switch (4), heating element (5), removable filter (6), bottom of base (7), lid bottom (8), lid release button (9), lid release spring (10), rubber seal (11), kettle control, switch and thermal fuse (12), LED (13), and bi-metallic disk (14). Apart from these shown in figure kettle also comes with cardboard packing box and usage manual.

Different stakeholders were involved in the use phase of the kettle; Retailer (who sells), Buyer (who buys), User (who uses), Maintainer (who maintains can be

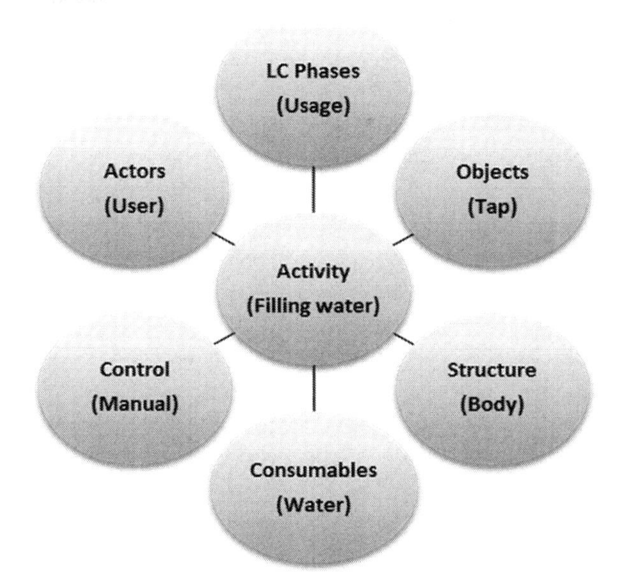

Fig. 2 Factors influencing *filling water activity* in use

user or a different person), Repairer (who repairs), Collector (who collects after use), which are all *actors* performing actions.

The use of the kettle need to be supported by other products (*objects*) like surface on which kettle is placed (table), place where kettle is cleaned (sink), (sponge) with which kettle is cleaned, (tap) from which water is filled, (cup) to which water is poured, (coffee machine) alongside which kettle is placed. Other elements which are consumed (*consumables*) in the process are water, energy, detergent, vinegar, tea which influence the usage.

The *activities* involved in use phase are: making place for kettle, removing packing, reading manual, unwinding the cable, placing the base, connecting the plug, removing any stickers/packing items (till here the activities corresponds to *installation*), opening lid, filling water, rinsing kettle, closing the lid, keeping kettle on base, pressing on/off switch, taking cup, taking tea packet, placing tea packet in the cup, pouring hot water from kettle after boiling, keep kettle on base, adding sugar, disposing tea packet, drinking tea, cleaning cup (these activities are in *usage*), rinsing, wiping, removing scales (corresponds to *maintenance*), fixing chord, replacing parts (*repair* subphase), taking back, donating, reselling (*dispose*). Some of the activities in use are dependent on the *control* (manual/programmable/ automatic) determined by the product's design.

Figure 2 shows an instance at which an activity is influenced by different elements; In usage sub phase, user is performing an activity of filling water into the body from tap and the activity is manually performed. The same activity can be

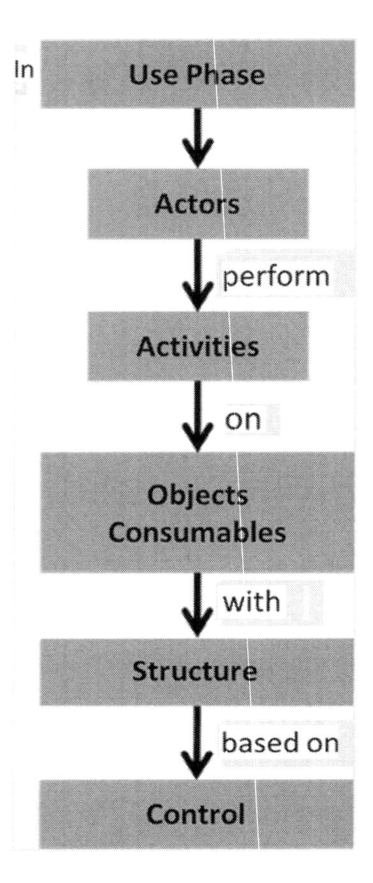

Fig. 3 Activity based model in use phase

differently performed based on the control and will be influenced by objects, structure and consumables. Figure 3 shows the activity based model in use phase; in *use phase*, *actors* perform *activities* on *objects*, and *consumables* with *structure*, based on *control* provided. These are defined as follows:

Life Cycle Phase: A distinctive phase of product life from raw material extraction through after use of the product under consideration.
Actors: Stakeholders involved in each phase of the product life cycle.
Activities: Actions performed by the actors throughout the life cycle of a product. Object: A non-consumable physical entity other than the product used in an activity.
Consumable: a consumable entity other than the product used in an activity.
Structure: Physical construction of the product throughout the life cycle.
Control: Way in which product can be operated.

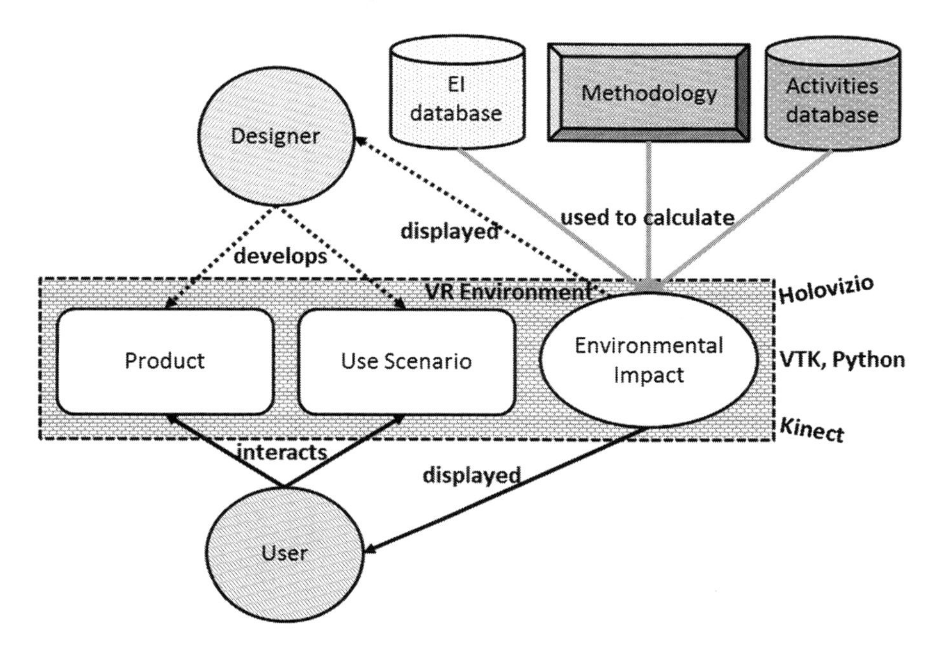

Fig. 4 Proposed platform architecture

4 Platform

The proposed platform will support designer to create and manipulate multiple concepts and scenarios for evaluation by users. It will also support users to use the product virtually for performing basic activities. Information regarding the status of objects, consumables and structure will be collected from the activities performed by user based on the control provided on the product. The support will also help in showing the impacts based on the above information to user and designer so that they can interpret and understand the consequences effectively. Illustrative 3D visualization and interaction of product, activities and the associated environmental impact by users and designers should help in better understanding of the product, users, and strategies during use phase.

The platform uses Holovizio 3D display [20] to show the product in virtual environment to the user and designer via user mode and designer mode. The interactive capability will be achieved by Kinect (a motion capture device) for performing basic activities by users. These two modes will have relevant characteristics based on requirements from users and designers. In Use phase the impact will be calculated based on activities performed and the severity of the results will be shown and explained per activity via intensity of the colour and depth in 3D.

The support can also provide facility to capture requirements, evaluate and select requirements, link between requirement & functions, help in performing activities, record the activities and their characteristics, calculate impact based on data from activities, life cycle databases and impact categories, and shows impact visually. Figure 4 shows the modules in the intended platform. The modules, people (designer and user), product, use, environmental impact visualization will be front-end of VR environment and other modules, like environmental impact database, methodology for assessment and database of activities which will be used in calculating environmental impact before showing to designer and user will be back-end. This platform is going to be implemented using visualization tool kit (VTK) in python programming language with Holovizio display unit and a motion capture setup.

5 Discussion

The platform will be evaluated by asking users to do set of tasks depending on different designs using the support and see how the designs and users contributing to the environmental impact based on different design characteristics and user activities. User feedback on interaction with the product using the proposed support will be collected by a questionnaire to evaluate the usability and effectiveness of the support.

The following will be evaluated using a questionnaire: Given a support to see how the different eco-design strategies actually (virtually) work instead of making prototype. See if providing the product and performing actions virtually help in determining the associated environmental impact (EI) effectively.

- Develop CAD model
  - Without temperature control (auto-off after boiling)
  - With temperature control (stepped/exact)
- Develop use environment
  - With interaction/Without interaction
- Capture details like
  - How many times device is used (allow user creation)
  - How long (show actual time in sec/min on kettle)
  - What capacity each time (Exact/Premeasured (1cup/2cups…etc.))
  - What temperature (Boil/Exact/Switched-off at will)
- Observe and measure
  - User satisfaction level

Can all the relevant actions be performed

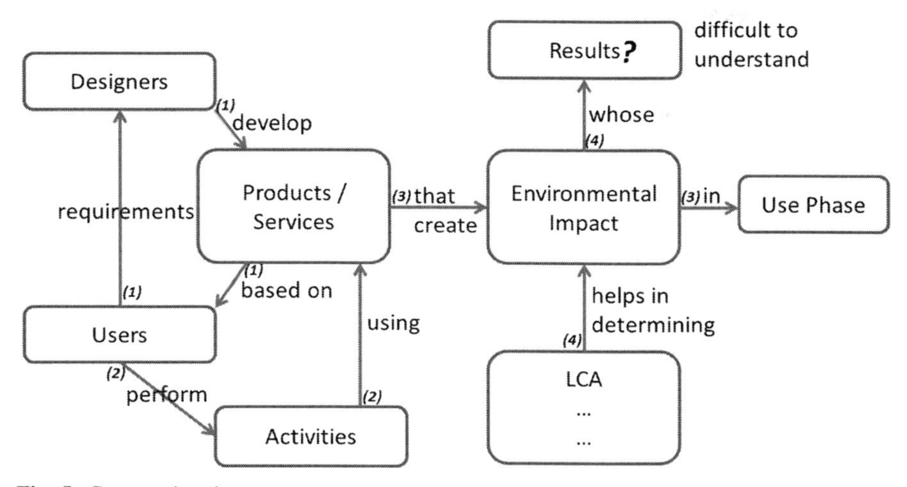

Fig. 5 Current situation

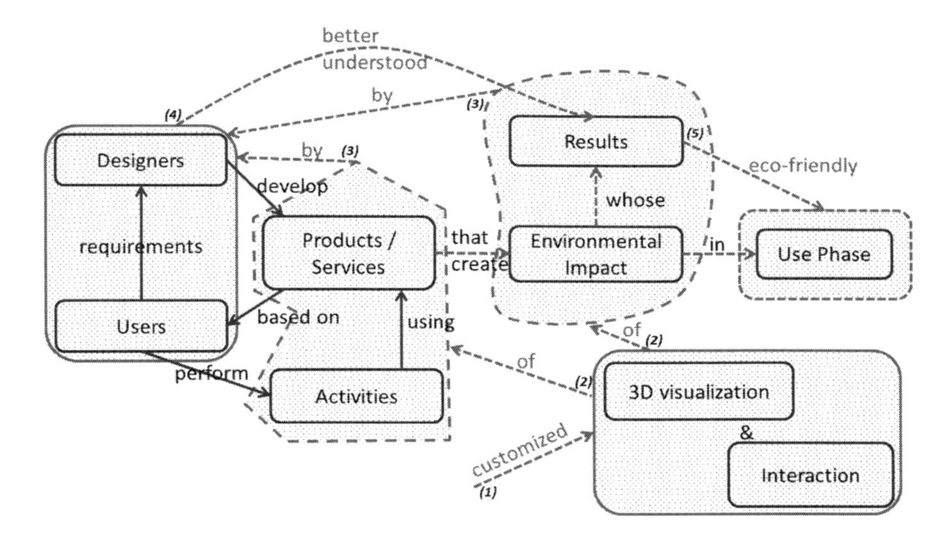

Fig. 6 Future situation

How satisfactorily each action can be performed

– What is the outcome

Planned (Full/Partial)/Un Planed (Good/Bad)

– Cleaning

After each use (water (hot water/normal water)/detergent)
After number of uses/After number of days

– On product information/display control labels/control positioning

What to display/How to display/where to display
What to give for control/How much control/When to give control/How to give control

– Control

Time
Materials/Energy
Desired output
Conditioned input or any input
Temperature control/sensor/Auto-cutoff temperature

– Use (hot water for (drinking / making (tea / coffee)))

Open tap
Fill the kettle with water (how much)
Press on/off switch (wait for water to boil (how much)
Take cup (clean/unclean)
Dispose tea packet (reuse/dispose)
Clean cup (detergent/no detergent/hot water/cold water)

– Maintain

Rinse with (water/detergent/vinegar) (daily/weekly/per number of uses)
Wipe with (damp cloth/dry cloth) (inner body/outer body) (after each use/daily/weekly)
Remove scales (weekly/bi monthly/monthly)

– Repair

Personal (broken cord etc.)
Service centre (parts broken/not working)

– Dispose

Take back/sell/donate/garbage
Reuse/recycle/incinerate/landfill

- Analyse to see the effect on environment

  Reduction in time / reduction in energy
  Exact temperature / exact amount of water
  User satisfaction in terms of reach / view / strategy

- Calculate impact based on "agreed" categories
- Show impact with and without strategy to user so that he can perceive the importance of the way activities done and steer towards eco-friendly usage

Figures 5 and 6 show the current situation and intended future situation using the platform. Designers develop products/services based on user requirements (1). Users perform activities using products/services (2) that create environmental impact in use phase (3). Life cycle assessment helps in determining the environmental impact whose results difficult to understand (4) that can be detrimental to producing eco-friendly products and their use. We want to better the understanding of the results by introducing customized 3D visualization and interaction of products, activities and associated environmental impacts and thus help in eco-friendly use phase.

The proposed support will lead to future situation in Figure 6 as customized 3D visualization and interaction (1) of products/services, activities and environmental impacts, results (2) by designers and users (3) will lead to better understanding of results (4) which helps in eco-friendly use phase (5)

6 Conclusions and Future work

It is important to bring together designer and user in design and assessment for creating environmentally friendly product usage. There are challenges associated with bringing different stakeholders together and need new methods and technologies to be adapted while developing support. 3D visualization and interaction can help in better understanding of the product and its use and will reduce the ambiguity in environmental impact results by presenting the relevant information effectively. The empirical study helped in identifying the elements and requirements in use phase that lead to development of the activity model. In use, impact is based on activity and it is influenced by actors, structure, objects, consumables and the control provided on the product. The knowledge generated will be used for implementing the conceptual support outlined here. The support will be evaluated using the proposed plan outlined in the discussion section in future work.

Acknowledgments This work is supported by Agence Nationale de la Recherche (ANR, The French National Research Agency) funding (project number: 1081C0212/ANR-10-ECOT-006-01).

References

1. Consoli F (ed) (1993) Guidelines for life-cycle assessment: a 'code of practice. Society of Environmental Toxicology and Chemistry (SETAC), Brussels
2. Wenzel H, Hauschild M, Alting L (1997) Environmental assessment of products, vol 1. Chapman and Hall, London

3. Reap J, Roman F, Duncan S, Bras B (2008) A survey of unresolved problems in life cycle assessment. Int J Life Cycle Assess 13:374–388
4. Cara DB, Rodney AS, Kelly F (2011) A novel mixed method smart metering approach to reconciling differences between perceived an actual residential end use water consumption. J Cleaner Prod doi:10.1016/j.jclepro.2011.09.007
5. Geppert J, Stamminger R (2010) Do consumers act in a sustainable way using their refrigerator? The influence of consumer real life behaviour on the energy consumption of cooling appliances. Int J Consum Stud 34(2):219–227
6. Zhou F, Xu Q, Jiao RJ (2011) Fundamentals of product ecosystem design for user experience. Res Eng Design 22:43–61
7. Sauer J, Wiese BS, Ruttinger B (2002) Improving ecological performance of electrical consumer products: the role of design-based measures and user variables. Appl Ergon 33(4):297–307
8. Sauer J, Wiese BS, Ruttinger B (2003) Designing low-complexity electrical consumer products for ecological use. Appl Ergon 34(6):521–531
9. Lu SCY, Shpitalni M, Gadh R (1999) Virtual and augumented reality technologies for product realization. Annals CIRP 48(2):471–494
10. Ronchi AM (2009) Date visualization and display technologies in ECulture: cultural content in the digital age. Springer, Berlin, pp 113–137
11. Sadeghi S, Masclet C, Noel F (2012) Gathering alternative solutions for new requirements in manufacturing company: collaborative process with data visualization and interaction support. In: 45th CIRP conference on manufacturing systems, 16–18 May 2012, Athens, Greece
12. Guidi G, Micoli LL (2011) A semi-automatic modeling system for quick generation of large virtual reality models. In: Proceedings of the ASME world conference on innovative virtual reality WINVR2011, 27–29 June, 2011, Milan, Italy
13. Lenne D, Thouvenin I, Aubry S (2009) Supporting design with 3D-annotations in a collaborative virtual environment. Res Eng Design 20:149–155
14. Tanriverdi V, Jacob RJK (2001) VRID: a design model and methodology for developing virtual reality interfaces. In: VRST'01, 15–17 Nov 2001, Banff, Alberta, pp 175–182
15. Weidlich D (2007) Virtual reality approaches for immersive design. Annals CIRP 56(1):139–142
16. Stark R, Israel JH, Wöhler T (2010) Towards hybrid modeling envrionments-merging desktop-cad and virtual reality technologies. CIRP Annals—Manuf Technol 59:179–182
17. Cabral M, et. al. (2007) An experience using X3D for virtual cultural heritage. In: Web3D, 15–18 April 2007, Perugia, Italy, pp 161–164
18. Hirte S. et. al. (2012) Data3—a kinect interface for olap using complex event processing. In: IEEE 28th international conference on data engineering, 1–5, April, 2012, DC, USA, pp.1297–1300
19. Chen LH, Lee CF (2008) Perceptual information for user-product interaction: using vacuum cleaner as example. Int J Des 2(1):45–53
20. http://www.holografika.com/Products/HoloVizio-128WLD.html

Design of Product Service Systems at the Base of The Pyramid

Santosh Jagtap and Andreas Larsson

Abstract The Base of the Pyramid (BoP) consists of about two-fifths of the world population. This population can be categorized as poor with income of less than 2 dollars per day. It is important to alleviate poverty. One of the promising approaches to tackle the wicked problem of poverty is business development combined with poverty alleviation. In this approach, integrated solutions are necessary in order to address the diverse issues in the BoP. These integrated solutions are in the form of product service systems (PSS) rather than the conventional product-oriented or service-oriented solutions. In this paper, we explore different issues that need to be addressed in the PSS design at the BoP. We have also explored strategies used in this PSS design. We have used a case study to explain these issues and strategies. In addition, we have identified salient characteristics of the PSS design at the BoP.

Keywords Product service systems · Design at the BoP · Design for sustainability

1 Introduction

The base of the world income pyramid, generally called the 'Base of the Pyramid' (BoP), consists of poor people. About two-fifths of the world population can be categorized as poor. Their income is less than 2 dollars per day. Many researchers

S. Jagtap (✉) · A. Larsson
Innovation Engineering, Department of Design Sciences, Faculty of Engineering, Lund University, Lund, Sweden
e-mail: Santosh.Jagtap@design.lth.se

A. Larsson
e-mail: Andreas.Larsson@design.lth.se

A. Chakrabarti and R. V. Prakash (eds.), *ICoRD'13*, Lecture Notes in Mechanical Engineering, DOI: 10.1007/978-81-322-1050-4_46,

prefer the poverty line of 2 dollars per day [1]. About a fifth of the world population is classified as extremely poor with income of less than 1.25 dollars per day.

Poverty is multifaceted, and has three intertwined characteristics as follows [1]: (1) Lack of income and resources required to satisfy basic necessities such as food, shelter, clothing, and fuel; (2) Lack of access to basic services such as public health, education, safe drinking water, sanitation, infrastructure, and security; and (3) Social, cultural, and political exclusion.

1.1 The Fight Against Poverty

It is important to alleviate poverty. Poverty is a trap—children born to poor parents are likely to grow up to be poor adults. Mahatma Gandhi often said—poverty is the worst form of violence.

Karnani [1] has analysed poverty reduction approaches. Since the 1990s, three different poverty reduction approaches have received attention. The first approach, namely 'microcredit', envisioned by Muhammad Yunus in Bangladesh, suggests that granting small loans to the poor can help them to grow their businesses, and thereby can help them to climb out of poverty [2]. The second approach of granting formal property rights to the poor was formulated by the Peruvian economist Soto [3]. The underlying principle of this approach is that property rights will give the poor access to credit, and thereby will help in poverty reduction. The third approach, popularised by the late Prahalad [4] proposes solutions involving business development combined with poverty alleviation. These solutions, consisting of business strategy, focus on the poor people as producers and consumers of products and services. The third approach can use the elements of the first approach (i.e., microcredit), and is powerful as it efficiently uses the resources of businesses.

1.2 Business Development Combined with Poverty Alleviation

In this paper, we focus on the third approach of business development combined with poverty alleviation. According to Prahalad and Hart [5], the most visible and prolific writers in the area of the BoP, this business strategy is important in "...lifting billions of people out of poverty and desperation, averting the social decay, political chaos, terrorism, and environmental meltdown that is certain to continue if the gap between rich and poor countries continues to widen." This suggests that such a business strategy offers a potential approach to meet the challenges of social, economic, and environmental sustainability at the BoP.

1.3 BoP People: Producers and Consumers

In this paper, we focus on the BoP people as producers and consumers of products and services. The businesses at the BoP design and develop products and services to serve the BoP producers, and to market these products and services to the BoP consumers. By focusing the poor as producers, their income can be raised. This can also help to generate employment opportunities for them, and can alleviate poverty. There are two ways to focus on the BoP people as consumers.

- First involves tapping BoP markets by selling products and services to the poor with the primary aim of earning profits. This approach may not be sustainable, and may not help to alleviate poverty. Karnani [6] has rigorously argued that this approach cannot alleviate poverty, and that it can exploit the poor.
- In the second way, businesses aim at the development of the poor and accordingly market appropriate products and services to them. In this approach, innovative solutions are devised to seek financial sustainability combined with the development of the poor.

In this paper, we focus on the second way when the BoP people are seen as consumers.

2 Research Aims and Research Methodology

In the approach of business development combined with poverty alleviation, some interventions are designed, developed and implemented in the BoP. These interventions need to address complexly intertwined issues (i.e. constraints)—such as poor physical infrastructure, lack of knowledge and skills of the poor, etc—in the BoP. In order to address these diverse issues in the BoP through the design and development of interventions, an integrated approach using knowledge from technical, social and management sciences is necessary [7, 8]. In this integrated approach, the interventions take the form of product–service systems (PSS) rather than the conventional product-oriented or service-oriented solutions. PSS consists of a set of products and services that jointly fulfill the needs of users. PSS can reduce the use of resources and the generation of waste as fewer products are manufactured.

While some authors have highlighted the need and importance of PSS at the BoP, much less work has been carried out in this area. In this paper, the following questions are explored.

- What are the different issues that need to be addressed in the PSS design at the BoP?
- What are the strategies used in the PSS design at the BoP?
- What arethe salient characteristics of the PSS design at the BoP?

We have explained these issues and strategies by using a case drawn from the study of United Nations Development Programme (UNDP) [9]. The UNDP led an initiative called 'Growing Inclusive Markets' (GIM). In this initiative, they analyzed several cases from different sectors (e.g., energy, healthcare, etc.) and countries. In order to explain the issues and strategies in the PSS design at the BoP, we selected one case from the UNDP study. This case is about a project where a for-profit company, Afrique Initiatives, used Information and Communication Technologies (ICTs) to monitor health conditions of children from low-income families in Mali.

Based on our literature review in the area of the BoP, we have identified some salient characteristics of the PSS design at the BoP.

3 PSS Design at the BoP: Issues and Strategies

In our prior research, we synthesized issues and strategies in the PSS design at the BoP [8, 10, 11]. We pulled together issues and strategies in this PSS design from the reviewed literature [5, 6, 12, 13]. We compared these issues and strategies with those identified in the UNDP study [9]. We used this UNDP study as a reference because the sample size of cases analysed in this study is large, and these cases are drawn from different sectors and countries. Our research identified that the issues and strategies of the UNDP study are comprehensive and include those identified in the reviewed literature.

3.1 Issues in PSS Design at the BoP

The issues in PSS design at the BoP are as follows.

- *Market information* This issue takes into account the knowledge of businesses regarding the BoP, for example, what the poor need, what capabilities the poor can offer, etc. Businesses often lack detailed information about the BoP markets and in particular about the rural BoP. The presence of intermediaries (e.g., market research, rating services) to consolidate or distribute information on the BoP cannot be assumed.
- *Regulatory environment* The regulatory frameworks are under- or un-developed in the BoP. In addition, enforcement of the existing rules is inadequate. Complying with the bureaucracy in developing countries can be time consuming and monetarily expensive. For example, in the Latin America and the Caribbean, opening a business takes about 73 days, and in Organisation for Economic Co-operation and Development (OECD) countries, it takes on average 17 days [9].
- *Physical infrastructure* This issue considers the inadequate infrastructure (e.g., roads, electricity, water and sanitation, hospitals, etc.) in the BoP.

In developed countries, the logistics system that is necessary for accessing consumers, selling to them, and servicing products exists, and only minor changes may be required for specific products. In the BoP, the existence of a logistics infrastructure cannot be assumed. PSS at the BoP need to work in hostile environment (e.g., noise, dust, abuse of products).

- *Knowledge and skills* The poor, generally, are illiterate and do not possess knowledge and skills regarding the availability of products, usage of products, etc. Furthermore, this lack of knowledge and skills inhibits them from starting their own businesses. PSS design at the BoP needs to take into account the skill levels of the poor. The heterogeneity of the BoP regarding language, culture, skill level, and prior familiarity with the functions or features of the PSS can be a challenging task in the PSS design.
- *Access to financial services* The poor lack access to credits, insurance products, and banking services. This puts limits to the purchases made by them. In addition, they cannot protect their meager assets from events such as illness, drought, etc. The PSS design at the BoP must take into account the price-performance relationship.

3.2 Strategies in PSS Design at the BoP

The strategies in PSS design at the BoP are as follows.

- *Adapt products and processes* This strategy includes product redesign, business model innovation, and technological adaptation. PSS design at the BoP can benefit from technological 'leapfrogging'—that is—avoiding intermediate steps to replace poor technology with the state of the art. While technology helps to deal with the daunting challenges in the BoP, it needs to go hand-in-hand with innovations in business models.
- *Invest in removing market constraints* This strategy includes investing for: educating consumers; enhancing or building capacities of the poor (e.g., supporting small producers who form a part of the supply chain); and building social marketing (e.g., health campaigns to increase demand of malaria nets).
- *Leverage the strengths of the poor* This strategy builds on the knowledge, networks, and abilities of the poor and their communities (e.g., developing cooperatives of the poor, employing the poor to fulfill some tasks of a business, leveraging the knowledge of the poor to design and develop PSS).
- *Combine resources and capabilities* Through collaborations and partnerships, this strategy combines resources and capabilities of different organizations such as businesses, NGOs, charitable sector, local governments, etc.
- *Engage in policy dialogue with governments* Businesses can overcome different issues in the BoP by engaging in dialogue with relevant governments, and this can help, for example, to formulate appropriate regulations, reduce bureaucracy, etc.
- PSS design at the BoP uses one or more of these above five strategies, which address one or more of the applicable issues.

4 Case Study: Pésinet's PSS in Healthcare Sector from Mali

In Mali, one child out of five dies before fifth birthday, and about 43 % of children are underweight. There is limited access to modern healthcare. The limited number of trained doctors and nurses worsen the problems. Furthermore, 40 % of the population lives more than 15 km away from a health facility.

More than 50 % of child mortality in Africa can be prevented. 55 % of children's mortality-causes can be detected easily by periodically checking basic symptoms, for example the evolution of the weight of a child, which is an accurate indicator of young child's health status. Patients in Africa usually come too late to a doctor. If diseases are detected and treated earlier on, mortality can be substantially reduced. This can also avoid risky and expensive emergency treatment, and health spending for households would be lower.

Afrique Initiatives, a for-profit company, focuses on investing in small- and medium-sized African businesses in order to promote sustainable enterprise and private sector development. Afrique Initiatives has the following main focus areas: education and training, nutrition, health, and information technologies (IT). Afrique Initiatives established an organisation called Pésinet with the aim of monitoring health conditions if children from low-income families. Pésinet implemented an intervention in Coura, a region near the capital city Bamako in Mali. The challenges faced by the people in this region are: malaria, low income, poor or no literacy, poor sanitation, and lack of adequate water supply infrastructure. Affordability is a crucial issue as the average income of the families in this region is less than 4 US dollars a day.

This intervention, implemented by Pésinet in the region Coura, monitored health conditions of children in that region, and helped to reduce child mortality rate. This intervention addressed the complexly intertwined issues in the region. The success of this intervention can be attributed to the fact that it was in the form of a PSS rather than the conventional product-oriented or service-oriented solutions. Cocreation played an important role in the design and development of this PSS. Pésinet cocreated the PSS with the following partners.

- A drug distributor, Medex from Mali
- The people from the region Coura
- An NGO, Kafo Yeredeme Ton from Mali
- Two French universities, ESSEC Business School and Ecole Centrale Paris
- Two major French telecommunications companies, Alcatel-Lucent and Orange

4.1 The Broader PSS Concept

In order to monitor health conditions of children in Coura, Pésinet used Information and Communication Technologies (ICTs). The weight of a child is used as an indicator of its health. The pattern in the weight-change is analysed by a doctor to identify anomalies, if any, in the child's health. The child's mother subscribes to Pésinet's services by paying nominal fees. A representative from the Pésinet weighs her child(ren) twice a week. The information on this weight is transmitted to a local doctor using SMS service of mobile phones. After reviewing the weight chart, the doctor requests the visit of the mother and child if any anomalies are identified. Pésinet implemented the project in 2007. This project benefited hundreds of children.

4.2 Elements of the PSS

The elements of the PSS, aimed at monitoring the health conditions of children from Coura, are as follows.

4.2.1 Products

In Mali, there are more mobiles phones than land lines, and this helps to contact remote areas. This fact was used by Pésinet in the PSS design. The products involved the use of ICTs. Required technical systems, consisting of the use of mobile phones and specific applications for data transmission, were designed by Alcatel-Lucent and Orange. The weight reading of a child is sent to a centralized database using these applications. A computer application that can be used by doctors processes this data on weight readings, and presents the evolution of the child's weight in a visual format. The doctor sends an SMS to the Pésinet representative when he/she identifies an anomaly in the evolution of the child's weight. The Pésinet representative then provides the mother with a consultation voucher.

Mali was selected because Alcatel-Lucent and Orange were already settled in the country. Alcatel-Lucent and Orange offered their support to Pésinet on their own, motivated by the 'digital divide' issue. While their involvement was as a part of their Corporate Social Responsibility, Orange gets some profits, as it charges for the SMS sent by the doctors.

4.2.2 Business Model

A team of students from the ESSEC Business School and the Ecole Centrale designed the business plan (e.g., subscription fee-structure to cover the operating costs and child sponsoring for low-income families). In Mali, the subscription fees are 500 Communaute Financiere Africaine (CFA) per month (about US$1.05) and include access to medicines. These fees entirely cover the operation costs, consisting of the Pésinet representatives' and doctor's wages, the scales renewal, the internet connection, medicines, and the salary of one manager. The doctors are employed by public hospitals, and are either directly compensated by Pésinet or indirectly through a financial contribution made to the hospitals.

4.2.3 Marketing and Awareness Program

Pésinet's staff organised small parties to inform the community in Coura regarding issues such as diarrhoea, cholera, etc. In the initial phase of the awareness raising program, the doctors went to the elementary schools and held their consultations with mothers and their children. Pésinet then kept promoting its services through word of mouth and District Associations.

The NGO, Kafo Yeredeme Ton from Mali, raised awareness by carrying out the marketing campaign together with Pésinet, by going door-to-door for a couple of months. The involvement of women community leaders in the programme helped to raise the awareness.

The government and local authorities were informed but not directly involved. The trust of local hospitals in Mali was achieved by informing them that the Pésinet's service would be a complementary service aimed at achieving early monitoring of children, and that the children with health issues would seek help from the regular healthcare system.

4.2.4 Services

Mothers subscribe to Pésinet's service by paying a nominal fee. This service consists of weighing her children once a week at her home (twice a week for children under one) plus advice and treatment (if required) by a doctor. Children are weighed by local women called Pésinet representatives (see Fig. 1). These representatives also register symptoms such as fever, diarrhoea, vomiting, and transmit the data to the doctors using SMS service of mobile phones. The doctor makes a decision regarding the visit of the mother and child based on the pattern in the change of the child's weight. One doctor can cover about 2000 children. When the doctor identifies an anomaly in the child's health, he/she sends an SMS to the Pésinet representative, who requests the mother and child to visit the doctor.

4.3 Transformation

The Pésinet's PSS helped to achieve the following changes of state. The Pésinet's service resulted in approximately 20 consultations per week for 400 children. This service helped mothers to get necessary advice and treatment for their children in a short time-scale.

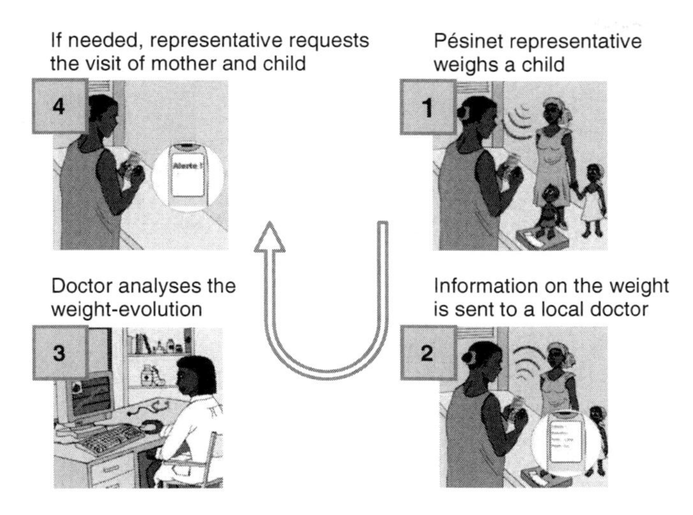
If needed, representative requests the visit of mother and child
4
Pésinet representative weighs a child
1
Doctor analyses the weight-evolution
3
Information on the weight is sent to a local doctor
2

Fig. 1 Service delivery in the case of pésinet's PSS—adopted from: [9], [14]

In September 2000, world leaders adopted the United Nations Millennium Declaration. This declaration is about committing to a new global partnership to alleviate extreme poverty and fulfilling time-bound targets with a deadline of 2015. These targets are known as the Millennium Development Goals (MDGs). One of the MDGs is to reduce child mortality. The changes of state, attributed to the Pésinet's PSS in Mali, can be interpreted as a step towards achieving the MDG of reducing child mortality.

In the case of the Pésinet's PSS in Mali, the UNDP study identified that the PSS addressed the following main issues—'physical infrastructure' (e.g., limited access to healthcare facilities) and 'knowledge and skills' (e.g., poor or no literacy). The study identified that the strategies used in the PSS design were: 'adapt products and processes' (e.g., use of mobile phones to transmit weight reading of a child), 'leverage the strengths of the poor' (e.g., involvement of local women as Pésinet representatives), and 'combine resources and capabilities' (e.g., partnership with Alcatel-Lucent and Orange, two French schools, a drug distributor and a local NGO).

5 Salient Characteristics of PSS Design at the BoP

Vasantha et al. [15] review of PSS design methodologies shows that the PSS design is still in initial stages of development, and that the PSS research is not mature. While most of the studies in the PSS are focused on developed countries, much less work has been carried out in the PSS at the BoP. Based on our literature review in the area of the BoP, we can note some salient characteristics of the PSS design at the BoP.

The PSS design in mature and developed country markets is driven by factors such as: customers want availability or capability rather than the purchasing of physical artefacts [16]; companies can establish long-term relationship with customers [17]; companies can obtain improved knowledge regarding the product use [18]; and growing concerns for the environment. Furthermore, a key criterion for evaluating the PSS in these markets is user experience. In contrary, at the BoP, addressing the complexly intertwined issues requires integrated solutions in the form of the PSS. One of the key aims of the PSS design at the BoP is to alleviate poverty through economic, environmental, and social development. The PSS at the BoP needs to be evaluated using criteria such as: satisfaction of un-met or underserved needs of the poor, increase in their income, a step towards the achievement of one or more of the MDGs, etc.

Cocreation and combining resources and capabilities of different partners including the poor people and non-traditional partners such as NGOs and charitable sector play a crucial role in the PSS design at the BoP (see Fig. 2). This cocreation helps in different stages of the PSS design and development such as: gaining information on BoP markets, sales, distribution, logistics, etc.

In tackling the challenges of poverty, the actors—businesses, governments, and civil society—have tended to view each other through the lens of negative stereotypes: businesses are exploitative, governments are corrupt and inefficient, and civil society is naive and ineffective [1]. The cocreation in PSS design at the BoP is in contrast to this negative stereotype. In this cocreation, the actors take a positive view as follows: businesses have resources and are efficient; governments have the power; and civil society has passion and energy.

Jagtap et al. [11]analysed the data available in the UNDP study to gain quantitative findings on the issues and strategies in PSS design at the BoP. They analysed the data on 48 cases from the UNDP study. One of the key findings of their analysis is that—in the PSS design, the strategy 'combine resources and capabilities' is predominantly used. 65 % of the 48 cases have used this strategy in PSS design.

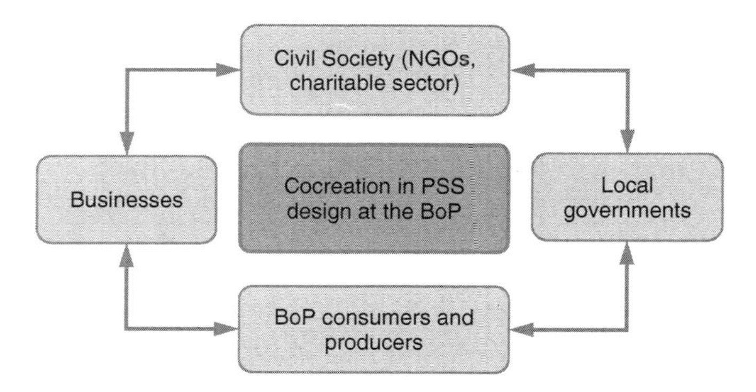

Fig. 2 Cocreation in PSS design at the BoP

Jagtap et al. [11]analysis of the data available in the UNDP study shows that the issue 'knowledge and skills' has frequently been addressed in the PSS design at the BoP (79 % of 48 cases). This suggests that this issue is ubiquitous in the BoP. This can be attributed to the prevalent lack of education in the BoP. Although, the field of education appears to be the responsibility of governments, the businesses have generally not used policy dialogue with the governments to address the issue 'knowledge and skills'. Instead, they have prominently used the strategies 'invest in removing market constraints' and 'combine resources and capabilities' to address this issue.

6 Summary and Conclusions

One of the promising approaches to tackle the wicked problem of poverty is business development combined with poverty alleviation. In this approach, integrated solutions are necessary in order to address diverse issues in the BoP. These integrated solutions are in the form of PSS rather than the conventional product-oriented or service-oriented solutions.

In this paper, we explored different issues that need to be addressed in the PSS design at the BoP. These issues are: 'market information', 'regulatory environment', 'physical infrastructure', 'knowledge and skills', and 'access to financial services'. We have also explored strategies used in this PSS design. These strategies are: 'adapt products and processes', 'invest in removing market constraints', 'leverage the strengths of the poor', 'combine resources and capabilities', and 'engage in policy dialogue with governments'.

We have explained these issues and strategies by using a case drawn from the UNDP study. This case is about a project where a for-profit company, Afrique Initiatives, used ICTs to monitor health conditions of children from low-income families in Mali.

In some aspects, the PSS design at the BoP appears to be different from that in the mature and developed country markets. The PSS design at the BoP is driven by the need to address complexly intertwined issues in the BoP markets. A key criterion for evaluating the PSS in mature and developed country markets is user experience. In contrary, the PSS at the BoP needs to be evaluated using criteria such as: satisfaction of un-met or under-served needs of the poor, increase in their income, a step towards the achievement of one or more of the MDGs, etc.

Cocreation and combining resources and capabilities of different partners including the poor people and non-traditional partners such as NGOs and charitable sector play a crucial role in the PSS design at the BoP. It is important to understand the process of PSSdesign at the BoP, and this can be an area of future research.The issue 'knowledge and skills' is prevalent in the BoP. This issue needs to be addressed in the PSS design, and can pose challenges is cocreation of PSS with the BoP people. There is a need of simple and easy to use methods and tools that can help in the cocreation of PSS with the semiliterate or illiterate BoP people.

Acknowledgments This work was partly financed by VINNOVA within the Product Innovation Engineering program (PIEp).

References

1. Karnani A (2011) Fighting poverty together: rethinking strategies for business, governments, and civil society to reduce poverty. Palgrave Macmillan, New York
2. Yunus M, Weber K (2007) Creating a world without poverty: social business and the future of capitalism. Public Affairs, USA
3. de Soto H (2010) The mystery of capital: why capitalism triumphs in the west and fails elsewhere, New York: Basic Books
4. Prahalad CK (2004) The fortune at the bottom of the pyramid: eradicating poverty through profits. Wharton School Publishing, NJ
5. Prahalad CK, Hart SL (2002) The fortune at the bottom of the pyramid. Strategy and Business 54–54.
6. Karnani A (2007) Misfortune at the bottom of the pyramid. Greener Manage Int 51:99–110
7. Kandachar P, Halme M (2008) Farewell to pyramids: how can business and technology help to eradicate poverty? In: Kandachar P, Halme M (eds) Sustainability challenges and solutions at the base of the pyramid. Greenleaf Publishing Limited, Sheffield
8. Jagtap S, Kandachar P (2010) Representing Interventions from the base of the pyramid. J Sustain Develop 3(4):58–73
9. UNDP (2008) Creating value for all: strategies for doing business with the poor. Available from: http://www.growinginclusivemarkets.org/reports
10. Jagtap S, Kandachar P (2011) Design for the base of the pyramid: issues and solutions. In: International conference on research into design (ICoRD '11), Bangalore, India
11. Jagtap S, Larsson A, Kandachar P (2012) Design and development of products and services at the base of the pyramid: a review of issues and solutions. Int J Sustain Soc (IJSSoc) 5(3)
12. Anderson J, Markides C (2006) Strategic innovation at the base of the economic pyramid [cited 2008 21 November]; Available from: http://www.jamieandersononline.com/uploads/ANDERSON_MARKIDES_SI_at_Base_of_Economic_Pyramid_FINAL.pdf
13. Keating C, Schmidt T (2008) Opportunities and challenges for multinational corporations at the base of the pyramid. In: Kandachar P, Halme M (eds) Sustainability challenges and solutions at the base of the pyramid. Greenleaf Publishing Limited, Sheffield
14. Pésinet [cited 2012 16 March]; Available from: http://www.pesinet.org/wp/
15. Vasantha GVA et al (2011) A review of product–service systems design methodologies. J Eng Desi 23:1–25
16. Military of Defence (2005) Defence industrial strategy. London: Her Majesty's Stationery Office
17. Vandermerwe S (2000) How increasing value to customer improves business results. MIT Sloan Manage Rev 42(1):27–37
18. Alonso-Rasgado T, Thompson G, Elfström B (2004) The Design of functional (total care) products. J Eng Des 15(4):515–540

Re-Assignment of E-Waste Exploring New Livelihood from Waste Management

P. Vivek Anand, Jayanta Chatterjee and Satyaki Roy

Abstract E-Waste is considered as one of the most difficult wastes to manage, not only to reuse and recycle them, but also to detoxicate and make them harmless to the environment. This not only has an effect on the immediate environment, but a lot of valuable metals in E-Waste are getting wasted in landfills. Even recycling, reusing and extracting valuable metals from E-Waste requires special equipment and consumes a lot of energy, contradicting the very reason for effective management of resources. In this paper, the author has explored product re-assignment as a method to handle E-Waste, i.e. making products with E-Waste as the raw material. The author has narrowed his focus to Printed circuit boards (PCBs) for this study and developed methods to handle PCBs and make products using constructive design method. These highly finished, pre-treated and safe products were designed not only to put PCBs to use, but also such that they can be made with basic hand tools (like a drilling machine, hand saw, Pliers, etc.), minimum training and minimum production time; appropriate for cottage industries and to make new and easy livelihood from waste management. Although this paper proposes only products made of PCBs, it concludes with discussing on employing such re-assignment designs on other parts of E-Waste (not only PCBs) and in managing other types of waste.

P. Vivek Anand (✉)
Design Programme Indian Institute of Technology, Kanpur 208016, India
e-mail: ar.anandvivek@gmail.com

J. Chatterjee
Design Programme/Department of Industrial Management and Engineering, Indian Institute of Technology, Kanpur 208016, India
e-mail: jayanta@iitk.ac.in

S. Roy
Design Programme/Department of Humanities and Social Sciences, Indian Institute of Technology, Kanpur 208016, India
e-mail: satyaki@iitk.ac.in

A. Chakrabarti and R. V. Prakash (eds.), *ICoRD'13*, Lecture Notes in Mechanical Engineering, DOI: 10.1007/978-81-322-1050-4_47, © Springer India 2013

Keywords Waste · Waste management · E-Waste · WEEE · Printed circuit boards · Reassignment · Curio · Beautility · Constructive design research · Probes

1 The Overview

"Electronic waste" may be defined as discarded computers, electronic equipment, phones, televisions, refrigerators, etc. This can also include used electronics, which are destined for reuse, resale, salvage, recycling, or disposal [2].

Out of 400 million tons of waste produced every year from over 60 identified types of wastes, E-Waste contributes 40 million tons, i.e. 10 % of the total waste produced (2011). And is estimated to reach 93 million tons by the year 2016 (5 years from 2011). [1].

Waste electrical and electronic equipment (WEEE) is one of the fastest growing special waste types with an estimated growth of 3–5 % per year. WEEE is a very heterogeneous waste type that contains many compounds that are considered to be harmful to both humans and the environment, as well as many metals that have the potential of being recycled and reused [2].

1.1 Where is All the E-Waste Going?

E-Waste including extremely hazardous waste like radioactive material, toxic heavy metals and poisonous components are regularly being transported across continents. Most of the E-Waste is being exported to developing countries like India, China, Pakistan and Malaysia [1], as shown in Fig. 1, where laws to protect workers and the environment are inadequate or not enforced. It is also cheaper to 'recycle' waste in these countries.

Landfilling, incineration, recycling, reusing and pulverizing few of the practices followed to treat E-Waste, where landfilling and incineration are highly practiced because they are cheaper and easier. Improper landfilling practices and incineration are extremely harmful for the environment and recyclers working with and around the recycling yards.

2 Understanding the Problem

It is apparent in the earlier paper that a problem exists, and there is a need to handle this effectively in these developing countries. Rather than attributing this problem to nations and laws, the root cause of the problem can be uncovered with reverse investigation, where the user plays a major antagonist in this scenario.

Fig. 1 World's E-Waste export sites [1]

Understanding the life cycle of electronics beyond their usage played a crucial step in the research to get insights to understand the human behavior and to approach the problem. This approach to research with interviews and investigation in constructive design research was conducted to retrieve vital inputs not only in analytical possibilities but also in procuring the raw materials for the constructive studies.

2.1 The Game of E-Waste

For a better understanding, a typical lifecycle of E-Waste was plotted in a board game structure, resembling 'Snakes and Ladders' shown in Fig. 2, where the ladders denote proper usage of the resources; and the snakes denote misuse and wastage of resources. The board game not only can be used as a part of this study but also can be used as a learning tool for all ages about E-Waste.

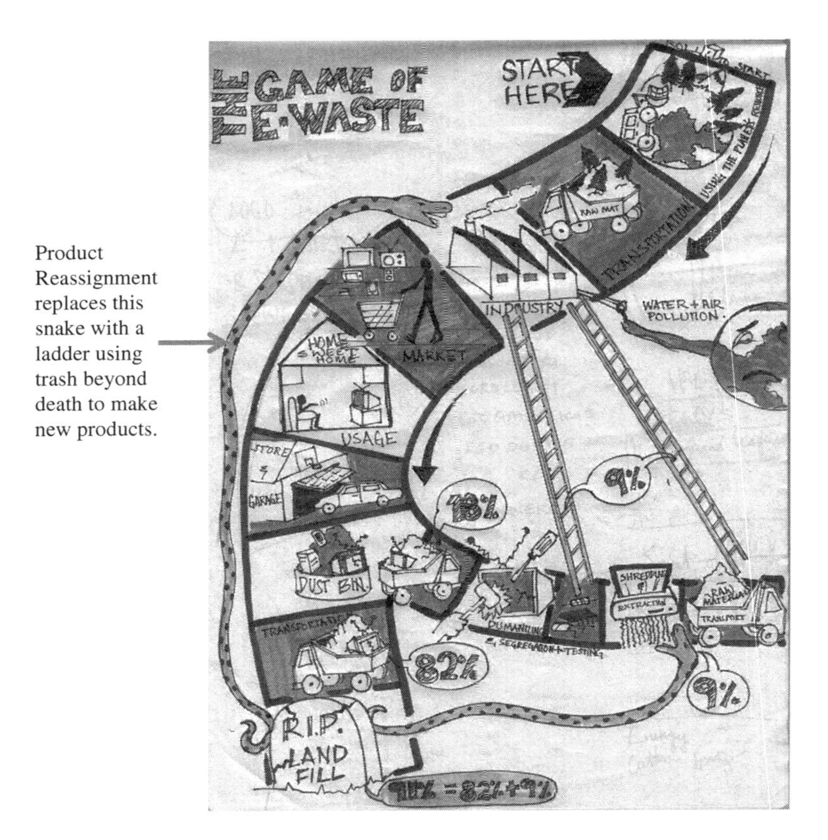

Fig. 2 The game of E-Waste

The board game based on interviews and constructive probing with the users, covered the story of E-Waste from the attaining of raw materials from nature, to usage patterns, recycling and land filling practices in detail. This exercise gave a considerable human and real perspective to the study, unlocked many opportunities for the ideation and solutions to flow.

2.2 How E-Waste can be Handled?

Using 'The game of E-Waste' as a design probe was used to initiate dialogue and discussion with the users followed by in-house brainstorming sessions. The tuning-in stage is for outlining the objective of design and understanding the phenomenon to be studied by making guesses about the causes and possible solutions. The tuning-in consists of preliminary study of the target group and phenomenon, recording the preconceived views, and examining the design challenge. These constructive design methods gave a better evaluation of the problem and fairly grouped handling methods at three stages,

1. At the manufacturing level
2. At the usage level
3. After the waste is produced

2.2.1 Handling E-Waste After the Waste is Produced

Based on the kind of measures to be taken to handle E-Waste, discarded electronics can be handled in three distinct ways. Firstly, ban hazardous waste imports and exports, secondly, proper detoxication and treatment of E-Waste before it is dumped in the landfills and thirdly, reusing and recycling.

Reuse and recycling as a huge domain include various measures that can be acted at many levels to make recycling and reusing much effective. These measures can be widely grouped in the following three groups.

- Better segregation, testing and collection
- Better reusing such that less goes to land fills, and
- Product reassignment

Extensive research was conducted to improve Reusing, testing and segregation of E-Waste. But product reassignment as a solution is not practiced widely to handle the problem of E-Waste, due to lack of design inputs and the gap between the market and production. In this study the author have attempted to design reassigned products to E-Waste, not only to put E-Waste to use, but also such that they can be made with hand tools (like a drilling machine, hand saw, pliers, etc.), minimum training and minimum production time; appropriate for cottage industries and to make new and easy livelihood from waste management.

2.3 Product Reassignment

Products in certain circumstances are transformed or reassigned from their original function and purpose into something entirely novel, a new service or utility not thought of by the original designer and manufacturer. Product 'reassignment' or 're-purposing' is a reconfiguration of an entire product through software or hardware addition that enables it to perform entirely new tasks. Not to be confused with recycled products, which are made from reprocessed materials.

3 Understanding E-Waste

E-Waste as a huge domain comprises of various types of devices and range of appliances from refrigerators to computers. This Huge domain was discussed and brainstormed on how to make it easy to handle and to contentedly apply design to reassign.

For effective and better results, the research was concentrated on computers, and further narrowed down to only Printed circuit boards (PCBs), as PCBs are a very common part of all kinds of E-Waste produced worldwide, and also one of the most difficult component to treat and handle. None-the-less this kind of constructive design approach can be applied to any province of E-Waste that needs to be reassigned.

3.1 Constructive Experiments on Printed Circuit Boards

Contemporary design research touches engineering, integrating design and research. It provides ways to justify methodological choices and understand these choices, through constructive methods. This type of research focuses on something far more concrete, which is, research that something is actually built and put to use. This not only includes concepts, but practical issues in manufacturing, making tangible opportunities for decisions [3]. The following are the practical experiments, observations and handling issues experienced working with PCBs.

Besides exploring general, physical and chemical properties, a series of constructive experiments were conducted to explore the best and optimum applications to handle PCBs. These experiments included various methods of cleaning, de-soldering, smoothening sharp edges, strategic patterning, cutting, finishing edges, protective coats and presentation techniques in the workshop. The following were the optimum methods, which were the easiest, quick and gave the best finish to the preliminary batch products (Fig. 3), which could make the product easy to produce, making opportunities for industries at cottage level, creating easy livelihoods.

3.1.1 Cleaning and Decontaminating

PCBs as internal parts of electronic devices are hardly cleaned and are usually dirty. These PCBs cannot be washed and were only cleaned with damp fabric and with a compressed air spray.

Although every PCB in the market comes with a moisture conformal protective coating, sealing all the organic and toxins used in their manufacturing, making the users almost able to be in physical contact with PCBs inside personal appliances like a phone, laptop or even kitchen appliances. However, it is the components like capacitors, which have poisonous dielectric liquid tightly sealed in them, which will be removed in the next stages carefully.

Fig. 3 Preliminary batch of products. *1* Printed circuit board (PCB). *2* Fridge magnets. *3* Paper clip with PCB grips.*4* Key chain/ring

3.1.2 De-Soldering

For further dismantling and de-soldering PCBs, extra care was taken by manually using pliers and de-soldering spike. The components are connected to the PCB in many methods, mostly by soldering, fittings and sometimes glued, sheer care was taken to keep the PCB is it's presentable state and also not to burst any capacitors.

3.1.3 Smoothening and Grinding Sharp Edges

Soldered connections to the PCB are usually untrimmed as they are intended to be inside the devices. These are usually sharp and were trimmed and ground smooth to make them suitable for usage in making products. This can be done on a drill with a grinding bit or with a sand paper.

3.1.4 Designing and Strategic Patterning

With so much of variation in the raw material, especially in E-Waste, every PCB was treated uniquely with a unique pattern plan of what is to be done with it.

Following were the things considered for patterning

1. Size
2. Evenness
3. Thickness
4. Placement of the visually interesting components
5. Color of the PCB
6. Thickness of the electronic circuits visible

Based on these constraints a cutting pattern was developed for every PCB to save time and maximize output.

3.1.5 Cutting

PCBs are non-elastic materials made of glass Epoxy, thus cutting them would leave the edges rough and chipped. The best way of cutting a PCB is grinding using either with an appropriate drill bit, or on the sander. Straight cuts for our convenience were done on an industrial trimmer but followed by further trimming on the belt sander and a grinding drill bit.

3.1.6 Finishing Edges

All the edges were trimmed and smoothened on a belt sander and with files. Any sharp edges of the products were checked and ground smooth. The edges after grounding were often left pale and were painted dark with any permanent ink marker or paint.

3.1.7 Toxin Protective Coat

A thin layer of easily available epoxy based transparent coat was painted to the finished pieces of PCBs and allowed to dry, not only to seal the toxins on the PCBs but also to add shiny finish. This epoxy protective coat can be preceded by a wash of Fenton's reagent (a solution of hydrogen peroxide and an iron catalyst that is used to oxidize contaminants) to treat any organics and metal ions from the freshly cut edges and the surfaces, which already have a conformal coating.

3.1.8 Finishing the Product

The finished pieces of PCB products were completed with other accessories needed to make the products usable and were sent to packing and retailing.

3.2 Preliminary Batch Products

While working on various constructive experiments to handle PCBs, a batch of products were made in the workshop shown in Fig. 3, which were very obvious and were used in the later stages of design as Design probes.

4 Probing and Brainstorming

4.1 Probing

Probes are used for applied research in design. These probes are equipped with instruments or objects selected to help answer research questions from the users.

A probe process intending dialogue emphasizes discussion and interpretation. Probes in Meetings, workshops and ideation sessions typically encourage the dialogue. As for sources of inspiration, they can be used without any particular outline for getting ideas [4].

Probing in this study was conducted in four stages, where the first two stages were conducted in isolation and the next two in group brainstorming sessions. In the first stage the users were left with the probes to interact with, while their comfort with the probes were watched. In the second isolated stage, the designer has interacted with the users triggering dialogue and discussion. The third and the fourth stages were conducted in groups, where the user was educated with information and constraints gradually; and the ideas were plotted, as shown in Fig. 4.

Fig. 4 Bain storming—idea mapping with basic constraints

5 Final Products

Of all the products plotted in the brain storming session, there was a need to follow a specific criterion, which can give scope for new livelihoods, starting at a cottage level using basic and cheap equipment. Following are few criterion based on which the final products were shortlisted for prototyping.

- Monolithic designs, minimizing finishing issues, maximizing finish and consistence.
- Products, which can be made with minimum number of operations and with basic tools, affordable by a cottage industry.
- Easy and quick to make, not compromising on the beauty, but saving on labor cost.
- Meeting the market needs and user preferences.
- Less skill required, thus less training.
- Beautility (Beauty + Utility) products, and total replacers of products in the market.

5.1 Products

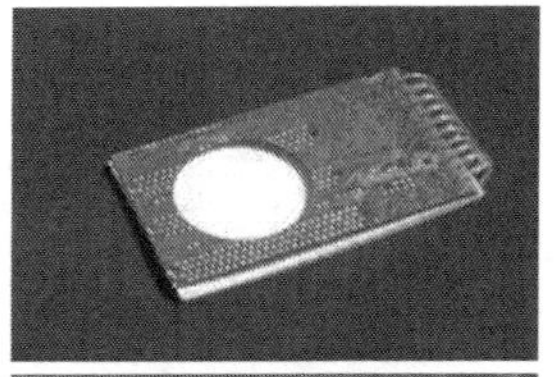	-A PCB has replaced the book cover with a circle cut in it, resembling an apple ipod Size—7 × 12 cm Cost of manufacturing INR 10
	-A PCB badge, circle was cut with a drill, and glued to safety pin at the back Size—4 × 4 cm Cost of manufacturing INR 2
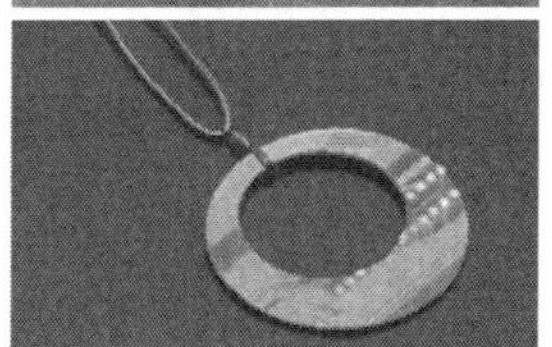	-Loose-hanging jewellery, crafted ring of PCB, attached by a small copper loop to a wire Size—3 × 3 cm Cost of manufacturing INR 3 each

(continued)

(continued)

-Loose-hanging earrings, crafted ring of PCB, attached by a small copper loop to a ear hook
Size—1.5 × 1.5 cm
Cost of manufacturing INR 2 each

-A PCB board replaced a common pad clip's wing
Size—4 × 5 x 4 cm
Cost of manufacturing INR 10

-Non-slip soap tray, made from three pieces of PCB, glued together. The rough undulations on the surface of the PCB are used to make it non-slip
Size—15 × 15 cm
Cost of manufacturing INR 4 each

-A famous product in market usually studded with PVC jewels, costs INR 30, was studded with PCB tiles, retaining it's beauty and reducing the cost
Size—1 × 2 cm × 9 studs
Cost of the new product INR 15

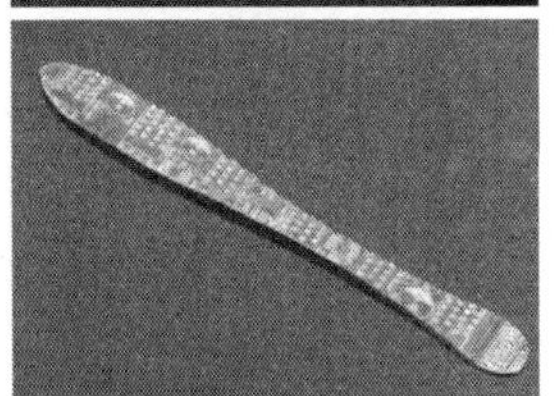

-Wooden paint stir sticks are commonly used to mix paints, epoxies and other general coating sold in small units or in large bulk for wholesale and costs 0.12–0.18 $ (ie. INR 6.5–9.5)
(www.jamestowndistributors.com, 16 May, 2012)
The product was reconstructed with epoxy coated PCB
Size—10 × 2 cm , cost- INR 1

-PCB spectacle frame glued to old spectacle temples and bridge
Size—10 × 5 cm
Cost of the new product INR 15

6 Conclusion and Future

The products prototyped at this stage have shown their possibility to create livelihoods using E-Waste, with optimum equipment and with great ease. And also can function in a cottage industry setup, fulfilling practical and startup issues with design intervention. There is still a lot of scope for improvement in design to use the full potential properties of a PCB. As this study was only targeted at startups, it can uncover more opportunities when the scale increases.

These prototypes were an outcome of Constructive design research methods, which can be applied to any waste that needs to be re-assigned by design opening many opportunities livelihood, using waste as a resource.

However, any product of recycled, reused or reassigned will have its end, so will these reassigned products be trashed in a regular landfill one day, along with regular non-recyclable trash. Although these "epoxy coated PCB reassigned products" are partially treated at the time of reassignment, because of the scattered distribution of E-Waste along with other non-recyclable trash dilutes the toxic effect of the PCBs on the landfill soil.

Because of these re-assigned products, the pressure on the raw materials used for making similar products will be relieved and shared. As these components have served a regular lifecycle of an electronic component and also an assigned product's life, the materials live a longer life than a material in a typical product's lifecycle.

References

1. Baker E, Bournay E, Harayama A, Rekacewicz P, Catelin M, Dawe N, Simonett O (2004) United nations environment programme (UNEP) and secretariat of the basel convention. Vital Waste Graph 8–9
2. Bigum M, Christensen TH (2011) Waste electrical and electronic equipment. Solid waste technology and management, Blackwell Publishing limited, pp 961–962
3. Koskinen I, Zimmerman J, Binder T, Redstrom J, Wensveen S (2011) Design research through practice—from the lab, Field and showroom, Elsevier, Amsterdam, pp 6–7
4. Tuuli M (2006) Probes seeking human life, design probes. Gummerus Printing, Finland, pp 39–65

Conflicts in the Idea of 'Assisted Self-Help' in Housing for the Indian Rural Poor

Ameya Athavankar, Sharmishtha Banerjee, B. K. Chakravarthy and Uday Athavankar

Abstract The paper seeks to establish the influence of government policy on housing for the rural poor in India. First, literature on the self-help approach to housing over the last 50 years is summarized. Then the paper proceeds to present a comparison of the macro level view of the Indira AwasYojana for rural poor with an understanding of ground realities of rural housing acquired through ethnographic studies in five areas of rural India. The studies reveal how the rural poor live and how IAY-funded houses fall short of meeting their housing requirements. The paper argues that the main shortcoming of the IAY is its focus on housing as a product rather than a process involving peoples' participation. It concludes that policies and schemes based on a limited understanding of housing processes are likely to meet with limited success and suggests ways in which architects and designers can contribute to housing in rural India.

Keywords Assisted self-help · Housing policy · Rural poor · Ethnographic studies

1 Housing the Rural Poor

Housing is a basic requirement for human survival. It provides a comfortable living environment and serves as a source of economic and social security. Yet, a large section of rural India continues to be houseless. In 1991, nearly 3.4 million households were houseless. This figure had reached 15 million by 2001 with almost 1 million being added annually [1]. The Working Group on Rural Housing

A. Athavankar (✉) · S. Banerjee · B. K. Chakravarthy · U. Athavankar
Industrial Design Centre (IDC), IIT, Powai, Mumbai 400076, India
e-mail: ameya@iitb.ac.in

A. Chakrabarti and R. V. Prakash (eds.), *ICoRD'13*, Lecture Notes in Mechanical Engineering, DOI: 10.1007/978-81-322-1050-4_48, © Springer India 2013

Fig. 1 Self-funded local (*left*) built using local and discarded reusable materials and IAY funded house (*right*) designed and built by a beneficiary himself

for the 12th plan has estimated the total housing shortfall between 2012 and 2017 at a little over 39 million houses [2]. Over 90 % of the total shortage pertains to below poverty line (BPL) households. In spite of the large requirement for low cost houses in rural areas, no private agencies have shown significant interest in the area, largely due to the low margin on these houses and the geographical spread, which makes development untenable. As a result rural India almost entirely consists of self-help housing built over time, sometimes generations using local and often discarded but reusable material as shown in Fig. 1.

Self-help housing has been a well-documented approach in housing literature since the 1960s. However, it must be noted, that for the most part, the discussion on self-help housing has been centered on housing for the urban poor. Nevertheless, it offers insights that are very relevant to this paper.

1.1 Self Help Housing: An Overview

The term 'aided self help' was coined by Jacob Crane in the 1940s and 1950s [3]. However, it was not until the 1970s and 1980s that self-help was propagated by practitioners and academics, following the path breaking work of self-help housing pioneers like Abrams [4], Mangin [5] and Turner [6, 7]. Based on their insights, in the early 1970s the World Bank started promoting self-help in projects in Africa, Asia and Latin America. Many Third World governments also realized the potential of self-help and this resulted in a major change of approach in housing. However, as Bredenoord and van Lindert [8] note, academic attention to self-help has been minimal since 1991 with little more than a few historical reflections.

1.1.1 Dwellers Versus Authority

The 'self help school' as Bredenoord et al. call it, rejected expensive standardized housing solutions. They made a convincing plea for a drastic reduction of government involvement in housing and a more active role for the local community [9]. Self-help, as Turner [10] stated did not imply self-home-building by untrained and ill-equipped people. To him, the central issue was that of control and the power to decide. It was the dwellers' control over the building process, he argued, which would help create settlements best suited to their needs. Turner and Mangin [11] showed how squatter settlements were well adapted to the needs and circumstances of their residents, and also improved over time.

1.1.2 Product Versus Process

The main contention of the self-help school was that people should be given the freedom to make their own decisions. They suggested that governments change their approach toward housing from building and managing houses, to the role of an enabler, providing people access to resources that they cannot provide themselves. The type of control, Turner [10] argued must be one that supports user participation and flexibility. In other words, setting limits to what may be done and giving individuals freedom to find their way within limits as opposed to mapping procedural lines that must be strictly adhered to. Turner suggested that housing should be viewed not as a noun but as a verb [12]. Housing as a noun refers to the physical shelter, a product or commodity. Housing as a verb, on the other hand deals with the activity of housing as an on-going process and focuses on what the house does rather than what it is.

1.1.3 Housing Paradigms: A Continuum

Turner's work became the foundation for what was later called the 'support paradigm' by Hamdi [13]. Hamdi developed Turner's ideas to identify what he called two conflicting paradigms of housing- provision and support. The provision paradigm, according to Hamdi, tries to deal with the housing problem by producing houses. The housing process is controlled by the provider, and it is he who is responsible for housing units being built to a certain standard. The support paradigm, on the other hand, seeks to assist users through finance, services or training helping them build their own houses, rather than controlling the production of houses. Providers typically see the housing problem as one that must be dealt with expertise in materials, design, technology etc. Supporters, on the other hand believe that solutions exist in the field and merely need to be 'recognized and the built on' [13]. In reality most housing processes are a combination of these conflicting paradigms. The idea of 'support structures' proposed by N. J. Habraken is one such example. Habraken breaks up the provision of housing into different components that can be tackled separately. In his solution, large physical infrastructure is designed and built by professionals working with the state while the 'infill' allow for users' participation in the design of their homes. Similar to Turner, the housing system is viewed as a process rather than a product, allowing users to make decisions according to their needs while respecting the larger structure.

Both Turner and Habraken, as Hamdi [13] states propagated the principles of flexibility, participation and enablement. However, while Turner promoted self-help, self-management and self-build, Habraken tried to incorporate industrial production to serve the interest of providers and users alike. Turner focused on housing politics and policies, while Habraken was more interested in the structure of the built environment.

It is this relationship between housing policy and the structure of the built environment that forms the focus of this paper. More specifically, the paper seeks to establish (a) how housing policy influences the structure of the residential built environment (b) in turn how an understanding of the built environment can inform the formulation of housing policy.

2 Housing by Financial Support: Indira Awas Yojana

The next section presents two views of an assisted self-help housing scheme for BPL households in rural India- the Indira Awas Yojana (IAY). The first half is a broad, macro level view of the scheme, its design its accomplishments and its shortcomings based on documented information. The second half presents an account of the situation on the ground as revealed by our field research focusing on houses created under the IAY.

2.1 The View From Above

The Indira Awas Yojana (IAY) was launched by the Government of India in 1985. The scheme was initiated as a part of the employment generation program, but eventually made independent in 1996 [14].The IAY is an assisted self-help scheme. It seeks to reduce rural homelessness by encouraging the rural BPL to build their own houses by providing them financial support. No type design is specified and the scheme beneficiary has complete freedom in the construction of houses, procurement of materials and in the employment of labor as shown in Fig. 1.

2.1.1 IAY: Accomplishments

The IAY has consistently reported high levels of satisfaction among households that were given grants and high levels of occupancy. Some reports suggest that about 86 % of beneficiaries expressed fairly good satisfaction levels with constructed houses. This has been attributed to two aspects of the scheme- (a) there are no architectural, material and layout requirements for the houses and beneficiaries are free to build the kind of houses they want and (b) the program grants a full subsidy and has no credit component [1].

The IAY has been receiving large fund allocations over the years. A total of Rs 35,450 crores has been allocated from the 7th to the 11th Plan periods. The Union Budget 2011 allocated Rs 11,075 crores for the IAY [15]. The scheme has also been effective in utilizing its allocations, meeting and often exceeding house construction targets.

2.1.2 IAY: Shortcomings and Quick Fixes

The IAY has been not without its shortcomings. It is unable to target the poorest of the poor landless laborers. Even in cases where it has succeeded, it has been unable to improve living conditions for the poor due to the lack of access to infrastructure. Moreover, a large percentage of beneficiaries have expressed their dissatisfaction with the adequacy of funds provided to build a house, leading to housing related indebtedness. The CDF [1] scheme brief on the IAY provides a detailed account of some of the shortcomings of the scheme. The government has been trying to address these issues through isolated, quick fix measures like homestead schemes, integrated habitat development schemes and upward revision mechanisms.

2.2 The View From the Ground

The discussion up to this point presented a broad view of the IAY as a housing scheme. This section is dedicated to presenting the other half of the picture, an account of how IAY funds are used by people to build their houses. Since no systematic studies of IAY funded houses and comparable local self-funded houses were found, an extensive field study was undertaken at five sites across rural India-Pohegaon-Khopoli and Nilshi-Khandi in Maharashtra, Wadi in North Karnataka and Nallur Vayal and Velliangiri in Tamil Nadu. The sites form a sample of convenience but do address variations in climate to some degree.

Data was collected through observation, unstructured interviews with inhabitants and measured drawings of houses. The focus was on understanding (a) how people go about the process of building their houses (b) how they inhabit their houses and (c) and how one affects the other. So typical data included a documentation of current status, history and future plans for their house and relationship of activities to individual spaces in and around the house.

The results reveal many common as well as unique patterns of practices across the five sites. These patterns has been expressed as 'insights'. The idea is somewhat similar to Alexander's 'patterns' [16]. Here, however an insight refers to a recurring building or living practice rather than a problem–solution set. The results are presented in the two parts. The first part is a comparative between the processes followed by an IAY funded house and a self-funded local house while the second deals with above questions.

Table 1 A comparative of typical local self-funded and IAY housing processes

Construction phase	Self-funded house	IAY-funded house
Initial phase	Minimum unit (Locally built, usually load bearing brick or stone walls, timber rafters, clay tiles, sheet roof, plinth built up with soling)	Core unit (IAY funded permanent construction, usually load bearing brick walls and sheet roof)
	1. *Enclosed multi-functional space*: Houses activities that require privacy including cooking, sleeping, some domestic work	1. *Strong room*: Houses safe, cupboards and storage units, good quality utensils, the television, anything valuable. Also used for sleeping
	2. *Proportionate outdoor space*: Effectively serves as living space, household activities spill over from inside.	2. *Bath*: Completely detached from strong room or attached but with access from outside.
Addition 1	Enclosure and isolation	Footprint addition and isolation
	3. *Kitchen*: Cooking and related activities separated from multi-functional space. (similar to minimum unit construction)	3. *Cooking area*: Only for cooking, does not contain house articles and dining
Addition 2	Footprint addition (Similar to minimum unit construction, precast fencing and lamp posts for columns)	Footprint addition (Locally built, usually timber posts, woven thatch or sheet for wall panels and roofs, reusable material. Size of spaces often determined by the dimensions of sheet roof components)
	4. *Additional kitchen*: Second kitchen built when a member gets married or starts a family	4. *Storage and cattle sheds*: For cattle, poultry and general storage. May abut the strong room or may be independent
	5. *Cattle, storage sheds, verandahs*: for cattle, poultry and general storage.	

2.2.1 IAY Funded Versus Self Funded

Generally, it was found that there were more shared insights in IAY houses as compared to local houses. A comparison of the typical process in an IAY dwelling as opposed to a traditional or self-funded dwelling is shown in Table 1 and Fig. 2. It must be noted that the example shown below is illustrative and does not account for the many regional and climatic variations in house forms. However, it is representative enough to demonstrate typical building and inhabiting processes.

2.2.2 Key Insights

Of the many shared insights and sub-insights, only two overlapping insight sets (a) multi-functionality (b) incrementality are presented here. Multi-functionality and incrementality have an inverse relationship but only to some extent.

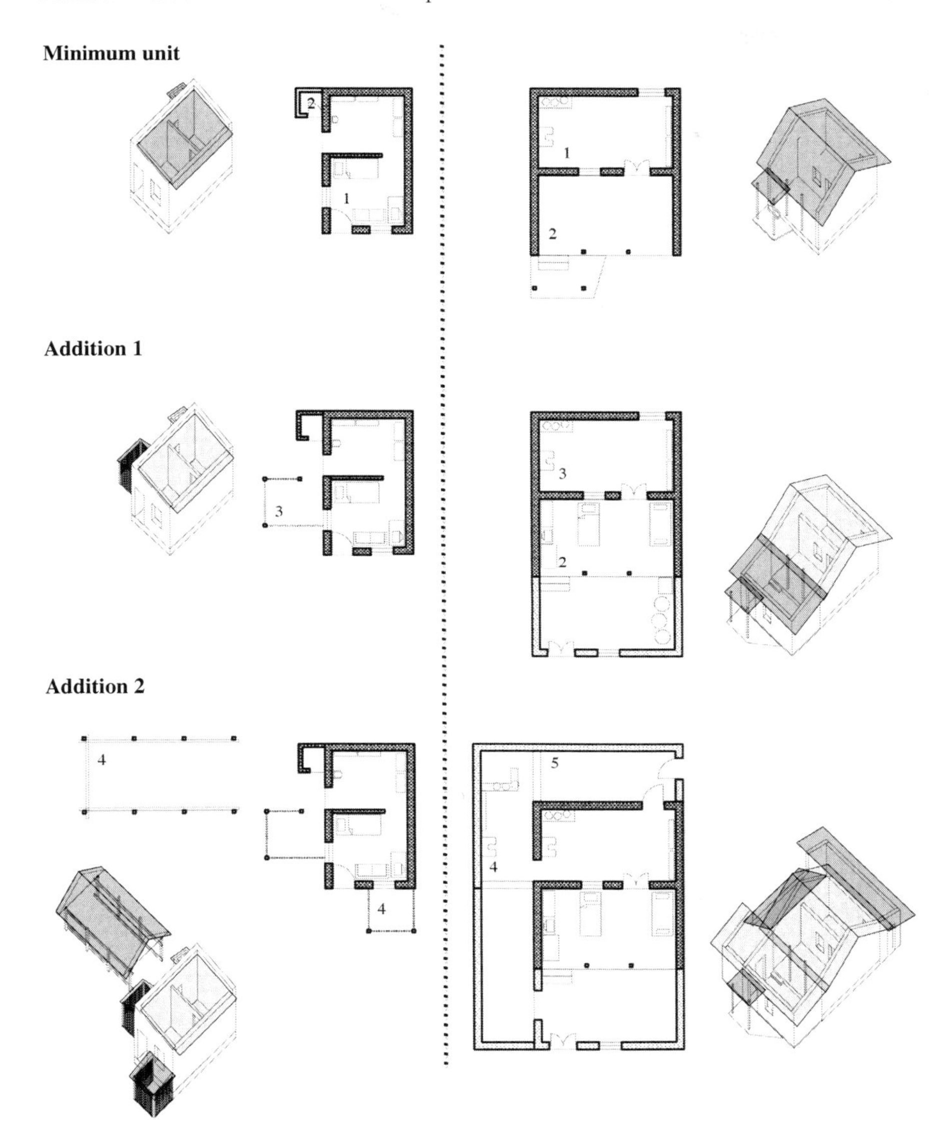

Fig. 2 Growth of a typical IAY funded house (*left*) compared to self-funded local house (*right*)

Incrementality, *Enclosure*, *Isolation*, *Footprint addition*: Rural households often do not have a regular income and this influences the way they build their houses. Initially, a household builds only the Minimum house needed to meet its requirements. The house is developed over a series incremental additions and extensions. Incrementality can be seen in gradual progression toward durable and modern materials as well as addition of extra spaces and internal lofts. Typical incremental changes include:

- *Enclosure* involving the complete or partial enclosure of a semi-open or open outdoor space for private use.
- *Isolation* involving separation or isolation of an activity from the multi-functional space, usually cooking from the rest of the house as this is primarily the woman's domain and needs to function relatively independent of the rest of the house. This separation is critical since cooking, washing and related activities start early in the day and need to be isolated from the rest of the household.
- *Footprint Addition* involves construction of adjacent or separate structures on previously unutilized ground, usually as the household acquires more assets, particularly cattle or poultry.

Multi-functionality, *Proportionate outdoors*, *dedicated kitchen*, *additional kitchens*: The use of space for activities is prioritized over its use for arranging related objects. Apart from the bath, the minimum house consists of a single multi-functional space with proportionate outdoors *See Enclosure*. As the house grows, *See Footprint addition* some activities are isolated from the multi-functional space *See Isolation*.

3 The Uncut Picture

With a more complete understanding of the IAY, it is now possible to examine the relationship between the IAY and the built form of the house. The first section discusses influence of the design of the IAY on the way people build and inhabit their houses. The second section discusses how this understanding can contribute to creating better rural housing policy.

3.1 IAY: The Conflicts

This section discusses the approach of the IAY as an assisted self-help housing scheme. It focuses on the adequacy, structure and method of fund disbursement and the very idea of financial support for the construction of a permanent house.

3.1.1 Funding What

The need to allocate more funds to the IAY beneficiary cannot be disputed. However, it is the IAY's focus on permanent houses that requires discussion. Clearly, the IAY funded 20 m^2 one room structure does not nearly meet the requirements of a rural BPL household. In fact, in most cases, a 20 m^2 room does not even constitute a minimum house. The overriding concern for building a permanent house with limited funds produces a scaled down house that uses far too much funding for the requirements it fulfills. The focus on the one room structure sidelines the need for a separate kitchen and finished outdoor space that are essential components of the minimum house.

The IAY presents a case where a user is being granted financial 'support' to deliver a product built to 'provider' standards at the cost of its ability to serve him. This conflict is one of the major reasons for the inadequacy of funds to build a good living environment.

3.1.2 Funding How

The incremental nature of the housing process does not find a corresponding response in the one time funding of the IAY. The lack of response to incrementality also extends to the funding structure of the scheme. IAY grants are disbursed in three to four installments. The release of these installments is conditional to the completion of pre-decided stages of construction, typically foundations, plinth, lintels and the roof. The method of disbursement is appropriate only for the construction of a finished structure. Again, the undue focus on finished house sidelines the need both for an initial structure that is supportive of incremental additions as well as funding for these additions.

The IAY's one time funding is characteristic of payment for a complete house. Even the stage-wise release of funds is suggestive of a monolithic structure instead of a series of connected structures built over time as per the need of the household. This is major reason for the asymmetry in quality and lack of cohesion between structures built under the IAY and additions made henceforth.

3.1.3 Only Funding

The IAY emphasizes the use of modern materials for the construction of the permanent house. This requires skills and training. For an IAY beneficiary, it is very difficult to employ skilled labor considering his location and the total value of work involved. The use of modern material by untrained workers results in poor quality construction prone to weather conditions. Off-the shelf components ensure some durability, but often create unusable spaces.

The rural poor do aspire for permanence and durability. This partially justifies the IAY's insistence on the use of modern materials. However, considering the amount of funding being provided, this must be accompanied by proper resources and training. Also, rejecting a wealth of local building skills seems rather unwise. The IAY, with its limited involvement of providing financial assistance cannot expect to create quality permanent housing.

3.1.4 Neighborhoods Versus Service Provision

In order to ease the provision of services such as electricity, water supply and sanitation, the government has been trying to propagate the 'cluster approach' to IAY housing. There is no doubt that cluster based development has distinct advantages over individual houses. It can not only facilitate provision of services, but also make it easier for beneficiaries to procure material or recruit skilled labor and share know-how on house construction. However, the cluster approach is often misconstrued as a colony of houses organized so as to save material or to ease the provision of services.

Cluster based development has potential to create rich neighborhoods and communities, provided the households participate in the process and share more than just services. If the IAY seeks to create good neighborhoods, it will need to develop a more participatory housing process and rethink its approach to the provision of services.

3.2 Challenges and Opportunities

The IAY, in many ways is characteristic of a scheme that seeks to reduce housing shortages and meet targets rather than create good living environments. This paper has gone to great lengths to show the limitations of the IAY's focus on the permanent house. This is perhaps the single largest shortcoming of the scheme, rooted in its view of housing as a product or a commodity rather than a process where the user is an equally, if not more, important participant.

The process of housing must be centered on the user and his capabilities. The government's role should be restricted to providing financial and technical support and setting 'limits' to how funds can be used. Architects and designers are well equipped to research and understand housing processes. They are also good judges of environmental quality and can make valuable contributions to the structure of financial packages, identification of areas that require technical support and drafting of context specific guidelines to define the limits of how IAY funds can be used.

The idea of 'support structures' proposed by Habraken [17] is another line of work that is worth exploring in rural India. Instead of building a finished house, IAY funds could be used to create a durable 'starter structure' that is weatherproof, offers good ventilation and allows additions and in-fills in local material. This could not only reduce cost but also utilize local building skills and materials. The role of the designer would be to conceive such a starter structure and its interaction with a variety of in-fills.

Building construction in India is yet to fully realize the immense potential of standardization and mass manufacturing, in bringing affordability and quality. Standardization does not always come at the cost of choice and individual preference. It is possible for architects and designers to create standardized building

Fig. 3 Precast fencing posts with random rubble in-fills (*left*) and a building material supplier showing palm leaf panels to be used as in-fills in local houses (*right*)

components that require simple skills to assemble. Such an approach can also effectively handle the incremental nature of rural housing. Rural India's faith in prefabrication and standardized components can already be seen in Fig. 3.

Architects and designers can also assume what Hamdi [13] called 'teacher's' or 'enabler's' role in the housing process. With their ability to understand local building processes, their benefits and limitations they can effectively disseminate knowledge, impart training and help create participatory design processes to bring about incremental improvements in houses.

These are just few of the many ways in which design can contribute to rural housing for the BPL. However, it is important to remember that people have been in-charge of building their own houses for the greater part of history while the participation of professionals in housing is a recent phenomenon. It is the professional class, as Habraken [18] pointed out that is participating in the age-old process of housing and not the other way around. The professional's role must involve providing expertise in areas that no one else is capable of handling better.

Housing the rural population of a nation of India's size is an enormous and complex task. But it is the conviction of this paper that architects and designers can play an important role in housing the Indian rural poor.

Acknowledgments We would like to extend our gratitude to ACC Ltd for supporting this research and to the many families of Pohegaon-Khopoli, Nilshi-Khandi, Wadi, Nallur Vayal and Velliangiri for welcoming us into their homes and openly sharing their views and experiences.

References

1. Raman NV (2012) Indira Awaas Yojana (IAY) scheme brief. Cent Dev Finan
2. Planning Commission Government of India (2012) Working group on rural housing for the twelfth five year plan (2012–2017). pp 7, 12, 15, 26
3. Harris R (1991) A burp in church: Jacob L. Crane's vision of aided self help housing. Plann Hist Stud 11:3–16
4. Abrams C (1966) Housing in the modern world. Faber and Faber, London

5. Mangin W (1967) Latin American squatter settlements: a problem and a solution. Lat Am Res Rev 2:67–98
6. Turner JFC (1967) Barriers and channels for housing development in modernizing countries. J Am Inst Planners 33:167–180
7. Turner JFC (1983) From central provision to local enablement, new directions for housing policies. Habitat Int 7(5/6):207–210
8. Bredenoord J, van Lindert P (2010) Pro-poor housing policies: rethinking the potential of assisted self-help housing. Habitat Int 34:278–287
9. Abrams C (1971) The language of cities: a glossary of terms. The Viking Press, New York, p 63
10. Turner JFC (1978) Housing by people: towards autonomy in building environments. Ideas in progress. Marion Boyars Publishers, London, p 133
11. Mangin W, Turner JFC (1968) Barrida Movement. Progressive Archit 37–56:62–154
12. Turner JFC, Ficher R (1972) Freedom to build. Macmillan, New York, p 151
13. Hamdi N (1995) Housing without houses: participation, flexibility, enablement. IT Publications Ltd, London, pp 27, 36, 45, 177–179
14. Ministry of Rural Development Government of India (2012) Rural housing. Web
15. HT Correspondent (2012) Indira Awas Yojana Receives Rs. 11,075 crore. Hindustan times
16. Alexander C, Ishikawa S, Silverstein M (1977) A pattern language: towns, buildings, construction. Oxford University Press, New York
17. Habraken NJ (1972) Supports, an alternative to mass housing. The Architectural Press, London
18. Habraken NJ (1986) Towards a new professional role. Des Stud 7(3):102–108

A Method to Design a Value Chain from Scratch

Romain Farel and Bernard Yannou

Abstract Value chain concept and methods has assumed a dominant position in studying industry from management point of view. Decision supports methods using value chain require the acquisition of data from various existing corporate databases or data warehouses. In design research discipline, the subject of value chain design is emerging. Only a few of published research took a wide scope comparable to theories used today in engineering design. As an effort in developing the methodology and as a result of research within a national industrial consortium, this paper proposes and discusses a general value chain design approach which opens up a promising perspective to provide a new direction for research and application of value chain from scratch for multi-stakeholder industrial systems. It introduces value chain design as a way to determine, model, and analyze and evaluate the industrial ecosystems, in order to generate future scenarios and provide evaluation criteria for decision makers. To illustrate its application, the establishment of end of life vehicle recycling subsidiary at national level is explored to identify potential values stakeholders.

Keywords Value chain · Structural analysis · System dynamics · Scenario · Matrix · Simulation

1 Introduction

Traditionally, organizations seek to streamline their processes and improve customer service by improving connectivity between both business processes and key operational units. The supply chain of the industry dominates the value generation chain

R. Farel (✉) · B. Yannou
Grande Voie des Vignes, Ecole Centrale Paris, Chatenay-Malabry 92290, France
e-mail: romain.farel@ecp.fr

A. Chakrabarti and R. V. Prakash (eds.), *ICoRD'13*, Lecture Notes in Mechanical Engineering, DOI: 10.1007/978-81-322-1050-4_49, © Springer India 2013

[1]. Porter has defined the value chain as a chain of activities for a firm operating in a specific industry [2]. Products pass through all activities of a chain from conception through the different phases of production, delivery and disposal, and at each activity the product gains some value [3]. Kaplinsky in [4] point out three key elements of value chain analysis: Barriers to entry, Governance, and Different types of value chains. Essentially, the primary returns (arising from design, production, marketing, etc.,) accrue to those parties who are able to protect themselves from competition. This ability to insulate activities can be encapsulated by the concept of rent, which arises from the possession of scarce attributes and involves barriers to entry. The economic rent arises in case of differential productivity, and takes various forms including technological capabilities, organizational capabilities, etc. Build on similar works [5] governance ensures that interactions between firms along a value chain exhibit some reflection of organization rather than being simply random. Value chains are governed when parameters requiring product, process, and logistic qualification are set which have consequences up or down the value chain encompassing bundles of activities, actors, roles, and functions.

Along with the increasing need of todays' enterprises for value chain analysis for integrating into new business models, industrial researchers study and explain adapted concepts, methods, and tools in different sectors. In IT and telecommunication [6, 7] and chemical industry [8] for example, the paradigm shift and new requirements in enhanced value chain design was discussed. Papazoglou proposed an integrated value chain model in order to create and enhance customer-perceived value by means of cross-enterprise collaboration [9]. However, the question of design a value chain from scratch seems to remain unsolved. First, the developed methods are based on existing workflows and their stakeholders. Second, the creation of a future multi-actor value chain has not been investigated from a design point of view, considering both technical and organizational issues. This paper proposes a first try on formalizing a methodology for design a value chain for multi domain industrial projects. The research question is formulated in what process is needed to design a large scale industrial system with stakeholders in different domains of activity, in which key decisions and strategic choices are supported with qualitative and quantitative data and analysis of the whole system. The detail of this method is illustrated using the data of an ongoing national level case of design the value chain for a non-existing End of Life Vehicle (ELV) material recycling. Data has been collected through interviews, field observation and investigation and developed models have been validated by project actors meeting and industrial experiments.

2 Methodology

In light of the evolving need and the limitations of existing approaches and methods, the main objective of this paper was to understand how to provide key decision indicators for stakeholders of a future value chain. The research behind this paper was conducted in a national consortium for establishing a future recycling subsidiary for End of Life Vehicle (ELV) materials. In a bottom–up approach, several industrial stake holders have been asked why they would be interested to take part in a future collaboration. They have also been asked what kind of information would help their decision making. The cost and benefit of the future activity, along with environmental impact was evoked. However, stakeholders needed to have simulated data, showing the requested variable (e.g., cost of recycling) in a long term future perspective, following the various possible situation. Those situations, called scenarios, should be generated intelligently from the crossing of dynamic evolution of key variables of the system. Thus, a system view should be established, and a method should be employed to identify key variables. The reason is in a systemic approach, generally a considerable number of variables can be found having interconnection, yet those relations remain unknown and impossible to be mathematically formulated.

A system view uses the basic flow model of the activities, characterized with variables, and determines the relation between those variables within the system and with external variables. Accordingly, it is necessary to model the physical flow (for production) or information flow (for service) using a modeling approach. What is important in modeling is to identify the operation activities, and parameterize each activity with requested variables such as economic (cost or benefit) and environmental (CO_2, waste). In the following each step is formalized and explained in detail. The method demonstration with a real case application is presented in the next section.

2.1 Modeling Material and Information Flow

Material flow can be defined as the description of the transportation of raw materials, pre-fabricates, parts, components, integrated objects and finally products as a flow of entities. In the same way, information flow is any tracking of referential information passing through operational units.

The interdisciplinary nature of the most of the today's consortium is fundamental to the task of modeling. It is required to:

- Identify the main process and sub-processes
- Define system boundaries
- Identify operational task alternatives
- Find/consider new innovative solutions that will result in value generation.

To perform the modeling task, one is free to choose or even define his own modeling approach. Hence, the body of literature in modeling approaches is simply immense. A valuable inventory of methods has been put together in [10]. Here, the most important issue is to parameterize input and output variables. At the end, variables value change in operation units are to be formulated mathematically, for example mass transfer in an operational unit of separation, or economic flow in a selling unit.

2.2 Establishing Value Network

The goal of this task is to identify and formalize the added value on each step of the process model, and for different stakeholders. While the main value is readily accepted to be the economic value or the cash balance, other form of measurable value types could be of the interests of the stakeholders. Environmental impact, such as carbon dioxide emission and waste production are two examples. To perform this task, firstly the value framework should be defined and be approved by stakeholders. Secondly, the value change should be formalized using the same variables and operation structure as of the previous step. For instance, a given operation of raw material purification through melting process would have a mass transfer equation between input and output material. The cost of this operation, the mass of by-produced waste, and the carbon dioxide mass in result of heating process are value measures.

2.3 Structural Analysis

Godet introduces the structural analysis as a method to determine the key questions concerning the future and to identify the influence of various stakeholders, in order to establish the relationships among them, as well as the stakes involved [11]. The reason of using structural analysis in this step is to add the stakeholders' point of view and interpretation of how the system works or would work to the technical model of previous steps. Structural analysis begins with a group made up of both internal personnel and outside expertise in the field under study. It includes three successive phases:

- Creating an inventory of variables
- Describing the relationships among the variables
- Identifying key variables.

The inventory of variables is of course closely linked to the value network. Establishment of system view on the process model using cause and effect relation proposed in Systems Dynamics [12] is helpful. The system view can also be a key enable in the necessary alignment of business logics, and is also helpful to address effectively the need for active and open-minded communication between stakeholders.

While a list of variables is obtained and their relationships are determined, it is essential from practical point of view to identify key variables to set up scenario generation and further simulation on variable dynamic estimation.

The key variable identification consists of identifying and re-ranking the key variables, i.e., those essential to the evolution of the system. Very few methods with similar matrix based techniques exist for this task. Here are two better known: MDM and MICMAC.

Multiple Domain Matrix (MDM) is an analyze method and tool which allows analyzing a system's structure across multiple domains. Matrix-based approaches integrating multiple views ("domains") become more and more accepted to manage several perspectives onto a system, especially when it comes to large structures. Detail on MDM can be found at www.dsmweb.org.

Matrix Impact Cross-Reference Multiplication Applied to a Classification (MICMAC) is method and tool that helps using direct classification (easy to set up from interviews with stakeholders) for primary analysis. Then through using indirect classification, obtained after increasing the power of the matrix, more structure analysis can be elaborated. Information on this method is available at http://en.laprospective.fr.

Both methods have their pros and cons. By MDM it is possible to describe and analyze a whole system, including multiple domains and several relationship types concurrently. This way MDM allows for systematic application of all available algorithms by compiling several matrices into aggregated views. On the other hand, comparing the hierarchy of variables in the various classifications (direct, indirect and potential) is a rich source of information. It enables one not only to confirm the importance of certain variables but also to uncover certain variables which, because of their indirect actions, play an important role (yet were not identifiable through direct classification).

2.4 Scenario Generation

Stakeholders always wonder the evolution of a given value system in future time horizon. Representing the future evolution in terms of a set of scenario is particularly effective strategy for achieving this objective. Scenario generation is a convenient tool with a broad application if different domains [13]. The more precisely the scenarios span the set of possible future events, the more accurate are the simulation calculated from the scenarios. The goal is to generate meaningful scenarios based on key variable evolution and their impact on the value chain. In this task, it is important to communicate the nature of a scenario set to stakeholders, because they should find a scenario set accessible. This task is simplified by focusing on single variables dynamics, for example increase of fuel price and consequently transportation cost, and generate a set of single variable scenario. Then, classify those single scenarios to generate new scenarios with multi-variable dynamics. An example is shown in the next section.

2.5 Simulation and Evaluation

Simulation is the simplest task in this method if previous modeling, formulation, and scenario generation are done correctly, while it is the most important one facing the stakeholders. Simulation of the value of the system in the future scenarios gives direct and tangible information to a decision maker. A parameterized and/or dynamic simulation can provide real-time simulated response to a decision alternative for each decision makers. Also, this provides a common understanding in the consortium of future trends of the collaboration.

In a former publication, we proposed a segmentation method for scenarios in order to generate a reduced number of multi-variable scenarios [14]. In this approach, a distinction should be made between endogenous and exogenous variables, and classify single scenarios according to the general effect of that scenario on the global value chain. Then cross those classes (for example best case and worst case) of both endogenous and exogenous to generate complex scenarios with two state dimensions. See the example in the following section for more detail.

3 Value Chain Design for ELV Recycling Network

Reuse, recycle and recovery of End of Life Vehicles (ELV) materials are increasingly of interest to researchers and industrial companies, mainly because of the application of an EU directive on ELV [15]. The directive sets the ultimate goal of reuse and recovery at a rate of 95 % of the ELV weight; furthermore, from 2015 on certain materials including glazing must be separated during dismantling from the ELV. Contrary to previous directives, this directive penalties car manufacturers if target reuse and recovery rate is not reached.

Within the boundaries of this study, recycling ELV glazing is defined as the process of dismantling, collection, storage and transportation, treatment, and ultimately reuse of recycled glass called cullet. Cullet is regularly used for making glass products, glass wool, or as a substitution for other raw materials. Beyond the price advantage, each 10 % increase in cullet usage in a glass furnace results in a 2–3 % energy saving in the melting process and each ton of cullet used saves 230 kg of CO_2 emissions. This has created an increasing demand for high-quality cullet by the glass industry. However, neither the demand for cullet, nor that of post-consumer recycled glass, is satisfied by current supply.

3.1 Modeling Material Flow

Figure 1 shows an ELV glazing recycling network in the form of a directed acyclic graph of value chains linking stakeholders dedicated to specific activities. The stakeholders of this network are the car manufacturers, the ELV dismantlers and shredders, the collecting and transporting companies, and the glass treatment companies. If the glazing network is considered as a whole, the cost and benefit can be stated as follows: the costs arise from dismantling, collection and

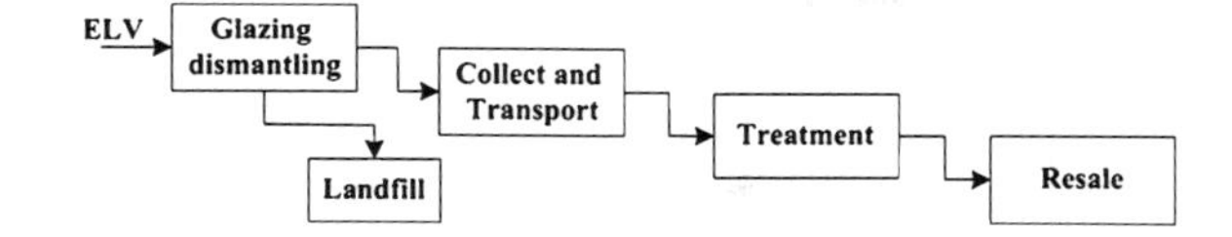

Fig. 1 ELV glazing recycling scheme

transportation, treatment, and the penalty paid by the producer in case of non-achievement of the directive target rate. The benefit of the network is from selling the produced cullet in the market, and the cost of not landfilling and shredding residual waste.

3.2 Establishing Value Network

For dismantlers, the interest to remove the glazing is questionable for two reasons. First, the dismantling procedure of glazing (itself) implies extra time and thus extra costs. After dismantling, the dismantled glass needs to be transported to a treatment unit, which costs more than the buyer can offer. Alternatively, the dismantled glass can be transported to landfill, which has its own costs. Meanwhile, none of these extra costs are compensated.

Second, with a proven market demand for cullet, treatment of the dismantled glazing and sale of the cullet could potentially make a profit for treatment units and for the glass production industry. However, in the absence of a large scale recycling network to share profit and finance the up-stream costs, it is not strategically justified for dismantlers to invest on a network with no future profit. Table 1 shows the motivations and disincentives of each stakeholder for a future ELV glazing recycling network.

Table 1 shows that the interests of stakeholders diverge. It is therefore difficult to establish the overall interest of the creation of this value chain. This is why we propose a Cost Benefit Analysis (CBA) of this value chain to make sure that the gains to the winners exceed the losses to the losers. This result will be employed to examine the possibility of a redistribution mechanism for benefits, assuring that being a member of recycling network is economically viable for each stakeholder.

3.3 Structural Analysis

We use Systems Dynamics approach to show the cause and effect diagram of the glazing recycling network (Fig. 2). There are two main mechanisms here, which have direct influence on the CBA of the network. The first mechanism concerns itself with the benefits of glazing recycling, composed of the sale of produced glass cullet and the savings in terms of landfill cost. The second mechanism deals with the glazing recycling cost, composed of the operational and logistic costs of the network, in addition to the potential penalty. The aforementioned CBA is the sum

Table 1 Individual motivation and disincentives of stakeholders for a future ELV glazing recycling network

	Individual motivation for recycling network	Disincentive for performing activity
Car manufacturer	Meet the 95 % target and avoid penalty	None
Dismantler	Attain a higher dismantling rate and use commercial benefits and potential government subsidies	Extra cost
Shredder	Less landfill cost	None
Collection and transportation	Improve the logistical operation	Volume issues
Glass treatment	Increase in feed and product flow	None
Cullet buyer	Satisfy the demand	Quality issues

of benefits minus the sum of cost. There are, however, a number of interconnections between two mechanisms. For example, the increase in benefits encourages more stakeholders to join the recycling network. Thus, the network coverage increases and this leads to a decrease in logistic costs, which then ripples into a decrease in total cost of the recycling network. The latter, influences motivation for joining the network, and results in a positive loop.

Creating a national recycling network and increasing the coverage rate is an endogenous change of the model. This change positively influences the saved cost of landfill, and decreases the logistic and the total cost of establishing this network. This aside, there are several time-varying elements from different exogenous sources that influence the behavior of the system. The first category of exogenous variables is the raw-material market evolution: change in cullet price and cullet demand. The dynamic of this change is explained in the following section. The second category is the politico-environmental decisions, which result in change in landfill tax, and in applying a penalty if the minimum recycling rate is not reached. The EU directive penalty on ELV reuse and recovery minimum rate of 95 % is one

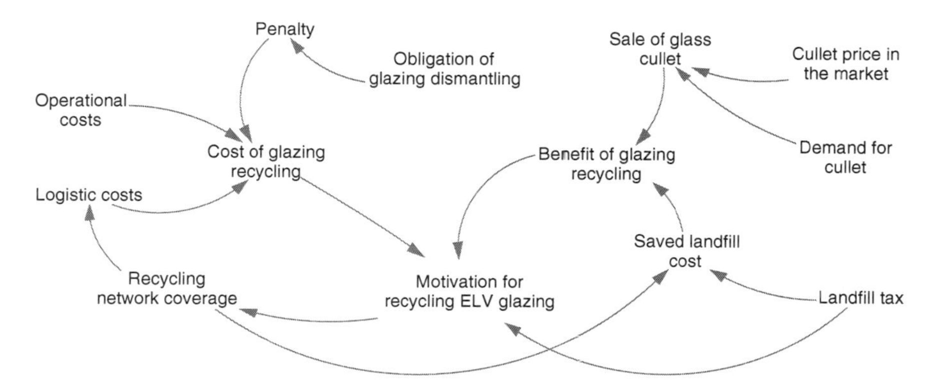

Fig. 2 System dynamics model of ELV glazing recycling network

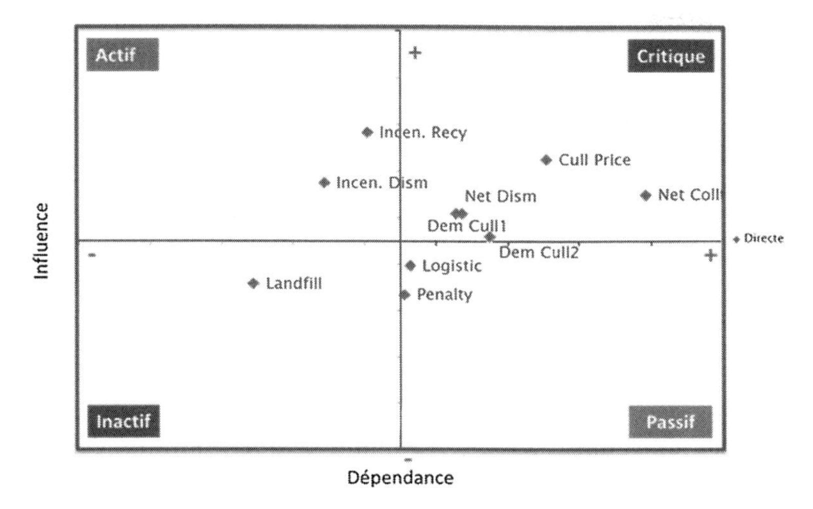

Fig. 3 Direct influence-dependence graph

example of this category. In this stage, with the inventory of variables, we used MICMAC method to identify key variables. In a series of interview, each stakeholder has been asked to fill a direct influence matrix, answering the following question: Is the variable A influence directly variable B? If yes, how?

The interview result has been transformed to numbered matrix with input standard of MICMAC software. With a considerable number of matrix calculations, several interesting demonstration of influence and dependency relation between variables was elaborated. Figure 3 shows a single frame result of the direct influence dependence graph.

3.4 Scenario Generation

Evolution of aforementioned system variables would have influence over the Future ELV glazing recycling network dynamics and consequently its cost and benefit. In a top–down approach, two trends can be imagined for the future network: first, generation of network revenue, and second decrease the operational costs. Cullet price and landfill cost as two exogenous variables, and logistics cost and network coverage as endogenous variables were chosen for the scenario generation. It is possible to consider at least three typical dynamics for each of these key variables, which produce three situations for the cost and benefit of the recycling network, namely best case, expected case, and worst case. Therefore 81 possible scenarios can be generated. Hence, in order to reduce the number of scenarios, a solution is to focus on how any of three situations (best, expected, worst) can happen for a variable category, here endogenous and exogenous. In result, nine scenarios can be generated as shown in Fig. 4.

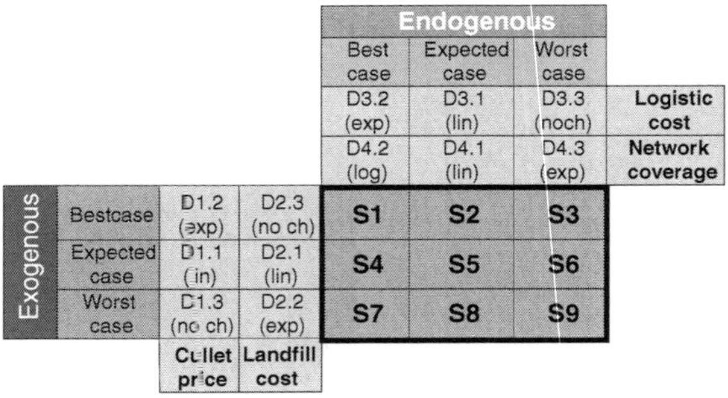

Fig. 4 Nine scenarios for the ELV glazing recycling network

				Endogenous			
				Best case	Expected case	Worst case	
				D3.2 (exp)	D3.1 (lin)	D3.3 (noch)	**Logistic cost**
				D4.2 (log)	D4.1 (lin)	D4.3 (exp)	**Network coverage**
Exogenous	Bestcase	D1.2 (exp)	D2.3 (no ch)	**S1**	**S2**	**S3**	
	Expected case	D1.1 (lin)	D2.1 (lin)	**S4**	**S5**	**S6**	
	Worst case	D1.3 (no ch)	D2.2 (exp)	**S7**	**S8**	**S9**	
		Cullet price	**Landfill cost**				

These two trends are explained as sub-systems in this section. Analysis of the cause and effect relationship between network variables reveals the importance of political decisions from outside of the system, and strategic decisions from within on the dynamic. As a result, a third trend emerges into identify the dynamics of internal and external motivation and decision variables on the network functions.

3.5 Simulation and Evaluation

The costs and benefits of the future ELV glazing recycling network in France depends on the values of time sensitive variable of the cost-benefit model. In order to estimate the cost and benefit in a time horizon, we build a simulator using the data obtained in the first and second tasks. The simulator uses the key variables

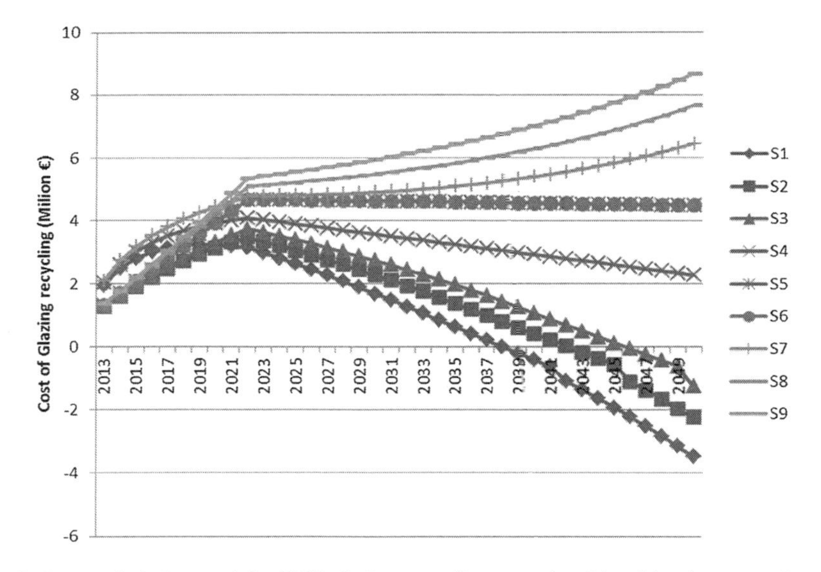

Fig. 5 Economic balance of the ELV glazing recycling at national level in nine scenarios

values change dynamic as input and reports total cost and benefit of recycling. Figure 5 shows the result of following the nine scenarios generated previously, and entering four variable dynamics for each scenario.

The figure demonstrates the cost of ELV glazing recycling in France for each scenario in a 50 years horizon. From a general perspective, three categories can be extrapolated: the first category, scenarios 1, 2 and 3 in which the CBA curve crosses the horizontal axe. There is an observable change in value from cost to profit. Scenario 1 is the best case scenario, in which the network revenue is at its greatest and the time it takes to reach break-point is at its least. In this category the external variables are experiencing in the most favorable conditions for the network. The performance of the network being at its best, expected or worst condition only modifies the slope of reaching the break point. In the second category, scenarios 4, 5 and 6 present a fixed scenario, albeit in the case of scenario 4, a decreasing cost for the network in the future. Scenario 4 would possibly reach the break point in the horizon of 100 years. In this category the external variables are in within the expected framework. The best case for internal variables present a significant cost savings, while S5 and S6 show no difference in cost and revenue of the network. In the third category, scenarios 7, 8 and 9, show an increasing cost for the recycling network. Scenario 9 is the most costly scenario where the external variables and internal variable are at their worst.

4 Conclusion

The need of developing methodology for value chain design is essential, and more and more increasing in multi domain industrial projects. The proposed method described in this paper can be summarized in three categories.

Stakeholders of an industrial consortium need concrete information about the future evolution of the value in the system by which they are concerned. The future changes are absolutely unknown to a design researcher, nonetheless using a method for generating scenarios and simulating the expected value in those scenarios is one of the best decision-making aids to the decision makers.

In a complex system with engineering, social and political disciplines, it is almost impossible to identify all variables and the exact relation among them. Hence, structural analysis approach brings techniques and tools to build a system view of the whole stakeholders and their operations and values, and identify the key variables out of all variables. Finally, in the first place, the value chain design is an engineering design problem, and like for other problems in this discipline, modeling is a fair choice. Last but not least, visualization is essential for providing transparent results. It is also important for communication between project stakeholders and having a common understanding of what the method provides.

References

1. Al-Mudimigh AS, Zairi M, Ahmed AMM (2004) Extending the concept of supply chain: the effective management of value chains. Int J Prod Econ 87(3):309–320
2. Porter M (1980) Competitive strategy: techniques for analyzing industries and competitors. Free Press, New York
3. Porter M (1985) Competitive advantage: creating and sustaining superior performance. Free Press, New York
4. Kaplinsky R, Morris M (2003) A Handbook for Value Chain Research: IDRC
5. Gereffi G (1994) The organization of buyer-driven global commodity chains: how U. S. retailers shape overseas production networks, in commodity chains and global capitalism. In: Gereffi G, Korzeniewicz M (eds) Praeger, London
6. Peppard J, Rylander A (2006) From value chain to value network: insights for mobile operators. Eur Manag J 24(2â€3):128–141
7. Dedrick J, Kraemer KL, Linden G (2011) The distribution of value in the mobile phone supply chain. Telecommun Policy 35(6):505–521
8. Vogt C et al (2005) Paradigm shift and requirements in enhanced value chain design in the chemical industry. Chem Eng Res Des 83(6):759–765
9. Papazoglou MP, Ribbers P, Tsalgatidou A (2000) Integrated value chains and their implications from a business and technology standpoint. Decis Support Syst 29(4):323–342
10. Heisig P, Clarkson PJ, Vajna S (2010) Modelling and management of engineering processes. Springer, London
11. Godet M, Durance P (2011) Strategic foresight for corporate and regional development, UNESCO
12. Sterman JD (2000) Business Dynamics: Systems Thinking and Modeling for a Complex World, Irwin/McGraw-Hill
13. Schwartz P (1996) The art of long view. Crown Publishing Group, USA
14. Farel R, Yannou B, Ghaffari A (2012) A cost and benefit analysis of future end-of-life vehicle glazing recycling in France: a system dynamics approach. Submitted to resources, conservation and recycling
15. EU-Directive, Directive 2000/53/EC of the European parliament and of the council. Official J Eur Commun, 2000

Part VI
Design Collaboration and Communication

Developing a Multi-Agent Model to Study the Social Formation of Design Practice

Vishal Singh and John S. Gero

Abstract This paper describes a computer simulation based approach to investigating the longitudinal patterns in social emergence of design practice. Legitimation code theory is adopted as the underlying framework to develop the model. The design practices in this model emerge and evolve under the influence of the social structure as well as the knowledge structure. This model simulates a society of designers with different design backgrounds, affiliated to different teams and organizations. Design agents interact with each other and the concepts associated with the different disciplines. Design agents within each discipline are modeled to be attracted towards concepts, i.e., knowledge mode, as well towards the other design agents, i.e., knower mode, which collectively influence design practice. The force of attraction towards the knower or concepts varies across disciplines. The emergent social pattern is plotted in a two dimensional space defined by the social and knowledge axes. The simulation environment allows studying the longitudinal emergence of design trends resulting from varied initial conditions and what-if scenarios that are difficult to study in the real-world. Exemplary results are presented.

Keywords Common ground · Legitimation theory · Identity · Influence · Emergence

V. Singh (✉)
School of Engineering, Aalto University, Espoo, Finland
e-mail: Vishal.Singh@aalto.fi

J. S. Gero
Krasnow Institute for Advanced Study, George Mason University, Virginia, USA
e-mail: John@johngero.com

A. Chakrabarti and R. V. Prakash (eds.), *ICoRD'13*, Lecture Notes in Mechanical Engineering, DOI: 10.1007/978-81-322-1050-4_50,

1 Introduction

Design involves multidisciplinary faculties, including technical and social know-how. The widespread use of the term 'Design' across different disciplines often leads to debates over 'art' and 'science' distinctions. While such debates may never have clear outcomes, a common understanding and acceptance of design typically emerges over time through social processes within a given discipline. Over an extended period of time, society develops mechanisms and processes to recognize and legitimise design practices within its community or discipline. Since this recognition and legitimation of design practice is a longitudinal process and as a consequence is difficult to study empirically, the understanding of social emergence of design practice is currently limited. Therefore, this research adopts a computer simulation based approach to investigate the longitudinal patterns in social emergence of design practice. Agent based studies have evolved as a powerful research method to conduct what-if studies for scenarios that are difficult to study in the real-world, and allow longitudinal studies with greater control of the parameters [1, 2].

This research reported in this paper adopts legitimation code theory [3] as the underlying framework to develop the model. LCT is based on five principal dimensions that include autonomy, density, temporality, specialization and semantics, of which the last two are most developed. According to the specialization principle of legitimation code theory, design practice and recognition within a social group are driven through both the knowledge and knower modes [4], i.e., the design practices emerge and evolve under the influence of the social structure as well as the knowledge structure. This model simulates a society of design agents with different design backgrounds affiliated to different teams and organizations. Design agents interact with each other and the concepts associated with the different disciplines. Following legitimation code theory, design agents within each discipline are modeled to be attracted towards concepts, i.e., knowledge mode, as well towards the other design agents, i.e., knower mode, which collectively influence the design practice. The force of attraction towards the knower or concepts varies across disciplines. The emergent social pattern is plotted in a two dimensional space defined by a social axis and knowledge axis such that design agents higher up the social axis exert higher knower force while the concepts higher up the knowledge axis exert higher knowledge force.

The simulation environment allows studying the longitudinal emergence of design trends resulting from varied initial conditions such as the number of design agents in the society and level of interdisciplinary interactions. The objective of this research is to investigate questions such as:

How does the design practice emerge over an extended period in a society? How do design trends vary across disciplinary and multidisciplinary communities?

This paper presents the theoretical basis for developing the computational model and provides preliminary simulation results to demonstrate the usefulness of an agent based approach in understanding complex social behaviors that emerge at global level from simpler local interactions.

2 Social Emergence of Design Practice

Design concepts and capabilities such as creativity have been argued to be social constructs [2]. Similarly, Bourdieu [5], Nonaka [6] and others have discussed the social creation and emergence of knowledge in a broader context. In this respect, legitimation code theory provides a useful conceptual framework to study the emergence and acceptance of knowledge in a social-cultural context.

Legitimation code theory is built on the premise that in any society the prevalent practices, beliefs and knowledge are driven towards something and or someone, such that there is an epistemic relation to an object (ER) and a social relation to a subject (SR) [7, 8]. The epistemic relation pulls the agents in the society towards knowledge, i.e., knowledge mode while the social relation pulls the agents towards the socially dominant agents, i.e., knower mode, Fig. 1.

Figure 1 represents the legitimation codes. The values (±) along the X-axis represent the strengths of epistemic relation, and the values (±) along the Y-axis represent the strengths of social relation. Each quadrant of the model corresponds to a specific LCT code. The knowledge code emphasizes concepts, while the knower code emphasises design agents. The elite code emphasises both the conceptual and social dispositions, while for the relativist code neither conceptual nor social dispositions are required.

These legitimation codes conceptualize the dominant basis of success in any particular social context. While there is always a knowledge and knower dimension in any social context, the knowledge or knower dimensions may dominate the other based on the established norms within the context. In design disciplines, even

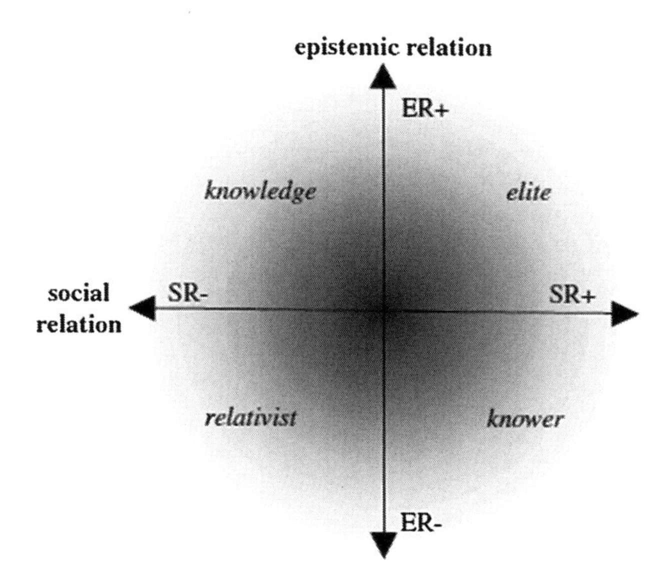

Fig. 1 Modes of legitimation of design practice (After [7])

Table 1 Legitimation codes across different design disciplines (based on [9])

Discipline	Epistemic relation (knowledge mode)	Social relation (knower mode)
Architecture	+	+
Fashion design	−+	++
Engineering design	++	−+

though such debates are common, there is little research in this area. More recently, [9] have compared the legitimation codes across different design disciplines including architecture, fashion design and engineering design, Table 1.

According to [9], while architecture tends to show greater balance between the knowledge and knower modes, in fashion design the knower mode tends to dominate, while in engineering the knowledge mode tends to dominate.

3 Why Computational Simulations as the Research Method?

Though [9] have shown that the relative contributions of the knowledge mode and the knower mode in legitimation and recognition of design practices varies across the studied disciplines, the understanding still remains subjective and abstract. For example, while [9] have established that legitimation of design practice in fashion is more knower driven than knowledge driven, it is currently difficult to establish how do these relative values compare, i.e., whether the relative contributions are linearly related or exponentially. The difficulty in establishing such relationships is further exacerbated because of the socially emergent nature of legitimation mechanisms, which are gradual and longitudinal processes, requiring observations and data collection over an extended period of time. In such a complex longitudinal scenario if the research were to rely entirely on real-world observations and case studies, the progress of our understanding of the legitimation practices and their relationships with associated parameters will remain rather slow and partial because of the time, cost and resources needed to build a comprehensive theory. Computational simulation test-beds using agent based models provide a complementary research method and infrastructure that can reduce the time, cost and resource constraints towards generating and testing the promising theories. Building such a research infrastructure, i.e., computational model may be a challenging task but once the initial infrastructure is created, it allows rapid extensions and explorations across different parameters and scenarios of interest.

Hence, this research adopts computational simulations to study the emergent social patterns across different design disciplines for what-if scenarios, based on informed assumptions derived from subjective understanding of the role of knowledge modes and knower modes across the different design disciplines. Since the underlying assumptions in the computational model and parameter values at

Table 2 Research questions, requirements and parameters

Research questions	Simulation requirements	
	Assumptions	Parameters
How does the emergent social pattern of design recognition vary across disciplines?	Forces between parameters	Design agents with different disciplinary backgrounds
How does the social pattern in multi-disciplinary design society compare to uni-disciplinary social environment?	Initial state (starting positions of design agents and concepts)	A set of concepts associated with different disciplines
		Demography, i.e., population mix, population size

time t = 0 are known in the simulations, the observed causal effects can be stochastically established with high confidence levels.

This research is planned bottom-up such that the initial simulations are conducted with as few assumptions and as few parameters as needed to generate and test meaningful hypotheses about the emergent social patterns across different disciplines. Table 2 lists the key research questions, and the corresponding minimal requirements to enable investigating these questions.

In order to compare the emergent social patterns resulting from differential knower and knowledge modes across different disciplines, we need to make informed assumptions about the relative knower and knowledge force of attractions across the different disciplines. Once force values are assumed they are kept constant across the different simulations, while the demography of design society is varied such that resulting patterns can be compared. Typically, designers work within teams and organizations, and affiliation to such teams creates a nested structure. These teams may have varied inclinations towards knowledge mode or knower mode and in the process affect the emergent patterns. Further, in order to understand the effects of scale, simulations are conducted with different population sizes, including more concepts and teams.

Once the findings of these simulations are known, additional parameters and assumptions can added to create complex scenarios. The research outcomes are intended to inform future empirical studies by identifying promising and potentially interesting hypotheses.

4 Computational Framework

The computational model is implemented in MASON [10], a java based multi-agent system. Each entity in the model that needs interaction, i.e., the designers, concepts and teams are implemented as agents within the simulation environment such that there are dynamic connections and forces of attraction between design agents, between design agents and concepts, between concepts, between design agents and teams, and between teams and concepts.

Representing each entity as agents gives them agency, which allows design agents, concepts and teams to move within the two dimensional space as per their interactions with other design agents, concepts and teams. Each entity has an influence radius such the force of attraction between any two agents is directly proportional to their influence radius.

Following the two modes in Legitimation Code Theory, the two dimensional space is defined by orthogonal axis with epistemic (knowledge) mode along the ordinate while the social (knower) mode is represented along the abscissa. As the interactions take place, the emergent social pattern including the knowledge and social dimensions of the design agents, concepts and teams are recorded and can be graphically observed. Following are the key assumptions and considerations in the model.

4.1 Forces of Attraction and Disciplinary Effects

Based on [9], three disciplinary backgrounds are considered to include architecture, fashion design and engineering. Design agents, teams and concepts are assumed to be associated to belong to one of these three disciplines such that the disciplinary background of a design agent determines how much it is influenced by knower mode and knowledge modes. Table 3 displays the assumed forces.

Following the findings in [9], forces corresponding to knower modes are highest for fashion disciplines and least for engineering disciplines. For example, constant **K** in agent–agent (knower) attraction is highest for fashion design agents and least for engineering design agents. On the other hand, forces corresponding to knowledge mode are highest for engineering design agents and least for fashion design agents. Accordingly, constant **E** in agent-concept (knowledge) attraction is highest for engineering design agents and least for fashion design agents. Similarly, other forces and constants are assumed using similar arguments.

4.2 Starting Conditions

At the start of the simulation, i.e., at $t = 0$, all the entities in the simulation environment including design agents, concepts and teams start with a pre-defined position the two dimensional space, defined by their social dimension and knowledge dimension. The initial scenario across all the entities, i.e., where the different agents, concepts and teams start from may influence the emergent social pattern. However, since the focus of the study reported in this paper is the effects of disciplinary backgrounds and not the initial conditions, in all the simulations the starting conditions remain the same.

Table 3 Assumed values for knowledge and knower attraction forces

Entities	Force	Discipline conditions
1 Agent (A^1)–Agent (A^2)	Constant **K** × (InfluenceRadius A^1 × InfluenceRadius A^2)/ (Square of social distance between A^1 and A^2)	For design agents IF discipline is architecture **K** = 100 IF discipline is fashion design **K** = 1000 IF discipline is engineering **K** = 1
2 Agent (A^1)–Concept (C^1)	Constant **E** × (InfluenceRadius A^1 × InfluenceRadius C^1)/(Square of distance between A^1 and C^1)	For design agents IF discipline is architecture **E** = 100 IF discipline is fashion design **E** = 1 IF discipline is engineering **E** = 1000
3 Concept (C^1)–Concept (C^2)	Constant **D** × (InfluenceRadius C^1 × InfluenceRadius C^2)/(Square of distance between C^1 and C^2)	IF C^1 and C^2 belong to same discipline **D** = 100 ELSE **D** = 1
4 Agent (A^1)–Team (T^1)	Constant **K** (Similar to 1)	Same as 1
5 Team (T^1)–Concept (C^1)	Constant **E** (Similar to 2)	Same as 2
6 Team (T^1)–Team (T^2)	Constant **K** (Similar to 1)	Same as 1

4.3 Analyzing the Outcomes

Given that there are limited prior study to benchmark the findings, this research poses challenges in analyzing the outcomes. Hence, visual representations of emergent patterns across different scenarios provide useful preliminary comparison across different cases. Developing the model requires iterations to calibrate the assumptions.

5 Simulation Results and Discussion

The first set of simulations was conducted to compare the effects of disciplinary backgrounds. A summary of the conducted simulations is presented in Table 4.

For each simulation case, 30 simulation runs were conducted. A snapshot of the typically emergent social pattern for the different cases is presented on the left hand side of Fig. 2, while the plot on the right hand side of Fig. 2 shows a pattern created by averaging the results from 30 simulation runs for each case.

Table 4 Simulations conducted to compare patters across different disciplines with population of 32, 6 teams and 12 concepts

Case	Distribution/Demography
1	All design agents, teams and concepts associated with architecture
2	All design agents, teams and concepts associated with fashion design
3	All design agents, teams and concepts associated with engineering design
4	Mixed population with equal distribution of architecture, fashion design and engineering design agents, teams and concepts

As seen in the first three cases of Fig. 2, the computational model simulates distinct patters of social emergence of design practice across different disciplines. A society with only architectural design agents (red), teams (black) and concepts (blue) grows towards the knowledge dimension but there is increasing pull for design agents towards the knower (social) dimension once higher levels of knowledge dimension is achieved. The knowledge leaders in architectural design society also grow along the knower dimension, becoming attractors for other agents to follow them through the knower mode. Though this pattern appears to be the result of assuming balanced knowledge and knower forces for architectural design agents, unlike the observed pattern, it was expected that the architectural design agents, teams and concepts would move at approximately diagonal across the two axes.

A society comprising of only fashion design agents, teams and concepts also shows a trend towards leaders and followers. However, unlike the architectural design society, in fashion design society agents at lower levels of knowledge tend to have greater diversion and attraction towards the knower mode. As a result, the knowledge divide between fashion design leaders and followers tends to increase over time because some of the followers at lower knowledge levels develop greater attraction for social dimension. While fashion design agents were assumed in the model to have greater attraction towards knower mode, we had not expected the emergence of differential knower level pulls corresponding to agents' relative knowledge levels. Furthermore, simulation results indicate that in the fashion design society the rate of growth of design agents (brown) may outpace teams (black).

A society of engineering design agents primarily follows the knowledge dimension, and the design agents (grey) and teams (black) grow at comparable rate. The society of engineering design agents shows patterns that were expected and are consistent with the assumptions.

Given that the three different design societies show distinct emergent patterns that broadly conform to the underlying legitimation code theory of design practice for the known disciplines, the findings from case 4 should provide insights into emergent social patterns of design practice in a multidisciplinary design society, which has not been empirically studied so far. It was expected that the multidisciplinary design society will show a greater balance between knowledge and knower driven design practice, with design agents distributed along both

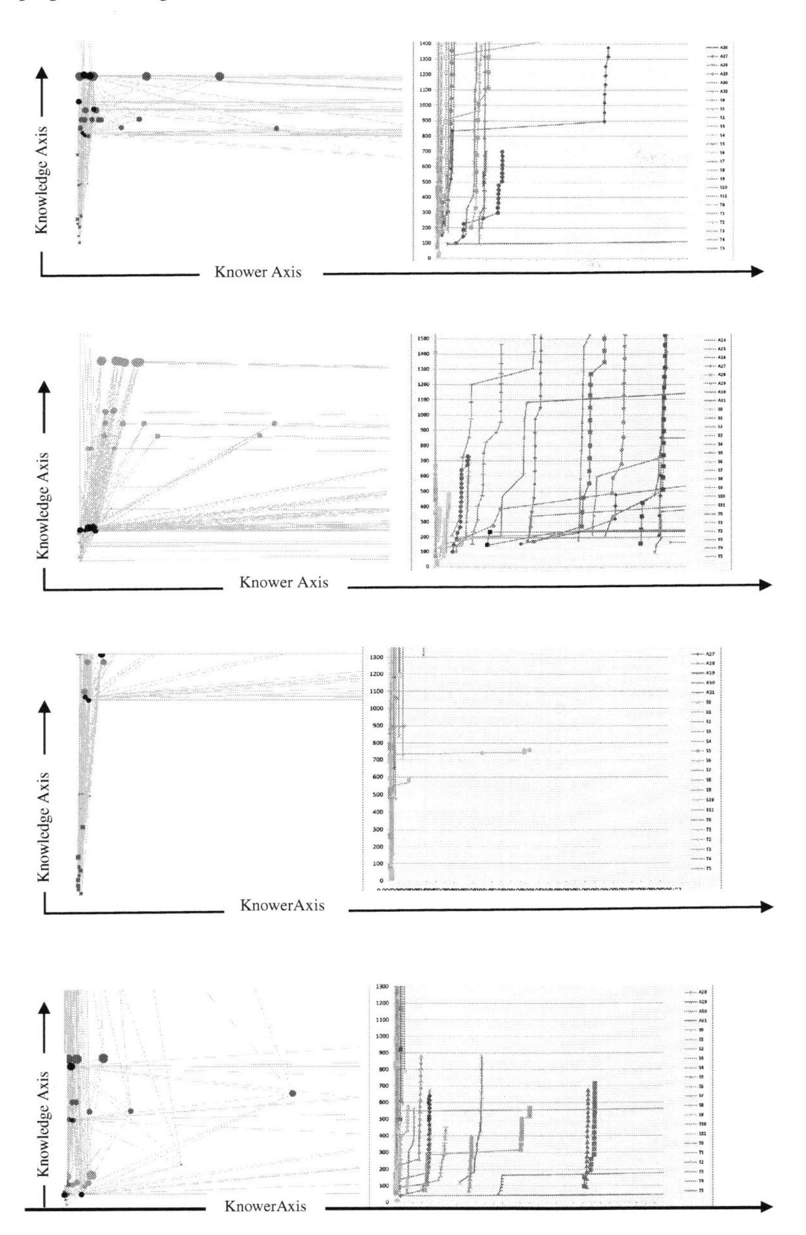

Fig. 2 Emergent social patterns across different simulation cases for cases in Table 4 *Case 1*: Only architecture design agents. *Case 2*: Only fashion design agents. *Case 3*: Only engineering design agents. *Case 4*: Multidisciplinary society with architecture, fashion and engineering design agents

dimensions. Though the emergent patterns in multidisciplinary design society does create a balance (compare right hand side of case 4 with the right hand side patterns of cases 1, 2 and 3), the design agents from different disciplinary backgrounds tend to cluster along different knowledge levels. Engineering design agents grow faster towards the knowledge dimension (being knowledge driven), during the same period the fashion design agents remain at lower knowledge levels as compared to the architecture design agents. Similarly, the growth of architecture and fashion design agents along the social dimension is greater than engineering design agents during the same duration. Thus, the simulation results in case 4 suggest that unless there is greater overlap in design concepts and practices across the different design disciplines, segregation of design agents across disciplinary groups is likely to emerge creating knowledge and social gaps that will widen over time.

In order to test whether the social pattern observed across the different cases depends on the number of design agents, teams and concepts, all the simulations listed in Table 4 were repeated with double the number of entities, i.e., 64 design agents, 12 teams and 24 concepts. The emergent social patterns in simulations with 64 design agents are similar to the observed patterns in simulations with 32 design agents. Findings suggest that the differential social pattern of design practice across different disciplines is potentially scalable, which can be tested further by varying the population sizes by higher order. Similarly, by changing the initial values of the parameters we can investigate how this pattern varies with the position of the influential design agents?

In the reported simulations, design agents are affiliated to teams and organizations, and the legitimation practice in the design society is mediated by this nested structure such that the teams and the design agents within these teams both influence the emergent legitimation practice. The role of this nested social structure in shaping the legitimation practice can be better understood by comparing these findings with results from simulations where the nested structure is not considered.

6 Conclusion

A computational model of legitimation of design practice is developed as a simulation test-bed, based on the specialization principles of legitimation code theory. The computational model is developed as a research infrastructure that supports generating and testing what-if scenarios with various design societies, starting with the comparison of the legitimation practices in uni-disciplinary design societies against multi-disciplinary design society. The emergent social patterns across the different cases involving design agents associated with architecture, fashion design and engineering are broadly consistent with the expected patterns, as known from the literature. The consistency of the simulation results with empirical data provides internal validity of the simulation platform, suggesting that the assumptions within the model can be relied upon for further studies. Findings from the simulation results suggest that irrespective of the social composition and disciplinary backgrounds, clusters of design agents are created at different knowledge levels. This clustering of agents in a multidisciplinary society is marked by disciplinary segregation. Though the assumptions and the simulations were based on relative epistemic and social code values for the studied disciplines, i.e., architecture, fashion and engineering design, the studies can be extended to study and compare the legitimation practices across other design disciplines and sub-disciplines.

Nonetheless, the primary challenge at this stage is to externally validate the model and simulation results because there are no empirical studies to benchmark the comparative patterns of legitimation practice across uni-disciplinary and multi-disciplinary design societies. Thus, the current scope of the computational model is limited to simulate what-if scenarios and generate potentially interesting hypotheses to be investigated empirically. The primary contribution of this paper is to describe the development, and demonstrate the usefulness, of an agent based simulation model in studying the legitimation of design practice as a socially emergent phenomenon.

Acknowledgments This work is supported in part by a grant from the US National Science Foundation under Grant no: NSF SBE-0915482. Any opinions, findings, and conclusions or recommendations expressed in this material are those of the authors and do not necessarily reflect the views of the National Science Foundation.

References

1. Carley K (1994) Sociology: computational organization theory. Soc Sci Comput Rev 12:611–624
2. Sosa R, Gero JS (2005) A computational study of creativity in design. AIEDAM 19(4):229–244
3. Maton K (2000) Languages of legitimation: the structuring significance for intellectual fields of strategic knowledge claims. Br J Sociol Educ 21(2):147–167
4. Carvalho L (2010) A sociology of informal learning in/about design. Ph D Thesis, Department of Architecture Planning and Design, The University of Sydney
5. Bourdieu P (1983) The field of cultural production, or: the economic world reversed. Poetics 12(5):311–356
6. Nonaka I (1994) A dynamic theory of organizational knowledge creation. Organ Sci 5(1):14–37
7. Maton K (2006) On knowledge structures and knower structures. In: Moore R, Arnot M, Beck J, Daniels H (eds) Bernstein: policy, knowledge and educational research. Routledge, London, pp 44–59
8. Maton K, Muller J (2007) A sociology for the transmission of knowledges. In: Christie F, Martin JR (eds) Language, knowledge and pedagogy: functional linguistic and sociological perspectives, Continuum, London pp 14–33
9. Carvalho L, Dong A, Maton K (2009) Legitimating design: a sociology of knowledge account of the field. Des Stud 30:483–502
10. Luke S, Cioffi-Revilla C, Panait L, Sullivan K, Balan G (2005) MASON: a multiagent simulation environment. Simulation 81:517–527

Improving Common Model Understanding Within Collaborative Engineering Design Research Projects

Andreas Kohn, Julia Reif, Thomas Wolfenstetter, Konstantin Kernschmidt, Suparna Goswami, Helmut Krcmar, Felix Brodbeck, Birgit Vogel-Heuser, Udo Lindemann and Maik Maurer

Abstract In collaborative engineering design research projects, the efficient and effective collaboration between the individual research groups is an important factor for the overall success of the project. However, different disciplines, research foci and personal backgrounds influence successful collaboration. The work presented in this paper focuses on the collaboration concerning models developed by individual subprojects and proposes a framework to support the necessary collaboration. Existing shortcomings are identified within a large scientific project in combination with an extensive literature review. The developed framework is based on two checklists in form of a model requirements document and model specification containing important aspects of models that need to be considered and communicated during the research project. Beyond the successful application for supporting collaboration in engineering design research projects, the insights and implications of this paper can be transferred to all other collaborative projects, where models are to be communicated between individuals or teams.

Keywords Modeling · Collaboration · Engineering design · Communication

A. Kohn (✉) · U. Lindemann · M. Maurer
Institute of Product Development, Technische Universität München, Munich, Germany
e-mail: kohn@pe.mw.tum.de

J. Reif · F. Brodbeck
Organisational and Economic Psychology, Ludwig-Maximilians-Universität München, Munich, Germany

T. Wolfenstetter · S. Goswami · H. Krcmar
Chair for Information Systems, Technische Universität München, Munich, Germany

K. Kernschmidt · B. Vogel-Heuser
Institute of Automation and Information Systems, Technische Universität München, Munich, Germany

A. Chakrabarti and R. V. Prakash (eds.), *ICoRD'13*, Lecture Notes in Mechanical Engineering, DOI: 10.1007/978-81-322-1050-4_51, © Springer India 2013

1 Introduction: Motivation and Problem Statement

In collaborative research projects, individual research groups work together in order to achieve common project goals. Efficient and effective collaboration plays an important role for the overall success of a project [1]. However, several barriers exist and have to be overcome to ensure the success. In this paper, we focus on the barriers, and the support mechanisms for overcoming these barriers in collaborative engineering design research projects. In general, the main goal of design research is to make design more efficient and effective in order to enable design practice to develop successful products [2]. The development of understanding (via models and theories) and the subsequent development of support are most important results of design research [2]. Several challenges emerging from the nature of design research occur in collaborative engineering design research projects (see Fig. 1). As design research is strongly linked to design practice, trends and changes in design practice influence design research. For example, today's products and corresponding innovation processes are characterized by increasing level of complexity [3]. Furthermore, a trend from "normal" mechanical products to a combination of products and services (Product-Service-Systems) is observed [4]. As a consequence of these trends, the research field splits up into different interrelated research objects which in turn increases the number of disciplines and research groups involved in research projects. Each of these research groups holds very different views (such as models) on certain research objects. As the research objects are interrelated, the need for communication and mutual adjustment of the emerging models is important. However, models evolve and change over time as the research progresses which might impede efficient collaboration. Additionally, different disciplines, research foci and diverse personal backgrounds influence collaboration concerning the emerging models. For example, one discipline builds a model which is misinterpreted by another discipline as the semantics of the model is unknown.

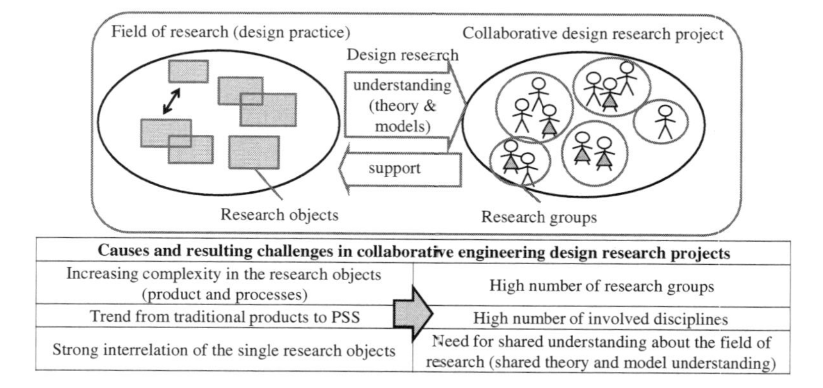

Causes and resulting challenges in collaborative engineering design research projects	
Increasing complexity in the research objects (product and processes)	High number of research groups
Trend from traditional products to PSS	High number of involved disciplines
Strong interrelation of the single research objects	Need for shared understanding about the field of research (shared theory and model understanding)

Fig. 1 Causes and resulting challenges in collaborative engineering design research projects

As models play an important role in design research, we propose a framework to support collaboration in design research projects by providing a more efficient and effective way of communicating emerging models among the various research groups. The framework is developed and tested within the collaborative research center 'SFB 768—Managing cycles in innovation processes—Integrated development of product-service-systems based on technical products'. In this paper, this collaborative research center serves as a use-case for deducing requirements and evaluating the proposed framework. The overall aim of the project is the understanding and design of innovation processes concerning the specific characteristics and interdependencies of cycles in innovation processes. The project consists of 14 subprojects (research groups) within 7 institutes at 4 faculties (mechanical engineering, psychology, economics and computer science). For details about the project SFB 768 please refer to [5] or www.sfb768.de.

This paper is organized as follows: First, the scientific method of our research is described. Second, we provide the theoretical and conceptual background for our work such as the necessary definitions and a state of the art overview concerning model theory, model quality and collaborative modeling. We then explain requirements for collaborative modeling in engineering design research projects based on our use-case. Next, we summarize our main findings and show the developed framework that bases on two checklists in the form of a model requirements document and a specification document. The framework is tested within the use-case and our main findings are presented in the next section. We conclude the paper with a discussion of our findings and provide an outlook for future research.

2 Method

We first conducted an extensive literature review in order to gain a deeper understanding about scientific modeling in diverse research disciplines, good modeling practices and the fundamentals of collaborative modeling. The results were structured and summarized. Second, we carried out three workshops in which participants discussed different aspects of 'modeling' in the context of the SFB 768. Participants (31 members of the SFB 768) circulated through the workshops so that everyone took part in each workshop. Participants' inputs were documented via written protocols. At the end of the workshops, a plenum discussion with all participants was held in order to integrate the results of the workshops and deduce requirements for the framework. The developed framework was tested within the SFB 768 iteratively. In a first step, it was tested for its applicability by 12 members of the SFB and revised accordingly. Then, it was applied by all the members of the project. Qualitative feedback was collected and potentials for improvement were derived.

3 Theoretical and Conceptual Background

3.1 Model Theory and Variety of Models Between Disciplines

In our context, we use the term "model" in a general definition according to [6]: "a model is a representation of an original. It represents relevant attributes for a model user and is used for a certain period of time to serve a certain purpose." Various types of models exist, differing in their purpose, representation, language, validation or application. Each discipline uses its own models and has a different interpretation of this term. For example, a scientific model as used in psychological research is "a set of representations, rules, and reasoning structures that allow one to generate predictions and explanations" [7]. In contrast, product models in engineering design always contain product-related information [8]. The modeling languages used to represent a model also differ between disciplines. A modeling language consists of a textual or graphical notation, called syntax, and the defined meaning of the notation, called semantics [9]. Basically every 'real' language can be used for a textual modeling. However, formal modeling languages avoid ambiguities, facilitate the computer-based model interpretation and make the models easier understandable for people of other disciplines [10].

3.2 Good Modeling practice

There are several rules and guidelines for good modeling. Perhaps the most important aspect about modeling is that a model has a clear purpose (e.g. to develop knowledge and understanding about a system) and not just to create the model itself [11]. The process of modeling should therefore be evolutionary since information about the system is obtained step by step during the act of modeling [12]. In this context modelers should start with simple assumptions and add complications only as necessary [13]. A basic trade-off in modeling is that models should be comprehensive but yet be sufficiently detailed to reflect important characteristics of a system in a realistic way. The amount of details should be based on the modeling objectives and only components that are relevant for decision-making need to be included [12]. When complex systems are to be mapped to a model one should refer to the divide-and-conquer principle. In general it is easier and also more useful to develop a set of simple interrelated models than to squeeze everything into a large and complex model. This applies especially to situations in which modeling is done by a team of people [13]. Different approaches for supporting modeling activity exist in various contexts. For example, [14] proposes that the development of simulation models can be supported by using a model requirements document and a model specification.

3.3 Collaborative Modeling: Challenges and Support

In order to use the knowledge and various perspectives from different experts to foster problem solving, good decision-making in complex contexts, and to innovate products and services, it is becoming more and more important to create multidisciplinary project teams [15]. A modeling project in an interdisciplinary team should involve close interaction and information sharing among the various project participants when formulating a problem and building a model. This interaction causes inaccuracies that need to be identified early and corrected efficiently [12]. Also, barriers concerning information and knowledge sharing as mentioned in [16] should be considered adequately. As support for collaborative modeling for distributed teams, IT-support can be a prime enabler for successful collaboration [17]. In order to facilitate distributed teamwork, so-called groupware tools should support communication, document sharing, group discussion and decision-making, as well as calendaring and scheduling. Furthermore teams need access to organizational and external information sources [18].

4 Deduction of Requirements for Collaborative Modeling in the Use-Case

Based on the above described background, we analyzed the needs of individual researchers and deduced requirements for supporting collaborative modeling in the context of our use-case—the transdisciplinary project SFB 768—through workshops. The workshops were grouped according to three main aspects of modeling that influence collaboration: The purposes of the individual models in respect to the overall goal of the project, the intended modeling languages to be used, and further boundary conditions for model integration were discussed. In each workshop, the requirements of the individual research groups were collected and discussed. At the end of each workshop, directives for supporting collaboration were deduced and combined through a plenum discussion.

In the first workshop, the interrelation between the purposes of the individual models and the overall purpose of the research project was discussed. It was identified, that different purposes of the models from the individual research groups as well as different views on the research field exist. This correlates with the initial situation in collaborative projects described above. Most importantly it was concluded, that each model clearly shows its connection to the overall purpose of the research project.

In the second workshop requirements for a modeling language in the single subprojects and the goals and benefits of a jointly used modeling language were discussed. As the different subprojects belong to different domains, and thus the modeling subjects differ strongly, a multitude of requirements for adequate modeling languages exist. While many requirements for a modeling language are discipline specific, it was identified that it would be beneficial to use a common modeling language in order to reduce complexity, understand models of other subprojects more easily and to present the entire research project more cohesively to other researchers and experts. Depending on the number of involved disciplines, it is possible that a single modeling language is not sufficient for all subprojects. In this case, model-clusters should be identified and within these clusters a common modeling language should be chosen.

The third workshop focused on model integration. The analysis of the discussion shows that efficiency is a basic purpose of model integration: redundancies may be reduced, mistakes can be eliminated, time and effort can be saved. A further purpose of model integration is transparency: modelers can develop a meta-understanding of the landscape of models, thus can broaden their knowledge of the whole model system which in turn fosters an advanced communication between the different disciplines. A third purpose of model integration is interconnectedness: relationships between models can be generated and dependencies can be illustrated. Finally, it is a purpose of model integration to enable a holistic perspective on the object of investigation. The following requirements concerning model integration were mentioned: models that are to be integrated should have a clear model type, a defined model context, internal and external interfaces and clear signals at the interfaces. Interfaces between models should be standardized so that they can be applied more easily. There should be a common or shared way of describing inputs and outputs and the models' level of abstraction should be clearly communicated.

The workshops showed that there is a strong need for supporting collaboration concerning the emerging models. In the three aspects a huge variety between the individual models of the subprojects concerning both the modeling process and the expected result was identified. As a conclusion of the workshops, the following aspects were identified as main requirements for supporting the collaborative modeling activities:

- Enable a continuous overview of the models in the individual subprojects.
- Enable a standardized description and documentation of the individual models.
- Provide a guideline and checklist for the modelers to identify which important aspects of modeling have to be considered.
- Identification of synergies between the single models of the projects. For example, common access to needed information or common evaluation scenarios can be identified to reduce the effort within the individual subprojects.
- Enable and identification of interfaces between the individual models.
- Improve deduction of requirements for possibly project-spanning modeling languages or tools.

5 Framework for Improving Common Model Understanding

This section shows the developed framework for supporting common model understanding in collaborative engineering design research projects. It was developed on the basis of the presented literature review and the identified requirements within the use-case SFB 768.

The core elements of the framework are two model checklists (a model requirements document and a model specification) that are exchanged between the researches via a data exchange platform (see Fig. 2). They contain aspects of the models that are most important for the communication between the research groups (see next sections for details). Those two checklists are to be filled out by the individual researchers for each emerging model. The model requirements document is to be filled in the early phase of the project; the model specification in later phases. Organizational aspects (meetings, involvement of experts, milestones, personal support and room for discussion) serve to support the successful application of the framework. Individual and team aspects (motivation, team processes and emergent states as trust or cohesion) complement the organizational aspects. The overall goal of the project (here: managing cycles in innovation processes) serves as "roof" for the framework and influences all tasks.

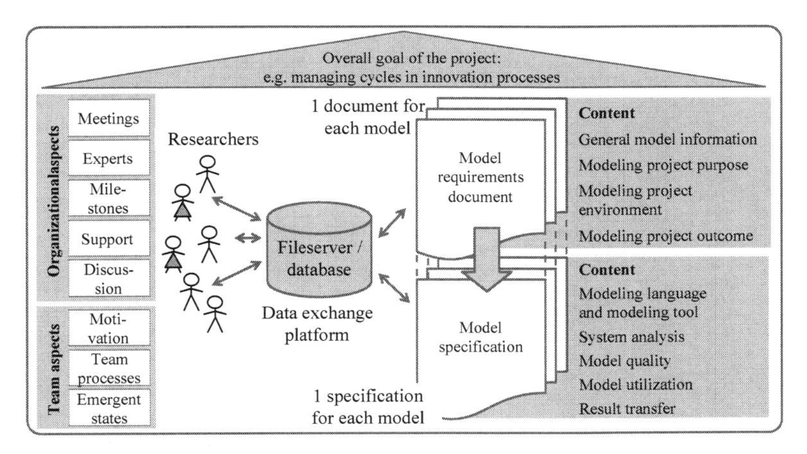

Fig. 2 Components of the framework for improving common model understanding

5.1 Model Requirements Document

The first of the two checklists is the "model requirements document" which is used to describe general specifications of the model and the modeling project at an early stage. The items are grouped based on their subject (Fig. 3). Basically, the

General model information	
Model name	A clear and unambiguous name should be chosen for each model
Subproject	Name of the subproject, which carries out the research
Modeling project purpose	
Purpose of the model	Which goal shall be achieved through the model? Are there any subgoals?
Original and system boundary	What is the original of the model (object of observation)? Can a model boundary be drawn already?
Classification of the model in the objective of the subproject	Which role plays the model in the objective of the subproject
User of the model	Who will use the developed model? (e.g. use in industry or science)
Relation to other models within the subproject	If other models exist or are planned within the subproject, which relations exist between this model and the other models?
Relation to models of other subproject s	Which relations between this model and existing/planned models of other subprojects already exist?
Requirements of the interfaces to other models	Which requirements for the interfaces to other models already exist?
Minimum requirements of the functionality of the model	When is the functionality of the model reached? What are the minimum requirements of the model concerning its functionality?
Timetable	When shall the model be completed? When shall the results of the model be used? What are important milestones during the development of the model?
Modeling project environment	
Institute/subproject	Which institute is responsible for the subproject/model?
Responsible person	Which persons are involved?
Previous knowledge	Which existing information/knowledge can be used for the modeling project? Hereby knowledge of the institute, previous projects and personal knowledge should be considered
Modeling tool/language	Which modeling languages and tools are available and which one shall be used? Are certain languages/tools already determined (reasons)?
Modeling project outcome	
Presentation of results	Where and in which way shall the results be presented? (Graphical/verbal…)
Documentation	In which way shall the model/its usage/experiments be documented?
Requirements for a further use	Which requirements have to be fulfilled if the model shall be used beyond its main purpose? (e.g. regular updates, transfer to other projects…)
Necessary effort for maintenance and care	Is a regular maintenance and care necessary for the use of the model?
Visualization	
Illustration of the model	An illustration of the model should be added for a better understanding. As many details are not yet defined in the early stage the illustration should focus on the modeling purpose or expected goals.

Fig. 3 Model requirements document

definition of requirements and constraints is conducted in the model requirements document while concrete activities are described in the model specification (see Sect. 5.2).

If necessary, the single aspects of the 'Project purpose', the 'Project environment' and the 'Project outcome' have to be specified during the project realization. Hereby the traceability of changes will be ensured by marking changes accordingly.

5.2 Model Specification

In the "model specification" checklist, detailed information about the elaboration of the model has to be provided. The focus lies on the description of the actions and execution of the necessary steps. The model specification can be complemented and added during the modeling project. It is structured according to the main steps of the modeling project in "modeling language and tool", "system analysis", "model quality", "use of the model" and "delivery of results" Fig. 4). Again, for each item within the three groups, a detailed description is given in the figure below.

Modeling language and modeling tool	
Selection of modeling language	How will the modeling language be selected? Are any adaptations of an existing modeling language necessary?
Selection of modeling tool	How will the modeling tool be selected? Are any adaptations necessary?
System analysis	
Information / data requirements	What are the requirements regarding information / data that are needed for modeling? (range/ quality/ format, etc.)
Data- / information transfer	How will existing and available data be used? Is there any format conversion necessary?
Data collection /execution of system analysis	In which way shall data/information about the original be acquired? Description of the implementation of experiments / studies to obtain data.
Presentation and documentation of information	How is the necessary information shown? Description of processing and documentation of the collected data.
Model quality	
Model verification	Is the model consistent to the mental model and the requirements? How shall the model be verified? (also cross-verification with interrelated sub-projects is necessary)
Model validation	Is the model consistent with reality? How shall the model be validated? Which limitations arise regarding the model scope?
Model applicability	How can the practical applicability be assured? (e.g. Case-Study)
Model utilization	
Determination of the experiment-plan / utilization-plan	Please provide a detailed plan stating: Who will use the model? How will the final model be used? What is the intended time frame?
Model adaption and improvement	How will the model be used, adapted and improved? (e.g. iteratively)
Evaluation of results	How will experiments / utilization of the model be evaluated? How will user experiences be captured?
Result transfer	
Working- and presentation documents	How will experiment results and the experiences regarding model utilization be documented? (Description of the intended approach)
Archival storage of model description, parameter sets and results	How will the model description, the necessary data and the results be documented or archived? (Description of the intended approach)
Model- and result transfer	How will the results be transferred to other sub-projects? How can be assured that the model is used in the designated way?
Compatibility of interfaces	How can the cross model functionality be assured regarding interfaces to other sub-projects?

Fig. 4 Model specification

5.3 Application of the Framework and First Experiences

After establishing the basis of the framework and the first drafts of the checklists, the members of the collaborative research project were asked to fill in the checklists. Within each institute, one "expert" was trained in the use of the checklists and the communication platform and served as multiplier for training the other members. So far, both checklists were filled in for 32 models. The content of the checklists was discussed within team meetings und the individual models were presented to the team. It was identified, that the grouping of the single aspects serves well for a structured discussion about the models. Two month after the checklists were filled in, a workshop with all project members was conducted. The positive qualitative impressions of the participants encourage the originally intended purposes of the framework. The usability of the data exchange platform was identified as a potential for improvement as there is currently no automatic versioning available. Furthermore, the individual entries in the checklists can be misinterpreted and sometimes needed personal explication. Despite some room for improvement, the above described requirements for advancing collaboration concerning the emerging models were commonly judged positively.

6 Discussion and Future Work

The problems that are faced by society and academia today are often not solvable by a single research discipline, and therefore call for the establishment of large multi-disciplinary research projects. However, researchers coming from different disciplines work with different research paradigms and this results in significant difficulties in collaborating with researchers from other disciplines. This can have a negative effect on the success of the overall research project. The objective of this paper was to provide a mechanism that facilitates collaboration and common understanding in trans-disciplinary research projects.

We achieved this by developing a framework that allows the diverse research groups to articulate and formalize their requirements regarding models, and a common platform for exchanging this formalized specification of models. Our framework, and in particular the proposed model checklists (requirements and specification documents) serve as initial step in facilitating collaboration in trans-disciplinary projects. Intensive workshops with a diverse group of researchers regarding what they understood and expected out of the models formed the basis for the framework. The usefulness of the checklists were further validated by asking researchers from various disciplines to fill out the checklists with their understanding and expectations regarding the model, and also to comment on the relevance and usefulness of the checklists.

While our use-case allowed us to identify the salient features that are important for a common understanding on models and modeling, the generalizability of the developed framework needs further research investigating its usability in the context of other collaborative projects. Further, the two checklists developed should be tested in different contexts and by different users and refined iteratively based on their feedbacks. This will enhance their completeness for capturing all requirements. In our use-case and evaluation, these checklists were filled in the initial stages of the project. It will be more interesting to see how the model requirements and specifications changes over time as the project progresses. Therefore, to be more useful, the checklists will be updated in regular intervals to reflect current state of understanding within the project or its various sub-projects. This can have significant implications on how the checklists help in creating a shared common understanding, as the understanding evolves over time.

Another potential shortcoming of the developed framework is, that it was both developed and tested within the same research project. The external validity of the framework therefore will be increased by testing it in the context of a different research project, or getting independent experts not involved in the project to review and evaluate the usefulness of the checklists. Moreover, it is planned to evaluate the framework through a longitudinal study. Participants working in collaborative projects fill out the checklists at the beginning of the project and then semi-structured interviews are carried on in later stages to measure the extent to which their overall understanding regarding models in the project have improved because of applying the checklists.

Acknowledgments We thank the German Research Foundation (Deutsche Forschungsgemeinschaft—DFG) for funding this project as part of the collaborative research center 'Sonderforschungsbereich 768—Managing cycles in innovation processes—Integrated development of product-service-systems based on technical products'.

References

1. Mostashari A (2011) Collaborative modelling and decision-making for complex energy systems. World Scientific, Singapore
2. Blessing LTM, Chakrabarti A (2009) DRM, a design research methodology. Springer, London
3. Clarkson PJ, Eckert CM (2005) Design process improvement: a review of current practice. Springer, London
4. Bitner MJ, Brown SW, Meuter ML (2000) Technology infusion in service encounters. J Acad Mark Sci 28(1):138–149
5. Hepperle C, Orawski R, Langer S, Mörtl M, Lindemann U (2011) Temporal aspects in lifecycle-oriented planning of product-service-systems. In: International conference on research into design (ICoRD11), Bangalore
6. Stachowiak H (1973) Allgemeine Modelltheorie. Springer Wien, New York
7. Schwarz CV, White BY (2005) Metamodeling knowledge: developing students' understanding of scientific modeling. Cogn Instr 23(2):165–205
8. Kohn A, Lutter-Günther M, Hagg M, Maurer M (2012) Handling product information—towards an improved use of product models in engineering design. In: 12th International design conference (DESIGN 2012), Dubrovnik

9. Harel D, Rumpe B (2004) Meaningful modeling: what's the semantics of "semantics"? Computer 37(10):64–72
10. Witsch M, Vogel-Heuser B (2012) Towards a formal specification framework for manufacturing execution systems. IEEE Trans Ind Inf 8(2):311–320
11. Crout N, Kokkonen T, Jakeman AJ, Norton JP, Newham LTH, Anderson R, Assaf H (2008) Chapter two good modelling practice. Dev Integr Environ Assess 3:15–31
12. Pritsker A, Henriksen J, Fishwick P, Clark G (1991) Principles of modeling. In: Proceedings of the 1991 winter simulation conference, pp 1199–1208
13. Pidd M (1999) Just modeling through: a rough guide to modeling. Interfaces 29:118–132
14. VDI 3633 (1997) Performance specification of the simulation study. Technical Rule, Germany
15. Brodbeck FC, Guillaume YRF (2012) Effective decision-making and problem solving in projects. In: Wastian M, Braumandl I, v. Rosenstiel L′, West MA (eds) Applied psychology for project managers: a practical handbook for successful project management. Springer, Berlin
16. Riege A (2005) Three-dozen knowledge-sharing barriers managers must consider. J Knowl Manage 9(3):18–35
17. Krcmar H (2010) Informationsmanagement, vol 5. Springer, Berlin u.a., p 658
18. Vick RM (1998) Perspectives on and problems with computer-mediated teamwork: current groupware issues and assumptions. J Comput Documentation 22(2):3–22

Issues in Sketch Based Collaborative Conceptual Design

Prasad S. Onkar and Dibakar Sen

Abstract Sketch is the primary mode of concept exploration and development in the early stages of design. In a global product development scenario, collaboration in the early fluidic stages could minimize design conflicts and enhance design understanding. In this work, the potential modalities of interactions among the collaborating designers in early stages using 3D articulated sketch based concept models are proposed. Depending upon the spatio-temporal characteristics, issues related to the hardware, software and network infrastructure in four possible collaboration scenarios are identified. It is argued that the 3D sketching techniques, earlier developed for haptics-enabled, articulated, conceptual designs, can be extended to support design collaboration using evolving concept sketches. The attributes of the 3D sketches and the requirements of the different collaboration scenarios seem to have an implicit and synergistic relationship. Results of initial experiments using TCP/IP based view sharing and interaction on an articulated 3D concept-sketch for elemental collaboration exercises are also reported.

Keywords Conceptual design · Collaborative design · 3D sketching

P. S. Onkar · D. Sen (✉)
Center for Product Design and Manufacturing, Indian Institute of Science, Bangalore, India
e-mail: dibakar@cpdm.iisc.ernet.in

P. S. Onkar
e-mail: prasad@cpdm.iisc.ernet.in

A. Chakrabarti and R. V. Prakash (eds.), *ICoRD'13*, Lecture Notes in Mechanical Engineering, DOI: 10.1007/978-81-322-1050-4_52,

1 Introduction

Product design is becoming pervasive and ubiquitous. Product design and development activities have become distributed. The products are being developed in different places by different people at different time [1]. Designing of products is becoming more complex to serve better to the increasing customer needs. Increasing complexities call for expertise from different domain. Collaborative design [2] is one of the important strategies which help to improve the overall productivity of the organization by enabling interaction between different stakeholders of the product being designed.

In the conceptual design process, interactions happen between the designers about the product concept and also with the design representations of the concept like sketches, CAD models and physical models/prototypes. Collaboration in the crucial phase of conceptual design has been envisaged to improve the design outcome [3].

In a global product development scenario, complex products are designed and developed in a distributed environment. It is, therefore, imperative to have effective collaboration among the participating teams which are often from diverse domains of expertise. Collaboration is about interaction among agents to work toward a common objective. It is characterized by leveraging the complementary core competencies of participating agents.

Collaborative creative exploration of the product concept is shared based on the domain expertise viz. structures, systems, controls, styling etc., of the participating parties. Further subdivision is limited by the meaning associated with a chunk of work. For example, when creating a concept sketch of a product, each designer in a team cannot be assigned the responsibility of drawing some strokes. Rather, each designer would individually explore a meaningful entity of the product, subassembly or component. A system for facilitating collaboration should support, record and assimilate the interaction among the agents. In the context of product design, product is the central theme for all interactions; clarifying and disambiguating the appearance, structure, behavior and function of the product. Thus, it is a peculiarity of collaborative design that we need to consider not only the interactions *about the design* but also that *with the design*. Queries about design require interaction among the designers; interaction with the design requires the representation to be responsive enough to support independent exploration. Such independent explorations are naturally supported by physical prototypes. However, the 2D rendered sketches of the product as presently being used as the reference, can barely support any query related to interaction with the design. Designers use sketching for different purposes in the early stages of design like, *visual externalization of ideas*, *capturing fleeting ideas* and for *exploration of ideas and concepts* [4]. Also, sketching is used as an aid for mental simulation of the physical behavior of the concept called as ***behavior simulation*** [5]. In this, the sketched components provide the visual feedback about the feasibility of the perceived combination and help a trained designer to visualize the component interaction and to evaluate if the product would work as desired. Apart from this, sketching is used as a medium of interaction in design ***collaboration***. In this, sketch forms a platform for negotiations and propositions on the ideas.

Concept sketching is a designer centric rather than a collaboration centric activity. Even today, designers prefer "paper and pencil" type sketching in the early stages of design for concept exploration because of their training, experience and skill. However, such 2D sketches of 3D products have inherent limitations such as (a) it requires special skills to represent the 3D objects on 2D paper with right view point and perspective (b) designers spend major portion of the time and skill in generating impressive *3D effects* and (c) sketch has no representation for the behavior that produces the *functionality*. Designers represent the *expected* motion and behavior of the components using *annotations* on the sketch which are *unverifiable*.

A CAD model, created from the concept sketches, overcome these drawbacks and facilitates the *engineers* to perform engineering analysis. Based on the analysis, engineers may suggest modifications. This cycle of "*designers generate* (*modify*), *engineers evaluate*" repeats until the required functionalities are achieved. Similarly, PDM systems support collaboration, but they require well-defined geometric models and other data related to different domains. However, such a crisp representation is not available in the early fluidic state of conceptualization. By the time the crisp representation of the product is available, majority of the design decisions are already made, and the cost of reversing them would be high. On the other hand, in the early fluidic state of conceptualization, in absence of the support for analysis, the designers may not be able to foresee if the product would deliver the required functionality! If one could intuitively explore the components, structure and behavior of a design systematically, with a little knowledge/expertise in the domain, many communications about the design would be redundant and collaboration would be more efficient and effective. Hence, the goal of the present work is to support design collaboration right from the conceptualization stage.

Methods developed for component identification [6], behavior simulation [5] and constraints specification [7] provides a tentative embodiment of the product concept which additionally supports easy and natural interaction with the design. These methods, though, developed for standalone concept exploration, have synergistic relationship with the requirements for collaborative conceptual sketching. Hence, in this work, we explore how these capabilities developed in-house enable a novel sketch based collaboration system to provide better means of interactions among collaborating designers.

2 Literature Survey

Kvan [2] gives a brief overview of the literature on collaborative design. Starting from the etymological perspective of the word *Collaboration*, he clarifies the distinction between words of similar meaning like co-operation and co-ordination. He also discusses its importance in the design process and computer based implementation. Maher et al. [8] studied the designer's behavior in a computer mediated collaborative design exercise. He observed that designers tend to document more information during non-collaboration and also exclusive collaboration gave more productive results.

Wang [3] reviewed the literature for collaborative conceptual design and mentioned several approaches and technologies for collaborative conceptual design. It also highlighted that the decisions made in the early stages have more impact on the design.

Sketching plays an important role in the early stage of design [4]. There has been growing interest to understand sketches in the context of design thinking and creative exploration. Some approaches focused on identifying geometric shapes and diagrams in sketches [9, 10]. Some other methods focused on generating 3D models from 2D sketches [11–13]. As the technologies developed to access 3D space, devices like Phantom, trackers and motion capture systems became popular. Direct 3D shape creation methods in 3D virtual environment have been proposed using such devices [14, 15]. Some of the issues related to feel of control, depth perceptions and visualizations have been addressed with the use of haptics and stereovision [7, 14].

Techniques are being developed to perform engineering analysis, which are typically carried out in the later stages of design, using sketches themselves viz. strength analysis [16], static analysis [17] and dynamic analysis [18] using a sketch. Concepts can also be evaluated for their behavior [5]. All of these systems are standalone applications and not used for collaboration. In a complex collaborative environment the creation and evaluation has to happen concurrently to reduce the concept development time. Huang and Mak [19] demonstrated the use of formal design methods in a web based collaborative design. NetSketch [20] is a sketch based collaborative system which supports distributed sketch creation. Fan et al. [21] has shown the use of freehand sketching and CAD modeling in distributed environment. Most of the existing literature focused only on the generation phase during the collaboration experiment.

However, the generative phase of the conceptual design is highly individualistic, involving introspection, creativity, visualization and understanding; wherein, the designers are unlikely to solicit collaboration. Interaction among designers is required for acceptance, conflict resolution, cross domain verification, etc. Hence, a sketch development environment should support the first (generation) and the collaboration environment should support the second (interaction). This paper explains the issues in developing an environment for collaborative conceptual design.

3 Requirements in Collaboration Scenarios

In collaborative product development, the aim is to support interaction with the designer and also with the design. As indicated in Sect. 1, interaction with design needs an embodied model of the product. In the initial phase of conceptualization, mostly physical principles and their interactions are explored which has limited scope for detailing. Sketches are rich in information because they contain data pertaining to the product's shape, behavior, functionality and usage details. Capturing these salient features of the sketch helps in better understanding of the design process. The techniques in [5–7], helps in constrained exploration with better insight into the product's behavior.

Designers use traditional methods to communicate like, talking with each other, write, makes gestures, etc. during their interactions which may not be enough when discussing *about a product*. Typically designers use some representation of the product model like sketches, CAD models, and physical models/prototypes to support the discussion. They enhance the communication by providing visual feedback and making the discussion more focused. Therefore, interaction *with the concept* model should be supported in a collaborative environment.

The requirement for collaboration in the early stages of design arises because product development needs knowledge from diverse domains. The expert designers may not be available at a single place at a particular time to resolve issues and conflicts of interest. They could be located at different geographic location and time zones. Based on the spatio-temporal relationships among the stake holders, collaborative interactions are categorized [22] into *co-located synchronous*, *distributed synchronous*, *co-located asynchronous* and *distributed asynchronous* modes. Approaches reported in literature aim at making two collaborating designers feel the presence of each other to perform a common task such as collaborative sketching [23]. The need for such tasks in the process of real designing is hardly elaborated. We believe that the types of interactions need to be supported for each category is different, and that different categories co-exist in a hierarchy.

3.1 Co-located Synchronous

In this scenario the designers are located at a common place simultaneously and are interacting with a common concept model. The designers discuss the proposed concept by communicating face-to-face, and interacting using white boards, sketches, CAD models and other supporting documents. In a sketch based collaboration interface, direct interactions with the proposed concept would clarify many doubts related to the form, structure and behavior of the product. The sketch is created with strokes but interpreted in terms of components and behaviors; direct access to the strokes therefore is irrelevant. Only the *functionally segmented* view of the sketch needs to be presented for view transformations and tangible kinematic *behavior exploration*. Additionally, the system should allow collaborating designers to scribble and comment on the sketch in an interactive environment to support natural interactions and future referencing. One challenge here is to manage the *identity* of individuals contributing to these *annotations*.

3.2 Distributed Synchronous

In this modality, the designers are located at different geographic location places are interacting in real-time. It is evident that the informal and natural modalities of communications such as gesture and eye-contact are not available. The interaction between the designers is facilitated through communication-tools like audio, video, e-mails and text messages. The deficiency can be supplemented by referring to CAD/physical models which are available at advanced stages of development. In collaborative concept development, CAD models are not available and hence concept sketches can be accessed simultaneously in the distributed environment. It should not only facilitate viewing, manipulation and annotations on the sketches by the designers but also maintain the integrity of the information presented on multiple (possibly diverse) display systems. The system could be augmented with technologies like audio/video chat with text messaging which provides effective communication among designers. The issues like network latency, bandwidth and group size have to be addressed to maintain concurrency.

3.3 Co-located Asynchronous

In this category, the collaboration happens at a single place, using the same infrastructure, but the information sharing happens at discrete interval of time. This is typical where the designers work in shifts to develop a common product, or the product is too complex to be solved in a single sitting by a single designer. The person working in the next shift can be the designer himself or any other designer who continues the exploration. The product development information related to one shift needs to be communicated to the designer working in the next shift. When starting, the designer need to know the current status of the design in terms of the issues encountered and changes made in the previous session and correspondingly he need to report similar information when a design is checked out for the day. A systematic procedure for monitoring, recoding and reporting significant modifications, developments and issues need to be in place for effective continuity of the multi-session design tasks.

When designers are sketching over *multiple sessions*, at the beginning of a session, designers may not recall the state of the thoughts at the end of the previous session. Hence, one of the important requirements for sketch based collaborative design is to support the continuity of the thought process of designers. For *multi-agent* explorations, identity of the person concerned and the dates need to be maintained for traceability of the development and accountability of the changes made.

For a sketch based collaborative system, it is required to support the saving the data of the concept generation process and also capture important events so as to restore the continuity of the thought process. Thus, a mechanism to maintain the temporal evolution of the product is necessary.

3.4 Distributed Asynchronous

In this scenario, one designer works on a design problem for some time and some other designer located at an alternate place might work on other aspects of the same design, or continue the design already initiated by others. Here, it is required to capture the design process, produce a compact summary of the major decisions taken during the development process and make it available to the other designer. Therefore, in addition to the features discussed in the previous categories, a facility to manage the design process information in a distributed environment is needed.

4 Issues in Sketch Based Collaboration

It can be observed that the requirements for the different collaboration scenarios are different. But the design in question is the same. The issues in all these different type of collaborations have to be addressed for development of a better design faster and also facilitate seamless transition into the subsequent review, concretization and analysis stages of the design.

Collaborative concept development is distinctly different from the collaborative sketching, as available in literature, wherein sketch strokes are created and viewed simultaneously by the collaborating designers. We believe that a sketch is the rendering of a mental imagery which therefore is very personal to the designer. The designer has the flexibility to create and modify the sketches as per the problem at hand. Each designer would explore based on his understanding and domain knowledge pertinent to the problem at hand. Collaboration has to enable assimilation and integration of the diverse views to improve the design. We envisage the following issues to be addressed in developing such an environment.

4.1 Data Storage

In a complex concept exploration a sketch needs to be created in multiple sessions. This necessitates for organizing and saving the sketch so as to facilitate the designer to create the sketch in multi sessions. Apart from the geometry described in the form of sequence of strokes which are represented using coordinates and the timestamp, the data required to support collaborative interaction need to be captured and stored. In a collaboration session, designers perform different activities like viewing creating other components, commenting on the existing components. For capturing the process of exploration, from the perceptive of the designer, it is necessary to record the view transformation with the sketch data. It is observed that when the designers are sketching they do not transform the view by pan zoom and rotate. Hence in the data file the geometry and transformation details are stored in alternate sequences. In collaboration, interactions are more focused on the components. Hence, information related to the associated strokes, accessibility and comments pertaining to different components have to be stored. In a distributed environment, different designers share and contribute to same data representation. Thus, it is important to record the individual identity and the location of the designer and their contribution toward the central database. Also, the sketches created from different devices needs to be identified separately because the device capabilities like working volume, number of actuators, magnitude of force and torque feedback, etc. will be different.

4.2 Data Manipulation

The designers can create strokes, components, constraints, articulation and animation of the components. On the other hand, during interactions for collaboration, view manipulation and annotations would be the primary manipulation activities. The data pertaining to a design exploration has to be created, accessed and modified in a controlled manner based on the permissions given to the designers. This philosophy of data management is similar to the version control in software development but requires custom changes to effectively implement for collaborative conceptual design.

4.3 Data Transmission

Data transmitted between the two terminals is pertinent mainly for the distributed synchronous collaboration scenario and it depends on the relationship between the two collaborating agents and the nature of interactions. One of them is to transmit the sketches created in one terminal to the other. Other interactions are typically cursor movements. View transformations are typically triggered through mouse events. Hence, communicating mouse position and button status could constitute the data which is decoded locally to display the information on the other terminal. For this the demand on the bandwidth is not expected to be very high but for sketch data transmission, load obviously depends on the complexity of the scene. In asynchronous mode, the sketch data has to be transmitted the centralized data server which houses the entire sketch data created across different terminals. For the implementation, different network topologies like peer to peer, client-server, and multicast can be employed depending on the requirement for different categories.

4.4 Event Logging

An event is an action performed by the designer in the design process. Every event initiated by the designer changes the state of the design. The event may be as simple as accessing a sketch in the repository and viewing or as drastic as deleting some components. Although a continuous recording of activities may be unnecessary, activities that lead to significant change in the design or view must be logged for future review and investigations. For example, simple interactions like click of a mouse button trigger some events like creating a sketch stroke or saving the sketched data. Therefore, logging of mouse events is important. A coding scheme for diverse actions of the designers is reported in [24] which may represent a high level event good for logging.

4.5 Data Viewing

In collaboration, designers are only allowed to view or comment on a particular section of the design which they are authorized. In such scenarios, designers should have restricted access to the data allowing the designers to view selectively. Domain specific simulations can be made accessible for viewing by the other designer. To facilitate viewing from diverse display devices like stereovision with head mounted display, desktop and mobile displays, layout and rendering methodologies have to appropriately design. In distributed asynchronous interactions, remote designers may be provided with recent development activities in the context. To support this, a quick view of the major events that happened in the previous session may be useful. Moreover, for multi-session concept sketch development, continuity of the thought should be facilitated; e.g. replay the sequence of the last few strokes created by the designer in real-time speed may help.

The event log represents the full history of the concept developed and viewed by multiple agents. Hence it is large and heterogeneous. A list view of all the events may not useful. To analyze and view different aspects of the development, meaningful queries and filters need to be supported. For example, “when was the component ‘X’ redesigned last and by whom?” is a meaningful query to be answered from the event log.

4.6 Hierarchy Control

In every organization, a hierarchy of people is involved in the product development activity where one has to follow the instructions of their higher authorities. In a sketch based concept exploration tool, this hierarchy need to be supported by providing control on the exploration process through the specification of constraints. These constraints may pertain to the space allocation, scope of view, access rights on data, time and other statutory requirements. The designers at the higher level have to make decisions based on the sketches made by the subordinate designers or guide the exploration by providing constraints. The designers can develop multiple concepts of the different components and the design managers can explore the design with the different components of the product. The system should help in the evaluation of the product concepts. As in case of a design of a car, multiple designers develop diverse concepts for different parts like body, interiors, dashboard, headlight cluster, front-grill etc. The manager can select the components from different designers and tryout different combinations to decide on a particular higher level concept. Though, we believe it would be of immense value, we are not aware of any such existing system that allows one to explore variety right from the early development phase.

5 Sketch Based Collaboration Experiment

Aim: To study the effectiveness and usefulness of sketch based collaboration in conceptual design.
Objective: (a) To develop a collaboration environment that support sketch based interaction, (b) demonstrate a distributed synchronous collaboration scenario, and (c) identify tools and methods used during the interaction.
Experimental setup: The experimental setup consists of two computer terminals with haptic devices and head mounted displays placed in two separate rooms to emulate a distributed interaction, and are connected to the Internet through LAN in a client-server configuration. Each system runs the client application and one of them acts as a server. The communication using TCP/IP protocols is enabled through network programming using Boost Pro™ ASIO library. The client application is the sketching program developed for direct 3D interaction [7]. A mobile-phone is used for audio communication.
Methodology: First, a designer creates a 3D concept sketch of a leisure-chair and stores it on the server. He then solicits the opinion from the co-designer in the remote system by making a phone call. Co-designer logs in into the sketch based collaboration system and load the shared sketch in the view synchronized mode. Both the designers now have identical images of the 3D sketch during the interaction. First the designer explains the design by showing the sketch from different views and asks the co-designer to respond. The co-designer suggests a change in the backrest angle for better comfort and indicates it by drawing 3D strokes over 3D sketch to explain what he meant (Fig. 1).
Observations: The time lag between user interactions to remote view updating was insignificant for the example tried. Verbal communication over mobile continuous but integrated with the interactions on the model. It is observed that the changes suggested to the concept are explicitly provided by direct interaction with

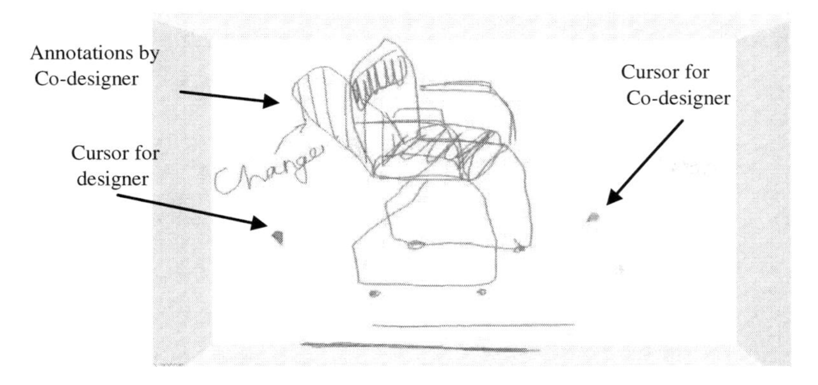

Fig. 1 Screenshot showing the annotations of one designer on a sketch created by other designer

the concept sketch supported by verbal communication through phones. Synchronized view transformation is used to explain the concept. The network bandwidth was sufficient to make the view transformations appear as real time.

Inference: Direct annotations on the sketch help in easy and natural communication. The strokes generated during interaction are only overlaid with the sketch and encapsulates the designer's identity. It acts as a record and helps in modification of the concept later on. Since after loading the model, only mouse interactions are exchanged over internet, which is a lean data, bandwidth issues did not arise.

Conclusions: Though the experiment carried out is elemental and indicative of the capabilities, it demonstrated that sketch based collaboration can be focused and effective in conveying the design intent. A designer centric *generation* and collaboration centric *evaluation* of product concepts is a viable sketch based collaboration model for the early stages of design.

6 Conclusions and Future Work

The work presented here shows that with small modifications to the existing 3D sketching interface, distributed collaboration can be facilitated. Further infrastructure based on information technology need to be developed around the existing methodologies to support different modes of collaboration as identified. The issues identified in this work form broad outlines for future enhancements. Further experiments have to be formulated and performed with real-time data to test the efficacy of such a system.

References

1. Eppinger SD, Chitkara AR (2006) The new practice of global product development. MIT Sloan Manage Rev 47(4):22–30
2. Kvan T (2000) Collaborative design: what is it? Autom Constr 9(4):409–415
3. Wang L et al (2002) Collaborative conceptual design—state of the art and future trends. Comput Aided Des 34(13):981–996
4. Onkar PS, Sen D (2009) A review of product sketching in early phases of design. International conference on research into design, pp 472–479
5. Onkar PS, Sen D (2012) Behaviour simulation in computer aided product concept sketching. 22nd CIRP design conference, Bangalore
6. Onkar PS, Sen D (2010) Functional segmentation of strokes for product sketch understanding. Int J Shape Model 16(1 and 2):9–38
7. Onkar PS, Sen D (2012) Constrained 3D sketching with haptics and active motion annotation. TMCE 2012, Karlsurhe
8. Maher ML, Cicognani A, Simoff S (1996) An experimental study of computer mediated collaborative design. In: Proceedings of 5th ETICE Workshop, pp 268–273
9. Jorge JA, Fonseca MJ (1999) A simple approach to recognise geometric shapes interactively. LNCS 3rd international workshop graphics recognition, vol 1941, pp 266–276
10. Kara LB, Stahovich TF (2004) Hierarchical parsing and recognition of hand-sketched diagrams. In: Proceedings of 17th ACM Symposium UIST '04, pp 13–22
11. Igarashi T, Matsuoka S, Tanaka H (1999) Teddy: a sketching interface for 3D freeform design. SIGGRAPH 26th CGIT, pp 409–416
12. Diehl H, Muller F, Lindemann U (2004) From raw 3D Sketches to exact CAD product models concept for an assistant-system. Eurographics SBIM
13. Lipson H, Shpitalni M (2007) Optimization based reconstruction of a 3D object from a single freehand line drawing. In: ACM SIGGRAPH 2007 courses
14. Keefe D, Zeleznik R, Laidlaw D (2007) Drawing on air: Input techniques for controlled 3D line illustration. IEEE Trans Vis Comp Graph 13(5):1067–1081
15. Schkolne S, Pruett M, Schröder P (2001) Surface drawing: creating organic 3D shapes with the hand and tangible tools. In: Proceedings of ACM-SIGCHI conference HFCS, pp 261–268
16. Murugappan S, Ramani K (2009) Feasy: A sketch-based interface integrating structural analysis in early design. In: Proceedings of ASME IDTC/CIE Conference
17. Atilola O et al. (2011) Mechanix: A sketch recognition truss tutoring system. ASME IDETC-design education conference
18. Alvarado C, Davis R (2001) Resolving ambiguities to create a natural computer-based sketching environment. IJCAI, pp 1365–1374
19. Huang GQ, Mak KL (1999) Web-based collaborative conceptual design. J Eng Des 10(2):183–194
20. Laviola J et al. (1998) Collaborative conceptual modeling using the sketch framework. IASTED International Conference of Computer Graphics and Imaging, pp 154–157
21. Fan Z et al. (2004) A sketch-based interface for collaborative design. Eurographics workshop on sketch based interfaces and modeling
22. http://en.wikipedia.org/wiki/Computer-supported_cooperative_work
23. Sallnas EL et al. (2000) Supporting presence in collaborative environments by haptic force feedback. ACM Trans Comput Hum Interact 7(4):461–476
24. Suwa M, Purcell T, Gero J (1998) Macroscopic analysis of design processes based on a scheme for coding designers' cognitive actions. Des Stud 19(4):455–483

Strategies for Mutual Learning Between Academia and Industry

Margareta Norell Bergendahl and Sofia Ritzén

Abstract The number of challenges facing companies in their development activities is numerous, some coming from new markets and technologies, and some more abstract, like conflicts between short term efficiency and long term innovativeness. Improving collaboration between industry and academia is considered critical—the aim of this paper is to contribute to the discussion on long term learning collaboration between academia and industry, being a core competence area in itself. Another purpose is to form a platform for experience sharing, and increased integration capability for sustaining common knowledge—and practice development. The paper includes an analysis of several collaboration programs between academia and industry conducted in Sweden, resulting in conclusions and advises concerning what to consider for collaborative work.

Keywords Learning collaboration · Product development · Academia and industry · Knowledge triangle

1 Background

The list of challenges facing companies in their development activities is almost endless, some concrete, like opportunities coming from new technologies, and some more abstract, like conflicts between short term efficiency and long term innovativeness. Product development is certainly today characterized by high

M. N. Bergendahl (✉) · S. Ritzén
Integrated Product Development, School of Industrial Engineering and Management, KTH Royal Institute of Technology, 100 44 Stockholm, Sweden
e-mail: maggan@kth.se

A. Chakrabarti and R. V. Prakash (eds.), *ICoRD'13*, Lecture Notes in Mechanical Engineering, DOI: 10.1007/978-81-322-1050-4_53,

complexity from several perspectives: strong interconnections between technologies and competence areas and consequently between people and organizations. Parallel to this, the need to develop learning relations to academic institutions is increasingly more articulated, and very much related to the perception of possible competitiveness, and delivery of competent employees. Deiaco et al. [1], expresses it as that the university plays a particular set of roles in the global knowledge economy by delivering research, education and applied problem solving. Collaboration that may appear of obvious importance at an overarching level, though need special attention to become successful at the level of an individual firm or research institution [2].

The goal of design research, according to Blessing and Chakrabarti [3] is to make the product development more effective and efficient in order to enable practitioners to develop successful products. Research not only aims to create knowledge, but is also assumed to cover implementation of tools, methods and procedures that improve the design process, and thus the competitiveness, in industry. Expected effects and implementation of research results in design practice have, however, been poor [3, 4], due to lack of mutual understanding between academia and industry and the way studies are performed. Improving collaboration between industry and academia is considered critical and is actively encouraged by authorities, e.g. the European Union [5].

Increasing complexity in product development makes integration capacity in development organizations a general and further increasing demand, especially when the need to be both effective and innovative is identified [6]. One used concept is Integrated Product Development, IPD, including different processes and working methods for improving efficiency and effectiveness [7]. *Integration* is to take into account the many aspects of how to optimize customer value (function, attractiveness, cost, environment etc.) in parallel and iteratively within the product development process. The integration requires collaboration between functions and competencies within and between organizations. There is also an increased demand for learning between companies and academic institutions. Thus, researchers in the fields of design, specially focusing on processes and methods for improved ways of working, have a double interest in this field, one to develop successful working procedures for integration for competitiveness, the other to develop the research based competence to collaborate for mutual learning.

This paper builds on empirical cases and experiences from collaborative initiatives in Swedish environments through the last 20 years, and forms a potential forum for discussion. We share some recent work on forming processes for strategic partnership between KTH management and important industrial and societal stakeholders. We will discuss experiences and reflections on the ENDREA program (Engineering Design and Education Agenda) in the 90s, including many IPD projects being action learning based; PIEp, the Product Innovation Engineering program running from 2006 including collaborations efforts both in research and more specifically in results dissemination; reflections from Industrial PhD students relations [8], and the work with Faculty for Innovative Engineering on University level at KTH, to further make the collaboration-learning effort stronger.

The aim of the paper is to contribute to the development of fruitful and long term learning collaboration between academia and industry, being a core competence area in itself. Another purpose is to form a platform for experience sharing, and increased integration capability for sustaining common knowledge—and practice development, addressing future challenges. The paper includes an analysis of several collaboration initiatives between academia and industry. The analysis combines experiences from each initiative, identification of strengths and challenges will be made.

2 Collaboration Between Academia and Industry: Theoretical Outlooks

Collaboration has been stressed as a key issue in product development for a long time. Several streams of intra-firm and inter-firm collaboration efforts can be found. One stream quite frequently addressed is the collaboration between developing firms and manufacturers/suppliers [9–11]. Another stream addressing collaboration is strategic alliances in R&D between firms of less mandatory dependencies than in the buyer–supplier relationship, exemplified by Uppvall [12] stating the importance of inter-organizational relationships when supporting innovation. Reoccurring issues concerning creating successful collaboration are trust between partners, commitment to the joint effort and a certain consistency in building and sustaining relations and thus collaboration.

Also collaboration between academia and industry is found as a stream of specific research interest. A rationale to academic-industry ventures is from academia to test and verify results and from industry to reach many more researchers than otherwise possible. Caravannis et al. [13] express many more reasons to such ventures and define benefits of collaboration between the sectors government, industry and academia to be sharing of risk and cost, access to complementary capabilities, access to specialized skills and access to new suppliers and markets. Beckman et al. [14] support these reasons and further express the significant benefits correlated to the different positions of academia: engaged in long term research and education, and industry: engaged in commercialization and applied research and development.

Barnes et al. [15] identified in a number of case studies factors for a successful University-Industry interaction, factors categorized into six main areas for attaining a good practice: Choice of partner, project management with progress monitoring and clear objective setting, management of environmental factors, building relations including securing trust and commitment, defining measures for keeping commitment and planning of tangible outcomes and achieving balance between academic objectives and industrial priorities. Further, their findings placed significant emphasis on the need to manage the inevitable cultural differences between academia and industry. The main cultural issues were related to the need to agree on priorities and timing.

Tartari et al. [16] studied barriers to academia-industry collaboration specifically from the perspective of academics and found it to be a 'non frictionless' process. Differences in timescales are the most important barrier they found (long term versus short term research) followed by timing of disclosure and conflicts regarding topics of research. They also saw that in order to overcome barriers related to timescales, difficulties in finding appropriate partners for mutual interests in research and mutual understandings about expectations, experience from former collaboration are crucial. They conclude in that an encouragement of academics to engage with industry requires an alignment of collaboration policies to individual incentives and that face to face meetings are key in this. Level or trust is reoccurring as a key for overcoming most barriers (ibid).

Since trust takes considerable time and effort to develop among partners the continuity of personal as well as organizational relations is important to support. The importance of facilitating tacit knowledge transfer (e.g. experiences, skills and mental models) was shown to be critical in a study investigating knowledge transfer from engineering research centers to companies [17]. Successful tacit knowledge transfer should focus on individuals involved in the process and the steps needed to ensure the development of close relationships. However, trust can face some challenges when differences in expectations are not aligned.

3 Experiences from Collaborative Initiatives

3.1 Endrea

In Sweden's manufacturing industry in the late 80s, there was a low recognition of the benefit of research educated people in the area of Industrial Product Development. At that time a strong initiative was taken within the national program Endrea (Engineering Design Research and Education Agenda) including the areas: Design Theory and Methods, Modeling and Simulation, and Engineering Management. Endrea was generously funded by the Swedish Foundation for Strategic Research; SSF and was designed as a Graduate school, based on collaboration between four major technical universities and a large number of industrial companies. The program built on the combination of technical disciplines and process-oriented, interdisciplinary research within industrial product development practice. The purpose was to increase the benefit of common knowledge development between industry and academy for long term competitiveness to be enhanced. A large number of engineering doctors were educated within the program (during 1995–2003), all with close relation to industrial practice, many of whom are now active in leading roles in industry and in different universities. A lot of experiences were drawn from Endrea [18], some of which are mentioned here:

Several large and medium sized companies were introduced to new forms of collaborative work with academia during the program; several of them (mainly large companies) started then strategic programs for Industrial PhD students as part of their long term development. More than 70 Engineering Doctors were examined, making a large impact in the companies where they were employed, one-third of the doctors made their careers in academia.

Today the Endrea-doctors are important ambassadors for collaborative work between industry and academia; they carry the understanding and bring interest in development of new collaborative and learning relations. Several of them are adjunct professors at a university; others have now upper management positions in manufacturing industry.

One of the learning is the time and effort consuming in building trustful and beneficial relations, not only between academia and industry, but also between the different, normally competing, universities. Endrea was funded for 5 years, plus the start and the close period. Due to a change in strategy at the funding body, the next period asked for programs where the universities was to compete, which turned to be negative for the well-grounded collaboration and trust built between the academic partners within Endrea.

It is eligible to say that the Endrea has brought a lot of learning to the Swedish engineering design and research system, by that forming a solid base for following initiatives, such as ProViking and PIEp (see below).

3.2 Product Innovation Engineering Program, PIEp

In 2006, PIEp was developed by KTH and Vinnova (the Swedish Governmental Agency for Innovation Systems), as a proactive initiative to develop innovation capability—to further develop an area of strength in Sweden. Increasing and further strengthening of innovation capabilities within existing organizations as well as among engineers educated at Swedish universities was defined as of utmost importance for long term competitiveness. An agenda for research and actions for change was developed jointly by academia and industry, led by academia. Vinnova agreed to fund the program, planned for 10 years. Five years later PIEp is a research and change program, constituted by 15 research projects, all in collaboration with industrial partners and partial funded by industry, including a Research School of about 50 doctoral students. In parallel actions to deploy research results, in order to induce change for increased innovation capability and actions for developing new education of engineers and designers are conducted. The program includes five academic partners in Sweden and several international partner institutions at Stanford, TUM, Cambridge, and Aalto University.

PIEp research projects build on collaboration where researchers in academia and engineers in industry have goals that mainly coincide, i.e. increasing innovation capability in organizations. However, goals are also somewhat diverging since researchers aim for general knowledge building, while industry aim for direct changes in their specific organization. Since PIEp as a program has the ambition to actually make change these somewhat divergent goals are of minor problem, though has to be dealt by researcher in relation to the research community as traditional measures rather refer to publishing than to action in industry. In this context two potential challenges emerge, based on the experiences within PIEp: researchers are many times too vague in defining the applicability of their research results and companies expect often research to be at a minimum of resources and cost for them.

Within PIEp a specific action for transferring research results has been performed during 2012, the Innovation Pilots. Carefully selected students have been trained in PIEp developed approaches for increasing innovation capability in R&D organizations and conduct an innovation screening as well as a few workshops initiating an action plan for increasing innovation capability in a company. The pilots have proven to be an effective means for inducing change in companies based on research results. A grand challenge for researchers coordinating this effort has however been to promote the pilots to companies. In order to reduce resources used and to build a sustainable structure, networks of companies has been approached; however, a striking finding is that the pilot projects build on direct personal relations between individuals in research and in companies. This experience further strengthens the need to consider and develop sustainable structures for long term collaboration.

3.3 Strategic Relations, Innovative Engineering

Most technical universities have strong habits and experiences of working relations with industry and surrounding society. Often these are developed on personal relations and trust built between professionals. Since, the last years KTH has increased the ambition for common pro-active knowledge development and defined the Faculty for Innovative Engineering including two major set of activities:

- To develop a strategic program to form long term learning collaboration with selected partners.
- To substantially increase the number of persons moving between KTH, industry and the surrounding society.

The strategic partnership is defined as a continuous dialog, the discussions are held on top management level, yearly meetings taking place between the Vice-Chancellor and the CEO of the partner. The implementation of the collaboration is about forming a Memorandum of Understanding for long term common development work in several fields, and to follow up a target document including specified goals for the coming year(s). The strategic partners all have several areas of interest of collaboration with KTH, from student recruitment, industrial PhD-commitments, adjunct professorships, and different areas of research interest.

Strategic partners in the Faculty for Innovative Engineering initiative today are large organizations: Scania, Ericsson, Stockholm City Council, Skanska, ABB—there are several more to come—and the development of a process for long term collaboration with SMEs and research institutes is in progress. The basic ideology behind the work with Innovative Engineering is the Knowledge Triangle: Education, Research and Innovation: increased quality in both research and education develops innovation capability for partners in a crucial collaboration.

The intention to increase the number of persons moving is based on the perception that the 'best way to exchange experience and knowledge is to move heads'. The goal is to double the number of adjunct professors, and the number of industrial PhD students in 3 years, and further more to increase the number of moving persons on mid-levels. Another interesting challenge is to increase the number of researchers/teachers to spend more time in environment outside the University, which will make important impact on teaching, research and innovativeness. To make this realistic the rules of the game are considered and changed, for example it must be possible to become an adjunct professor with somewhat different merits than an ordinary professor, and it must be valued and rewarded to move between academy and industry/society.

3.4 Industrial PhD-Positions

A specific kind of collaboration between academia and industry is the industrial research student. Industrial PhDs are employed by a company, associated to a research institution and conduct scientific research of interest both for the company and for academy. Cultural differences become surficial in this kind of collaboration as PhD students have clearly defined requirements with regard to their research work, which hence constrain their role within industrially-orientated projects. Experiences from this collaboration setting reveal that this is an important dialog to be held before starting an Industrial PhD. The industrial partner has a specific research interest that must be integral to the research institution; however, their educational goal must also be aligned to the interest of the company. This is most often the case as assigning a person for a PhD position has long term goals.

Being an Industrial PhD Student is an opportunity, but also often means being in a conflict, since the Industrial PhD student is expected to handle the research work, and at the same time work in the employing company: long term research, along with short term project work with tight deadlines at the companies. Lack of understanding for what constitutes doctorate level research at some industrial companies is a challenge, as is the understanding from academy for the industrial reality. Possible conflicts regarding publishing and IPR must be clarified as publishing is mandatory for research, however might interfere with opportunities to patenting, which could be a key in doing business.

The overall experience shows that Industrial PhD-positions is a very powerful instrument to ensure collaborative work lasting for several years, if handled in a proper way. In a former work [8], some conclusions regarding Industrial PhD student relations are drawn:

- Be careful (and take the effort needed) to reach common understanding for academic freedom and quality, rules for publication etc.;
- Discuss balancing of time between operative industry projects and research project/education—time for 'reflexivity';
- Search ways to anchor research results in the industrial organization;
- Select an academic advisor with experience from close work with industry;
- Plan for exchange of information and communication between the partners;
- Place the PhD student in the functional engineering organization (not staff) at the company, plan for a substantial proportion of time in the academic organization.

4 Experiences from Collaborative Initiatives

The work to develop sustainable learning relations between academy and industrial companies is as difficult as to create win–win integrative work procedures in any human relation. Not surprisingly it is very much about how to develop trust between people and partner organizations as also is stated in [17].

From our experiences with the initiatives presented in this paper some highlights can be made: It is of major importance to build a sustainable financial infrastructure, not to say that all funding must be front-loaded, but a relatively safe 5 years start is a strong recommendation. The risks and responsibilities should be shared, to motivate carefully delivery and reflection.

A successful collaboration relies many times on long term relationships with partner organizations, therefore the organizing of the collaboration has to be relying on several persons, and expected deliveries and roles should be defined. Also commonly defined processes for continuation are of highest value to make the collaboration less person dependent. Normally it is helpful to have a management organization not only including the active researchers/teachers, but also people giving administrative support—effective supporting structures as defined in [16].

An increased mobility, people from a company acting within a research group at an university and vice versa, is both a means for building long term relationships and for reaching the actual goals of collaboration: increased quality of research and education, and following on that innovation capacity. We have already mentioned adjunct professors from industry to university, other positions are affiliated faculty and adjunct expert as well as industrial post-docs. Often a trust-building relation can build on one or a number of master thesis work performed by students, advised by an industrial PhD student etc. Mobility could be further widened as experiences reported here reveal: both extended to include also people at doctorate level and increased concerning academics spending (limited) time in companies. In mobility settings as well as in other settings the three corners tones of the Knowledge Triangle should define a common vision.

It is of major importance to build not only trust, but understanding and respect from both parties concerning expectations on what are relevant deliveries, such as publications, change management practices, continuous education, or other applicable deliveries. A general discussion on how to value impact on practice is needed (recently introduced in a major Research Assessment Exercise, RAE, at KTH in June 2012 being one of three headlines), and how this can be understood as a merit in the academic career.

Researchers also need to develop skills to communicate research and package it in a way that makes research results applicable in practice, or to find the right collaborative partners that can be the ‘carrier’ or the ‘implementer’—this being part of a more general discussion on what are the skills to expect from an academic person.

One underestimated challenge is how to choose the ‘right’ partner, and how not to continue a non-successful partnership. Selecting partner is also emphasized in [17]. First in this is to try to be brave, realistic and focused—there has to be real potential winnings of some magnitude for both sides if the work to form a strategic alliance with a partner is worth doing.

5 Conclusive Advices

Here we summarize a number of advises as ‘what to do, and what not to do’, based on the initiatives presented and related research efforts and our identification of strengths and challenges from these experiences. It is our hope that this could be a platform for an interesting experience exchange.

Opportunities—to develop:

1. *Articulated terms and concepts*

   Clearly articulate and define a common view on the terms, concepts, and possible results as well as how these will be followed up (long- and short term);

2. *Role definitions*

   Define and discuss the roles and the responsibilities of all the engaged persons to build commitment;

3. *Trustful relations*

   Develop personal relations and trust, start with smaller projects and/or exchange—follow up—expand the common work and fields of collaboration;

4. *Search a sponsor*

   For long term relations—search a high level sponsor in each partner organization, take time needed to anchor the collaboration and the interest in the organizations;

5. *Planning and delivery*

   Plan realistically, always deliver on time & budget;

6. *Reflection and learning*

   Include all partners in feedback, reflection and how to implement learning;

7. *Long term relation with several instruments*

   Design the strategic relation on several different kind of instruments and level of actions, develop a bunch of collaboration activities, including education, research, innovation as well as exchange of experts;

8. *Maintenance*

   Organize for operation and survival of the relation, be aware of the budget needed and the budget processes for partners;

9. *Common vision*

   Define a common vision for the collaboration, where to be in 1-3 years;

10. *Communicate*

    Plan how to communicate and implement findings to develop the relation and results

Challenges—not to do:

1. Start a 'collaboration' without a clear management sponsoring;
2. Underestimate the understanding of difference in industrial goal, research goal and education goal, as well as expectations on results;
3. Promise short term winnings, but communicate about mutual winnings;
4. Compete with consultant companies;
5. Start with finance talk, if there is perceived benefit to come, financing will come;
6. Start without any thoughts from the partners on resources and budget needed;
7. Mix the duties and responsibilities from the different partners;
8. Start the process without a clear commitment and general plan for the continuity and for the coming years.

Our aim is to invite to discussion, we think that sharing experiences and results of learning collaborations will both increase the benefit of our research, and also make this area of research interesting for academics as well as practitioners. Further work in the area is emphasized, specifically on the complexity on building strategic long term relationships and in parallel launching operational projects delivering results beneficial in academia and industry.

Acknowledgments We would like to acknowledge The Strategic Foundation of Research (SSF www.ssf.se) and Vinnova (www.vinnova.se), funding partners making the research and collaborative initiatives possible. We would also like to express our gratitude to partner companies and all the people in these organizations that have been open to challenging joint effort, the actual base for mutual learning. Specifically we would like to mention Scania, Ericsson, Volvo Cars and St. Jude Medical.

References

1. Deiaco E, Huges A, McKelvey M (2012) Universities as strategic actors in the knowledge economy. Cambridge J Econ 36(3):525–541
2. Liyanage S, Mitchell H (1994) Strategic management of interactions at the academic-industry interface. Technovation 14(10):641–655
3. Blessing LTM, Chakrabarti A (2009) DRM a design research methodology. Springer, London
4. Cantamessa M (2003) An empirical perspective upon design research. J Eng Des 14(1):1–15
5. European Commission (2007) Improving knowledge transfer between research institutions and industry across Europe, EUR 22836 EN, Luxemburg
6. Teece DJ (2007) Explicating dynamic capabilities: the nature and micro foundations of (sustainable) enterprise performance. Strateg Manag J 28(13):1319–1350
7. Norell M (1999) Managing integrated product development. In: Mortensen NH, Sigurjónsson J (eds) Critical enthusiasm, Trondheim/Lyngby
8. Kihlander I, Nilsson S, Lund K, Ritzén S, Norell Bergendahl M (2011) Planning industrial PhD projects in practice: speaking both 'Academia' and 'Practitionese'. In: Proceedings of ICED11, international conference on engineering design, Copenhagen
9. Monczka RM, Petersen KJ, Handfield RB, Ragatz GL (1998) Success factors in strategic supplier alliances: the buying company perspective. Decis Sci 29(3):553–577
10. Eneström P, Ritzén S, Bergendahl MN (1999) Co-operation between supplier: buyer in integrated product development. In: Proceedings of 12th international conference on engineering design, Munich
11. von Corswant F, Tunälv C (2002) Coordinating customers and proactive suppliers: a case study of supplier collaboration in product development. J Eng Tech Manag 19(3–4):249–261
12. Uppvall L (2010) The collaborative challenge of product development: exploring sustainable world systems through critical incidents in R&D alliances, PhD Thesis, industrial economics and organizations, KTH, Sweden, ISSN 1100-7982
13. Caravannis EG, Alexander J, Ioannidis A (2000) Leveraging knowledge, learning, and innovation in forming strategic government-university-industry (GUI) R6D partnerships in the US, Germany and France, Technovation, vol 20, pp 477–488
14. Beckman K, Coulter N, Khajenoori S, Mead NR (1997) Collaborations: closing the industry-academia gap. IEEE Softw pp 49–57
15. Barnes T, Pashby I, Gibbons A (2002) Effective university—industry interaction: a multi-case evaluation of collaborative R&D projects. Eur Manag J 20(3):272–285
16. Tartari V, Salter A, D'Este P (2012) Crossing the Rubicon: exploring the factors that shape academics' perceptions of the barriers to working with industry. Cambridge J Econ 36(3):655–677
17. Santoro MD, Bierly PE III (2006) Facilitators of knowledge transfer in University-industry collaborations: a knowledge-based perspective. IEEE Trans Eng Manag 53(4):495–507
18. ENDREA, Final report, 2004, published for the Swedish Foundation for Strategic Research, Stockholm 2004

Participatory Design for Surgical Innovation in the Developing World

Florin Gheorghe and H. F. Machiel Van der Loos

Abstract The field of surgery is very much a technology-mediated practice. Unfortunately, locally appropriate medical equipment is largely unavailable, or in the case of Western donated devices, is non-functional in much of the developing world. This paper presents two critical challenges that face medical device manufacturers and designers looking to innovate in international surgery, and proposes a methodology to address these concerns. First, designers approach the process with a set of embedded assumptions and biases that are rooted in their experience of traditional markets, thus delimiting the solution space too narrowly. Second, designers working cross-culturally with expert users face numerous difficulties in understanding the problem space. Through a reflective process within both a Canadian and Ugandan context, this study proposes that the assumptions Western designers hold can be challenged to co-create and uncover innovative technology solutions in international surgery.

Keywords Medical device · Design · Emerging markets · International surgery

1 Introduction

The global burden of disease from trauma and injury is estimated at 5 million deaths annually and contributes upwards of 20 million disabilities, with over 90 % of those deaths taking place in the developing world [1, 2]. This problem is largely

F. Gheorghe (✉) · H. F. M. Van der Loos
Department of Mechanical Engineering, University of British Columbia, Vancouver, Canada
e-mail: gheorghe.florin@gmail.com

H. F. M. Van der Loos
e-mail: vdl@mech.ubc.ca

A. Chakrabarti and R. V. Prakash (eds.), *ICoRD'13*, Lecture Notes in Mechanical Engineering, DOI: 10.1007/978-81-322-1050-4_54,

aggravated due to the lacking technical capacity to safely and efficiently manage surgical cases across these regions.

This epidemic is expected to grow in the coming decades, with injury from traffic accidents alone estimated to move from 8 to 4th place as leading cause of disability in the world by the year 2030 [3]. This untreated burden of disease unfortunately has a disproportionally negative role in overall social and economic development, as those killed and disabled for life are often the breadwinners on whom families and entire communities depend.

While much of this disability and mortality can be prevented through timely access to surgical intervention, there is currently a major gap in provision of surgical care in the developing world [1]. Hospitals in these regions lack the technical resources to manage and effectively treat the vast numbers of patients who present with trauma every day [4].

Efforts over the past decades to equip developing world hospitals with donated medical technology have largely failed. In some cases, developing countries depend on foreign donations for up to 95 % of their equipment, though studies from 16 countries across Asia, Africa, and South America have found that 80 % of these donated technologies fail within the first year [4]. As many as 39 % of all donated technology never worked at all once arriving in the recipient country. The reasons for failure are sometimes technical, as in the case of an inappropriate power supply, but often arise from a lack of foresight into the appropriate design, use, and maintenance of technology. Technology that is designed for use in a highly developed Western medical context simply does not fit that of a developing world hospital. The WHO's report *Medical Device Mismatch* identified that, although medical technologies of all sorts are created by industry to treat and manage a wide variety of disease states, a significant mismatch exists between what is needed in the developing world and what is being designed currently for Western markets [5]. This gap is one of availability, accessibility, affordability, and appropriateness for local needs of developing world users.

This paper examines two major challenges of designing medical technology for international surgery, which we define as taking place in a developing world context. Firstly, it is argued that there is a Western bias inherent when models of medical device design are applied in non-traditional markets. This issue arises from assumptions that designers may hold about the design space. Secondly, there is a difficulty in this field of becoming familiar with and understanding the nuances of the problem space, due to the expertise required to understand the field of surgery. This, combined with power dynamics that often accompany cross-cultural projects in low-resource settings, makes for great difficulty in understanding and defining the problem. A method is proposed for uncovering the assumptions that are inherent in traditional medical device design, validating, and then moving past them to co-create technologies that can benefit international surgical practice.

2 Understanding the Mismatch

It is critical to ask why such a mismatch exists between what the medical device industry is producing and what billions of people around the world need desperately.

The Bottom of the Pyramid (BOP) market, defined by Prahalad as the four billion people in the world living on under $2 a day, presents one of the greatest challenges to medical device innovators [6]. This gap in access to technology-mediated health care is not only an opportunity to bring the benefits of modern day technology to underserved populations, but also to gain an economic foothold in some of the fastest growing economies of the world. Although the United States currently makes up 40 % of the global market for medical technology, emerging markets in Brazil, Russia, India, and China (BRIC nations) are expected to more than triple their health expenditure between 2010 and 2020 [7, 8]. India and China alone are predicted to increase health spending by a compounded annual growth rate of 15.5 and 14.5 % respectively, compared with only 4.0 % for the G7 nations during this same period. Though Asian giants have taken the spotlight of emerging market projections, African economies such as Ghana and Rwanda are ranked impressively as the 4 and 13th fastest growing economies in the world, with 2011 estimates of 13.6 and 8.8 % annual GDP increase respectively [9]. Medical device manufacturers, like other sectors, are beginning to take note of this trend, with companies such as GE Healthcare, Johnson & Johnson, and most recently Covidien opening R&D centers within these target markets and investing heavily to find appropriate solutions [10–12].

Despite these efforts, many established manufacturers are finding great challenge in reaching the true Bottom of the Pyramid. The socio-economic stratification and geographic separation of the poor majority within developing countries presents two consumer tiers. The wealthiest within these populations are not much different from their Western equivalents. They have access to international standards of healthcare, trained physicians, and cutting-edge technology. For this reason, they are a familiar and welcome sight to Western medical innovators, and have typically been the primary target of companies seeking to enter emerging markets [13]. In contrast, the poor majority of these populations have little access to healthcare. Despite spending a disproportionately higher percentage of their income on health [14], the system that serves them is lacking even the most basic of technology. This second tier of consumers is greatly underserved by the medical device industry to date.

To understand this problem further, it is necessary to look at the way in which medical device design takes place in the wealthy, developed world. Exploring the basis of these methods that lead to success in the first world may reveal why these steps cannot achieve a similar result in emerging markets. At its root, the problem lies not only in how we approach the design process, but also in how we understand the problem space. It is these two challenges that this paper will further address.

2.1 Uncovering Assumptions that Guide Traditional Design

Western biomedical engineers and designers are trained to navigate a rigorous, step-wise process of design for medical technology, based on directives set out by regulating bodies such as the United States FDA [15].

Although there are a number of guidelines and representations of this process, including the waterfall model and the stage-gate model [16], these examples may not deliver successful products in BOP markets. Such models have been developed through the lens of Western designers, and with experience within the complex Western medical device environment. Designers themselves apply these steps and procedures within a particular technology design paradigm, based on a set of expectations and assumptions about the users and context of use that are unique to traditional markets.

This can include, for example, assumptions around the information flow throughout a surgical procedure. In a Western setting, a great deal of information is available to the operating team about the patient and the progress of the surgical intervention. Real-time patient monitoring and past medical records allow operating staff to proceed through the surgery with confidence, and to anticipate any complications they may face. Not having this information means that a surgeon must rely more heavily on clinical skill, while delaying the operation and practicing more conservatively due to the increased risks.

Visualization of injuries through imaging is a key aspect of information flow that is often missing as well. The procedures, implants, and specific techniques that have advanced the field of orthopedic trauma are all dependent on certain information that is commonly accessible in a Western operating suite. These problems of information flow can highlight opportunity areas for innovation, and can become leverage points rather than deficiencies. The starting point is to understand how practice is currently adapted to manage this uncertainty, and how it could be supported with a locally appropriate technological approach. A further summary of systemic differences and assumptions is described in Sect. 5: Expected Results.

The paradigm we are designing for in the Western medical world features a different set of characteristics than those of resource-constrained countries. What is needed then, is a shift in this technology design paradigm, and a change in how designers and engineers approach the design of medical technology for the developing world. In order for this shift to take place, we must begin to understand and catalog the differences present in each system, specifically how Western assumptions surrounding technology design may or may no longer apply in a developing world context. Shedding light on our assumptions about the design, delivery, and use of medical technology can also uncover the values and priorities that inform designers on appropriate trade-offs that can be made in the new medical technology paradigm.

2.2 The Challenge of Expert Users

A further challenge present in medical device design exists in understanding the complex needs of expert users, a task made more difficult than designing in common consumer focused sectors such as home electronics or kitchenware.

There is a distinct asymmetry of information present between user and designer. Expert medical users have a depth of understanding of the practice and associated challenges that is rooted in their own experience and education. Designers on the other hand typically have a depth of knowledge in the solution process and method, but have only a common understanding of the problem space. Von Hippel labels this a problem of "sticky information", which has a high cost of transfer from users to the designers who need it in order to work effectively [17]. The information is hard to transfer for a variety of reasons, but primarily due to the complexity and nuances that exist in practice. To the users, the necessary information may seem second nature and obvious among their peer group, thus resulting in their not sharing of certain critical details with designers. Other times the information may not be encoded and the users themselves may not recognize their own habits and needs, such as the manner in which surgeons use bracing strategies to improve manual dexterity and precision during an operation. Sure terms these valuable but unconscious adaptations to one's environment as "thoughtless acts" [18].

Manufacturers try to move past this challenge of understanding the problem space by maintaining close relationships with doctors, nurses, and other medical users. Methods borrowed from the field of human centered design are used to understand the user's context, challenges, and subtle needs. However, applying many of these techniques, such as contextual inquiry, where users are interviewed and use a think-aloud approach to verbalize their decisions, are often not appropriate for the trauma setting, where users are intensely focused on a life or death situation. Some work on developing techniques for needs-finding in trauma surgery has been done. In the field of human factors, Brown presents a process specifically for use in trauma settings consisting of expert interviews, observation, as well as getting feedback on prototypes through usability testing and heuristic evaluation [19].

Despite progress made on the development of trauma-specific methods for the early stages of design, this practice within a low-resource setting provides an additional challenge in the power dynamic that exists between designer and user. Evident in international development projects, the dichotomy of the rich, Western donor and a poor, suffering beneficiary intensifies problems in communication, trust, and partnership [20]. A similar perception may exist in a design context where the medical users, working in a resource-constrained environment and struggling both personally and professionally, may perceive a designer to be arriving with money, technology solutions, and the significant backing of a foreign entity or company. The real or imagined incentives in this scenario may then become a barrier to hearing truthful comments from users and require a designer to enter with a heightened sense of skepticism and situational awareness.

3 Background on the Study Approach

Overcoming these two issues—a Western-biased design approach and an asymmetry of information among expert users—is the focus of this study into the medical device design mismatch.

This study proposes that it is possible to address such barriers, and that the methods used to do so actually have a complementary effect on the two challenges.

While they may be hidden to the individual, Adams identifies various conceptual blocks that can hamper the designer's ability to see important information [21]. Perceptual blocks may cause a designer to stereotype—to see only what they expect to see. They may have a difficulty in isolating the problem, or a tendency to delimit the problem area too closely. In the case of a medical device designer, these blocks result in an embedded bias toward the characteristics of a Western medical system that we have come to know and take for granted, such as culture, infrastructure, as well as human and technical capacity. All of these expectations, which manifest as blocks, are implicit in the approach and perspective brought to a design activity.

It may be possible to overcome these perceptual blocks by becoming aware of the design assumptions that we hold. Surfacing such assumptions by examining past and current design practice allows one to then question and validate them against the realities of a design space. Understanding whether these same ideas apply in a new context allows one to step back and look at the broader system, possibly uncovering new opportunities for innovation. Indeed, Stefik suggests that, "for routine problems, previous experience helps us work through the problems effectively. For novel situations, however, our experience can get in the way of having a breakthrough" [22]. Thus, by stepping back and uncovering pre-existing biases, it is possible for an expert mind to approach a situation with beginner's eyes. This is especially important when approaching the design of medical technology for a complex and unfamiliar environment.

Through the process of validating preconceptions, a designer begins to also uncover priorities among different design attributes. This type of prioritization is required in the model of frugal innovation. Sehgal suggests that consumers of products in the developing world are not seeking low-cost, low-quality alternatives to what is available to more wealthy customers. Rather, they are a value-conscious group that demands—and deserves—affordable alternatives that still meet a competitive performance standard [23]. Understanding and validating design assumptions inform designers on which compromises are acceptable in a given context, and where trade-offs can be made to still achieve a reasonable outcome.

The approach proposed in the following section not only addresses the first problem of an inherent bias in the process of needs finding and design, but also partially begins to address the information asymmetry problem faced when designing for expert users. Identifying and questioning the built-in assumptions in a traditional system and examining them in the context of a novel and lesser-known system is a way of gaining deep insight on the new problem space too. This process of reflection and critical thought enables a shift in the source of information from users to designers, thereby partially overcoming, but not negating the need and difficulty in learning directly from users. This, combined with the use of methods borrowed from the field of human centered design, such as observation and cultural probes, can contribute to a better understanding of the problem space and life context of an expert user.

4 Proposed Methodology

In order to shed light on the challenges posed in this paper, the authors are collaborating with the Uganda Sustainable Trauma Orthopedic Program (USTOP).

This project of the University of British Columbia in Vancouver, Canada conducts annual trips to Mulago Hospital in Kampala, Uganda to provide capacity building training for local medical staff. This partnership provides access for investigators to study the environments, people, and interactions in both a Canadian medical setting at Vancouver General Hospital (VGH), and also in a Ugandan setting at Mulago Hospital.

In the first phase of this study, which is in progress, data are being collected at VGH in order to catalog the assumptions present in how technology currently in use was designed. The following six questions guide the search for embedded design assumptions:

1. What technology interdependencies exist and are built into design?
2. Who is the technology intended for and what is assumed about them?
3. What skills are implicitly required for the use of technology?
4. What assumptions are revealed from marketing and the imagery used?
5. What assumptions are made about the context and use environment?
6. What supporting systems or infrastructure are assumed available?

Investigators will use a Grounded Theory approach for conducting this research [24]. In this method, data from various sources including media, observation, and interviews are collected and documented. This information is then coded and interpreted, and becomes the source of emergent patterns in an iterative process of theory building.

The next phases of this study will take place in the fall of 2012 at Mulago Hospital as investigators join the USTOP surgical team for several weeks. The data and insights gathered about assumptions from VGH will be validated in an alternate context at Mulago through observations and interviews with medical device users.

In addition to validating assumptions on design from VGH, new perspectives will be sought on the values and needs of users at Mulago through the use of cultural probes. This technique introduced by Gaver equips users with a reflection toolkit consisting of a journal, questionnaires, and a photo camera [25]. Users are instructed to collect photographs and reflections, and to answer specific questions that serve to inform designers about their life context. Currano demonstrates the value of 'reflection-out-of-action', which the cultural probes tools would afford by allowing users a self-guided reflection process without the pressure of a formal interview setting [26]. This method is expected to uncover important differences among this user group and their environment as compared to their Western counterparts, and will partially address Von Hippel's sticky information problem and the power dynamics involved.

After validating the design assumptions from VGH and gathering new insights at Mulago, these can then be prioritized to identify which attributes are of greatest concern in design, whether they be human, technology, or systemic in nature. Focus group discussions with medical users at Mulago will identify which results are of priority in the local context, placing each attribute on a spectrum from absolute to more flexible. This exercise will inform what trade-offs are acceptable and desired in a local context, which previously held assumptions would have hidden from view. The areas of most flexibility will provide for designers a rich set of opportunities for innovation.

Finally, users will be engaged in a creative thinking and imagination exercise in order to co-create locally appropriate solutions. Using the method of Outcome Driven Innovation (ODI), opportunity areas are mapped as a series of tasks, or jobs, that the user performs, each leading to an overall desired outcome [27]. Armed with insights about local priorities and acceptable trade-offs, users will be engaged to re-imagine how their desired outcomes could be achieved in innovative ways. Such innovations that reduce the need for inaccessible inputs or remove steps that don not fit the local context can radically change the face of technology-mediated surgical care in the developing world.

5 Expected Results

Background discussions with Canadian surgical staff involved in the USTOP program have revealed a series of valuable insights and serve to highlight preliminary areas of exploration for design requirements and potential design assumptions. These examples will be explored further in all phases of this study.

Vital signs monitoring: An example of a problem facing surgical staff in the developing world is the information gap within their practice, as compared to Western counterparts. Due to the lack of technology, especially for patient vital signs monitoring, surgeons and anesthetists are, in a sense, operating blindly to the status of the patient.

Patient medical history: Another information gap is in the lack of medical records for patients, both in terms of longitudinal patient history, past procedures, and potential reactions, but also lacking details of the interventions and test results from the current hospital stay. These unknowns may lead the surgeon to feel a heightened sense of risk, which affects the decisions that will be made during a procedure and the type of care a patient can receive.

Logistics and tracking: Similarly, a lack of logistics and tracking information for equipment inventory and sterile indicators for quality control may affect the confidence of practice and willingness to move forward with certain decisions. These unique differences in information availability, as well as the resulting risk tolerance that impacts decision-making, affect the way in which technologies will be used or not within a medical system. As such, they may have great implications for how design is to be approached in such a context.

Infrastructure and technology: The availability of various infrastructure and supporting systems for the practice of surgery is often quite different. Designers in traditional markets may take for granted the availability and consistency of power and water supply. Thus they may not take into consideration the immense voltage spikes and variability that cause nearly 30 % of donated device failures in developing world hospitals [4]. The lack of appropriate technology within the operating suite may also have implications for the design and use of co-dependent technologies. For long, the ability to repair long bone fractures using internal fixation was unattainable due to the unavailability of intra-operative fluoroscopic imaging. This supporting technology is an essential tool in the accurate placement of any modern day intramedullary nail, which is an internal supporting rod that keeps the broken bone in alignment during healing. Recently, designers of the SIGN Intramedullary Nail were able to overcome this challenge by creating a purely mechanical system that uses a jig for accurate placement of a nail without the need for fluoroscopy [28]. By stepping back to question the basic assumptions about long bone fixation procedures, they were able to turn constraints into an opportunity for innovation. Users could then reach the same health outcome but with a change in the steps and inputs of the process.

Time constants: Other examples of potentially embedded design assumptions include the time-constants assumed in a surgical setting, which are often not the same in international surgery due to resource gaps, inefficiencies, and differing priorities or values.

Sterilization: Sterility is a factor assumed critical and near-absolute in a Western surgical suite, but similar standards and reliability simply can not be met in a developing world hospital. Technologies then cannot be assumed to depend on such standards.

Human resources: The human resource capacity in a hospital is assumed to be above a certain standard. Even in the Western medical setting, designers often assume technical competence with information technology as a given, but many older nurses and doctors face difficulty using new generations of devices.

This human challenge is compounded in a developing world context where other factors such as literacy, education level, culture, motivation, and confidence in practice can come into surprise designers.

Visualization: Finally, certain characteristics of a procedure are assumed. A tourniquet is one of the most basic pieces of equipment in a trauma department, used to contain blood loss in a limb during surgery. The result is that surgeons typically operate in a bloodless field with a high degree of visibility of the tissue. Technology designed for use in such working conditions may not take into account the visualization issues in a time-critical and stressful surgery where blood cannot be controlled due to the lack of a tourniquet.

These are a sampling of assumptions built into the design of medical technology, and part of the reason why technology developed in the West stands little chance of sustainable use in a developing world hospital. Each of these has implications for how designers must proceed in their work, and will be explored in further detail throughout this study, along with new assumptions that surface.

It is expected that this study will result in insights that help ground in reality the design of medical devices intended for use in the developing world. Assumptions will be explored with users in order to prioritize and better understand where trade-offs could pave the way for novel approaches that lead to potential innovations. Design workshops using Outcome Driven Innovation techniques will generate alternative solutions for achieving acceptable surgical outcomes in the field of orthopedic trauma surgery.

6 Conclusion

The field of international surgery is greatly lacking in technological capacity, with Western-designed medical technology so far failing to improve this outcome. As surgery is a highly technology-dependent field, it is critical that designers become familiar with innovating products for this context if both social and economic benefits are to be realized.

This study proposes that the challenges faced in achieving success are twofold. First in the ability to understand the problem space, and second in the biased approach that Western designers bring to the solution space.

Through a process based on reflection and user feedback, the authors propose that these two problems can be addressed and a step forward taken in the design of medical technology for international surgery. Assumptions and biases can be surfaced in Western medical settings and validated within a developing world context in order to enable a more appropriate approach to design. Uncovering and then challenging the attributes assumed important in a design space can allow for prioritization and trade-offs to be made in order to achieve an acceptable outcome with limited resources.

This approach allows designers to account for the constraints in an unfamiliar system, while surfacing insights, which may lead to disruptive new health technologies.

Acknowledgments The authors would like to thank the Engineers in Scrubs program, as well as the Branch for International Surgery at the University of British Columbia for their generous support of this study.

References

1. Mock CN, Jurkovich GJ, nii-Amon-Kotei D et al (1998) Trauma mortality patterns in three nations at different economic levels: implications for global trauma system development. J Trauma 44(5):804–814
2. World Health Organization (2004) The global burden of disease: 2004 update
3. Mathers CD, Loncar D (2006) Projections of global mortality and burden of disease from 2002 to 2030. PLoS Med 3(11):e442
4. Malkin RA (2007) Design of health care technologies for the developing world. Annu Rev Biomed Eng 9:567–87
5. World Health Organization (2010) Medical devices: managing the mismatch
6. Prahalad CK (2003) The fortune at the bottom of the pyramid: eradicating poverty through profit. Wharton School Publishing, Upper Saddle River
7. Holtzman Y (2012) The U.S. medical device industry in 2012: challenges at home and abroad. Med Device Diagn Ind 17, http://www.mddionline.com
8. Das R (2012) Asia pacific healthcare outlook 2012–2015. Frost & Sullivan, USA
9. CIA, CIA World Factbook (2012) http://www.cia.gov
10. GE Healthymagination (2010) 2010 Annual report, http://www.healthymagination.com
11. Johnson & Johnson (2011) 2011 Annual report, http://www.jnj.com
12. Covidien (2012) Covidien opens US$45 million R&D facility in China, http://investor.covidien.com
13. Deloitte (2012) Securing the next level of growth: second tier emerging markets, http://www.deloitte.com
14. Duflo E, Banerjee AV (2011) Poor economics: a radical rethinking of the way to fight global poverty. Public Affairs, New York
15. Zenios S, Makower J, Yock P (2009) Biodesign: the process of innovating medical technologies. Cambridge University Press, Cambridge
16. Pietzsch JB (2009) Stage-gate process for the development of medical devices. J Med Devices 3:021004
17. Von Hippel E (2005) Democratizing innovation. MIT Press, Massachusetts
18. Suri JF (2005) Thoughtless acts?. Chronicle Books, San Francisco
19. Brown DS, Motte S (1998) Device design methodology for trauma applications. Proc Comput Hum Inter (CHI), pp 590–594
20. Dudley E (1993) The critical villager: beyond community participation. Routledge, London
21. Adams JL (2001) Conceptual blockbusting: a guide to better ideas. Perseus Publishing, Cambridge
22. Stefik M, Stefik B (2005) The prepared mind versus the beginner's mind. Design Management Review, Winter, pp 10–16
23. Sehgal V, Dehoff K, Panneer G (2010) The importance of frugal engineering. Strategy + Business 59
24. Glaser BG, Strauss AL (1967) The discovery of grounded theory: strategies for qualitative research. Transaction Publishers, New Jersey

25. Gaver B, Dunne T, Pacenti E (1999) Design: cultural probes. Interactions 6(1):21–29
26. Currano RM, Steinert M, Leifer L (2011) Characterizing reflective practice in design: what about those ideas you get in the shower. Int Conf Eng Des (ICED11)
27. Ulwick AW (2002) Turn customer input into innovation. Harvard Business Rev 80(1):91–97
28. Sekimpi P, Okike K, Zirkle L, Jawa A (2011) Femoral fracture fixation in developing countries: an evaluation of the SIGN intramedullary nail. J Bone Joint Surg 93(19):1811–1818

Co-Web: A Tool for Collaborative Web Searching for Pre-Teens and Teens

Arnab Chakravarty and Samiksha Kothari

Abstract This paper presents a design case of a Rich Internet Application Co-Web: a collaborative work tool for pre-teens and teens to help collect, collaborate, organize and format information from the web for their school projects more holistically by enhancing their in-search and post-search experience. A user-centered design methodology was followed to help identify how teens and pre-teens use search engines and other technology tools in their daily life to collaboratively work on school projects and recognize the key real-world interactions, which help support this activity. This design solution proposes that integrating collaboration with family, friends and visualizing search keeps the user group interested and helps them retain and use the information they seek more cohesively within the existing search paradigms. This paper describes the design methodology, findings from co-design with participants and reflects on designing new web search interfaces that provide tools for collaborating.

Keywords Web search · Collaboration · Teenager · Computer-supported cooperative work

A. Chakravarty (✉) · S. Kothari
Industrial Design Centre, IIT Bombay, Adi Shankaracharya Marg, Ramabai Nagar, Vikhroli West, Mumbai, MH 400706, India
e-mail: chakravarty.arnab@gmail.com

S. Kothari
e-mail: kothari.samiksha@gmail.com

A. Chakrabarti and R. V. Prakash (eds.), *ICoRD'13*, Lecture Notes in Mechanical Engineering, DOI: 10.1007/978-81-322-1050-4_55,

1 Introduction

Children and young adults are one of the biggest adopters of computers and Internet in India and across the world [1, 2]. The primary use of Internet by secondary and high school children in India is for education and entertainment purposes [3]. Due to increase in exposure to these technologies in schools and classrooms, more and more children are using the Internet and computer based media as primary sources of information for academic related activities [3]. Search engines and online libraries require proper articulation of query, analysis of results to find out the most important result or involve an understanding of abstract concepts which are beyond a child's still- developing skills [4–6] especially in India, where English is not the first language. Searching for information on the web is usually seen as a solitary activity, which often results in incomplete searches, unsatisfactory results and discouraged users [7]. There exists a big need to understand the real-world interactions that children turn to for support in their information seeking activities and to develop applications to support the ecosystem. We conducted interviews with educators, parents and others who frequently observe how children in high schools use computers and Internet for their academic purpose to understand the extent of use and behaviors of the stakeholders. We followed this up with a contextual inquiry with 8 school students and observed them while they carried out a collaborative project in their regular environments. Based on these interviews and observations, we identified several limitations of current collaborative information gathering, search and collaboration practises. Informed by the findings of this formative study, we propose the design of Co-Web, a system for facilitating collaborative Web search among secondary school students working at different locations on a common topic. We provide a detailed overview of the Co-Web system, including a sample usage scenario. We also report some early experiences in co-designing and exploring the design space offered by Co-Web, highlighting on future possibilities. Finally we conclude by highlighting the need of integrating aspects of collaboration and teamwork to support users' desires to engage in active, small-group collaborative searching while finding information on the web.

2 Related Work

Co-Web is built upon several areas of research, including studies of people's information retrieval habits, systems that support synchronous/asynchronous web searching, 'passive' collaboration systems, and systems supporting multi-user searching.

Commercial search engines and Web browsers focus on single-user scenarios [8]. However, there are some systems that enable collaborative work around web searching but most of them focus around co-located collaboration on Web searching like Co-Search, which allows collaborative web search in a co-located setting, i.e., when several people are gathered around a single computer [8]. Teamsearch supports co-located search of digital photo collections by groups seated around an interactive tabletop display [9]. WebSplitter [10] generates personalized views of Web pages for multiple co-located users based on currently available devices with the users at a given point of time.

There has been some prior work which has explored in way to support remote Collaboration on Web tasks. GroupWeb is a browser that allows group members to visually share and navigate World Wide Web pages in real time using tele-pointers for enacting gestures [11]. Search Together enables groups of remote users to synchronously or asynchronously collaborate when searching the Web [12]. Search Together [12] focuses on supporting collaboration during the process of searching the Web, including formulating queries, exploring search results, and evaluating the information that has been found. S3 allows users to asynchronously share useful sites found during a Web search by representing search results in a persistent file format that can be sent to and augmented by several people [13].

In contrast to these systems, Co-Web's design is specifically contextual to secondary and high school students and it has been designed by keeping their unique needs in mind. It supports two or more synchronous or asynchronous searchers using a variety of Web-based search services, and provides integrated support for information collection, collaboration, assimilation and presentation.

3 Formative Studies

In order to verify the prevalence of collaborative web-based information projects and to understand the needs of the stakeholder, we conducted a series of semi-structured interviews with 10 people who work in settings where computers were used for projects in schools. We interviewed 3 teachers, 2 librarians, 3 mothers and 5 children. The teachers included a primary school teacher teaching in Dahanu, a town in Thane District, India, a primary school teacher teaching in a medium-income public school in Mumbai and a secondary school teacher teaching in Kandivali, a medium-income suburb in Mumbai, India. Of the 2 librarians, one worked in a school library in Mumbai and one worked in a small public library. Of the 2 mothers, one was computer-literate and could guide and assist her children at home, one was non computer-literate and had a computer at home and one was not computer-literate and had children who used it at school. In each interview, a set of open-ended questions were asked which were customized to each individual. The questions were designed to investigate about how high school children searched for information in schools, libraries for school projects, the frequency of collaborative search, participation of each collaborator, the within-group roles that emerge, the type of search tasks, the reservations against use by the parents/teachers, the motivations of searching collaboratively and the physical setup of resources needed.

4 User Study

We recruited 8 students and the children were observed in their natural settings to perform pre-defined search tasks on their computers. The studies were conducted in two phases:

In phase 1, Participants were asked to 'think aloud' during the task and tell the researchers about any issues or problems they faced. The participants were also encouraged to avail the help of other people in case of difficulties to observe how they traditionally completed tasks, which they could not do independently. Finally, we conducted semi-structured interviews with participants after they had completed the task.

In phase 2, the same 8 students were divided into 2 groups and were asked to do a project on 'Surface Tension', which was a topic from their curriculum but had not been covered by the teachers yet. The project was to be submitted within five days and the final deliverable was a MS Power-point presentation, which is the most common form of presenting content in most schools (as conveyed to us by the participants in formative study). The participants were all acquainted with each other. They all had previously used internet and computers for school projects but had not used online collaborative tools. Each group member had a computer connected to the Internet and they were free to use other standard software like the Microsoft Office suite as well as well as take notes.

During both phases, we observed the participants' search strategies, usage of tools, and creation of any electronic and non-electronic artifacts during the search. Also, we saved all the information found by participants during the task and the phase 1 individual searching was video recorded. This two-phase study design gave us a chance to observe information gathering not only during synchronous collaboration but also when collaboration was asynchronous. We analyzed the data using affinity mapping and looked for themes related to how group members understood the information (e.g., websites) found during the task as well as information related to division of labor, group members' task strategies, and task state.

5 Insights

The formative study revealed several themes regarding the challenges of finding information and collaborative Web search for students, teachers and the parents.

Prevalence of internet based information gathering: The librarians and parents both reported that students frequently perform online searching for their project purposes. Teachers pointed out that it is also mentioned in the teacher's curriculum to encourage students to find additional information on the internet after a lesson is taught in class to arouse their curiosity to know things and break beyond the information provided in textbooks. Librarians mentioned that due to lack of books in the library about specific topics taught in classroom, children often use the library computer which has access to internet and try finding relevant books and in turn ask the librarian to order these books for their school library. Students and teachers both mentioned that more and more classroom assignments related to exploring subjects and making presentations were expected to be done on the computer.

Participation and enthusiasm in group projects: Teachers also stated that collaborative activities in classroom prove to be very helpful for the students. Students also enjoy it as they like to work and play together. Parents and children also reported that the energy levels of the children would peak up during collaborative and group projects if they had selected the group themselves. Children would see it more as a fun activity compared to usual homework. Also, competing with other groups in the class would make them do little more than required at times, to prove themselves as the best group. There were reservations about the lack of visibility of individual contribution in group projects. If a member was not working, the entire group had to suffer. The intent of the project was clearly defined to the users as to try and explore the subject and gain more information apart from what is currently there in the textbook to get a better understanding of the subject, but it was observed that for some of participants, working on the project merely was a task completion activity and due to which the motivation and enthusiasm in other participants got affected. The initial division was soon forgotten and work was done on an on-need basis. The work kept shuffling between the members of the group as per need.

Locations of computer use: The librarians stated that they rarely saw students in her library using computers in isolation. Librarian also reported that students would use the computer lab for collaborative tasks during free periods, as that would require discussion, which would not be possible in a library atmosphere. Parents and students reported that although most of the first sessions would start with a physical meeting with all the teammates either in school or some collaborators house, most of the other coordination would happen over telephonic conversations. While students preferred working in a co-location setting, day-to-day logistics issues prevented them from physically meeting teammates and do group project.

Resolution of breakdowns: The librarians stated that students of secondary school often come with problems that they are unable to find information of particular topics on the Internet and need guidance on how to frame the query to get relevant information. One parent reported that her son's online searches are mediated either by his father or elder brother as he generally needs help as he is unable to find information by himself. The help provided is typically guiding the search by making query suggestions (verbally) or navigation suggestions (by pointing). In the individual task, 6 out of the 8 users gave up when they could not locate the desired information. Students reported that they gave up more easily when they were working alone on a project. They had to either wait to ask a family member or request help from a class-mate. In collaborative tasks, it was easier as everyone was working on the same project so they did not face any hesitation.

Minimal Sense-making: During collaborative Web searches, the interviewees reported that students would divide the projects in multiple tasks and choose these tasks as per their skill sets and facilities available with each one of them. There would be students who would be good at making presentations and finally after all the other collaborators have gathered information, one student would compile all of it into a beautiful looking presentation. In some cases, the presentation would be divided into stages and each one would contribute some slides.

Lack of Awareness of Search Engines and Boolean searches: The problem seen with students searching for information on the web was that Children are poor at representing their needs in form of a query. Most of the children found it difficult to articulate what they were looking for. On the other hand, search engines are poor at responding to vague queries. They were not aware of the various techniques and tricks available to articulate their query in a search engine friendly manner for the best results.

Minimal Awareness: In the collaborative task, there were communication problems observed among the users as they were not at the same pace. Most of the things were communicated over telephonic conversations. There were misunderstandings among the team mates as they couldn't see each other's screen and visualizing what exactly is being viewed by the teammate became difficult.

Plethora of programs: Users generally coordinated and worked using a multitude of different programs/websites for different tasks. e.g.: MS-Word for editing content, Skype, Instant Messengers for co-ordination with other team mates, PowerPoint for formatting and presenting content etc. The users kept switching between multiple applications which were pointed out by users as being irritating and distracting. Most of the files were shared over email or instant messenger file transfers which added to their confusion and a lot of time was spent on keeping track of different files spread over different locations on the computer. A lack of version control also ensured that mismatches in data crept up especially when the data was being assimilated for presentation.

Video as a distraction: Users generally got distracted while using video chat of instant messengers. Due to poor broadband speeds, video generally did not stream well and much time was spent in trying to fix it.

Parental Control: Parents often enforced a time-limit of anywhere between 45 min to 2 h for a single session. Parents often could not identify whether the computer was being used for casual purposes or for project work.

6 Design Goals

It was quite clear from the user studies and interviews that collaboration on projects remotely was quite prevalent. Also, students collaborating on finding information together had the potential to improve search experiences and outcomes even within the existing search paradigm. Since these activities are not supported by current paradigms, people were employing workarounds and multiple applications to communicate and coordinate. It was postulated that systems and tools were designed to support such user behavior will benefit students and help them. With the above considerations in mind, the envisioned design goals of Co-Web were as follows:

- To create a collaborative environment to reduce redundancy and repetition in search queries thereby increasing productivity.
- To provide tools to save/search results and create new visualizations of search results to elicit more user involvement in the information search rather than it just being a mechanical activity.
- To provide a sense of comfort and reinforce user confidence in completing a search by turning it into a group activity yet not a competitive one.
- To help users articulate their queries better by incorporating advanced search techniques in a visual manner.
- To ensure that users complete their search activities with higher confidence/correctness by involving family, friends and teachers in the search activity.

7 Co-Web

Based on our investigations of students' collaborative search practises and needs, Co-Web is designed to enable either synchronous or asynchronous remote collaboration between students. The application is a RIA (rich internet application), which ensures that the reach of the internet can be leveraged with a compelling user interface. In the next section, we shall illustrate the experience of using Co-Web by describing its features. In the most basic Co-Search usage scenario, a group of students can login to the application and run an instance of co-web on their individual computers. They can collaboratively search information and collect relevant pieces of information in the shared space. Each user is represented by a unique color and icon. Every piece of collected information displays the identity of the person who created it. Audio and text chat functionality is in-built so that participants can collaborate. Special search categories like Encyclopedia and summary search are integrated so that participants can find relevant information quickly. Related search and Boolean operations are also included. Basic text editing tools are also integrated so that basic formatting and presentation can be done in the application itself. The application has a separate interface for the teachers where they can have an overview of the project.

7.1 Features

The first screen is sectioned into three main sections:

- Start individual search (Fig. 1a)
- Start collaborative search (Fig. 1b)
- Continue an ongoing project (Fig. 1c)

Clicking Collaborative search on (Fig. 1b) takes the user to the screen (Fig. 2) where he/she can choose to either browse through previous projects (Fig. 2a) or start a new project (Fig. 2b). The user can choose to invite friends, family and other people from his contacts to join the session, merely by dragging and dropping the contacts (Fig. 2c) on the project book (Fig. 2b). (In case of a group project, the friends added to the project become the collaborators).

Dynamic Cloud and Booleans: A tag cloud (Fig. 3a) displays a network of words related to the inputted query word. On clicking a related search terms from the interactive cloud, a Boolean operator 'AND (+)' (Fig. 3b) is added along with the initial query term to narrow down search and get more accurate results to help ease the problem of articulating search queries for the best results.

Encyclopedia and Project Search categories: Two new categories of 'Encyclopedia' (Fig. 4) and 'Project search' (Fig. 4d) have been created especially considering the search requirements of the user group: 'Encyclopedia' category returns results of the search term from a set of pre-selected online encyclopedias. It was noticed during the user studies that people were not using other encyclopedias for content and defaulting to Wikipedia since it was the only encyclopedia they had used regularly and were aware of. However, there was a latent aspiration to use information from other resources. 'Project' category summarizes different categories (Fig. 4a, b, c) of information traditionally related to an academic project (namely encyclopedia, experiments, diagrams, videos, definition, images) on a

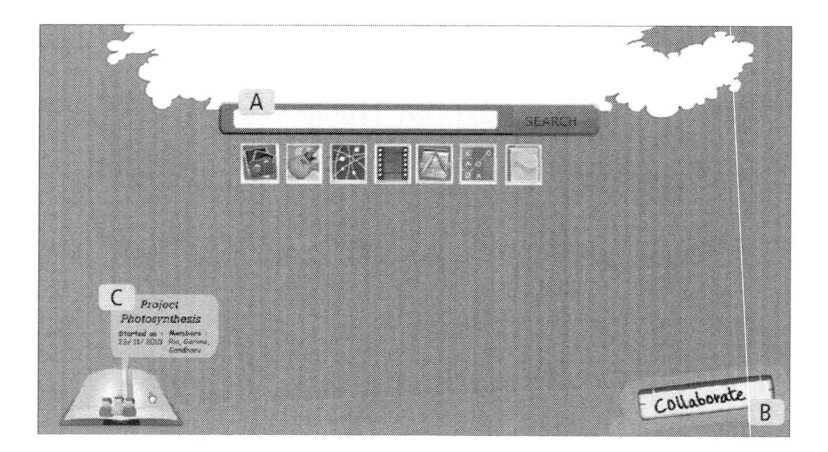

Fig. 1 First screen

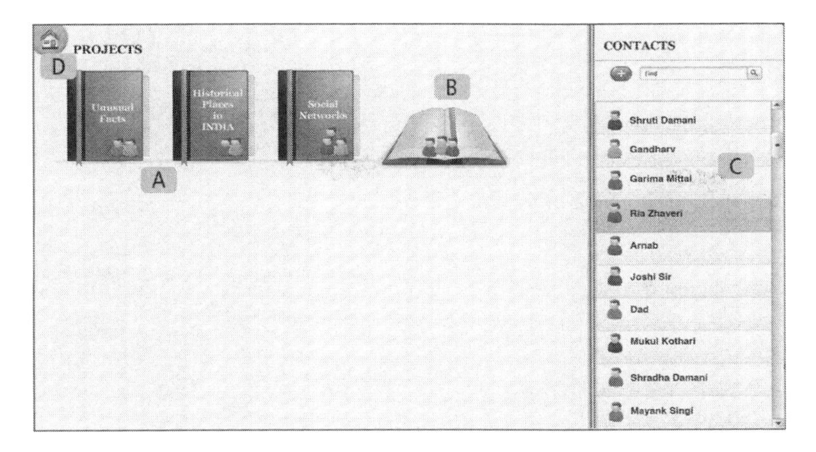

Fig. 2 Project view

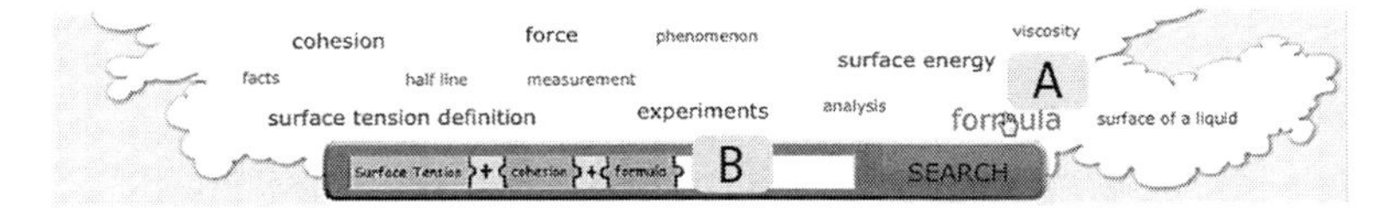

Fig. 3 Related query terms

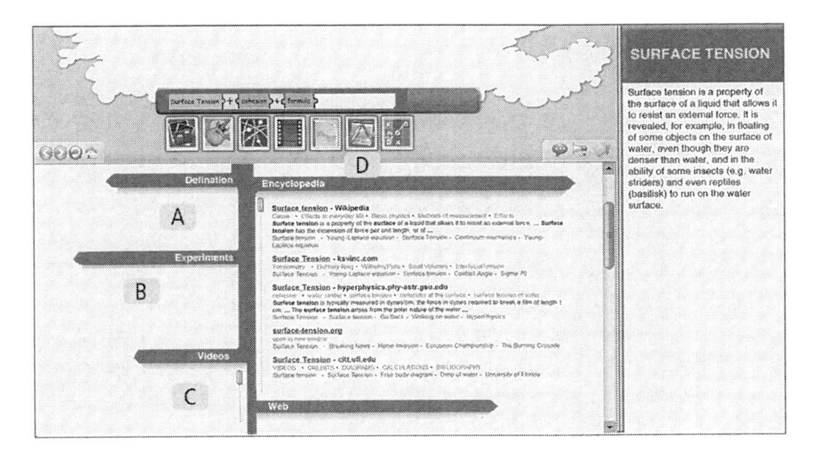

Fig. 4 Project search category visualization

single screen. Clicking the button (Fig. 4d) fires multiple queries to get all project related information at one go and it's visualized in a common space. Users don't have to keep swapping between different categories every time while searching for information and can access it from a common screen.

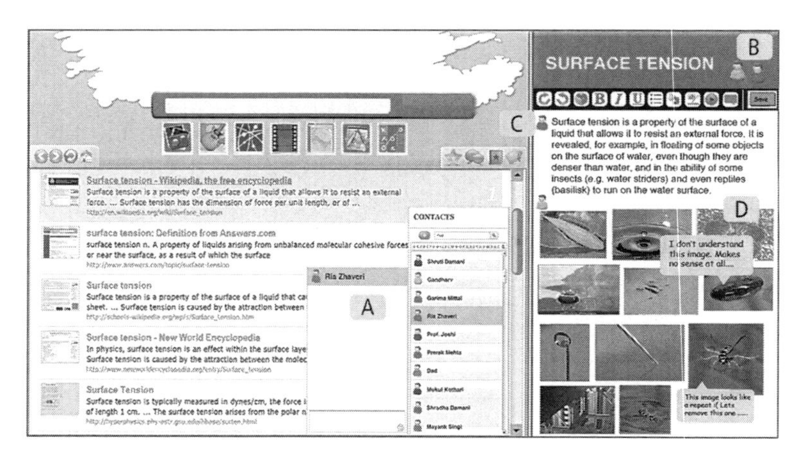

Fig. 5 Project tools for documentation

Shared Space: (Figure 5d) The application provides a shared space for all the team members to keep dumping all data collected by them. This data can be viewed by all the collaborators and anyone who has been invited to join the session. The icon (Fig. 5c) of members who dumped the data is shown to avoid confusion and for further referencing. Also, the application gives the users an indication when the collaborators join/leave (Fig. 5b) the session by showing their icon in the shared space.

Collapsible and flexible space: The shared space is a flexible space, which can be adjusted to any size to avoid using too much space while searching. During the search activity the shared space can be collapsed and while editing the content it can be expanded into full screen. This provides a common ground between the users to carry out discussions, view, share and comment (Fig. 5d) on the data being accumulated by team members. Users can collectively take a decision on what can be finally chosen for the final presentation. Changes made by the group members are reflected in real time. Discussion and expressing view-point leaves a scope for reflection—a vital component of learning.

Timeline: The timeline option in the project tools (Fig. 5c) shows a replay of how the shared document has been edited from the start. This is a valuable aid to view the chronological progression of the actions of all group members in editing the information gathered during the entire search session. This can be valuable aid to recognizing dead-ends, reorganizing and adopting a different search strategy also prevents redundancy.

Integrated chat: (Figure 5a) The application helps the team to communicate by integrating a text and audio chat feature in the application. This is to ensure that searching and discussing information can happen simultaneously. A combination of a singular shared space and audio/text chat helps a lot in preventing miscommunication.

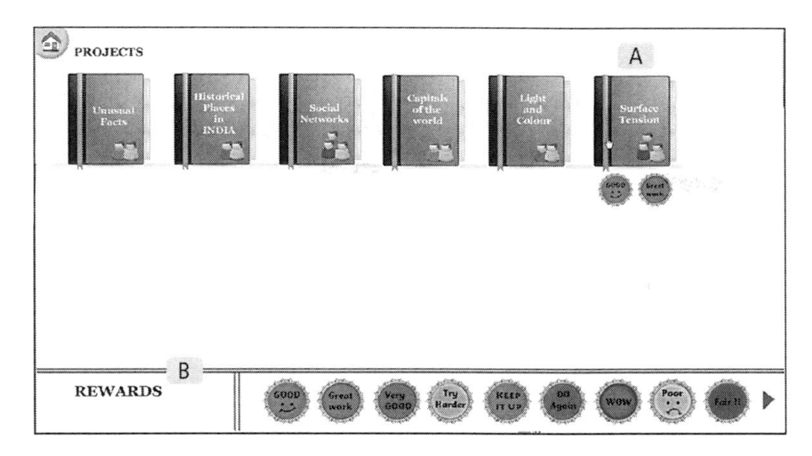

Fig. 6 Project tools for documentation

Project Tools: (Figure 5c) The application provides basic editing tools in the shared space, so that a rich documentation of the project can be completed in the application itself. The final edited summary can also be exported into different file formats.

The application has a separate interface (Fig. 6) for the teachers (to make sure they are not directly involved) where they can view, comment, suggest corrections and reward (Fig. 6b) the projects to encourage the team.

8 Co-Designing and Experiencing Co-Web

A high-fidelity prototype created in Action script and deployed as an AIR application was used to evoke the participation of multiple students working on finding information for a common project. A group of students and teachers were asked to try out the sketch and then discuss the possibilities that opened up during the sketch. The students believed that the overall idea of carrying out informational searches in a common space was helpful but stressed the need of more ways to simplify articulation search queries. Features visualizing search (timeline, shared space, dynamic tag cloud) were among the most highly rated aspects of the prototype. Users expressed a desire to directly print the edited summaries and use it in their project scrapbooks. This suggests that providing rich documentation/post searching experiences is an important aspect of supporting collaborative Web search and is something that the users enjoy and cherish. Teachers expressed the view that it would not be comfortable for them to add any person to the project (esp. family members of students) and also wondered if there was enough sense-making of the data that was would be collected by the students using this application. One student expressed that she could use this to even do collaborative searches with her siblings and use it to teach her mom who did not know how to use internet search. Thereby in addition to evaluating the specific features of the sketch, the process facilitated a move forward in co-articulating what it means to design and use digital technology to facilitate collaborative information searching in everyday life.

9 Future Work and Possibilities

Technologically, the first step would be to develop a completely working application which is tested with a bunch of students on a real project. This will be taken to the homes and schools of users to discuss the design space of possibilities. Future efforts should be taken in the direction of increasing sense-making of the information collected collaboratively by the participants. It may also be interesting to expand the formation of search queries through different Boolean operators apart from 'AND' and also visualizing other advances search techniques in an intuitive and usable manner. Therefore, future research plans include not only further evaluation of Co-Web interface, but further development of the interface to support various collaborative scenarios, increasing leaning and ensuring better search articulation.

10 Future Work and Possibilities

In the above sections, we detailed how the design space offered by collaboration in online searches and information retrieval in secondary school students was explored. We further detailed the interviews and user studies carried out to flesh out this design space and how Co-Web emerged out of the pain-points and opportunities presented by the design space. We finally conclude by reflecting on the contributions of the paper: firstly, providing a high-fidelity prototype in the form of Co-Web and secondly this initial exploration opening up and adding to the emerging design space of collaborative web searches and information gathering online for children.

As described above, the design of Co-Web evolved from user studies and interviews to early software manifestation in a co-design setting where the students and teachers evaluated the ideas and specific features as the process moved forward. Adding to this, the process was set in the actual site of their individual homes of the students, thereby brining forward more direct and relevant evaluation of the concept. We end by discussing the future directions and the possibilities that emerge out of this exploration and the direction further research can take for creating unique collaborative interfaces used by children for better information retrieval.

References

1. The “Internet Habits of High School Children”. http://kraran.com/internet-habits-of-high-school-children/
2. “UNICEF-India- Statistics”. http://www.unicef.org/infobycountry/india_sta tistics.html
3. PricewaterhouseCoopers (2010) Information and communication technology for education in India and South Asia
4. Moore P, St. George A (1991) Children as information seekers: the cognitive demands of books and library systems. Sch Libr Media Q 19:161–168
5. Druin et al. (2001) Designing a digital library for young children: an intergenerational partnership. In: Proceedings of ACM/IEEE joint conference on digital libraries
6. Walter VA, Borgman CL, Hirsh SG (1996) The science library catalog: a springboard for information literacy. Sch Libr Media Q 24:105–112
7. Gayo-Avello D, Brenes DJ (2009) Making the road by searching—a search engine based on swarm information foraging CoRR abs/0911.3979
8. Amershi S, Morris MR (2008) Co-search: a system for co-located collaborative web search. CHI 2008, pp 1647–1656
9. Morris MR, Paepcke A, Winograd T (2006) Team-search: comparing techniques for co-present collaborative search of digital media. IEEE Tabletop, pp 97–104
10. Han R, Perrett V, Naghshineh M (2000) WebSplitter: a unified xml framework for mutli-device collaborative web browsing. CSCW 2000, pp 221–230
11. Greenberg S, Roseman M (1996) GroupWeb: a WWW browser as real time groupware. CHI 1996 conference companion
12. Morris MR, Horvitz E (2008) SearchTogether: an interface for collaborative web search. In: Proceedings UIST 2008, pp 3–12
13. Morris MR, Horvitz E (2007) S3: storable, shareable search. Interact 2007, in press

Printed by Publishers' Graphics LLC